Lecture Notes in Computer Science 14038

Founding Editors

Gerhard Goos
Juris Hartmanis

The series Lecture Notes in Computer Science (LNCS), including its subseries Lecture Notes in Artificial Intelligence (LNAI) and Lecture Notes in Bioinformatics (LNBI), has established itself as a medium for the publication of new developments in computer science and information technology research, teaching, and education.

LNCS enjoys close cooperation with the computer science R & D community, the series counts many renowned academics among its volume editors and paper authors, and collaborates with prestigious societies. Its mission is to serve this international community by providing an invaluable service, mainly focused on the publication of conference and workshop proceedings and postproceedings. LNCS commenced publication in 1973.

Fiona Nah · Keng Siau

Editors

HCI in Business, Government and Organizations

10th International Conference, HCIBGO 2023
Held as Part of the 25th HCI International Conference, HCII 2023
Copenhagen, Denmark, July 23–28, 2023
Proceedings, Part I

 Springer

Editors
Fiona Nah
City University of Hong Kong
Kowloon, Hong Kong

Keng Siau
City University of Hong Kong
Kowloon, Hong Kong

ISSN 0302-9743   ISSN 1611-3349 (electronic)
Lecture Notes in Computer Science
ISBN 978-3-031-35968-2   ISBN 978-3-031-35969-9 (eBook)
https://doi.org/10.1007/978-3-031-35969-9

This Springer imprint is published by the registered company Springer Nature Switzerland AG
The registered company address is: Gewerbestrasse 11, 6330 Cham, Switzerland

Foreword

Human-computer interaction (HCI) is acquiring an ever-increasing scientific and industrial importance, as well as having more impact on people's everyday lives, as an ever-growing number of human activities are progressively moving from the physical to the digital world. This process, which has been ongoing for some time now, was further accelerated during the acute period of the COVID-19 pandemic. The HCI International (HCII) conference series, held annually, aims to respond to the compelling need to advance the exchange of knowledge and research and development efforts on the human aspects of design and use of computing systems.

The 25th International Conference on Human-Computer Interaction, HCI International 2023 (HCII 2023), was held in the emerging post-pandemic era as a 'hybrid' event at the AC Bella Sky Hotel and Bella Center, Copenhagen, Denmark, during July 23–28, 2023. It incorporated the 21 thematic areas and affiliated conferences listed below.

A total of 7472 individuals from academia, research institutes, industry, and government agencies from 85 countries submitted contributions, and 1578 papers and 396 posters were included in the volumes of the proceedings that were published just before the start of the conference, these are listed below. The contributions thoroughly cover the entire field of human-computer interaction, addressing major advances in knowledge and effective use of computers in a variety of application areas. These papers provide academics, researchers, engineers, scientists, practitioners and students with state-of-the-art information on the most recent advances in HCI.

The HCI International (HCII) conference also offers the option of presenting 'Late Breaking Work', and this applies both for papers and posters, with corresponding volumes of proceedings that will be published after the conference. Full papers will be included in the 'HCII 2023 - Late Breaking Work - Papers' volumes of the proceedings to be published in the Springer LNCS series, while 'Poster Extended Abstracts' will be included as short research papers in the 'HCII 2023 - Late Breaking Work - Posters' volumes to be published in the Springer CCIS series.

I would like to thank the Program Board Chairs and the members of the Program Boards of all thematic areas and affiliated conferences for their contribution towards the high scientific quality and overall success of the HCI International 2023 conference. Their manifold support in terms of paper reviewing (single-blind review process, with a minimum of two reviews per submission), session organization and their willingness to act as goodwill ambassadors for the conference is most highly appreciated.

This conference would not have been possible without the continuous and unwavering support and advice of Gavriel Salvendy, founder, General Chair Emeritus, and Scientific Advisor. For his outstanding efforts, I would like to express my sincere appreciation to Abbas Moallem, Communications Chair and Editor of HCI International News.

July 2023   Constantine Stephanidis

HCI International 2023 Thematic Areas
and Affiliated Conferences

Thematic Areas

- HCI: Human-Computer Interaction
- HIMI: Human Interface and the Management of Information

Affiliated Conferences

- EPCE: 20th International Conference on Engineering Psychology and Cognitive Ergonomics
- AC: 17th International Conference on Augmented Cognition
- UAHCI: 17th International Conference on Universal Access in Human-Computer Interaction
- CCD: 15th International Conference on Cross-Cultural Design
- SCSM: 15th International Conference on Social Computing and Social Media
- VAMR: 15th International Conference on Virtual, Augmented and Mixed Reality
- DHM: 14th International Conference on Digital Human Modeling and Applications in Health, Safety, Ergonomics and Risk Management
- DUXU: 12th International Conference on Design, User Experience and Usability
- C&C: 11th International Conference on Culture and Computing
- DAPI: 11th International Conference on Distributed, Ambient and Pervasive Interactions
- HCIBGO: 10th International Conference on HCI in Business, Government and Organizations
- LCT: 10th International Conference on Learning and Collaboration Technologies
- ITAP: 9th International Conference on Human Aspects of IT for the Aged Population
- AIS: 5th International Conference on Adaptive Instructional Systems
- HCI-CPT: 5th International Conference on HCI for Cybersecurity, Privacy and Trust
- HCI-Games: 5th International Conference on HCI in Games
- MobiTAS: 5th International Conference on HCI in Mobility, Transport and Automotive Systems
- AI-HCI: 4th International Conference on Artificial Intelligence in HCI
- MOBILE: 4th International Conference on Design, Operation and Evaluation of Mobile Communications

List of Conference Proceedings Volumes Appearing Before the Conference

1. LNCS 14011, Human-Computer Interaction: Part I, edited by Masaaki Kurosu and Ayako Hashizume
2. LNCS 14012, Human-Computer Interaction: Part II, edited by Masaaki Kurosu and Ayako Hashizume
3. LNCS 14013, Human-Computer Interaction: Part III, edited by Masaaki Kurosu and Ayako Hashizume
4. LNCS 14014, Human-Computer Interaction: Part IV, edited by Masaaki Kurosu and Ayako Hashizume
5. LNCS 14015, Human Interface and the Management of Information: Part I, edited by Hirohiko Mori and Yumi Asahi
6. LNCS 14016, Human Interface and the Management of Information: Part II, edited by Hirohiko Mori and Yumi Asahi
7. LNAI 14017, Engineering Psychology and Cognitive Ergonomics: Part I, edited by Don Harris and Wen-Chin Li
8. LNAI 14018, Engineering Psychology and Cognitive Ergonomics: Part II, edited by Don Harris and Wen-Chin Li
9. LNAI 14019, Augmented Cognition, edited by Dylan D. Schmorrow and Cali M. Fidopiastis
10. LNCS 14020, Universal Access in Human-Computer Interaction: Part I, edited by Margherita Antona and Constantine Stephanidis
11. LNCS 14021, Universal Access in Human-Computer Interaction: Part II, edited by Margherita Antona and Constantine Stephanidis
12. LNCS 14022, Cross-Cultural Design: Part I, edited by Pei-Luen Patrick Rau
13. LNCS 14023, Cross-Cultural Design: Part II, edited by Pei-Luen Patrick Rau
14. LNCS 14024, Cross-Cultural Design: Part III, edited by Pei-Luen Patrick Rau
15. LNCS 14025, Social Computing and Social Media: Part I, edited by Adela Coman and Simona Vasilache
16. LNCS 14026, Social Computing and Social Media: Part II, edited by Adela Coman and Simona Vasilache
17. LNCS 14027, Virtual, Augmented and Mixed Reality, edited by Jessie Y. C. Chen and Gino Fragomeni
18. LNCS 14028, Digital Human Modeling and Applications in Health, Safety, Ergonomics and Risk Management: Part I, edited by Vincent G. Duffy
19. LNCS 14029, Digital Human Modeling and Applications in Health, Safety, Ergonomics and Risk Management: Part II, edited by Vincent G. Duffy
20. LNCS 14030, Design, User Experience, and Usability: Part I, edited by Aaron Marcus, Elizabeth Rosenzweig and Marcelo Soares
21. LNCS 14031, Design, User Experience, and Usability: Part II, edited by Aaron Marcus, Elizabeth Rosenzweig and Marcelo Soares

47. CCIS 1836, HCI International 2023 Posters - Part V, edited by Constantine Stephanidis, Margherita Antona, Stavroula Ntoa and Gavriel Salvendy

https://2023.hci.international/proceedings

Preface

The use and role of technology in the business and organizational context have always been at the heart of human-computer interaction (HCI) since the start of management information systems. In general, HCI research in such a context is concerned with the ways humans interact with information, technologies, and tasks in the business, managerial, and organizational contexts. Hence, the focus lies in understanding the relationships and interactions between people (e.g., management, users, implementers, designers, developers, senior executives, and vendors), tasks, contexts, information, and technology. Today, with the explosion of the metaverse, social media, big data, and the Internet of Things, new pathways are opening in this direction, which need to be investigated and exploited.

The 10th International Conference on HCI in Business, Government and Organizations (HCIBGO 2023), an affiliated conference of the HCI International (HCII) conference, promoted and supported multidisciplinary dialogue, cross-fertilization of ideas, and greater synergies between research, academia, and stakeholders in the business, managerial, and organizational domain.

HCI in business, government, and organizations ranges across a broad spectrum of topics from digital transformation to customer engagement. The HCIBGO conference facilitates the advancement of HCI research and practice for individuals, groups, enterprises, and society at large. The topics covered include emerging areas such as artificial intelligence and machine learning, blockchain, service design, live streaming in electronic commerce, visualization, and workplace design.

Two volumes of the HCII 2023 proceedings are dedicated to this year's edition of the HCIBGO conference. The first volume covers topics related to advancing technology and management in public sector organizations and governance, user experience and business perspectives in the context of mobile commerce and e-commerce, as well as the use of disruptive technologies to enhance customer experience. The second volume focuses on topics related to exploring the intersection of robotics and autonomous agents in business and industry, applications of AI in business and society, as well as case studies and empirical research in the domain of exploring human behavior and communication in business.

Papers of this volume are included for publication after a minimum of two single-blind reviews from the members of the HCIBGO Program Board or, in some cases, from members of the Program Boards of other affiliated conferences. We would like to thank all of them for their invaluable contribution, support, and efforts.

July 2023

Fiona Nah
Keng Siau

10th International Conference on HCI in Business, Government and Organizations (HCIBGO)

The full list with the Program Board Chairs and the members of the Program Boards of all thematic areas and affiliated conferences of HCII2023 is available online at:

http://www.hci.international/board-members-2023.php

HCI International 2024 Conference

The 26th International Conference on Human-Computer Interaction, HCI International 2024, will be held jointly with the affiliated conferences at the Washington Hilton Hotel, Washington, DC, USA, June 29 – July 4, 2024. It will cover a broad spectrum of themes related to Human-Computer Interaction, including theoretical issues, methods, tools, processes, and case studies in HCI design, as well as novel interaction techniques, interfaces, and applications. The proceedings will be published by Springer. More information will be made available on the conference website: http://2024.hci.international/.

General Chair
Prof. Constantine Stephanidis
University of Crete and ICS-FORTH
Heraklion, Crete, Greece
Email: general_chair@hcii2024.org

https://2024.hci.international/

Contents – Part I

Mobile Commerce and e-Commerce: User Experience and Business Perspectives

Use of Disruptive Technologies to Enhance Customer Experience

Contents – Part II

Applications of AI in Business and Society

Exploring Human Behavior and Communication in Business: Case Studies and Empirical Research

Advancing Technology
and Management in Public Sector
Organizations and Governance

Virtual Reality for Smart Government – Requirements, Opportunities, and Challenges

Matthias Baldauf$^{(\boxtimes)}$ (ID), Hans-Dieter Zimmermann (ID), Pascale Baer-Baldauf (ID), and Valmir Bekiri

OST – Eastern Switzerland University of Applied Sciences, Institute for Information and Process Management, Rosenbergstrasse 59, 9001 St. Gallen, Switzerland
{matthias.baldauf,hansdieter.zimmermann,
pascale.baer,valmir.bekiri}@ost.ch

Abstract. Virtual Reality (VR) might enable promising novel applications to involve citizens and businesses in governmental processes. As the body of literature in the field of VR in digital government is limited, we conducted an exploratory interview study with experts from the fields of e-government (both academia and practice) and VR technology to gain more knowledge about the potential utilization of the technology for smart government. The semi-structured interviews were encoded around thematic topics that emerged from the data. The findings cover important requirements, chances and opportunities, risks and challenges as well es potential application scenarios. Among other factors, authentication has been identified both a requirement and a core challenge to apply VR in governmental services. Despite recent decreases in prices of VR devices, our experts considered current costs still high for general interest. Nevertheless, promising potential applications such as community gatherings or the immersive presentation of urban planning projects have been identified.

Keywords: Virtual Reality · E-government · Smart government

1 From E-Government to Open and Smart Government

The term e-government generally refers to the use of information and communication technologies (ICT) in the context of government agencies, either within or between agencies, or between agencies and citizens, or between businesses and agencies. There are two distinct themes in the e-government debate: (1) the general development of digitization and digitalization also reached government agencies and public administration. (2) Beyond this more technological perspective we must also consider the developments of the so-called New Public Management (NPM), which was coined in the 1990s. Its basic ideas, together with the opportunities offered by ICT, have led to a vision of modern government in a digital society, often referred to as open government. ICT enable government agencies and public administrations to engage citizens and businesses in a wide range of activities. In the course of the "open" movement also government

© Springer Nature Switzerland AG 2023
F. Fui-Hoon Nah and K. Siau (Eds.): HCII 2023, LNCS 14038, pp. 3–13, 2023.
https://doi.org/10.1007/978-3-031-35969-9_1

Fig. 1. Example of a VR application: Using a head-mounted display with handheld controllers (left) to explore a remote building via immersive 360-degree panorama images (right).

agencies engage citizens as well as private sector companies in participatory processes to utilize the "collective wisdom" and to develop solutions to solve social and societal problems (cf. [9,13]).

When one looks at open government, social software in particular comes into play. The term describes software tools that serve to support people in the areas of communication and collaboration, and generally the establishment and maintenance of social relationships. And as technological developments emerge, new technological solutions should be considered for utilization for interactions between public agencies and citizens resp. businesses, or more general for human-computer interaction (HCI) in the public services environment.

Related research has been studying HCI trends such as conversational interfaces for digital government (cf. [1,2]), for example. In this work, we focus on the application of Virtual Reality (VR) technology for the use in e-government. VR refers to computer-generated 3D environments (either through rendered 3D graphics or 360°C photos/videos) immersively experienced in multi-projected spaces or via head-mounted displays. Users naturally interact in and with these environments, e.g., through head and body tracking and/or using handheld controllers (see Fig. 1).

After years of basic VR research, affordable VR goggles and corresponding software platforms have become available on the mass market recently. While we see a current application focus on entertainment, VR might also provide a novel promising channel for governmental applications. Because of the novelty of the research field, we decided for an exploratory approach to (1) uncover requirements, opportunities, and challenges of applying VR technology for governmental applications and (2) identify potential uses cases and applications.

2   Related Work

The existing body of knowledge in this area has been limited. Among others, the literature quite often discusses the use of VR in e-government within the context of e-participation, which is a major field of open government.

Leible et al. [9] conducted a literature review analyzing ICT applications of e-participation. The authors identified VR as one of six application types utilized by e-participation. Other authors consider VR as an extension to e-participation and define "the domain of VR participation as a multi-modal, convergent, immersive communication extending existing e-participation paradigm" [11]. The authors argue that a VR-enhanced dialogue may help to involve practitioners in e-participation to better "gain an additional perspective on contemporary online digital communication issues and the opportunities for e-participation in employing cutting-edge Virtual Reality technology" [12]. In their research the authors emphasize the dialogue as the core of e-participation. From the literature, they identify two distinct affordances of human interaction in VR to be considered when discussing the human perception of the virtual world and collaborators: immersion and presence. In addition, they complement these two characteristics with the "sense of community", which focuses on the social aspects and is based on immersion and presence [12]. Beyond the opportunities for VR-enhanced dialogues the authors also point out some challenges, which are the availability of hardware equipment and the available skill set to operate the devices. Finally, the authors propose a "VR dialogue framework" to support VR-participation complementing traditional e-participation and provide some practical guidelines for building VR-participation.

Based on a design sciences-based research study Fegert et al. came up with 20 meta requirements and design principles for an Augmented Reality (AR)- and VR-based e-participation application organized in five different categories: access, information, motivation, transparency, and data security [4]. Beside e-participation, Buljat [3] investigated the use of AR and VR for environmental communication and for a motivation for environmentally-friendly behavior, e.g., by providing a simulation of normally unobservable long-term negative consequences of environmental threats, but so far, the topic has barely been addressed by information systems (IS) literature. Based on a comprehensive literature study the authors conclude that "the field is still young, as there is a lack of AR/VR research for environmental sustainability in the leading IS journals" [3]. Pfeiffer et al. [10] presented an overview of how immersive systems might help addressing sustainably goals through "addressing challenges regarding awareness, motivation, information transfer, and educating citizens to act in a sustainable manner". They propose the following areas of application: Immersive systems as a tool to increase awareness, to increase motivation to participate, for smart information transfer, and for learning to act.

Already in 2013, Tozsa addressed the use of VR technology in public administration [15]. While referring to the virtual world of "Second Life" the author suggested that 3D type of applications also might be accepted in the public administration area. Thus, he proposed 3D graphic models of the office building and the customer management desks in order to allow real time interactivity within a virtual space where representatives as well as customers are being virtually present through their avatars. The most important areas of application, according to the author, are virtual administrative visits and internal meetings.

Taking into account the development of open government and the increasing involvement of citizens and businesses in governmental decision processes on the one hand, and the increasing digitalization of government agencies on the other hand, it can be concluded from our literature review that additional research is needed. In particular, the use of new technologies for low-threshold integration of citizens and businesses in government processes needs to be explored. VR technology could offer exactly this potential.

3   Method

As the body of literature in the field of VR in e-government is very manageable, we decided to go for an exploratory interview study to gain more knowledge about the potential utilization of VR in digital government. To cover a broad view, we decided to include e-government experts from academia, e-government professionals as well as VR experts. Overall, 13 potential interview partners in Switzerland (with respective knowledge and several years of relevant professional experience) were identified and invited to interviews. Finally, 8 of them agreed to participate in this study. 5 experts have a background in e- and digital government, 2 of them are from academia, 3 from practice, and 3 experts are VR specialists. Table 1 shows an overview of the experts involved:

Table 1. Overview of interviewees.

Id	Profession
P1	E-government practitioner
P2	E-government practitioner
P3	E-government IT professional
P4	E-government researcher
P5	E-government researcher
P6	VR expert
P7	VR expert
P8	VR expert

Each interview was conducted by two research assistants online via MS Teams in the period of March to May 2022. Each interview took between 45 and 60 min. The interviewers followed a semi-structured interview guide. For the participants' preparation, the main questions were sent to the participants one week before the interview. These included key requirements for applying VR technology for governmental applications as well as opportunities and challenges of VR applications for smart government.

The interviews were conducted in German and recorded using the respective MS Teams feature. For post-study analysis, the interviews were transcribed. Based on the transcriptions the interviews have been encoded around thematic topics which emerged from the study of the transcripts. In a next step, common themes in the interviewees' responses were identified and related statements clustered.

4   Results

During the interviews the following overall topics have been addressed: requirements for VR-based government services, chances and opportunities as well as risks and challenges of such services, and potential governmental VR applications. In this section, we present the results of the expert interviews, summarize common themes, and illustrate them with participants' statements.

4.1   Requirements

The experts identified several important requirements for applying VR technology for e-government cases.

Identity Management. Ensuring and verifying the identity of a user/citizen was emphasized by several interview partners as a crucial requirement and challenge for e-government applications in VR. P2 illustrated this requirement with an example of a wedding in VR: "How can you make sure that the people behind these avatars are really Ms. Miller and her husband to-be?" As long as the identity of person is not verifiable, only trivial non-personalized simple services are possible.

Data Security. Since many governmental applications access and process personal privacy-relevant information, the security of the data involved was another key requirement identified by the interviewees. They suggested that a respective VR application must be transparent about the personal data involved and the authorities having access. P3 pointed out the corresponding challenges when providing governmental services within the *Metaverse* or a similar platform by a major tech company: "Offering the whole thing in Metaverse would give the impression that a tech giant is listening and watching".

Usability. Several interviewees considered good usability a key acceptance criterion for governmental VR applications: "It must be as easy as using a mobile phone" (P2). P7 and P8 referred to the concept of "plug and play" and had concerns that VR technology is not immediately ready for use. Since VR glasses and corresponding applications take some time to launch, they questioned the acceptance for spontaneously used governmental applications. In addition, P2 suggested starting with simple cases and avoiding complex scenarios.

Affordability of Devices. For the widespread usage and acceptance of governmental VR applications, the interviewees mentioned affordable consumer hardware, i.e., stand-alone VR glasses, as another key requirement. Despite recent decreases in device prices, the interview partners considered the costs for such devices still too high to be of general interest.

Return on Investment. All e-government professionals involved brought up the subject of the costs of respective VR applications for the municipality. "Return on investment is always a big issue for the government", as P1 put it. Since taxpayers' money is used for the implementation of such applications, their usefulness and cost-effectiveness must be thought through and evaluated thoroughly at an early state. Furthermore, potential cost savings in comparison to an established analog or digital process must be analyzed.

4.2   Chances and Opportunities

In the following, we cluster and summarize the interview partners' remarks on chances and opportunities of VR technology for governmental applications.

Informing Citizens. Many interviewees considered VR an alternative touch point and channel for municipalities to communicate with citizens. Two participants imagined advanced information and consultancy services for citizens via VR technology. Advantages mentioned include less travel time and travel costs for citizens and, possibly, less waiting time in contrast to physical counters. P3 emphasized administration's ongoing efforts towards sustainability and saw VR as an opportunity for further decreasing the number of printed information material.

Involving and Connecting Citizens. In addition to providing information, interviewees identified government-related social gatherings and participation in VR as a major opportunity and emphasized the social dimension of VR technology. As main advantage the involvement of citizens over large distance and with mobility constraints was seen. For example, P2 questioned the necessity of physical presence for official marriage processes and the civil registry office and imaged: "You can do that online, then everyone puts on their VR goggles". Similarly, social gatherings in VR for participating in political processes were considered an opportunity. However, several interviewees point out political regulations for formal community assemblies. For first experiments with virtual citizen gatherings, e-government practitioners recommend unbureaucratic and informal exchanges suitable.

Visualizing Spatial Development. One core feature of VR, the interactive immersive presentation of virtual worlds, was considered a major opportunity for governmental cases: Public buildings in the planning phase or related spatial development projects of a municipality can be demonstrated realistically to citizens at early stages. In combination with participatory processes mentioned above, such visualizations can be utilized for decision-making in referendums. P5 also suggested VR technology to enable digitally experiencing recent building projects financed with taxpayers' money. While only a small number of citizens might be able to visit and view the building in reality, VR technology would allow many more people "access" to such a site. One interview partner also deemed the retrospective exploration of spatial development over a longer period promising for governmental purposes, e.g., to enable tracking the development of a city for citizens.

Supporting Municipal Utilities. Finally, several experts saw great potential of VR technology for municipal utilities. For example, P4 mentioned efficiency gains in maintenance work through the visualization of broken pipes and utilities. In addition, P6 and P7 pointed out training scenarios relevant for municipal utilities. As an example, VR professional P6 mentioned a recent VR application to learn how to use a high-voltage control box. While in real life mistakes obviously can have devastating consequences, the application helped trainees to build up muscle memory safely in virtual space.

4.3   Risks and Challenges

Besides the opportunities summarized above, the experts also found several risks and challenges when applying VR technology for governmental use cases.

Lack of Acceptance. With reference to the novelty and citizens' overall unfamiliarity with VR technology, several interview partners suspected a lack of acceptance for prospective VR applications for governmental purposes. P1 described his concerns to the main type of prevailing VR applications, gaming and entertainment: "Many still see VR as fun and associate it with less serious applications". P6 and P7 referred to recent government-related applications for Covid tracking and digital citizen IDs and pointed out an overall skepticism about exchanging confidential information with the state via digital means.

Technical Implementation Challenges. The interviewees identified several challenges when it comes to implementing governmental VR applications. P5 pointed to the manifold skills that are required to implement a sophisticated VR application: In addition to professional experts for the complex integration of services and data from several departments of a municipality, experts in 3D modelling, computer graphics, or animation are needed. P1 added, that so far, many e-government application have been implemented with a pure focus on technical functionality, not on usability. As one example, P3 emphasized that media breaks must be avoided: "Who buys VR glasses to do their tax return when they have to print something out later?" While already introduced as key requirement by many interview partners, the implementation of authentication mechanisms was considered a major challenge. P1 referred, for example, to the lack of the possibility to take a fingerprint in VR (as done when ordering a passport).

Health-Related Concerns. Several interview partners stressed that VR experiences are very tiring and that even 15 min spent in a VR world, for example, can be very exhausting. P4 imagined the case of an information desk in VR but considered the application for both employees and citizens not feasible due to health concerns. Also, P7 suggested keeping potential governmental VR application short and argued that "[...] interaction with the state and the administration is an expense that I want to minimize".

Legal Uncertainties. P4 stressed current legal uncertainties as a challenge. He related to VR examples in professional settings: "As an employer I am responsible for occupational safety - what if someone collapses while using a VR application?". Other participants referred to former discussions regarding suitable authentication processes and emphasized legal issues for governmental cases if the real person behind a digital VR avatar is unverified.

4.4 Potential Applications

In the following, we summarize the participants' ideas for potential applications, which also reflect on some of the previous statements made above.

Urban Planning. Various areas of use in urban planning were frequently mentioned as potential applications. VR could be utilized to illustrate how buildings fit into cities, e.g., in the design of new urban neighborhoods or the planning of future buildings. A further field of application may be checking building codes. Building codes are strict and complex, including distance rules, light incidence angles, etc. So, one could put a proposed project online and check those rules in virtual reality.

E-Participation. VR could support e-participation in the field of urban planning. Various aspects of urban planning, as mentioned above, could be presented, and illustrated to the residents in order to improve the basis of participation. Finally, also community meetings in virtual spaces are possible in the near future. Thus, the technology might level the way to more smart government.

Municipal Services. It has also been mentioned several times that VR either might be utilized for cooperation within and between different governmental agencies, but also for advising citizens. E.g., citizens could meet the responsible employee of the authority in virtual spaces to create the feeling of sitting in a room for interaction and exchange, e.g., to work on complex forms.

Training and Education. Different field of application for training and education have been mentioned by the interviewees. Especially for emergency services such as police, fire services, or emergency responders, but also in the field of military, several application areas can be imagined training the handling of dangerous situations. VR also could make sense in the use of public utilities such as gas, water, electricity. Here, damage to infrastructure could be visualized either for training or to support the repair of damage.

Arts and Culture. One expert mentioned the application of VR in the area of arts and culture. Here, interested people could be enabled to participate virtually in performances as well as exhibitions.

Weddings in "Metaverse". One e-government professional also mentioned organizing virtual weddings: "So for me anyway, there is no reason to go to the civil registry office to get married. [...] I actually still think it's pretty cool."

5   Discussion

Referring to requirements mentioned by the experts for applying VR technology for governmental applications, we see many general prerequisites for e-government services such as strong data security and reliable user authentication. In particular, authentication has been identified both a requirement and core challenge for other advanced e government channels such as conversational government [1]. Authentication in VR is a heavily investigated research topic with a lot of innovative approaches proposed by researchers (cf. [6,7]). However, only few authentication techniques have been implemented so far in mass-market VR platforms. In addition, authentication for governmental applications requires a digital ID to verify a user's identity.

Costs for developing and consuming governmental VR applications have been a major topic for several of the experts. They recommended careful analysis of the return of investment for such applications. Additionally, they considered current prices of recent VR goggles still too high for widespread use and acceptance of such devices. For first implementations of selected use cases, we thus consider web-based approaches for implementing VR applications highly suitable. Example technologies include *WebXR*[1] and related VR framework such as *A-Frame*[2]. Resulting VR applications can be viewed both on dedicated hardware such as VR goggles and in modern Web browsers, thus are accessible for a very large number of citizens.

During our interviews, the experts identified potential VR applications in several areas, from typical government tasks (information, participation), over use cases supporting municipal utilities (maintenance, training) to less government-related areas (arts, culture). The VR application they considered most promising, participatory urban planning and spatial development, is covered by a decent body of literature (cf. [5,8,14]). In contrast, citizen gatherings, town meetings, and referendums in VR are less investigated, yet might offer new opportunities for including diverse populations. While advantages of VR representations in the spatial development processes (such as realistically visualizing buildings in the planning stage) are obvious, the added value of VR for citizen meetings in digital space over video-based online meetings, for example, is less clear.

6   Conclusion and Outlook

In this paper, we presented an exploratory study on requirements, chances and opportunities, risks and challenges as well es potential application scenarios for

---

[1] https://www.w3.org/immersive-web/.
[2] https://aframe.io/.

VR-based e-governmental services. To cover a broad view in this not yet well researched field, we interviewed eight experts: two e-government practitioners, an e-government IT professional, two e-government researchers and three VR experts.

While having identified potential use cases to involve citizens and business in government processes using VR technology, our results also show that important prerequisites must be met for a successful diffusion of the technology in governmental agencies and public administration. Today VR technology is mainly used in the gaming and entertainment industry. For e-governmental application specific adaptions are needed to enhance acceptance and dissemination. Furthermore, we shall keep in mind, that a young generation is growing up, which deals much easier with new technologies. Their demands, i.e., what they experience in private, might increase pressure on administrations and thus trigger new applications and developments.

Future research could study the added value of VR applications in governmental services in depth, more specifically the added value of VR for including diverse population in citizen gatherings, town meetings, and referendums through meetings in a digital space. Research needs to investigate both the usefulness from a citizen perspective and the cost-effectiveness of VR applications for specific use cases and can provide the data basis justifying the investment of taxpayer's money in VR-enabled smart government services.

Acknowledgements. The authors would like to thank Admir Durmo, Aleksandar Obradovic, Melanie Baumgartner, and Veton Rasaj for conducting the interviews.

References

1. Baldauf, M., Zimmermann, H.D.: Towards conversational e-government. HCI in Business, Government and Organizations. HCII 2020, pp. 3–14 (2020). https://doi.org/10.1007/978-3-030-50341-3_1
2. Baldauf, M., Zimmermann, H.D., Pedron, C.: Exploring citizens' attitudes towards voice-based government services in switzerland. Human-Computer Interaction. Design and User Experience Case Studies, pp. 229–238 (2021). https://doi.org/10.1007/978-3-030-78468-3_16
3. Buljat Raymond, B.: When information systems address environmental sustainability challenges: the role of immersive technologies. In: AIM2021 (Association Information et Management), June 2021
4. Fegert, J., Pfeiffer, J., Peukert, C., Golubyeva, A., Weinhardt, C.: Combining e-participation with augmented and virtual reality: insights from a design science research project. In: ICIS 2020 Proceedings, pp. Paper- Nr.: 1521. AIS eLibrary (AISeL) (2020), 12.02.04; LK 01
5. Jamei, E., Mortimer, M., Seyedmahmoudian, M., Horan, B., Stojcevski, A.: Investigating the role of virtual reality in planning for sustainable smart cities. Sustainability **9**(11), 2006 (2017). https://doi.org/10.3390/su9112006
6. Jones, J.M., Duezguen, R., Mayer, P., Volkamer, M., Das, S.: A literature review on virtual reality authentication. IFIP Adv. Inf. Commun. Technol. **613**, 189–198 (2021). https://doi.org/10.1007/978-3-030-81111-2_16

7. Kürtünlüoğlu, P., Akdik, B., Karaarslan, E.: Security of virtual reality authentication methods in metaverse: an overview (2022). https://doi.org/10.48550/arxiv.2209.06447. https://arxiv.org/abs/2209.06447v1
8. van Leeuwen, J.P., Hermans, K., Jylhä, A., Quanjer, A.J., Nijman, H.: Effectiveness of virtual reality in participatory urban planning: a case study. In: Proceedings of the 4th Media Architecture Biennale Conference, pp. 128–136. MAB18, Association for Computing Machinery, New York, NY, USA (2018). https://doi.org/10.1145/3284389.3284491
9. Leible, S., Ludzay, M., Götz, S., Kaufmann, T., Meyer-Lüters, K., Tran, M.N.: ICT application types and equality of e-participation - a systematic literature review (2022). https://aisel.aisnet.org/pacis2022/30
10. Pfeiffer, J., Fegert, J., Greif-Winzrieth, A., Hoffmann, G., Peukert, C.: Can immersive systems help address sustainability goals? insights from research in information systems. Market Engineering, pp. 135–150 (2021). https://doi.org/10.1007/978 3 030-66661-3_8
11. Porwol, L., Ojo, A.: Harnessing virtual reality for e-participation: Defining VR-participation domain as extension to e-participation. In: ACM International Conference Proceeding Series, pp. 324–331 (2019). https://doi.org/10.1145/3325112.3325255
12. Porwol, L., Ojo, A.: Virtual reality-driven serious communication: Through VR-dialogue towards VR-participation. Inf. Polity **26**, 501–519 (2021). https://doi.org/10.3233/IP-210331
13. Simonofski, A., Snoeck, M., Vanderose, B., Crompvoets, J., Habra, N.: Reexamining e-participation: Systematic literature review on citizen participation in e-government service delivery. AMCIS 2017 Proceedings (2017). http://aisel.aisnet.org/amcis2017/eGovernment/Presentations/3
14. Stauskis, G.: Development of methods and practices of virtual reality as a tool for participatory urban planning: a case study of Vilnius City as an example for improving environmental, social and energy sustainability. Energy Sustainability Soc. 4(1), 1–13 (2014). https://doi.org/10.1186/2192-0567-4-7
15. Tozsa, I.: Virtual reality and public administration. Transylv. Rev. Admin. Sci. **9**, 202–212 (2013). https://rtsa.ro/tras/index.php/tras/article/view/120

Measuring the Effectiveness of U.S. Government Security Awareness Programs: A Mixed-Methods Study

Jody L. Jacobs(✉) [iD], Julie M. Haney[iD], and Susanne M. Furman[iD]

National Institute of Standards and Technology, Gaithersburg, MD 20899, USA
{jody.jacobs,julie.haney,susanne.furman}@nist.gov
https://csrc.nist.gov/usable-cybersecurity

Abstract. The goal of organizational security awareness programs is to positively influence employee security behaviors. However, organizations may struggle to determine program effectiveness, often relying on training policy compliance metrics (e.g., training completion rates) rather than measuring actual impact. Few studies have begun to discover approaches and challenges to measuring security awareness program effectiveness within compliance-focused sectors such as the United States (U.S.) government. To address this gap, we conducted a mixed-methods research study that leveraged both focus group and survey methodologies centered on U.S. Government organizations. We discovered that organizations do indeed place emphasis on compliance metrics and are challenged in determining other ways to gauge success. Our results can inform guidance and other initiatives to aid organizations in measuring the effectiveness of their security awareness programs.

Keywords: security awareness · training · government · effectiveness · metrics · mixed-methods

1 Introduction

The goal of organizational security awareness programs is to help employees recognize and appropriately respond to security issues, improving the overall security posture of organizations [28]. Various public and private industry sectors require or recommend annual security awareness training. For example, the Federal Information Security Modernization Act (FISMA) - a security law for U.S. Government organizations - mandates the implementation of security awareness training for all employees [2].

Organizations collect metrics about their security awareness programs to satisfy mandatory training requirements, show return on investment, or demonstrate overall program success and value to management [5,14]. The success of security awareness programs is often measured by the number of organizational employees completing or attending the training (i.e., compliance to the training mandates) [5,16]. However, these compliance metrics tell little of how employee

© Springer Nature Switzerland AG 2023
F. Fui-Hoon Nah and K. Siau (Eds.): HCII 2023, LNCS 14038, pp. 14–33, 2023.
https://doi.org/10.1007/978-3-031-35969-9_2

security behaviors and attitudes have been positively changed [3]. Indeed, prior literature and industry surveys have revealed that security awareness programs often fall short in changing behaviors, in part because they struggle with how to measure program impact [4,9,22,24]. Without insight into impact, security awareness programs may not be able to identify and plan for improvements necessary for facilitating behavior change and adjusting to ever-changing threats and organizational needs [5].

Few studies have begun to discover the approaches and challenges to measuring the effectiveness of organizational security awareness programs within compliance-focused sectors like the government [17,21]. To address this gap, we conducted mixed-methods research involving U.S. Government (federal) professionals who implement or oversee security awareness programs. Focus groups with 29 individuals informed the development of a survey completed by 96 participants. While the research looked at multiple aspects of government security awareness programs, this paper focuses on a subset of research questions (RQs) about measuring program effectiveness:

RQ1: How do U.S. Government organizations determine the effectiveness of their security awareness programs?
RQ2: How do government security awareness teams use program effectiveness data?
RQ3: Which types of effectiveness data do managers find most valuable?
RQ4: What are the challenges government organizations face when trying to measure effectiveness?

Our study makes several contributions. We provide new insights into how security awareness programs approach and struggle with measuring effectiveness within a yet-to-be-explored context (the U.S. Government). This understanding can inform government security awareness professionals, organizational decision makers, and policy makers in their efforts to improve security awareness programs. Results are also contributing to the development of a publication to guide organizations in building effective security awareness programs [18]. While our study is U.S. government-focused, findings may be transferable to other sectors and countries.

2   Background and Related Work

To better contextualize our study results, we provide a summary of prior literature related to measuring the effectiveness of security awareness programs and background information on security awareness mandates.

2.1   Security Awareness Mandates

Security awareness programs are meant to provide employees with an understanding of security risks and the knowledge and tools to help them take appropriate action, with a goal of achieving long-term behavior change [28]. The cornerstone of security awareness programs is awareness training, most often conducted online and annually. U.S. Government agencies are mandated to conduct this annual training for all employees and contractors in accordance with several directives, including Office of Management and Budget Circular A-130 *Managing Information as a Strategic Resource* [20] and FISMA [2]. Beyond federal organizations, some U.S. state governments have also adopted security awareness training as a part of their information security program. For example, the Massachusetts data security law requires ongoing training focused on internal and external risks to data records containing personal information [27]. Other countries, such as Canada [10], also have training directives.

Organizations in the private sector may also be subject to security awareness training mandates. For example, organizations in the healthcare sector are required to conduct training under the Health Insurance Portability and Accountability Act [7], and the financial sector must adhere to the Gramm-Leach-Bliley Act, which requires similar training to ensure the protection of sensitive client and financial information [1].

Beyond annual, mandated training, programs – though not required – may integrate other security-related activities and communications throughout the year, including newsletters, emails, speaker events, posters, and even novel approaches, (e.g., virtual reality and escape rooms) [11]. These additional activities intend to reinforce learning and dynamically to address new security threats, policies, and processes as they arise.

2.2 Evaluating Security Awareness Programs

Measuring success is a critical, but challenging aspect of security awareness programs, with many organizations failing to adequately gauge program effectiveness [4,9]. In fact, in an industry survey of 600 organizations, less than half reported that their organizations attempt to measure the effectiveness of their awareness programs [16]. This shortfall may in part be due to reliance on metrics focused on compliance to awareness training policies (e.g., FISMA) as indicators of success [16]. However, compliance metrics fail to capture overall program impact (i.e., employee behavior change) and ignore the influence of additional awareness efforts throughout the year.

Several research and industry groups developed frameworks for measuring the effectiveness of security awareness programs. Manifavas et al. developed a tool to automate and formalize the deployment and maintenance of security awareness assessment, including metrics to measure changes in workforce knowledge, attitude, and behavior [14]. The European Network and Information Security Agency (ENISA) defined four categories of measures for security awareness evaluation: process improvement, attack resistance, efficiency and effectiveness, and internal protections [8]. The guidelines further recommended that organizations continually measure and monitor program performance and automate metrics gathering as much as possible. Rantos et al. developed a methodology for assessing the effectiveness of organizational security awareness programs [22]. They identified two major issues that must be considered when measuring effectiveness: 1) whether the information has reached the target audience (e.g., if and how information was delivered) and 2) whether the information was absorbed by the target (e.g., if learning and behavior change has been achieved). In their survey research, the security training institute, SANS, found that organizations that assess their programs against peers tend to have greater leadership support for security awareness training, and, therefore, more success [24]. To provide this peer benchmark, they developed the five-level Security Awareness Maturity Model [23].

In a more recent effort, Chaudhury et al. conducted a systematic literature review towards defining metrics for measuring the success of a security awareness program [5]. The resultant metrics framework consisted of four overarching categories of indicators measured by quantitative (objective) data:

- *Impact indicators* measure changes in security knowledge, attitude, and behavior and can be measured by quantitative surveys, web-based tests, simulated attacks, or analysis of passive data (e.g., audits, risk assessments, security incidents).
- *Sustainability indicators* measure the value-added and impact on organizational policies and regulatory frameworks and can be assessed via changes in program funding and resources, cost-benefit analysis of the program, and percentage of awareness processes in organizational policies and processes.
- *Accessibility indicators* measure topic relevance, quality of training materials, and the reachability and usability of awareness dissemination channels. Indicators can be collected with quantitative surveys and analysis of passive data (e.g., attendance logs and training material hit counts).
- *Monitoring indicators* gauge the workforce's interest and participation in the security awareness program and leadership support. These indicators can be collected via quantitative surveys and analysis of passive data (e.g., attendance logs, training material hit counts).

The value of quantitative versus qualitative data to help measure effectiveness has been a topic of debate. Some argue for the use of only quantitative metrics (e.g., quantitative surveys, percentages of employees performing an action, analysis of incidents), saying that these are preferred since the data are more objective, repeatable, and can provide benchmarks for future evaluations [5,14]. However, others (e.g., [8,22]) suggest collecting a combination of quantitative and qualitative data. Qualitative measures (e.g., observations, detailed reports from employees, open-ended feedback forms) can be used to gauge audience satisfaction with the program, obtain ideas for improvement, and provide context and root-cause analysis to quantitative data.

Our study sought to position these prior research findings and frameworks within a new context: the U.S. Government. We also wished to gather data from the perspective of those working in security awareness programs, an approach that is in contrast to the majority of prior studies aimed at measuring the awareness levels of users [17,21].

Table 1. Focus group composition

Focus Group	# Participants	# Unique Organizations
Department #1	3	3
Department #2	3	3
Sub-component #1	3	3
Sub-component #2	5	4
Sub-component #3	3	3
Independent #1	4	4
Independent #2	4	4
Independent #3	4	4
Total:	**29**	**28**

3 Methodology

From December 2020 - July 2021, we conducted exploratory, sequential mixed-methods research consisting of focus groups followed by a survey. The National Institute of Standards and Technology (NIST) Research Protections Office approved the study. Focus groups provided an understanding of security awareness approaches and the concepts and challenges viewed as most important by participants. These insights informed a follow-on survey distributed to a larger population.

3.1 Focus Groups

We first collected qualitative data via focus groups. We selected a multiple-category design [13] with participants from three categories of organizations: 1) department-level organizations (e.g., U.S. Department of Labor[1]), 2) sub-component agencies, which are organizations under a department (e.g., Bureau of Labor Statistics under Department of Labor), and 3) independent agencies, which are not in a department (e.g., Federal Trade Commission). In the Executive Branch of the U.S. Government, there are 15 departments, over 200 sub-components, and just over 100 independent agencies. Participants were federal employees who had security awareness duties or were managers or executives who oversaw the programs within their organizations. We identified participants via: recommendations from security awareness colleagues; our professional contacts; security-focused government online mailing lists; and internet searches.

We conducted eight virtual focus groups with 29 total participants, representing 28 unique government organizations. Table 1 shows the composition of each focus group. Participants provided informed consent and completed an online survey to collect demographic and organizational information. Focus groups lasted 60–75 minutes and were audio-recorded and transcribed.

---

[1] Organization names are for illustrative purposes only and do not signify the organizations' participation in the study.

Following an analysis methodology informed by Grounded Theory [6], each member of the research team independently coded a subset of three transcripts (one from each category of focus group) using a preliminary code list based on the focus group questions. We added new codes as needed and met several times to discuss codes and develop a codebook. Coding continued until all remaining transcripts were coded by two researchers, who met to discuss code application and resolve differences. In accordance with the recommendation of qualitative methodologists [15], we focused not just on agreement but also on how and why disagreements in coding arose and the insights afforded by subsequent discussions. When disagreement occurred, we discussed as a group to reach consensus. In rare cases where agreement could not be reached, the primary coder made the final decision. The entire research team convened to discuss overarching themes identified in the data and areas of interest to include in the subsequent survey.

3.2 Survey

Focus group insights informed the development of an anonymous, online survey. The final survey included questions about security awareness approaches and challenges. This paper focuses on a subset of questions related to measuring program effectiveness.

Recruitment methods and participation criteria mirrored those in the focus groups. The survey was open for 18 days, with 96 survey responses in the final dataset. Survey participants represented a diverse range of organizations of different types and sizes. Table 2 shows the organizations represented in both the focus groups and the survey. As indicated in the table, participants reported their organizations' type and size (number of government employees), the number of people (government and non-government contractors) covered by the organization's security awareness program, and the number of individuals tasked with implementing the security awareness program (team size). We calculated descriptive statistics of quantitative responses. We also calculated inferential statistics to look for potential differences among organizations of different types, program sizes, and security awareness team sizes (Kruskal Walls H Test for ordinal dependent variables and Chi-square tests for categorical dependent variables). We only report significant results. For open-ended responses, two researchers performed qualitative data coding similar to the method employed for the focus group data.

3.3 Limitations

Although we recruited participants from organizations of varying sizes and types, our participants may not represent the full range of government security awareness programs. Our investigation is also limited to the U.S. Government, which may have different security awareness training policies and pressures as compared to other sectors. However, given that security awareness training is common in many sectors, our findings may be transferable, at least in part, to other organizations. Similar studies with other populations would be valuable.

Table 2. Represented organizations

		Focus Groups (n = 28)	Survey (n = 96)
Type	Independent	42.9%	35.4%
	Department	21.4%	32.3%
	Sub-component	35.7%	31.3%
*Size**	Less than 1,000	7.1%	17.7%
	1,000-4,999	32.1%	29.2%
	5,000-29,999	28.6%	25.0%
	30,000+	32.2%	25.0%
	Don't know	0%	3.1%
*Program size***	Less than 1,000	0%	22.1%
	1,000-4,999	25%	25.3%
	5,000-29,999	28.5%	26.3%
	30,000+	12.8%	24.2%
	Don't know	3.6%	2.1%
Team size	1 - 2	25%	33.8%
	3 - 4	53.6%	29.7%
	6 - 10	10.7%	14.9%
	11+	10.7%	21.6%

*Size = number of government employees. **Program size = number of government and contract employees covered by the security awareness program.

4   Results

Since participants had the option of skipping survey questions, we report the number of responses (n) for each survey question. Direct quotes from the focus groups and open-ended questions in the survey are included to further expand upon quantitative survey results. We attribute focus group quotes with identifiers D01-06 for participants from departments, S01-11 for sub-components, and N01-12 for independent agencies. Survey participants are indicated with Q01-96.

4.1   Measures of Effectiveness

We asked participants how their organizations try to measure the effectiveness of their security awareness program. Response frequencies are shown in Fig. 1. Sixty-four percent used at least five different measures, and only 4% selected just one measure of effectiveness or did not measure effectiveness. Indicators of compliance to training mandates (training completion rates and audit reports) were common across both survey and focus group participants, with completion rates being the most-selected measure in the survey (84%). In the survey, organizations also frequently utilized phishing simulation click rates (72%) and reporting of simulated (62%) and real-world phishing (53%) to gauge effectiveness of phishing-related training.

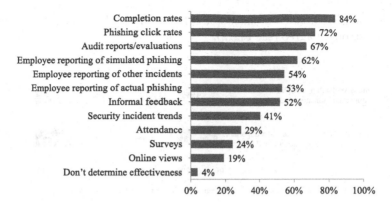

Fig. 1. Measures of effectiveness (n = 79)

Some participants looked for demonstrated employee behaviors, for example, monitoring trends in user-caused security incidents (41% in the survey) or employee incident reporting (54%) to determine whether certain security topics were being translated into action by the workforce. For example, a program lead in the focus groups said, "I interact with our SOC [security operations center] to see what types of events and incidents are being reported to see if there's any way that I can incorporate some sort of training if the incident is the result of user behavior within the agency" (N09).

Other participants made use of employee feedback to determine if their security awareness efforts were perceived as valuable: 52% of survey participants gauged success via informal feedback, and 24% used surveys. A focus group participant remarked:

> "For all of our virtual events and at the end of our training, we have surveys... It gives them a rating scale and asks them, was the training effective?... Was the delivery or the presenter's delivery effective? And we use that feedback to measure our training" (D06).

Other measures, such as event attendance and views of online materials (e.g., newsletters or videos) were also used, although by fewer survey participants (less than 30%). Several focus group participants mentioned that their organizations routinely track attendance as an indicator of reach across the organization: "We keep tally of whenever we have a speaker, we make sure that we determine all the people that are there and sort of use those as some rough stats as to success with the campaign" (S04).

4.2   Compliance as Indicator of Success

To determine if compliance with government mandatory training requirements (e.g., as measured by training completion rates) was regarded as the most important indicator of program success, in the survey we asked participants to rate their agreement with two statements on a five-point scale ranging from strongly disagree to strongly agree (see Fig. 2).

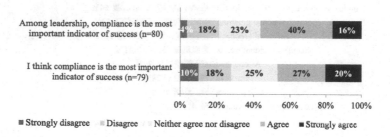

Fig. 2. Agreement that compliance is the most important indicator of success

Leadership Perspective on Compliance: In the first statement, participants were asked to indicate whether they believed their organization's leadership thinks compliance is the most important indicator of security awareness program success. Over half of responding participants (56%) agreed or strongly agreed with this statement, and 22% either disagreed or strongly disagreed.

In the focus groups, several participants commented on how compliance metrics garner leadership attention, regardless of how meaningful those might be. A security awareness program lead commented, "We have found that, yes, management pays attention to things with compliance... Now, that doesn't identify effectiveness,... but it does help increase management awareness and attention to supporting these programs" (S11).

Participant Perspective on Compliance: In the second statement, participants were asked to rate their agreement related to their own opinion on compliance being the most important indicator of program success. As compared to the leadership perspective, fewer (47%) agreed with this statement and more (28%) disagreed.

Despite almost half of survey participants believing compliance is the most important indicator of success, many participants in both the focus groups and survey voiced a concern that compliance metrics in the form of training completion rates, although required, do not demonstrate long-term attitude or behavior change: "Completion of training is one statistic, but that doesn't really tell you whether anything's sunk in. It tells you that they got through the course" (N11).

4.3   Using Effectiveness Data

We asked participants how their security awareness program uses program effectiveness data (see Fig. 3). Most commonly, programs use the data to demonstrate training compliance (78%) or to improve or inform the program (70%). Over half use the data to demonstrate the value of their program to leadership (58%). Less than a quarter provide the data to employees to provide transparency about the security awareness program or pass on the data to inform the efforts of other groups in the organization.

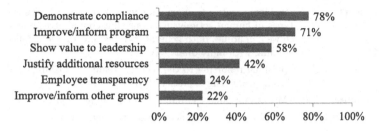

Fig. 3. How programs use effectiveness data (n = 72)

Participants provided further explanations on how they use effectiveness data. A security awareness program lead at a sub-component agency used data to inform leadership and employees within the organization: "We do have compliance metrics that we report. Management does pay attention to that, and it does heighten awareness with staff" (S11). A survey participant suggested that security awareness professionals "capture metrics to show where you started (e.g. phishing susceptibility, training rates, incident data), inform your program's strategy and tactics, and show progress" (Q43).

4.4   Manager Preferences

We asked survey participants who were managers an open-ended question about what data would help demonstrate the value and effectiveness of the program to them. Twenty-nine participants answered this question (see Table 3). Security incidents were most frequently mentioned as valuable (59% of those responding to this question). However, in the previous question on measures of effectiveness, only 41% said that their program uses security incident data, possibly demonstrating a gap in current measures. Phishing data (31%), training completion rates (24%), employee feedback (21%), and other demonstrations of employee behaviors (21%) were among other frequently-mentioned data.

4.5   Program Support

Perceived support in the form of direct feedback, actions, and allocated resources can be another effectiveness indicator. While we were not able to collect direct evidence of support, we were able to gauge how supported our participants felt.

Perceptions of Support. Participants rated their level of agreement for two statements about perceived support for the security awareness program. Figure 4 shows the agreement percentages.

Table 3. Manager perspective - Data demonstrating security awareness program value (n = 29). # indicates number of participants mentioned that data.

Type of Data	Example Responses	#
Security incidents	"incidents more granularly analyzed and categorized as to the types of human actions/inactions that contributed, and who, so we can adjust both general training and targeted follow-up training with individuals." (Q43)	17
Phishing data	"phishing reporting to the security team or phishing clicks during a phishing exercise." (Q74)	9
Completion rates	"Metrics for timely completion of training" (Q38)	7
Employee/user feedback	"We also review feedback of the training." (Q39)	6
Other demonstrations	"Adhering to the rules of behavior" (Q39)	6
Data relationships	"annual CSAT [cybersecurity awareness training], IT Professional/Role Based Training, and Phishing Click data graphed with the Network Monitoring data and Helpdesk reporting data" (Q17)	4
Training topics	"Categories of questions pertaining to each area of operations... Topical areas help to identify the practical application of cybersecurity across the organization." (Q38)	4
Employee reporting	"The number of staff who actually recognized an incident, report them, and follow recommended practices." (Q30)	3
Participation	"Event attendance" (Q24)	3
External data	"peer agency metrics" (Q83)	3
Knowledge testing	"Exam scores, number of times a course is repeated, most likely failed questions" (Q38)	3

Leadership Support for the Security Awareness Program: A large majority (88%) thought their leadership was supportive of their program. Only 5% disagreed or strongly disagreed. When participants were asked what advice they would provide to their colleagues, gaining leadership support was the one of the most frequently mentioned topics in both the focus groups and survey. A focus group participant commented on their leadership's support: "I would say we have good support from our management and executives. They seem to give us a lot of flexibility. If we want everyone to have a phishing exercise, they give us a little leeway to do so. If we draft a newsletter or a poster or something, they'll send it out to the user population agency-wide" (S07). However, several lamented the lack of support. A program lead said, "I don't think management would do it unless it was mandated by law... At least every few years, I have to quote the legal basis for delivering this required training" (D05). A survey participant commented, "Anything that was done in the past was personal initiative. I've done newsletters, websites, tried to get Hollywood movies regarding security shown with discussion afterwards, tried to get a security game (I was told that was insulting). Management just didn't care" (Q34).

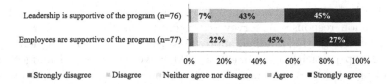

Fig. 4. Agreement that workforce supports the security awareness program

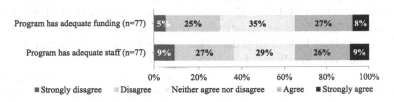

Fig. 5. Agreement about having adequate resources

Employee Support of the Security Awareness Program: Seventy-two percent of survey participants agreed that employees were supportive of their program. Just 6% disagreed/strongly disagreed. When asked about successful aspects of their program, some participants commented on employee engagement and interest in the program. One said that their program was "popular with the people, encouraging engagement and behavior change" (Q56). In contrast, others remarked that employees may lack the time or motivation to engage with security awareness information or activities: "A lot of times we'll find that sometimes our users aren't engaging with the message, or they may delete it, or they don't report it the way that we want them to" (N01).

Program Resources. Survey participants indicated their level of agreement about two statements about whether the security awareness program has adequate resources in the form of funding and staff. Figure 5 shows the results.

Adequate Funding for the Security Awareness Program: Only a little over one third of participants (35%) agreed/strongly agreed that their security awareness program has adequate funding, with 30% disagreeing or strongly disagreeing. There were statistically significant differences between: very small and large teams ($z = -2.445$); and small and large teams ($z = -2.925$). For very small teams (n = 22), 32% disagreed with the statement and 23% agreed (the remainder were neutral). For small teams (n = 19), 47% disagreed/strongly disagreed and only 16% agreed/strongly agreed. In contrast, large teams (n = 12) were more likely to agree/strongly agree that they had adequate funding (67%).

When asked what could help their programs be more successful, more funding was a common response. A survey participant remarked, "Finding content is not the problem- getting funding/approval to purchase it is the problem" (Q43). A focus group participant commented, "We have a very small budget for our cybersecurity awareness program. I've seen some products in the private sector that are very slick and customizable, but they're also expensive" (S06).

Adequate Staff Dedicated to the Security Awareness Program: When asked about program staffing, 35 agreed/strongly agreed that they had adequate staff, while 36% disagreed/strongly disagreed. Staff resources were closely related to perceived lack of funding. There were statistically significant differences between: very small and large teams ($z = -2.198$); and small and large teams ($z = -2.758$). While a large number of participants with very small and small teams did not think they had adequate staff (45% and 58%, respectively), only 8% (one) participant in a large organization disagreed with the statement.

In qualitative remarks, participants often discussed needing more staff to improve their programs. For example, a survey participant expressed the need for "Additional staff/SMEs [subject matter experts] to help create content other than only myself" (Q74). The fact that most security awareness team members were part-time and had other duties also contributed to the staffing shortage: "The team...who perform the security-related operations for our network, they're the same team that helps create and manage the training. So, if we have an issue or a series of issues, sometimes we may have to either delay training or make a lighter version of training" (N07). Staff turnover was also viewed as a disruption for programs: "Frequent staff turnover, including CISO and CIO positions decrease the long-term success of a program because ideas, funding and priorities change and ultimately limit program strength and growth opportunities. Meaning you can't build a great house, if you keep ripping up the foundation every year or two" (Q24).

4.6   Challenges

We asked participants to rate their programs' challenges related to determining program success on a five-point scale ranging from "very challenging" to "not at all challenging" with a "does not apply" (N/A) option (Fig. 6). The remainder of this section provides details on survey results for each challenge and includes example supporting quotes from focus group and survey participants.

Determining what and how to Measure: Forty-four percent of survey participants rated determining what to measure and how to measure program effectiveness as very or moderately challenging. Only 14% rated it not at all challenging. Although most programs make at least some attempt to determine success, almost half of focus group participants expressed uncertainty about how. A program lead remarked, "How do we determine whether or not it is effective?... How are we making a difference when we educate our workforce?" (N04)

Participants expressed a desire for more government guidelines and standards on how to measure program effectiveness, including what variables to measure

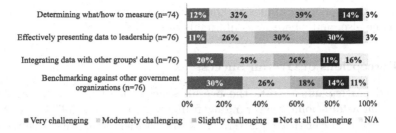

Fig. 6. Challenges determining program effectiveness

and how to interpret training metrics. For example, a participant saw the poten tial benefit of having "something standard that all the departments and agencies could actually end up measuring" (S01).

Effectively Presenting Data to Leadership: Presenting program data to leadership in an effective way was rated very or moderately challenging by 37% of survey participants. One focus group participant expressed frustration with not being able to convince their leadership to help solve challenges faced by the security awareness team: "I have no idea how to solve the issues and challenges as, even though I have expressed challenges to the Department, it appears they all fall on deaf ears" (S09). Other participants recommended developing a robust plan to garner leadership support:

> "Write up some type of training and awareness program plan so that you can document what it is that you want the program to do and how you want it to work and all of the players that would be involved so that you can brief senior leadership on that. Because if you don't have their buy-in, then your program is probably not going to go anywhere" (D02).

Integrating/Correlating Security Awareness Data with Data Collected by Other Groups in My Organization: Being able to bring together data from multiple groups to inform the security awareness program was rated as very or moderately challenging by 48% of survey participants. Only 11% rated this as not challenging at all. Focus group participants commented on how their organizations were not currently connecting security awareness data with security incident data. A program lead said, "Ideally, you'd be able to track the incidents and see based on your security awareness and training and if your incidents are going down. We are not doing that, probably due to lack of resources" (S06).

Benchmarking My Organization Against Other Federal Organizations: Over half (56%) of survey participants rated benchmarking (comparing) their organization's security awareness program against programs in other government organizations to be challenging. Several participants expressed a desire to have more government-specific information as a comparison point:

> "With our phishing exercise results, I would love to have... a standard way of looking at our agency or across agencies or across departments. We

could judge apples to apples to know where we are, how we stand up to someone else, and where we could focus our training" (S08).

4.7   Perceptions of Overall Success

We asked participants to rate the overall success of their security awareness program on a four-point scale ranging from "very unsuccessful" to "very successful" (Fig. 7). Over three-quarters (77%) rated their programs as moderately or very successful. None rated their program as unsuccessful, and only 4% rated their programs as very unsuccessful. Using Kendall's Rank Correlation for ordinal data, we found a statistically significant association between the number of measures of effectiveness employed by organizations and the ratings of program success ($\tau = 0.36$, p = 0.001); the more measures of effectiveness used, the higher the rating of success.

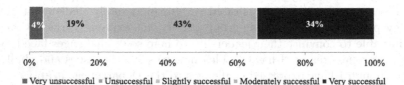

Fig. 7. Ratings of overall security awareness program success (n = 80)

During the focus groups, participants differed on the ultimate indicator of success for their security awareness program. Some emphasized compliance: "I was at 99.9% last year, which is pretty hard when you have between 38 and 45 thousand employees" (S09). However, others saw overall success as being grounded in a tangible reduction in incidents. One focus group participant remarked, "That is really the number of incidents that we end up having and tracked throughout the year and ultimately, not to be on the five o'clock news for some type of compromise or breach" (S01). Another explained:

> "It is...the elimination or...mitigation of all those threats and vulnerabilities, those incidents that have to be reported and even those that don't have to be reported. Just you want to make sure that we have smooth sailing as far as our daily operations, that there's no impact to...the service that we're supposed to provide for the federal government" (S08).

5   Discussion

In this section, grounded in our results and situated in prior research literature, we offer suggestions on how organizations can be supported in effectively measuring program impact. We also discuss areas for future work.

5.1   Development of Guidance and Standards

The following are suggestions for supporting organizations (e.g., sector oversight, standards, and training institutions) that develop security awareness guidance and policies. Organizations can also individually document their own standards and lessons learned to aid in repeatability and continuity when program staff changes.

Develop Standards and Share Lessons Learned. Most study participants said that their security awareness programs were successful. However, this raises the question of how participants know their programs are successful given their expressed challenge with determining what and how to measure and a lack of government guidance and standards, as also confirmed in prior literature focused on the private sector [17, 21]. To address these challenges, guidance could include concrete advice on deliberate planning of measures of effectiveness and standardized measures. An upcoming U.S. Government document entitled "Building a Cybersecurity and Privacy Awareness and Training Program" [18], which was informed by our study, will incorporate many of these suggestions. Guidance could also include how to correlate data from multiple sources (viewed as challenging by almost half of survey participants). Additionally, because the meaning of effectiveness metrics can be contextual [5], guidance can include suggestions for how to tailor baselines of measures to the needs and risk levels of an organization.

Emphasize Impact Over Compliance. Since U.S. Government organizations are required to conduct security awareness training, it was not surprising that training completion rates were the most common measure of effectiveness in our survey and viewed by many as being the most important indicator of success. However, as compared to findings by other researchers in the private sector [3,4,24], we observed a substantial disconnect between the emphasis on compliance and the actual purpose of security awareness: facilitating better employee security behaviors. To combat this issue, guidance documents should emphasize the importance of assessing behavioral impacts.

Provide Guidance on Presenting Data to Leadership. We also observed a disconnect in how security awareness professionals present effectiveness data to organizational leadership. While security incident trends were less commonly utilized by our surveyed organizations, participants who were also managers listed incidents as the measure of effectiveness most preferred for helping them make decisions about the security awareness program, placing less emphasis on compliance metrics. We also found a dissonance between the high levels of perceived leadership support and the high percentages saying that their programs lacked adequate funding and staff. This leaves one to wonder why leadership support had not been translated into resources. Therefore, we see a need for guidance documents to provide examples of what kind of data is most relevant to organizational decision makers to garner both support and needed resources. Suggestions on how to effectively present that data to leadership (e.g., using visualizations and ensuring data is contextually specific [26]) can also help address participants' challenge in that area.

Facilitate Benchmarking and Information Sharing. Given that over half of our participants rated benchmarking as challenging, oversight organizations could also aggregate and share sector-specific data to allow comparisons across programs. Also helpful will be the encouragement of security awareness professionals to utilize maturity models for benchmarking their programs (e.g., the SANS Security Awareness Maturity Model [23]) and forums for sharing experiences related to measuring effectiveness with their peers (e.g., via the Federal Information Security Educator's forum [19] and the SANS Security Awareness online community [25]).

5.2 Collect Holistic Measures of Effectiveness

Collect Data from Multiple Sources for Multiple Purposes. For a holistic perspective, organizations should not rely on only one metric. Rather, they can leverage and combine a variety of different types of metrics - both quantitative and qualitative – as suggested by prior research [3,8,9,12,22,26]. Ultimately, measures should be part of an iterative feedback loop to continually identify areas of concern, refocus, and improve security awareness initiatives. Situating our findings within the metrics framework suggested by Chaudhury et al. [5], we observed an emphasis on impact and monitoring indicators, but suggest collecting the following, more comprehensive indicators:

Impact Indicators: More than half of participants measured program effectiveness with phishing click rates, audit reports, and reporting of security incidents (real and simulated phishing and other incidents). In addition to these, programs could look at further demonstrations of employee behaviors, such as the use of secure authentication mechanisms, user-generated security incidents, and security policy violations.

Sustainability Indicators: Sustainability indicators were only addressed tangentially in the survey in the expression of challenge programs encountered when trying to present meaningful and influential data to leadership. To remedy this current shortfall, programs could better track changes in program resourcing and influences on organizational policies.

Accessibility Indicators: In addition to who and how many employees were reached by security awareness training and other communications, programs can track which types of employees or organizational groups seem to have the most security-related issues or are less likely to receive or pay attention to awareness information. These program teams could then put additional effort into reaching those populations. Accessibility indicators could also be collected via workforce surveys (which were utilized by less than 25% of participants) to gauge topic relevance and perceived quality of materials. Furthermore, informal usability evaluations of security awareness information could be valuable in determining whether security awareness communications are properly tailored to the various workforce audiences and actionable.

Monitoring Indicators: While most organizations collected training completion rates, other types of data could help assess the workforce's interest and engagement in the program. Event attendance and online views of awareness materials were less popular but could be valuable for demonstrating effort in accessing awareness information. Also helpful is the collection of both informal and formal feedback from employees about what is working or not working for them (e.g., via anonymous surveys and focus groups). Feedback from organizational leadership could also help assess impact and organizational attitudes towards security awareness initiatives.

Automate Metrics Collection. Deliberate planning of *what* measures to collect should be followed by deciding *how* to collect those measures. For efficiency and consistency, quantitative metrics should be automated as much as possible [9,14,26]. For example, organizations can leverage existing technology, such as learning management systems, automatic phishing reporting buttons on email clients, or security operations data queries.

5.3   Areas for Future Research

While quantitative data can be especially helpful in identifying issues for managers, unlike Chaudhury et al. [5] and Manifavas et al. [14] who advocated for the exclusive usage of quantitative metrics, our results indicate that qualitative indicators may also be complementary as this data can expand upon quantitative indicators and get at the root cause of workforce challenges and behaviors [8,22]. Additional research is needed to develop recommendations on how programs can gather robust qualitative data and to explore how quantitative and qualitative data can be most effectively and efficiently synthesized. We also do not address potential ethical implications of the collection of effectiveness indicators, especially if used punitively against employees [5]. Additional investigation is needed to determine how data can protect the privacy of employees while still being meaningful and actionable to the organization.

6   Conclusion

Through focus groups and a survey, we provide additional evidence towards developing standards on how to evaluate security awareness programs, a current and important gap [5]. We extended prior research focused on the private sector by exploring the approaches and challenges of U.S. Government organizations in measuring security awareness program effectiveness. We found that compliance metrics were viewed as a primary indicator of program success as opposed to impact on workforce behaviors. Organizations were particularly challenged in determining what to measure due to a lack of standards, management support, and resources across the government. Our results are informing guidance and other initiatives to aid organizations in measuring the effectiveness of their programs.

Disclaimer

Certain commercial companies or products are identified to foster understanding. Such identification does not imply recommendation or endorsement by the National Institute of Standards and Technology, nor does it imply that the companies or products identified are necessarily the best available for the purpose.

References

1. 106th Congress: S.900 - Gramm-Leach-Bliley Act (1999). https://www.congress.gov/bill/106th-congress/senate-bill/900
2. 113th Congress: Federal information security modernization act of 2014. Pub. L. 113–283, 128 Stat. 3073 (2014). https://www.govinfo.gov/app/details/PLAW-113publ283
3. Alshaikh, M., Maynard, S.B., Ahmad, A., Chang, S.: An exploratory study of current information security training and awareness practices in organizations. In: 51st Hawaii International Conference on System Sciences, pp. 5085–5094 (2018)
4. Bada, M., Sasse, M.A., Nurse, J.R.: Cyber security awareness campaigns: Why do they fail to change behaviour? (2019). https://arxiv.org/ftp/arxiv/papers/1901/1901.02672.pdf
5. Chaudhary, S., Gkioulos, V., Katsikas, S.: Developing metrics to assess the effectiveness of cybersecurity awareness program. J. Cybersecur. 8(1), tyac006 (2022)
6. Corbin, J., Strauss, A.L.: Basics of Qualitative Research: Techniques and Procedures for Developing Grounded Theory, 4th edn. Sage, Thousand Oaks, CA (2015)
7. Department of Health and Human Services: The HIPAA privacy rule (2021). https://www.hhs.gov/hipaa/for-professionals/privacy/index.html
8. European Union Agency for Cybersecurity (ENISA): The new user's guide: how to raise information security awareness (en) (2010). https://www.enisa.europa.eu/publications/archive/copy_of_new-users-guide
9. Fertig, T., Schütz, A.E., Weber, K.: Current issues of metrics for information security awareness. In: European Conference on Information Systems (2020)
10. Government of Canada: Directive on security management (2019). https://www.tbs-sct.canada.ca/pol/doc-eng.aspx?id=32611&section=procedure&p=H
11. Haney, J., Jacobs, J., Furman, S., Barrientos, F.: NISTIR 8420A approaches and challenges of federal cybersecurity awareness programs (2022). https://nvlpubs.nist.gov/nistpubs/ir/2022/NIST.IR.8420A.pdf
12. Jaeger, L.: Information security awareness: literature review and integrative framework. In: 51st Hawaii International Conference on System Sciences, pp. 4703–4712 (2018)
13. Krueger, R.A., Casey, M.A.: Focus groups: a practical guide for applied research. Sage (2015)
14. Manifavas, C., Fysarakis, K., Rantos, K., Hatzivasilis, G.: Dynamic security awareness program evaluation. In: Proceedings of the 16th International Conference on Human-Computer Interaction, pp. 258–269 (2014)
15. McDonald, N., Schoenebeck, S., Forte, A.: Reliability and inter-rater reliability in qualitative research: norms and guidelines for CSCW and HCI practice. In: ACM on Human-Computer Interaction, p. 72 (2019)

16. Monahan, D.: Security awareness training: it's not just for compliance (2014). https://www.enterprisemanagement.com/research/asset-free.php/2734/pre/Report-Summary--Security-Awareness-Training:-It's-Not-Just-for-Compliance-pre
17. Muronga, K., Herselman, M., Botha, A., Veiga, A.D.: An analysis of assessment approaches and maturity scales used for evaluation of information security and cybersecurity user awareness and training programs: a scoping review. In: 2019 Conference on Next Generation Computing Applications (NextComp), pp. 1–6 (2019)
18. National Institute of Standards and Technology: pre-draft call for comments: Building a cybersecurity and privacy awareness and training program (2021). https://csrc.nist.gov/publications/detail/sp/800-50/rev-1/draft
19. National Institute of Standards and Technology: FISSEA - Federal Information Security Educators (2022). https://csrc.nist.gov/projects/fissea
20. Office of Management and Budget: Circular a-130 managing information as a strategic resource (2106). https://www.whitehouse.gov/omb/information-for-agencies/circulars/
21. Rahim, A., Hayani, N., Hamid, S., Kia, M.L.M., Shamshirband, S., Furnell, S.: A systematic review of approaches to assessing cybersecurity awareness. Kybernetes 44(4), 606–622 (2015)
22. Rantos, K., Fysarakis, K., Manifavas, C.: How effective is your security awareness program? an evaluation methodology. Inf. Secur. J. Global Perspect. 21(6), 328–345 (2012)
23. SANS: Security awareness maturity model (2018). https://www.sans.org/security-awareness-training/blog/security-awareness-maturity-model-kit
24. SANS: 2021 SANS security awareness report: Managing human cyber risk (2021). https://www.sans.org/security-awareness-training/resources/reports/sareport-2021/
25. SANS: SANS security awareness resources (2022). https://www.sans.org/security-awareness-training/resources/
26. Spitzner, L.: Security awareness metrics - what to measure and how (2021). https://www.sans.org/blog/security-awareness-metrics-what-to-measure-and-how/
27. State of Massachusetts: Title 201 CMR 17.00 - Standards for the protection of personal information of residents of the commonwealth (2017). https://casetext.com/regulation/code-of-massachusetts-regulations/department-201-cmr-office-of-consumer-affairs-and-business-regulation/
28. Wilson, M., Hash, J.: NIST Special Publication 800–50 - Building an information technology security awareness program (2003). https://nvlpubs.nist.gov/nistpubs/Legacy/SP/nistspecialpublication800-50.pdf

Stakeholder-in-the-Loop Fair Decisions: A Framework to Design Decision Support Systems in Public and Private Organizations

Yuri Nakao(✉) and Takuya Yokota

Fujitsu Limited, Kawasaki City, Japan
{nakao.yuri,yokota-takuya}@fujitsu.com

Abstract. Due to the opacity of machine learning technology, there is a need for explainability and fairness in the decision support systems used in public or private organizations. Although the criteria for appropriate explanations and fair decisions change depending on the values of those who are affected by the decisions, there is a lack of discussion framework to consider the appropriate outputs for each stakeholder. In this paper, we propose a discussion framework that we call "stakeholder-in-the-loop fair decisions." This is proposed to consider the requirements for appropriate explanations and fair decisions. We identified four stakeholders that need to be considered to design accountable decision support systems and discussed how to consider the appropriate outputs for each stakeholder by referring to our works. By clarifying the characteristics of specific stakeholders in each application domain and integrating the stakeholders' values into outputs that all stakeholders agree upon, decision support systems can be designed as systems that ensure accountable decision makings.

Keywords: Machine Learning · Fairness · Explainability · Decision Support Systems

1 Introduction

Decision support systems using machine learning have expanded in private and public organizations and they have a great influence on our society. The stakeholders of the decision support systems are diverse and many of them are non-experts in information technology, who have difficulty fully understanding the mechanism of technology. Despite that, each stakeholder needs to know how and why decisions that s/he is responsible for or that affect her or him, are made. Hence, there is a requirement for interfaces that encourage smooth communication between the systems and their users based on the knowledge of human-computer interaction (HCI) techniques.

1.1 Lack of Explainability

To develop the interfaces, we have to overcome the social issues that machine learning provokes. Especially the issue of the lack of explainability and the issue

© Springer Nature Switzerland AG 2023
F. Fui-Hoon Nah and K. Siau (Eds.): HCII 2023, LNCS 14038, pp. 34–46, 2023.
https://doi.org/10.1007/978-3-031-35969-9_3

of discriminatory bias are well-known. The lack of explainability occurs due to the opacity of machine learning to its users or even to engineers [4]. Because, with complex methods such as deep learning, machine learning models are trained through a multilayered network, it is unable for human users to trace the whole process of the training. This leads to that no one can explain the reason for the output from the trained models. This issue of lack of explainability is fatal to high-stakes social decisions such as medical diagnosis or prediction of recidivism. To overcome this issue, the interpretability or explainability of machine learning has been explored. In the conventional research, there are approaches in which, after a machine learning model is trained, a different model to explain a local situation that its users want to know about is trained [25], and that try to train machine learning models in understandable forms for human users [14].

In addition to the technology of interpretable and explainable machine learn ing, what kind of explanation is needed by the target users should be investigated. The decision support systems used in private and public organizations have a variety of stakeholders. Hence, the systems' explanations should be changed so as to make themselves understandable for each stakeholder. For example, engineers in organizations can understand the statistical results explained quantitatively by the decision support systems. In contrast, non-expert users inside and outside the organizations do not understand such results. For non-expert users, the system should express the same results using natural languages. How to express the appropriate explanation for each group of users is a major research topic that HCI community should investigate.

1.2   Lack of Consensus on Fair Decisions

On the other hand, the issue of discriminatory bias in machine learning processes has also been pointed out by academia and industry [17]. This issue occurs because, in the training data, there remain discriminatory biases in decisions in the past, such as gender bias or racial bias, and intersectional bias where such biases combine [13]. The machine learning models trained with biased data output discriminatory results. To remove the discriminatory bias from the training data or trained models, many technologies have been developed [17].

One major concern about the fairness issue is the lack of consensus on the definition of fairness. There are major two definitions of fairness: group fairness and individual fairness [5]. Group fairness is ensured when there is not any difference in the acceptance ratio or performance metrics between the protected group, i.e., discriminated groups such as a minority, colored people in Western countries, or people with disabilities, and the other group. On the other hand, individual fairness is ensured when people who have the same ability or condition are treated the same. It is known that these two definitions of fairness are not sometimes compatible. For example, when a person in a minority group is poorer than a person in a majority group, the person in a minority group will have less education and s/he will have a skill level that is lower than a person in a majority group. In this case, when we follow the definition of individual fairness, we should give preferential treatment to the person from a majority group. However, when we follow the definition of group fairness, we may be ought to give preferential treatment to a person from a minority group.

In addition to this kind of trade-off among the fairness definitions, there are different perceptions of fairness in different cultures. For example, Kim and Leung [12] clarified the cultural difference in the situations considered fair by people. According to them, employees in Japan and the U.S. tend to consider interactional fairness, which is the degree to which the people affected by a decision are treated with dignity and respect as fair. On the other hand, employees in China and Korea tend to consider distributive fairness, which is the fairness related to how rewards and costs are distributed across group members, as fair. Hence, when we try to reach a consensus on fairness, we should consider to what cultural areas stakeholders belong.

In the context of the decision support systems used in public and private organizations, we should consider what are fair decisions for the stakeholders beyond the discussion of discriminatory bias. According to Oxford Learners' Dictionary[1], the word 'fair' means to be *"acceptable and appropriate in a particular situation."* The acceptable and appropriate conditions differ among the local situations of stakeholders. For example, one group of stakeholders might think of the accurate decisions based on the historical data as fair, while another group might consider the decisions where discriminatory bias is removed as fair. Despite that, a social decision should ideally be what is considered fair by all stakeholders. Hence, how to extract the condition stakeholders consider fair and how to integrate the perceived fair condition into the final decision is one of the important research topics of HCI.

2   Our Focus

In this paper, we focus on how to consider appropriate explanations and fair decisions output by the decision support systems used in public and private organizations. We classified the stakeholders who are related to the decision support systems into four groups: experts in organizations, direct recipients of decisions, indirect recipients of decisions, and regulators. Then, we discuss how to consider the explanations and decisions the stakeholders will agree upon. And, finally, we summarize the requirements for the explanations and decisions as the framework of stakeholder-in-the-loop fair decisions for further discussion. Through the framework, we contribute to the field of HCI to provide the foundation of discussion about how to design decision support systems in organizations.

3   Explanation and Fairness for Each Stakeholder

3.1   Experts in Organizations

Here, we consider the preferable explanation and fair results of the decision support systems to the stakeholders in the organizations. First, there are two

[1] https://www.oxfordlearnersdictionaries.com/.

different types of stakeholders in the organization: information technology (IT) experts and domain experts. IT experts are those who operate IT systems in organizations. Some of them manage the dataset and train machine learning models and adjust the models in their daily practice. Although IT experts sometimes exist in outsourced companies, here we consider the outsourced companies are a part of the organization for convenience. On the other hand, domain experts are those whose roles are not related to IT, e.g., loan officers in a bank, who decide if a loan is approved for a customer or doctors in hospitals. Although they do not usually operate decision support systems, they use the results from the systems to execute their roles. Because they usually have responsibilities for their role, they have to understand the results based on their own contexts.

Additionally, there are two different types of domain experts in organizations [20]. First, there are domain experts who work on the deliverables that are used in different departments in the same organization. The deliverables become the resources on which the domain experts in other departments in the organization work based. Usually, there are various departments in one organization or company such as the customer service or citizen relation section, accounting section, and contract audit section. The final decisions for the outer customers or citizens are based on the integration of the decisions made by such various departments. To make the final decision accountable, domain experts in each department have the responsibility to make the accountable deliverables. Hence, they have to know the reason for their decisions. When decisions are made by an IT system, the reason for the decisions should be explained to the domain experts in a form that is reasonable and can be understood by the experts in other departments. The second type of domain experts is those who work for the outer customers or citizens directly, such as doctors in hospitals, and loan officers in banks. Their works have effects on the people outside the organization. Works done by public organizations, such as local and national governments, affect the citizens' lives and works done by private companies affect the behavior of their customers.

It has been pointed out that different types of explanations are needed for different types of experts [2]. For IT experts, statistical explanations are appropriate since IT experts have specialties in understanding statistical outputs. On the other hand, for domain experts, the systems' outputs should be explained in natural languages because the experts do not usually have the skills to understand the statistical outputs. At the same time, for the domain experts, the explanation should be made to meet the responsibility that the experts have. For example, of course, the explanation for the doctors and that for loan officers should be different. Moreover, the appropriate explanations for the stakeholders inside the organizations have to be investigated to make HCI research match the daily practice of experts. The appropriate explanation for the experts who communicate with outer customers or citizens should be investigated because the experts' criteria are related to the accountability to the customers or citizens. Additionally, not only that, the accountability of the work done by the domain experts working for the experts in other departments inside the same organization should also be ensured because ensuring the accountability of the final decisions requires ensuring accountability in each phase of the decision process done in each department [20].

In addition to the different kinds of explanations required for different types of experts, different criteria for fair decisions are needed for the different types of experts. For example, an IT expert might consider accurate decisions meaning that the result from a decision support system matches historical data are fair. On the other hand, a domain expert, e.g., a loan officer, might consider the decisions that match the expert's intuition based on her/his experiences, e.g., workers in big companies tend to be approved in loan decisions, are fair. Accordingly, the appropriate explanations and the criteria for fair decisions for the experts in organizations have to be investigated in each domain.

To investigate the similarity and differences inside a bank, we did research that explores a design space of user interfaces to support data scientists, i.e., IT experts, and loan officers, i.e., domain experts, to investigate the fairness of machine learning models [21]. Using loan applications as an example, we held a series of workshops with loan officers and data scientists to elicit their requirements. As a result, for example, only data scientists need the information on sensitive attributes while loan officers need to feedback to data scientists on "questionable" attributes that should not be used for decision-making. This result indicates that the data scientists consider decisions without discriminatory bias as fair while the loan officers consider the decisions that match their intuition as fair. In the paper [21], we proposed a test case of how to investigate the experts' viewpoints about explanations and fair decisions.

3.2 Recipients of Decisions Outside Organizations

Next, there are two types of stakeholders outside the organizations who are affected by the decisions based on the outputs of decision support systems: direct recipients of decisions, and indirect recipients of decisions. Here, we discuss how to consider the appropriate explanation and fair decisions for them.

Direct Recipient of Decisions. First, there are direct recipients of decisions. They receive and are influenced by the decisions made by domain experts in the organization directly. The direct recipients are, for example, defendants in recidivism predictions, patients in medical diagnosis, recipients of investment in finance, and job candidates in job matching. In the artificial intelligence ACT (AI ACT) [7], which is a draft of regulation to AI systems proposed by the European Commission (EC) in April 2021, AI systems except for the minimal risk AI have the obligation of transparency such as the obligation for chatbot systems to inform human users of s/he is interacting with AI systems. Additionally, according to General Data Protection Regulation (GDPR) [6], data subjects, which are similar to the users of AI systems, *"should have enough relevant information about the envisaged use and consequences of the processing to ensure that any consent they provide represents an informed choice. (Article 6(1)(a))"* Therefore, decision support systems in organizations have to be designed with a clear understanding of what explanations are necessary and sufficient for the direct recipient and what decisions are considered fair.

To clarify such explanations and criteria for fair decisions, now we discuss the types of direct recipients. There are both recipients who have interests that are the same as and different from the experts in organizations have. For example, in the medical context, patients, the direct recipients, might consider accurate decisions, i.e., diagnoses, in light of historical data as fair decisions. In this case, the patients consider accuracy, one of the performance metrics of machine learning, as the most important indicator for fair decisions. This preference for performance metrics is the same as doctors, the domain experts in organizations who want to judge the remedies accurately. On the other hand, there are also direct recipients who have different interests from the experts. For example, in the context of recidivism predictions, judges, the domain expert in the court, want to make accurate decisions on the likelihood of recidivism. Hence, they will prefer to have the result from a decision support system whose accuracy is maximized. On the contrary, defendants, the direct recipients of the decisions, do not want to be mistakenly judged as a person likely to re-offend. Therefore, they will prefer to maximize the false-positive rate than the accuracy. By clarifying if the interests of the direct recipients and the experts are the same or different, we can tell if the same or different explanations have to be expressed by the decision support systems and if the fair decisions are coherent or not between the experts and the direct recipients.

Moreover, when considering the direct recipients, we sometimes have to take global cultural diversity into account. When a private company such as a big bank has its branches globally, the decision support systems have effects on customers all over the world. In such cases, the concepts of fair decisions might be different based on cultural differences. For example, Geert Hofstede [10], a social psychologist, developed a six-dimension model of national cultures based on global research and advocated that different countries have different tendencies in such dimensions as power distance, individualism, uncertainty avoidance, etc. If we follow the argument, for example, a person in a country which has a high individualism score might have a tendency to consider the denial of a loan application because of the arrest record of the customer's family member as unfair although a person who lives in a country whose individualism score is low might consider the loan decision is fair. Hence, when there are stakeholders globally, the difference in the cultural context should be taken into account.

To explore methods to investigate appropriate explanations and fair decisions for the direct recipients that exist globally, we did a research consisting of a series of workshops and crowdsourcing study [22]. Through workshops with end-users, we co-designed and implemented a prototype system that allowed end-users to see why predictions were made in a machine learning model of loan decisions, and then to change weights on features to debug fairness issues. We evaluated the use of this prototype system through a crowdsourcing study. To investigate the implications of diverse human values about fairness around the globe, we also explored how cultural dimensions might play a role in using this prototype.

From this research, we found that cultural differences explained differences in assessing and improving fairness. The cultural dimensions that seemed to matter most were Masculinity, Uncertainty Avoidance, and Indulgence in the Hofstedes' model [10]. This research [22] is also a test case of how to investigate the global direct recipients' preference related to explanations and fair decisions.

Indirect Recipients of Decisions. Outside the organizations, there are also indirect recipients, who are indirectly affected by the decisions about the direct recipients. The indirect recipients are affected by decisions due to their relationships with the direct recipients. For example, in the domain of recidivism prediction, the members of the local community that a defendant belongs to are indirect recipients because they will be harmed if the decision, the prediction of recidivism, is wrong. Similarly, insurance companies in the domain of medical diagnosis, recipients' business partners in loan decisions, or personnel placement agencies in job matching are the indirect recipients of decisions.

To consider appropriate explanations and fair decisions for the indirect recipients, we need to discuss the similarity and differences in the interests between the indirect recipients and other stakeholders. For example, in the medical context, an insurance company, which is an indirect recipient, does not want to pay the medical expenses for the erroneous diagnosis. Hence, when we consider the diagnosis that a patient has a disease as a positive instance, the company wants to minimize the false positive rate, which is the rate of patients diagnosed wrongly. On the other hand, in the domain of job matching, a personnel placement agency, an indirect recipient, wants to maximize its profits by receiving commissions for recruiting from the company where the candidate decides to be employed. Hence, they do not want a job candidate to be wrongly judged as an unqualified person. In this case, when we consider the decision to hire a candidate as a positive instance, the personnel agency wants to minimize the false negative rate. This interest is the same as the job candidate, who does not either want to be judged wrongly as an unqualified candidate.

Based on the examples we discussed above, we can tell that the appropriate explanations and fair decisions for the indirect recipients change according to their interests. While that is the same as for other stakeholders, the obligation about what has to be considered also changes according to the domain of the decisions. When the public organization such as national or local governments possibly must not fail to consider the indirect recipients such as the local community in the case of recidivism predictions when generating explanations and fair decisions. This is because public organizations generally have to be accountable to the public. On the other hand, private companies, such as banks might not have to consider the indirect recipients when they make fair decisions due to the trade secret. Hence, for the decision support systems used in organizations, especially in public ones, there is the necessity to output that explains that their decisions are fair for the indirect recipients. The HCI community should explore how to generate the appropriate explanations for indirect recipients.

3.3   Regulators Outside the Organizations

The final stakeholder we consider is the external regulators outside the organizations. The regulators check if the organizations obey the laws, regulations, or constitutions regarding human rights. While they check the organizations that make social decisions entirely, some of the regulators are paying attention, especially to the algorithmic decision support systems used in the organization. In the famous example, ProPublica, which is known as a non-profit organization (NPO) for investigative journalism, pointed that the existence of racial bias in a recidivism scoring system used in the US court called COMPAS [1]. Although there are some criticisms of this report because the way of evaluating the discriminatory bias is not appropriate, the report was so influential that academia and industry started to focus on the fairness issues in the decision support systems. Other than NPO, various countries have proposed their regulations on AI systems and have tried to control the decision support systems used in the organization to ensure the transparency, fairness, and accountability of social decisions to protect human rights [8,27]. Moreover, since GDPR and AI ACT, the regulations developed by EC have or will have effects globally, even if a country does not set any regulations on AI systems or data processing systems, companies, and organizations that try to operate globally are subject to control under those regulations. In this case, EC is a regulator.

Appropriate explanations and fair decisions for the regulators change depending on the laws or regulations on which the regulator is based. An explanation is appropriate if the explanation provides the outer regulators with information enough to audit the process of decisions' lawfulness. For fair decisions, what is needed by the regulators are fairness which means that there are not any discriminatory bias in the process or results of decisions. This kind of fairness can be ensured by using the conventional methods of fairness-aware machine learning [17], which remove the bias based on sensitive attributes, such as race or gender, from training data or machine learning models. However, since there are various types of fairness that should be ensured, such as group fairness or individual fairness [5], fairness in acceptance rate [3] or in performance metrics [9], and which sensitive attributes should be focused, the stakeholders should reach a consensus about the fairness that will be ensured.

To be responsible for the regulators' requirements, many companies declared their own AI Ethics guidelines [18,23]. With them, the companies try to show their attitude that they use AI technologies in ethical ways. Moreover, several companies proposed methods to check ethical issues in the process of machine learning [11,16,24]. With these methods, people can come up with the potential ethical issues that can be evoked when machine learning and human users or society interacts. Through these activities, the organizations which are mainly private companies try to ensure accountability for the outer regulators.

4 Integration of Diverse Concept of Fairness

Now, the authors are working on the research to design understandable and fair results of the decision support systems considering the preference of diverse stakeholders. As we explained above, there are diverse stakeholders in decision support systems used in organizations. And different stakeholders have different skills and concepts of fair decisions. Since the decision support systems in the public and private organizations have a great impact on society, the difference in the preference for explanations and fairness should be intermediated. For the purpose of this intermediation, there are some previous studies that explore co-creation methods using workshops where multiple stakeholders meet in one place and discuss the preferable results of algorithmic systems [15,28]. However, since making the diverse stakeholders get together in one place is difficult, and the number of people who participate in a workshop is limited, the workshop approach to extract preference for the machine learning models has limitations in terms of its scale.

Hence, we are now taking the crowdsourcing approach. Via crowdsourcing platforms such as Amazon Mechanical Turk[2] and Prolific[3], we can access more diverse people than those who can meet in one place. We have explored what kind of situations are considered fair by diverse people via crowdsourcing using binary search method [19], interactive systems with which users can evaluate the fairness of the decisions [22,26], and choice of the preferable model seeing the performance metrics of machine learning models [29]. Although each method has its strength and drawbacks, we are now continuing to explore the best way to extract the diverse stakeholders' preferences for machine learning systems and integrate them in an agreeable form for all stakeholders.

5 Framework of the Stakeholder-in-the-Loop Fairness

Here, we summarize the requirements for the explanations and decisions we discussed above as the framework of stakeholder-in-the-loop fair decisions for further discussion. In Fig. 1, blue icons and frames indicate they are or belong to a public or private organization, and red icons indicate that they exist outside of the organization. The narrow arrows between icons indicate the relationship between stakeholders. And the bold arrows colored blue indicate the considerations for each stakeholder by the decision support system in the organization. The meanings of blue arrows as described as follows:

a Considerations on the appropriate outputs for each expert. Each expert has her/his own skill and responsibility. Decision support systems' output should be accountable based on the skills and responsibilities.

---

[2] https://www.mturk.com/.
[3] https://www.prolific.co/.

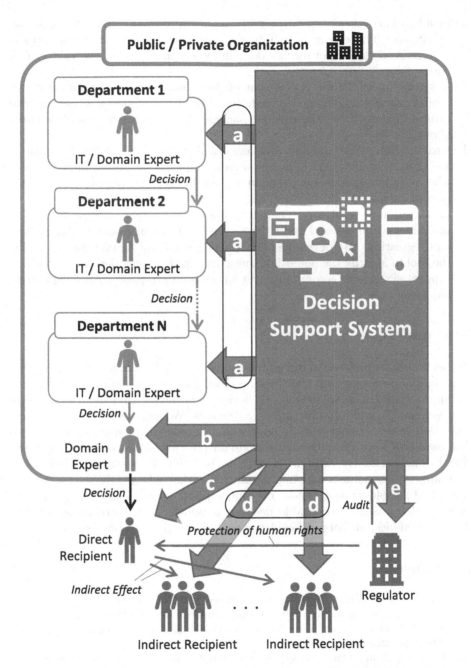

Fig. 1. Framework of the Stakeholder-in-the-loop fairness. This figure shows the relationships between stakeholders, and a decision support system in an organization that considers the stakeholders.

b Consideration of the domain expert who makes decisions for the outer customer or citizens. The expert is mainly the direct user of the decision support system. And the output for this domain expert should cover the values of all stakeholders.

c Consideration of the direct recipient of decisions. Decision support systems' outputs should be designed considering the direct recipients' situations and values. In some cases, the cultural background of the users should be cared about.

d Consideration of the indirect recipients of decisions. The indirect recipients are those who are affected by the direct recipients somehow. In some cases where the decisions are highly public, the indirect recipients, who are the part of citizens should be cared about.

e Consideration of regulators. The regulators try to protect the human right of the direct recipients and audit the activities of organizations based on laws and regulations. By developing the guidelines for the AI systems and using the tool to identify the potential issues, organizations can design the decision support systems as accountable systems that can respond to the regulators' requirements.

6   Conclusion

In this paper, we proposed a discussion framework that we name stakeholder-in-the-loop fair decisions. This framework is developed to consider the requirements for appropriate explanations and fair decisions obtained from the decision support systems in public and private organizations. We identified five stakeholders that need to be considered to design accountable decision support systems and discussed how to consider the appropriate outputs for each stakeholder by referring to our works. By clarifying the characteristics of specific stakeholders in each application domain and integrating the stakeholders' values into outputs that all stakeholders agree upon, decision support systems can be designed as systems that ensure accountable decision makings. To achieve accountability decision makings, authors will continue to work on this line of research.

References

1. Angwin, J., Larson, J., Mattu, S., Kirchner, L.: Machine bias: there's software used across the country to predict future criminals. and it's biased against blacks, May 2016. https://www.propublica.org/article/machine-bias-risk-assessments-in-criminal-sentencing

2. Arya, V., et al.: One explanation does not fit all: a toolkit and taxonomy of AI explainability techniques (2019). https://doi.org/10.48550/ARXIV.1909.03012, https://arxiv.org/abs/1909.03012

3. Barocas, S., Selbst, A.D.: Big data's disparate impact. Calif. L. Rev. **104**, 671 (2016)

4. Clinciu, M.A., Hastie, H.F.: A survey of explainable AI terminology. In: 1st Workshop on Interactive Natural Language Technology for Explainable Artificial Intelligence 2019, pp. 8–13 (2019)

5. Dwork, C., Hardt, M., Pitassi, T., Reingold, O., Zemel, R.: Fairness through awareness. In: Proceedings of the 3rd Innovations in Theoretical Computer Science Conference. ITCS 2012, pp. 214–226, New York, NY, USA. Association for Computing Machinery (2012). https://doi.org/10.1145/2090236.2090255, https://doi.org/10.1145/2090236.2090255

6. European, C.: Regulation (EU) 2016/679 of the European parliament and of the council of 27 April 2016 on the protection of natural persons with regard to the processing of personal data and on the free movement of such data, and repealing directive 95/46/ec (general data protection regulation) (text with eea relevance), April 2016. https://eur-lex.europa.eu/eli/reg/2016/679/oj

7. European, C.: Proposal for a regulation of the European parliament and of the council laying down harmonised rules on artificial intelligence (artificial intelligence act) and amending certain union legislative acts, April 2021. https://eur-lex.europa.eu/legal-content/EN/TXT/?uri=CELEX%3A52021PC0206

8. Government of Canada.: Responsible use of artificial intelligence (AI), November 2022. https://www.canada.ca/en/government/system/digital-government/digital-government-innovations/responsible-use-ai.html

9. Hardt, M., Price, E., Srebro, N.: Equality of opportunity in supervised learning. In: Proceedings of the 30th International Conference on Neural Information Processing Systems. NIPS 2016, Red Hook, NY, USA, pp. 3323–3331. Curran Associates Inc. (2016)

10. Hofstede, G., Hofstede, G.J., Minkov, M.: Cultures and Organizations: Software of the Mind, vol. 2. Mcgraw-Hill, New York (2005)

11. IBM: AI design ethics, December 2022. https://www.ibm.com/design/ai/ethics/

12. Kim, T.Y., Leung, K.: Forming and reacting to overall fairness: A cross-cultural comparison. Organizational Beh. Hum. Dec. Process. **104**(1), 83–95 (2007). https://doi.org/10.1016/j.obhdp.2007.01.004, https://www.sciencedirect.com/science/article/pii/S0749597807000076

13. Kobayashi, K., Nakao, Y.: One-vs.-one mitigation of intersectional bias: a general method for extending fairness-aware binary classification. In: de Paz Santana, J.F., de la Iglesia, D.H., López Rivero, A.J. (eds.) DiTTEt 2021. AISC, vol. 1410, pp. 43–54. Springer, Cham (2022). https://doi.org/10.1007/978-3-030-87687-6_5

14. Lakkaraju, H., Bach, S.H., Leskovec, J.: Interpretable decision sets: A joint framework for description and prediction. In: Proceedings of the 22nd ACM SIGKDD International Conference on Knowledge Discovery and Data Mining. KDD 2016, New York, NY, USA, pp. 1675–1684. Association for Computing Machinery (2016). https://doi.org/10.1145/2939672.2939874, https://doi.org/10.1145/2939672.2939874

15. Lee, M.K., et al.: Webuildai: participatory framework for algorithmic governance. Proc. ACM Hum.-Comput. Interact. **3**(CSCW) (2019). https://doi.org/10.1145/3359283

16. Madaio, M.A., Stark, L., Wortman Vaughan, J., Wallach, H.: Co-designing checklists to understand organizational challenges and opportunities around fairness in AI. In: Proceedings of the 2020 CHI Conference on Human Factors in Computing Systems. CHI 2020, New York, NY, USA, pp. 1–14. Association for Computing Machinery (2020). https://doi.org/10.1145/3313831.3376445, https://doi.org/10.1145/3313831.3376445

17. Mehrabi, N., Morstatter, F., Saxena, N., Lerman, K., Galstyan, A.: A survey on bias and fairness in machine learning. ACM Comput. Surv. **54**(6) (2021). https://doi.org/10.1145/3457607, https://doi.org/10.1145/3457607

18. Microsoft: Responsible AI principles from Microsoft (2022). https://www. microsoft.com/en-us/ai/responsible-ai?activetab=pivot1%3aprimaryr6
19. Nakao, Y.: Toward human-in-the-loop AI fairness with crowdsourcing: effects of crowdworkers' characteristics and fairness metrics on AI fairness perception (2022)
20. Nakao, Y., Shigezumi, J., Yokono, H., Takagi, T.: Requirements for explainable smart systems in the enterprises from users and society based on FAT. In: Trattner, C., Parra, D., Riche, N. (eds.) Joint Proceedings of the ACM IUI 2019 Workshops co-located with the 24th ACM Conference on Intelligent User Interfaces (ACM IUI 2019), Los Angeles, USA, March 20, 2019. CEUR Workshop Proceedings, vol. 2327. CEUR-WS.org (2019). http://ceur-ws.org/Vol-2327/IUI19WS-ExSS2019-3. pdf
21. Nakao, Y., Strappelli, L., Stumpf, S., Naseer, A., Regoli, D., Gamba, G.D.: Towards responsible AI: a design space exploration of human-centered artificial intelligence user interfaces to investigate fairness. International Journal of Human-Computer Interaction, pp. 1–27 (2022)
22. Nakao, Y., Stumpf, S., Ahmed, S., Naseer, A., Strappelli, L.: Toward involving end-users in interactive human-in-the-loop AI fairness. ACM Trans. Interact. Intell. Syst. **12**(3) (2022). https://doi.org/10.1145/3514258
23. Nakata, T., et al.: Initiatives for AI ethics: formulation of Fujitsu group AI commitment. Fujitsu Sci. Techn. J. **56**(1), 13–19 (2020)
24. Nitta, I., Ohashi, K., Shiga, S., Onodera, S.: AI ethics impact assessment based on requirement engineering. In: 2022 IEEE 30th International Requirements Engineering Conference Workshops (REW), pp. 152–161 (2022). https://doi.org/10. 1109/REW56159.2022.00037
25. Ribeiro, M.T., Singh, S., Guestrin, C.: "why should i trust you?": explaining the predictions of any classifier. In: Proceedings of the 22nd ACM SIGKDD International Conference on Knowledge Discovery and Data Mining. KDD 2016, New York, NY, USA, pp. 1135–1144. Association for Computing Machinery (2016). https://doi.org/10.1145/2939672.2939778
26. Stumpf, S., et al.: Design methods for artificial intelligence fairness and transparency. In: Glowacka, D., Krishnamurthy, V.R. (eds.) Joint Proceedings of the ACM IUI 2021 Workshops co-located with 26th ACM Conference on Intelligent User Interfaces (ACM IUI 2021), College Station, United States, April 13–17, 2021. CEUR Workshop Proceedings, vol. 2903. CEUR-WS.org (2021). http://ceur-ws. org/Vol-2903/IUI21WS-TExSS-13.pdf
27. The White House: Blueprint for an AI bill of rights, October 2022. https://www. whitehouse.gov/ostp/ai-bill-of-rights/
28. Woodruff, A., Fox, S.E., Rousso-Schindler, S., Warshaw, J.: A qualitative exploration of perceptions of algorithmic fairness. In: Proceedings of the 2018 CHI Conference on Human Factors in Computing Systems. CHI 2018, New York, NY, USA, pp. 1–14. Association for Computing Machinery (2018). https://doi.org/10.1145/ 3173574.3174230
29. Yokota, T., Nakao, Y.: Toward a decision process of the best machine learning model for multi-stakeholders: A crowdsourcing survey method. In: Adjunct Proceedings of the 30th ACM Conference on User Modeling, Adaptation and Personalization. UMAP 2022 Adjunct, New York, NY, USA, pp. 245–254. Association for Computing Machinery (2022). https://doi.org/10.1145/3511047.3538033

Theoretical Model of Electronic Management for the Development of Human Potential in a Local Government. Peru Case

Moisés David Reyes-Perez[1]([✉]) [iD], Jhoselit Lisset Facho-Cornejo[2] [iD],
Carmen Graciela Arbulú-Pérez Vargas[3] [iD], Danicsa Karina Carrasco-Espino[4] [iD],
and Luis Eden Rojas-Palacios[5] [iD]

[1] General Studies Unit, Norbert Wiener Private University, Lima, Peru
moises.reyes@uwiener.edu.pe
[2] San Martin de Porres University, Pimentel, Peru
[3] Cesar Vallejo University, Pimentel, Peru
[4] Señor de Sipan University, Pimentel, Peru
[5] Tecnológica del Perú University, Chiclayo, Peru

Abstract. The present investigation had a general objective to build a theoretical model of electronic management based on the humanist paradigm and labor competencies that will optimize the processes for the development of human potential in a local government of the Lambayeque region, the present investigation is relevant because it will allow to analyze the administrative systems proposed by SERVIR in a local government of the Lambayeque Region, in addition, a proposal is proposed that is characterized by developing people, above any financial interest, seeking human fulfillment, using technology. The sample consisted of 80 public servants from a Municipality of the Province of MPCH, the methodology used is a quantitative - positivist approach. Concluding that 70% of the collaborators surveyed, refers to the fact that the municipality executes the employment management dimension at a very low level, which means that hardly the collaborators have knowledge about the selection, induction and personnel rotation processes.

Keywords: e-HRM · E-Recruitment · E-Performance management y E-Learning

1 Introduction

1.1 Problematic Reality

The quality and performance of an entity depend on the development of the human potential of its collaborators, in this way suitable and efficient servers are guaranteed, which contribute to the improvement of administrative processes, speed of response to the demands of citizens and public value (Abas & Imam 2016; Harky 2018).

The processes of calls, attraction, evaluations, and job interviews are operational activities that demand a high percentage of the budget, which was initially allocated for the development of people in public entities (Johnson et al., 2019). Currently, in these post-COVID-19 pandemic circumstances, organizations cannot continue working with traditional methodologies and tools that allow physical contact, that is, in person (Luballo & Simon 2017).

With the advancement of Artificial Intelligence and the creation of new digital platforms, they have transformed the ways of managing an organization in the financial and commercial areas; In the same way, it could be used in the areas of human potential management, streamlining operational activities that are cumbersome in their processes, allowing collaborators to get involved in tactical and strategic functions (Roshchin et al. 2017).

The evolution of ICTs has generated a review of traditional administrative methods and processes in Public Entities, with the aim of transversally inserting new electronic tools, using an electronic management model in the development processes of human potential (profiles of position, recruitment, selection, training, performance evaluation, labor welfare) will allow optimizing and reusing some budgeted resources in operational activities that were carried out in a traditional way; in the same way; the calls will have a long-range diffusion, attracting the right people; In addition, the officials who make decisions will have real and reliable data, through reports prepared by the implemented systems (Nurlina et al. 2020).

In this sense, the problematic reality arises, where it is appreciated, in a Local Government of the Lambayeque region, breaches of the regulations established by the Peruvian State modernization law that indicates electronic government as a transversal axis, however, in said institution, human potential continues to be developed with traditional methods and in the face-to-face modality, limiting the adequate flow of the processes established by the Responsible Body.

In addition, there is a lack of a computer system that allows managing the development of human potential, allowing to attract the best talents, systematized performance evaluations, standardized feedback, providing real data that helps the authorities to execute the promotion and promotion processes of transparent way; in the same way, with performance evaluation and training virtually (Simonova et al. 2020)., as proposed by the transversal axis: electronic government, promoting the digitization of a set of administrative activities in local governments (Government of Peru 2017).

The present investigation is justified in the contribution of new contributions to manage the development of human potential under a modern theoretical framework based on labor competencies, allowing to know the link that exists between human talent and the continuous improvement of the modernization of public management, optimizing current resources, with low costs and benefits in public services, providing an approach to citizens as well as timely attention in an equal manner for users and beneficiaries. This research is necessary because a theoretical model of electronic management will be built to optimize the processes of developing the human potential of local government servers so that they can provide users with a modern, efficient and quality service.

The problem formulated is: How a theoretical model of electronic management based on the humanist paradigm and labor competencies will optimize the processes for the development of human potential in a local government in the Lambayeque region?

1.2 Literature Review

Electronic Management of Human Potential Development. Electronic management groups a series of actions such as planning, coordinating, monitoring, directing, controlling and obtaining resources to execute activities at the administrative level in the virtual modality, with the objective of achieving strategic goals of an Entity (Hernández-Sampieri & Mendoza 2018).

Electronic documents are produced by electronic systems and equipment, with legal support; these documents have to be reliable and true, such as physical documents (paper with signature and seal, prepared by human hand), the fundamental characteristics that electronic documents must have are: Available (locate as soon as possible); reliable (count on exact content, for which it was created); integrity (not be altered or tampered with by an unauthorized user); authenticity is being able to prove the time and that user created, modified, sent and received it (Ardila 2020).

The concept of electronic management of human potential development has evolved and expanded over decades of research on the intersection between human potential and technology. There are several definitions, but one of the most cited is that of Giri et al. (2019). Who states that it is a broad term that includes contents of self-development, improvement, personal growth, streamlined by the support of technology, allowing to provide public value and differentiation with other entities.

This definition corresponds to four critical aspects of e-HRM: HRM practices, technology implementation, the actors in those processes, and the consequences of implementation. Other definitions provided by (Blom et al. 2019; Wirtky et al. 2016) also highlight the application of technology to support the performance of human talent activities. Issues related to organizational actors using technology, organizational practices, and the impact of technology are not only central to e-HRM, but to almost all studies on the intersection of technology and organizational studies (Bortnikas 2017).

Currently, human potential development strategists innovate new ways to predict the work performance of employees, which is why they need digitalized systems designed to issue reports with data that managers need (Dulebohn & Johnson 2013). The decisions of financial managers have improved because they have platforms that provide them with metrics to make the best decisions, modern managers of human potential must accept artificial intelligence as a strategic ally to automate those processes that are essential to face a trend globalized post-pandemic, for this you must have electronic recruitment, activities that contribute to motivation, climate and job satisfaction (Bortnikas 2017).

Artificial intelligence would add public value to the services provided by a local government, it would allow the development of systems that report the work situations of collaborators, entry time, time required to perform a function, how many people are involved in meeting the goal, level job performance, level of training gaps, predictive data for promotions, promotions, leadership, feedback and metrics (Karwehl & Kauffeld 2021; Johnson et al. 2020). In Europe, the public management of human potential has had innovations in its personnel selection processes, carrying out the entire process digitally, from the call, evaluation of files, evaluation of skills and signing of contracts (Simonova et al. 2020; Ramkumar 2018; Dayarathna et al. 2020).

The post-pandemic trends of human potential in the public sector of Latin America, is to implement software that reduces the operational burden on employees in the human talent area, helping them with CV evaluation actions in a massive way, extending the job offer through social networks, virtual platforms, university job bank (Kwok & Muñiz 2021; Chandratre & Soman 2020; Roshchin et al. 2017).

The challenges of human potential allied to ICTs and artificial intelligence is to be a key piece in the development of all Public Entities, through its intelligently electronic execution (Wolff & Burrows 2021).

Evidencing itself in its performance evaluation processes, to then virtually build a training plan to strengthen labor skills, reducing gaps and providing an efficient and electronic service to users (Colonnelli et al. 2020; Forsyth & Anglim 2020; Krichevsky & Martynova 2020). With a friendly interface for the collaborators responsible for the processes of human potential, motivating and generating an organizational culture from the electronic perspective, as proposed by transversal axis number three of the modernization law of the Peruvian State.

Humanistic Approach. This approach shows that an entity that is constituted under the humanist gaze cares about developing people, above any financial interest; it seeks the fulfillment and satisfaction of the basic needs of the collaborators, in the same way, it is concerned about human freedom, the way to transcend, the dignity and the legacy that the collaborators wish to leave to the new generations. Human beings not only want to work to survive, but also want to build a better society based on their abilities (Urdang 2013; García de la Torre et al. 2014).

The management of human potential has become a relevant aspect, because the existence of entities depends on the actions carried out by public servants, so investing in people brings long-term benefits; from being just an operational process, it is now a competitive advantage that aims to generate value in its services (Strohmeier et al. 2012).

In non-profit entities, such as a district municipality, its competitive advantage is to be able to efficiently fulfill the services that citizens demand, for this leadership is required, which guides the staff to fulfill the mission and strategic objectives according to their job skills (Johnson et al. 2020).

Competency Management. The people management approach based on professional skills allows for the reduction of disability, ineptitude and deficiency gaps; this approach; It has been generating changes in private companies, it has its scoop on the relevant factor that are competencies (knowledge + skills + attitudes) so that people can have above-average job performance and performance (Karwehl & Kauffeld 2021).

The competency-based management approach is a process of updating job profiles (competences required according to the demand for the job) in order to achieve the vision for which the entity was created. This process covers methodologies, tools, and procedures to classify and develop competences in the positions according to the demands of the users (Stanley & Aggarwal 2019).

Objectives. Build a theoretical model of electronic management based on the humanist paradigm and labor competencies that will optimize the processes for the development of human potential in a local government in the Lambayeque region.

Analyze the Human Resources Management Administrative System proposed by the Peruvian government.

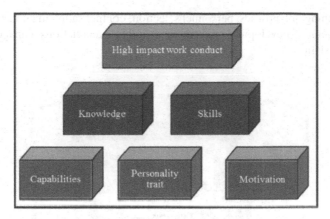

Fig. 1. Construction of work behavior in organizations

Establish the indicators that electronic management must meet to improve the development of human potential based on the humanist paradigm and labor skills in a local government.

Compare the results of the pre and post test.

2 Method

Electronic management model of the development of human potential to improve in a local government in the Lambayeque region, is a study, which is part of the quantitative method, detached from the positivism paradigm (Hernández - Sampieri & Mendoza 2018), likewise, this study of this basic type is created from an existing theory without changing the nature of the variables (CONCYTEC 2018). In addition, the problematic manifestations of the population were objectively observed and analyzed.

It is a design that responds to the classification of a non-experimental study, the categories and variables are not intentionally altered, the phenomenon is identified, appreciated and its current state (diagnosis) is described, in order to later elaborate and build a proposal. That allows the established goal to be achieved (Hernández-Sampieri & Mendoza 2018).

The population consisted of 260 public servants, after a statistical analysis, a sample of 80 people was obtained.

3 Results

Theoretical Model of Electronic Management Based on the Humanist Paradigm and Labor Competencies that Will Optimize the Processes for the Development of Hu-man Potential in a Local Government in the Lambayeque Region

The management of the human potential of a local government in the future is to migrate to a digitally intelligent management, for them it is first necessary to implement the basic administrative processes of the development of human potential, such as human potential planning, job profiles, personnel selection, compensation management, performance management, development management and human relations management in the electronic modality.

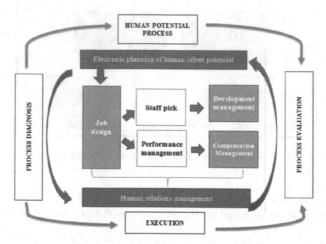

Fig. 2. Human potential process diagram

The strategy comprises five phases: The first phase is the diagnosis of the main administrative processes for the development of human potential, which has already been executed, evidencing the absence of artificial intelligence (Information and communication technologies), in the analysis and job description, induction process for new staff, performance evaluation process to identify gaps for improvement, likewise, there is no annual training plan based on a diagnosis of needs and finally the compensation process based on the performance of each server public.

The second phase is the sensitization of the work team of the human resources area and the technology office of the local government, in addition, in this stage they will be trained on the preparation of the technical documents that must be prepared at the time of carrying out each process.

The third phase is the work schedule, where each collaborator, both from the human resources area and from the technology area, must assume the commitment and responsibility for the execution of the administrative processes of human potential, according to the bureaucratic limitations of the local government.

The fourth phase is the execution of the processes electronically; In the job analysis process, formats must be developed to complete a job profile that meets the minimum requirements, such as academic degrees, skills, years of experience, and relevant functions; this information will serve as input for the call and the selection process of suitable, meritocratic and transparent personnel; in the same way; The profiles allow performance evaluation, because according to the execution of their functions we will know the level of performance of each collaborator, the report provided by said evaluation will help to build a training diagnosis to strengthen the capacities of the servers to be being efficient in the fulfillment of its functions; in addition; Being able to provide recognition according to the performance of each collaborator, for which it is necessary to have a warm and motivating work environment.

The fifth phase is the evaluation of the administrative processes, identifying gaps, providing feedback and an improvement plan so that the theoretical model of electronic management for the development of human potential in a local government in the Lambayeque region can remain over time and meet your goal.

Levels of the Dimensions of the Development of Human Potential Variable

Table 1. Levels of the dimensions of the development of human potential variable

Dimensions	Execution level	Pre test		Pos test	
		F	%	F	%
HR policy planning	Does not execute	0	0	0	0
	Very low	7	8.8	1	1,25
	Low	20	25	5	6,25
	Average	53	66.3	42	52,5
	High	0	0	32	40
Work organization	Does not execute	17	21.3	0	0
	Very low	45	56.3	2	2,5
	Low	18	22.5	6	7,5
	Average	0	0	37	46,25
	High	0	0	35	43,75
Employment management	Does not execute	0	0	0	0
	Very low	56	70	28	35
	Low	21	26.3	12	15
	Average	3	3.8	10	12,5
	High	0	0	30	37,5
Compensation Management	Does not execute	0	0	0	0
	Very low	64	80	32	40
	Low	15	18.8	4	5
	Average	1	1.3	29	36,25
	High	0	0	15	18,75

(continued)

Table 1. (*continued*)

Dimensions	Execution level	Pre test		Pos test	
		F	%	F	%
Development and training management	Does not execute	11	13.8	0	0
	Very low	59	73.8	22	27,5
	Low	10	12.5	5	6,25
	Average	0	0	35	43,75
	High	0	0	18	22,5
Human relations management	Does not execute	0	0	0	0
	Very low	6	7.5	2	2,5
	Low	13	16.3	5	6,25
	Average	61	76.3	32	40
	High	0	0	41	51,25
Performance management	Does not execute	77	96.3	0	0
	Very low	2	2.5	0	0
	Low	1	1.3	18	22,5
	Average	0	0	38	47,5
	High	0	0	24	30

Note. Table 1 shows the significant effect generated by the theoretical model of electronic management in the development of human potential, in a local government in the country of Peru.

4   Conclusions

In the Local Government, before the proposal of the theoretical model, the subsystem of training, employment management and performance evaluation was not executed, however, after the proposal, all the subsystems of human potential were implemented.

The proposal is based on the humanist theoretical approach and management by competence, which are characterized by developing people, above any financial interest, it seeks human fulfillment, in the same way, it is concerned with human freedom, the dignity of the collaborator, and the legacy that he wishes to leave to the new generations from the beginning of building a better society from his abilities.

The updating of the organizations and functions manual and the preparation of the job profile by competence are relevant indicators to develop human potential, since the competence approach is built on the premise of knowledge (know), skills (know how), and the attitudes (wanting to do) required by the work activities that are carried out daily.

The theoretical model of electronic management based on the humanist paradigm and labor competencies that optimizes the processes for the development of human potential was validated using the Delphi method.

References

Abas, M., Imam, O.: Graduates; competence on employability skills and job performance. Int. J. Eval. Res. Educ. **5**(2), 119–125 (2016). https://eric.ed.gov/?id=EJ1108534

Ardila, E.: Electronic management and its application in the current environment: case study at the Agustín Codazzi IGAC Geographic Institute [University of the Salle] (2020). https://ciencia.lasalle.edu.co/cgi/viewcontent.cgi?article=1994&context=sistemas_informacion_documentacion

Blom, T., Du Plessis, Y., Kazeroony, H.: The role of electronic human resource management in diverse workforce efficiency. SA J. Hum. Resource Manag. **17**, 118 (2019). Scopus. https://doi.org/10.4102/sajhrm.v17i0.1118

Bortnikas, A.: Human resources management modernization of contemporary organization. Publ. Policy Administ. **16**(2), 335–346 (2017). Scopus. https://doi.org/10.13165/VPA-17-16-2-12

Chandratre, S., Soman, A.: Preparing for the interviewing process during Coronavirus disease-19 pandemic: Virtual interviewing experiences of applicants and interviewers, a systematic review. PLoS ONE **15**, 243415 (2020). Scopus. https://doi.org/10.1371/journal.pone.0243415

Colonnelli, E., Prem, M., Teso, E.: Patronage and selection in public sector organizations. Am. Econ. Rev. **110**(10), 3071–3099 (2020). Scopus. https://doi.org/10.1257/aer.20181491

Dayarathna, V.L., et al.: Assessment of the efficacy and effectiveness of virtual reality teaching module: a gender-based comparison. Int. J. Eng. Educ. **36**(6), 1938–1955 (2020). Scopus

Dulebohn, J.H., Johnson, R.D.: Human resource metrics and decision support: a classification framework. Hum. Resource Manag. Rev. **23**(1), 71–83 (2013). Scopus. https://doi.org/10.1016/j.hrmr.2012.06.005

Forsyth, L., Anglim, J.: Using text analysis software to detect deception in written short-answer questions in employee selection. Int. J. Select. Assess. **28**(3), 236–246 (2020). Scopus. https://doi.org/10.1111/ijsa.12284

García de la Torre, C., Portales, L., Arandia, O.: The formation humanistic at the tecnológico de monterrey: the citizen of the future. In: Lupton, N., Pirson, M. (eds.) Humanistic Perspectives on International Business and Management, pp. 256–268. Palgrave Macmillan, New York (2014)

Giri, A., Paul, P., Chatterjee, S., Bag, M., Aich, A.: Intention to adopt e-HRM (electronic—Human resource management) in Indian manufacturing industry: an empirical study using technology acceptance model (TAM). Int. J. Manag. **10**(4), 205–215 (2019). Scopus. https://doi.org/10.34218/IJM.10.4.2019.020

Government of Peru approves Law No. 30057, Modernization Framework Law of state management (2017). https://busquedas.elperuano.pe/normaslegales/decreto-legislativo-que-modifica-la-ley-n-27658-ley-marco-decreto-legislativo-n-1446-1692078-21/

Harky, Y.: The significance of recruitment and selection on organizational performance: the case of private owned organizations in Erbil. North Int. J. Contemp. Res. Rev. **9**(02), 422 (2018). https://doi.org/10.15520/ijcrr/2018/9/02/422

Hernández - Sampieri, R., Mendoza, C.: Metodología de la investigación. Las rutas cuantitativa, cualitativa y mixta. (McGRAW-HILL) (2018). https://up-pe.libguides.com/c.php?g=1043492&p=7612751

Johnson, R.D., Stone, D.L., Lukaszewski, K.M.: The benefits of eHRM and AI for talent acquisition. J. Tourism Futures **7**, 40–52 (2020). Scopus. https://doi.org/10.1108/JTF-02-2020-0013

Karwehl, L.J., Kauffeld, S.: Gruppe. Interaktion. Organisation. Zeitschrift Für Angewandte Organisationspsychologie (GIO) **52**(1), 7–24 (2021). https://doi.org/10.1007/s11612-021-00548-y

Krichevsky, M.L., Martynova, J.A.: Multifactor selection of personnel based on fuzzy logic. Int. Multi-Conf. Ind. Eng. Modern Technol., FarEastCon. (2020). Scopus. https://doi.org/10.1109/FarEastCon50210.2020.9271197

Kwok, L., Muñiz, A.: Do job seekers' social media profiles affect hospitality managers' hiring decisions? A qualitative inquiry. J. Hospital. Tourism Manag. **46**, 153–159 (2021). Scopus. https://doi.org/10.1016/j.jhtm.2020.12.005

Luballo, W., Simon, C.: Human resource management practices and service delivery in county government of Siaya, Kenya. Hum. Resource Manag. **1**, 95–113 (2017). http://www.ijcab.org/wp-content/uploads/2017/07/Human-Resource-Management-Practices-and-Service-Delivery-in-County-Government-of-Siaya-Kenya.pdf

National Council of Science, Technology and Technological Innovation CONCYTEC (2018) regulation RENACYT. https://portal.concytec.gob.pe/images/renacyt/reglamento_rcnacyt_version_final.pdf

Nurlina, N., Situmorang, J., Akob, M., Quilim, C.A., Arfah, A.: Influence of e-HRM and Human Resources Service Quality on Employee Performance. J. Asian Finan. Econ. Business **7**(10), 391–399 (2020). Scopus. https://doi.org/10.13106/jafeb.2020.vol7.no10.391

Roshchin, S., Solntsev, S., Vasilyev, D.: Recruiting and job search technologies in the age of internet. Foresight STI Govern. **11**(4), 33–43 (2017). Scopus. https://doi.org/10.17323/2500-2597.2017.4.33.43

Ramkumar, A.: A conceptual study on how electronic recruitment tools simplify the hiring process. Indian J. Pub. Health Res. Develop. **9**(6), 136–139 (2018). Scopus. https://doi.org/10.5958/0976-5506.2018.00537.5

Simonova, M., Lyachenkov, Y., Kravchenko, A.: HR innovation risk assessment. En Kudriavtcev, S., Murgul, V. (eds.), E3S Web Conference, vol. 157 (2020). EDP Sciences; Scopus. https://doi.org/10.1051/e3sconf/202015704024

Stanley, D.S., Aggarwal, V.: Impact of disruptive technology on human resource management practices. Int. J. Business Continuity Risk Manag. **9**(4), 350–361 (2019). Scopus. https://doi.org/10.1504/ijbcrm.2019.102608

Strohmeier, S., Bondarouk, T., Konradt, U.: Editorial: Electronic human resource management: Transformation of HRM? Zeitschrift fur Personalforschung **26**(3), 215–217. Scopus (2012). https://doi.org/10.1177/239700221202600301

Urdang, E.: Human behavior in the social environment: Interweaving the inner and outer worlds. Routledge, New York (2013)

Wirtky, T., Laumer, S., Eckhardt, A., Weitzel, T.: On the untapped value of e-HRM: a literature review. Commun. Assoc. Inf. Syst. **38**(1), 20–83 (2016). Scopus. https://doi.org/10.17705/1CAIS.03802

Wolff, M., Burrows, H.: Planning for virtual interviews: residency recruitment during a pandemic. Acad. Pediatrics **21**(1), 24–31 (2021). Scopus. https://doi.org/10.1016/j.acap.2020.10.006

Enhancing Transparency for Benefit Payments in the Digital Age: Perspectives from Government Officials and Citizens in Thailand

Saiphit Satjawisate[(✉)] ⓘ and Mark Perry ⓘ

Department of Computer Science, Brunel University London, Uxbridge, UK
{1930700,Mark.Perry}@brunel.ac.uk

Abstract. Government operations and services are undergoing a rapid shift towards digitalization, but for this to be successful, it is essential that such government systems are transparent for citizens and government officers to effectively utilize and make sense of them. This study, therefore, examines system mechanisms and citizen engagement with Social Security Office (SSO)s' benefit payment services to enhance operational transparency. We use a case study of the SSO in Thailand, where digitalization has had a high priority across all areas of government administration. Social Security Benefits (SSB) payment form a large portion of government spending relative to public revenue, highlighting the need to increase the transparency of government action. It remains unclear, however, *how* and *in what ways* digital technologies in the benefit payments can improve this situation. We explore the transparency indicators of the benefit payment system by interviewing public officers and insured persons using a set of transparency frameworks. We examine the relationship between how individuals engage with digital technology and its consequences on the system's transparency. Results show that designing systems transparency in SSB payments is challenging due to complex socio-technical issues. We recommend that HCI designers and system developers are concerned not only with technological aspects but also organizational, human, and social aspects, enhancing transparency for the benefit payments.

Keywords: Transparency · Openness · Open government data · Benefit payments · Citizen engagement · Socio-technical systems

1 Introduction

New forms of interaction made possible by information and communication technologies (ICTs) are considered crucial to implementing social security programs, allowing performance and service quality enhancements in Social Security Administration (SSA). Consequently, many public social security agencies have attempted to develop or improve their electronic government (e-government) services to achieve this. In addition, numerous governments have acknowledged that the widespread availability and usage of ICTs

© Springer Nature Switzerland AG 2023
F. Fui-Hoon Nah and K. Siau (Eds.): HCII 2023, LNCS 14038, pp. 57–73, 2023.
https://doi.org/10.1007/978-3-031-35969-9_5

in public sector organizations and institutions remains a key tool for increasing transparency [24] and reducing corruption [7]. However, the advantages of deploying ICTs may be accompanied by several obstacles, such as the digital divide, data quality, and data privacy [16, 17], making this set of developments both societally and interactionally complex.

To respond to public perceptions and expectations of the Social Security System (SSS), new approaches to enhance service quality and become more user-centered are being explored by attempting to develop policies that: 1) find a balance between privacy and openness; 2) allow public involvement and interaction with service users; and 3) support digital transparency as they move away from paper records [17]. In the case of SSS in Thailand, the Social Security Office (SSO) has supported using digital technology in its financial systems to reduce corruption, limiting the use of photocopied documents, and expanding citizen engagement with the public via the office's website and mobile devices [30–32]. When compared to other countries, the SSA's direction in Thailand is typical in many ways; however, we question how the government officials promote transparency in benefit payments with interaction through a back-end system and how citizen increase the use of open government data (OGD) via SSO e-services. A close examination of this topic allows us to develop design guidelines that will enhance transparency in the SSO's payment system.

To answer these questions, our study focuses on transparency as a system mechanism [4, 14, 27] and transparency occurring when citizens engage with OGD [13, 25, 26]. We examined how two groups of participants interact with one another and digital technology: 1) public officers who use the SSO's back-end system and 2) insured persons who engage with OGD via SSO e-services. Analysis of these two groups revealed that they shared their perspectives on transparency improvement, enabling us to evaluate the strengths and weaknesses of existing transparency both within and without the SSO. We then discuss the findings and draw implications for socio-technical systems design that offer opportunities to enhance benefit payment transparency.

2   eGovernment, User Interaction, and Transparency

E-Government is broadly defined as the fundamental concern of changing government services through the use of ICTs to improve cost savings, accessibility and delivery of government service, and operations for the benefit of citizens, businesses, and government officers [1, 35]. The ability to transfer money online plays a significant part in e-government activities that are frequently mentioned in the context of financial transactions involving procurement or revenue collection, such as paying fees, taxes, or fines [36]. According to Csáki et al. [9] and Lochan et al. [21], the trend in welfare payment systems is to eliminate paper-based payment and replace it with electronic payment (e-payment); the intention of this is that by moving toward digital payments, there will be an increase in transparent payment trails [21]. There are two main reasons for this: 1) recipients have digital records of their payments and minimize the number of people 'touching' the money, reducing the number of potential leakage points; 2) financial service providers require more stringent identity proof for digital payments beneficiaries, making it difficult to hide the recipient's identity [18]. These reasons demonstrated that the signs of transparency would come from the back-end system if the internal process is transparent, which not only prevents money loss but also protects financial information since officials and citizens can track and monitor all financial transactions at every stage.

The provision of open data is recognized as having many advantages for enhancing digital transparency in government. By sharing their information with the public, government agencies can create societal benefits such as monitoring government activities, improving the quality of public policy, increasing public trust in government, and corruption prevention [7, 13, 26]. According to Purwanto, Zuiderwijk, and Janssen [26], citizen engagement has been crucial for the success of many OGD systems. One of the central motivations for engaging with OGD systems is that the data should be seen as relevant and important to citizens. For instance, citizens in developing countries suffering from low transparency are likely to place a high value on government spending data. In contrast, some residents from a different country are concerned about the openness of profession and life affairs such as certificated examinations, property acquisition policies, and house demolition [37]. Consequently, diverse populations require various and different forms of government information. However, e-government has been limited by barriers such as data quality, internet speed, and limited access (to data, process and decisions) for citizens [7, 26]. Indeed, this explains why the index of information transparency is low particular in developing countries [38].

It is in this context that we examine the benefits and barriers of e-government services in meeting citizens' demands. We do so by addressing human-computer interaction (HCI) aspects in order to improve citizen users' access to and interaction with OGD. This requires understanding users' prior experience, expectations and needs while interacting with information, technology, and the diverse tasks that they are trying to undertake [33]. Traditional methods of systems investigation first create the technical system and then fit people to it [2], which typically achieve their technical requirements, but these often fail to support the real work being done [5] because of its social embeddedness. However, the use of socio-technical system methods provides a way of allowing HCI researchers to consider both technical and non-technical factors are together [5, 29]. We therefore address how to enhance transparency in the SSO's benefit payment system through socio-technical system methods. The outcome of applying these methods is a more integrated understanding of how organizational and technological factors affect the way that the SSO system is used. In addition, we show how human, social, and technological factors interact and influence OGD through the SSO e-services functionality and usage.

3 Transparency Framework

The literature on transparency principles in public sector systems design tends to consider either the "back-end" information system around data management or the "front-end" system that increases transparency with citizen engagement through access to OGD. Accordingly, these studies make it impossible to reflect the "overview system": an organization's internal and external perspectives on generating transparency. Based on this gap, we examined transparency from two high-level perspectives. These factors reflect on how the government officials promote data, process, and decision/policy transparency in benefit payments and how citizens' motivation, their social environments and ICTs infrastructure impact their decision of use OGD. In our analysis, both transparency types (see Fig. 1) are used as main themes for exploring what we frame as "transparency work" in the SSO's payment system:

3.1 Transparency in System Mechanisms

A government operation in the back-end system generates information with a bottom-up structure. Thus, information from government officials dealing with the operational system of benefit payments can reflect their work processes which should be transparent to create accurate and verifiable data for users. By providing meaningful transparency are considered, covering three aspects of transparency [4, 14, 27]:

- *Data transparency* relates to answers primarily what information is required in the system, who may be included as stakeholders, when it will be used, and where it will happen, as appropriate.
- *Process transparency* is the dissemination of information on the steps through which government processes progress. This process addresses how information is handled in the context of transparency, when task units and sub-units are completed, and how users can track the status of a process.
- *Decision/Policy transparency* concerns the government's intentions, decision-making, and policies. This includes explaining why and how a decision or policy is decided.

3.2 Transparency Through Open Government Data

HCI designers need to understand a collection of social, organizational, and political elements before designing systems. To do so, these systems and applications can boost citizens to engage and collaborate with modern ICTs, which disclose information about government policies and procedures [20] so that citizens can track and monitor the government's performance through these tools. Three elements including motivation [13, 26, 34], social influence [25, 26, 39], and technical factors [6, 25, 26] are used to identify the indicators that influence citizens' participation in OGD to increase government transparency:

- *Motivation* covers citizen motivation to engage with OGD, such as its benefits (extrinsic motivation) or doing something for fun and enjoyment (intrinsic motivation)
- *Social influence* relates to influence from important people who may affect a citizen's intention to engage with OGD, such as family, friends, colleagues, and social media communities.
- *Technical factors* refer to data classification, format, and description that are clear and easy to understand (readability), regularly updated (timeliness), and easily accessible to the public (accessibility).

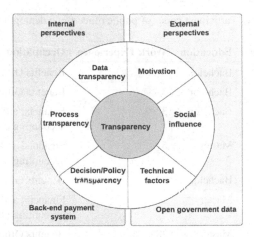

Fig. 1. Transparency framework for analysis

4 Research Methodology

An initial analysis of the SSO back-end system and e-services such as the official website and mobile app was undertaken to understand the nature of the interactional task for users, and how visible information was to those users (i.e., the transparency of content and of any changes made). We then embarked on a process of empirical data collection using qualitative methods.

4.1 Data Collection

Detailed online semi-structured interviews were conducted with 10 public SSO officers and 12 insured persons in Thailand about their understanding and use of the SSO back-end and front-end systems between June to September 2022, each lasting, on average, around 45 min. Demographics are shown in Tables 1 and 2 below; the participant codes have the following structure: < I > represents interview data, < PO > identifies public officer, and < IP > identifies the insured person, with the running number serving as a unique identification.

For public officers (see Table 1), we used a judgmental sampling approach [3, 11] to select participants responsible for operating the benefit payment system from "benefits department" and "financial and accounting department" to share their knowledge and experience of this system. Participant ages ranged from 30–59 years old. They were composed of two men and eight women. Gatekeepers explained that there are a larger proportion of female officers than male officers (in a 10:1 female to male ratio) is a common occurrence for officers in the SSO of Thailand. During these interviews, we recruited operational and professional-level personnel to obtain a broad range of views and experience using the SSO back-end system for benefit payments.

Table 1. Participant codes of public officers and demographics

ID	Gender	Age	Education	Work Experience	Occupation	Level
iPO01	M	38	Bachelor	3	Benefits Officer	Operational
iPO02	F	42	Bachelor	3	Benefits Officer	Operational
iPO03	F	36	Bachelor	3	Financial and accounting officer	Operational
iPO04	F	30	Master	5	Financial and accounting officer	Operational
iPO05	F	55	Bachelor	29	Benefits Officer	Professional
iPO06	F	50	Master	20	Financial and accounting officer	Professional
iPO07	F	54	Master	28	Benefits Officer	Professional
iPO08	M	59	Bachelor	29	Financial and accounting officer	Professional
iPO09	F	52	Bachelor	2	Benefits Officer	Operational
iPO10	F	52	Bachelor	24	Financial and accounting officer	Professional

For insured persons (see Table 2), we employed the techniques of judgmental and snowball sampling [3, 28]: we initiated an online interview with three individuals who then connected us to nine other claimants with experience of the SSO's benefit payment system. Participants ranged in age from 27–51 years old, with two not disclosing their age. Five men and seven women from varied demographic backgrounds were interviewed, all of whom had recently utilized the SSO e-services to engage with OGD.

Table 2. Participant codes of insured persons and demographics

ID	Gender	Age	Family Members	Career	Education
iIP11	M	35	3	University staff	Doctoral
iIP12	F	Prefer not to say	4	Hotel staff	Bachelor
iIP13	F	34	4	Pharmacist	Bachelor
iIP14	M	34	3	Factory worker	Bachelor
iIP15	F	33	4	Project developer	Master
iIP16	M	33	5	Senior full stack developer	Bachelor
iIP17	F	29	3	Purchasing staff	Bachelor

(continued)

Table 2. (*continued*)

ID	Gender	Age	Family Members	Career	Education
iIP18	F	27	3	Marketing staff	Bachelor
iIP19	M	40	4	IT Manager	Master
iIP20	F	Prefer not to say	4	Product owner	Bachelor
iIP21	F	51	3	Farmer	Bachelor
iIP22	M	28	4	Programmer	Bachelor

4.2 Data Analysis: Hybrid Approach to Thematic Analysis

We used a hybrid method [10] to analyze data: data were evaluated both using the transparency framework through a process of deductive thematic analysis and using inductive thematic analysis to reveal previously underdeveloped themes arising from the interview data. Deductive themes were drawn from the academic literature and preliminary engagements with the SSO e-services to populate the transparency framework as theory-driven codes with the six themes shown in Fig. 1. By doing this, we can scope the analysis with the participant's behavior and the environment surrounding their claims to delimit the scope to only cover issues of transparency within the SSO. Initially, the 22 interview audio recordings were transcribed and coded using the NVivo software application. We then applied an inductive approach to render 107 codes. Finally, these codes were grouped under the six themes of transparency framework.

5 Results

Based on our findings, we coded main themes following the transparency framework; the findings were classified around the two groups of participants: 1) Perspectives from public officers who were involved in the SSO's back-end payment system known as "Sapiens", and 2) Perspectives from insured persons who had the experience of benefit claims and who had engaged with OGD around their claim information.

5.1 Perspectives from Public Officers

The primary responsibilities of public officers involved in the benefit payments took place in two departments: the benefits department, and the finance and accounting department. These officers are crucial in promoting transparency within the payment system, given that they work with "*data*" from the insured person's personal information, employment history, and claim transactions; "*process*" in diagnosing benefits and payment process, and "*decision/policy*" around the approval of benefit payments and money transfers to citizens that we required to follow the SSO's policies.

Data Transparency: Promoting data transparency is as simple as allowing the public to explore information online themselves [4], in contrast to public officers who do not have equal access to information. Access to the Sapiens menu varies between public officers based on their role and responsibilities within the organizational structure, so for instance, registration officers are not authorized to access information in the benefits department. This apparent separation of duties allows employees to do their duties with full performance based on their own experience and expertise. As explained by iPO03, *"in the case of refunds approval to the insured person, if the amount exceeds 100,000 baht (approx. USD$2,994.01), an officer with professional level authority is required. However, if the amount is less than 100,000 baht, the operational level can handle it"*. Each public officer has a unique user identification (User ID), verified at login into Sapiens and the scope of the user's responsibility is checked. If a public officer tries to act beyond his/her authority the *"Sapiens system will present a notification that you cannot approve since your authority has been exceeded"* [iPO09]. User ID makes it easy to identify ownership and contact for more information when other public officers face accusations of suspicious information or data entry errors. One participant explained that the financial officer has to validate the insured person's personal information, diagnosis results, and benefit payments amount before making payment. If they find any manual data entry errors in the Sapiens system, for example, that the insured person's sick leave differs from the number of days recorded on the medical certificate, then this can be tracked backwards: *"The financial officer accessed the inquiry menu by entering the citizen ID number. The system then presented the name and last name of the benefits officer who was responsible for this case in order to report this problem to that benefits officer"* [iPO05]. Here, benefit payments accuracy is improved by rechecking data between both departments as any data entry or benefit calculation mistakes will be corrected before paying money to the insured person.

Process Transparency: In the Sapiens system, a public officer can observe how the transaction is progressing while it is ongoing and when it will be completed. In this way, each processing step shows a letter code as its status: *" When the benefits officer has completed recording the data, the screen displays the letter "O" and then the information is sent to the head of department, at which point the screen changes to the letter "A" when he/she approves this claim"* [iPO05]. This authorization is transferred to the financial officer responsible for approving the payment: *"I fill in data like the cheque number, amount, and the insured person's book bank number. After that, I approve by entering my username and password, and the screen will display the letter "Y" when the transaction has been completed"* [iPO10]. However, if the financial officer discovers a data error, such as misidentifying the month: *"I log in, fix the month, and press Enter. The screen then displays a "Y," symbol, I know immediately that the data has been stored"* [iPO04]. As we show here, decisions made by public officers in the claim process are recorded and linked into a chain of prior decisions, and every step provides data for traceability because the Sapiens system keeps a record of all transactions, accessed via an "Inquire" menu. In order to monitor the behavior of personnel involve in the benefit payment process, SSO headquarters need to validate their actions at various process steps. iPO01 noted that *"if someone looks at the same company's information frequently or accesses several companies strangely in one day. In that case, Headquarters IT center personnel will report the situation to the head of the Provincial SSO to decide whether your public officer is likely to be corrupt or misuse, such as selling data to third parties"*. In this way, the SSO considers the security of the insured person's information and has an established procedure to monitor the public officer's behavior, reflecting process transparency that can be validated at every stage of data processing.

Decision/Policy Transparency: Decisions around benefit payments rely on the SSO policy as the condition for advancing each step in the process. The rules and requirements are stored in the Sapiens system. That way, it would be complicated for the public officer to change the regulations or avoid the requirements in this manner. Practically, the system cannot make 100% of decisions on behalf of people so sometimes decisions will depend on interaction between human judgment, known rules, and policy documents. Public officers need to use extreme caution when making decisions, particularly regarding complex medical expenses: *"I have to review the medical record to help me decide to assess the cost of treatment, room, board, and so on. I need to attend training in reading medical records first to see if that case is eligible for a claim or not"* [iPO02]. As a result, augmenting the public officer's knowledge is critical at this stage, such as through receiving information via the Line app, a freeware communication program: *"I regularly access the SSO's Line app group across the country to see what has changed or if other provinces have any problems. So how do they solve the problem?"* [iPO09]. This shows that public officers develop their understanding of policies through conversations with others. Elsewhere in the payments process, the financial and accounting department's decisions are based mainly on rules and required documents. Although the system shows claim approval information from the benefits department, the financial officers are unable to pay anything to the insured person until they have reviewed those documents. At this stage in the money payment process at the service counter with insured persons, financial officers are directly involved. As iPO06 explained, *"I focus on safety first and try not to pay cash at the counter, but I pay by bank transfer. Except in some cases, such as the elderly, I will use my discretion to pay in cash."* The SSO has a policy requiring staff to pay with digital transfers through internet banking. However, citizens who have informed the public officers that they do not have a bank account, especially the elderly require payment with cash or a cheque at the SSO.

5.2 Perspectives from Insured Persons

Citizen engagement is essential to the successful and sustainable implementation of OGD, which involves multiple activities relate to retrieving and transferring data into OGD-based applications. These tools in the SSO consist of 1) SSO website, which discloses information about the SSO news, conditions for receiving benefits, insured person's registration report, the SSO annual performance report, and the insured person's claims history, and 2) SSO connect mobile related to mobile function regarding insured person information such as important news, hospital for treatment, and historical claims. However, it is recognized that there is a lack of understanding into what motivates citizens to participate in OGD activities [20]. Based on our literature review, we studied three factors, namely *"motivation," "social influence,"* and *"technical factors,"* on how these elements promote insured persons to engage with the SSO e-services to raise transparency.

Motivation: All participants were extrinsically motivated, with the primary motivators being to determine the type of benefits to be involved, money to be received, and the need to learn in more detail about the relevant regulations to use the SSO e-services through the official SSO website and SSO mobile application. One participant remarked that the SSS offers citizens a safer living, health insurance, and a place to recover from illness: " *I always get a toothache because it wears down with time as we get older. But at least I can reimburse this treatment [...] I save 900 bahts (approx. USD$26.95) a year, even if it's a small amount* " [iIP21]. In this study, legal sanction does not relate to the extrinsic motivations to engage with OGD: "*If I ignore to get money following the rules, I'm at a loss benefit by myself and not be punished from the SSO*" [iIP19]. We did not find other common intrinsic motivations such as fun, enjoyment, and contributing to society to be a driving factor for OGD, as our participants were mainly concerned in how much money they could receive from the SSO. Some participants suggested that the SSO should proactively promote public relations through social media channels to increase the motivation to search for government data because they often use social media for communications, and easy to access. Moreover, [iIP20] stated that "*I've worked for many companies; however, the HR staff never informed me on orientation day about fundamental social welfare, including our rights and conditions on how to claim*". This point implies that the SSO should be issued as a rule requiring the human resources (HR) staff to educate employees on the fundamental of the SSB within one day or on the first day of employment.

Social Influence: The social influence of participants mainly comes from their workplace, important persons, or influences on their behavior such as family members, friends, colleagues, and internet communities that share the common interest of SSB. In our study, the influence of family members had both effects and did not affect driving citizens to use OGD services. For example, after having their first child, some participants became concerned with understanding SSB from various sources by consulting with family members, colleagues, and HR staff, as well as searching on the internet. During this time, family and friends often assisted participants in gaining a basic understanding, such as what child allowance benefits were available and what documents or evidence were required, and then confirmed the accuracy of this information by using the SSO e-services. Some participants' experiences indicated that families did not contribute to their access to SSB data because none of the family members had any previous claim experience. In addition, family members could provide ambiguous information: "*my relatives gave me unclear answers [...] I asked the questions about sickness benefit to my company's HR staff again. Then, I got my answer*" [iIP17]. We noted that colleagues have the greatest impact on participants' SSB comprehension. For instance, the opinion in [iIP22] explained that "*I often discuss problems with my senior at work [...] she is my close friend and I feel comfortable when talk with her [...] she once told me about the experience of using SSO mobile app and how useful it is*". In the internet communities, our participants mentioned the benefits of social media, such as Facebook and YouTube, in sharing SSB information with friends and the SSO staff. Although this content is short and usually easy to understand, such information on regulations may too generic so that insured persons need to review additional details on the SSO website or mobile app, which have more complete and current SSB information.

Technical Factors: Most participants expressed satisfaction with the convenience of using e-services to search for information and ensure its accuracy because government agencies are providing the information. They can access the SSO official website; however, the SSO mobile application regularly fails, and users need to switch to the website instead. Barriers to accessing OGD are not only due to system failures or the unavailability of digital devices, but also what was generally referred to as age constraints. Some participants were concerned about the elderly due to their unfamiliarity with computers and lack of computer skills. As the experience of elderly friends and family one described: " *they can't register to login into the mobile app or government website [...] even if they can, they wouldn't dare use it on their own*" [iIP17]. Another obstacle was the formal language of the policy wording, which was difficult to comprehend and interpret. Some participants resolved the issue by calling the public officer to further explain. The SSO is aware of this and has created an infographic to describe the process of submitting a claim for each sort of benefit which has received a very positive response from the insured persons: " *I like a visual communication because it is easy to understand [...] I am more knowledgeable about the procedure of preparing documentation for a claim*" [iIP12]. Information on the SSO e-services was regularly updated with social security news and information in the claim history section; most insured persons retrieved their claim history normally. However, if the claim was still pending, the insured person could not track a claim's status, so that the *"mobile app doesn't show the progress status. I don't know where my required documents are, HR didn't send them, or the public officer lost them. Finally, I didn't receive the money for the accident reimbursement at that time"* [iIP17]. This is a major gap in supporting technical transparency. Finally, security was a concern to our participants as a technical factor, with iIP15 concerned that personal information may be leaked, including their medical care information.

6 Discussion

This paper discusses the socio-technical aspects that must be considered when building a transparent benefit payment system. If we focus only on the technological issues in the back-end system and e-services as the front-end system. This is insufficient for enhancing transparency in the case of the SSO in Thailand. For this reason, it is essential for HCI designers and system developers to consider and understand how human, social, and organizational elements influence the decisions around their work and in the utilization of technology [5, 8] in order to establish service policies that can satisfy the demands of citizens and will also foster greater confidence in government services.

6.1 Collaborative Work in Information Transparency

Creating digital transparency is not restricted to technical challenges involved with system development, but also concerns organizational conditions for digital transparency [23]. Based on our results, organizational and technological factors impact the back-end payment system's design requirements. The organizational structure of government agencies is evident as organizational factors that each public officer has unique roles, responsibilities, and authorizations. This demonstrated that the SSO has appropriate data transparency because they can answer questions from the public regarding who is involved, when and where data happens, and what information is required at the various points of processing a claim [4, 27]. Additionally, segregation of duties between officers also helps to manage monitoring and controlling activities. For instance, the financial and benefits officers conduct a collaborative audit of financial transactions regarding benefit payments with paper-based and data in the Sapiens system. Here, we see that the system's process transparency is functioning as it allows participants to track all previous transactions [27].

Another form of collaborative work that the system performs is sharing data between the SSO and hospital. If public officers require additional medical evidence, the insured person does not have to waste time traveling to send these documents at the SSO. Importantly, access to medical records should not be provided to all public officers and should be limited to those involved in the claim process. According to a recent International Social Security Association report [17] sharing individual data between public agencies is necessary for implementing more efficient and proactive public services. However, policy challenges must be overcome, such as establishing a balance between privacy and openness. Whenever the SSO changes process or policy, these changes should therefore be documented [23] and the explanations published for public dissemination so that they can enhance decision/policy transparency [14]. The changes in policies and regulations will affect the benefit payment system adjustments. This is why some insured persons call public officers frequently during this time because they do not understand and are familiar with new practices. Consequently, all public officers should have the opportunity to attend regular training from the SSO headquarters in order to advise the most appropriate solutions to the insured persons.

After completing the claim, the IT department at the SSO headquarters will be responsible for data storage. Typically, a database administrator is responsible for system governance to provide a suitable level of transparency. People holding this role will need to be familiar with the relevant privacy legislation: they must have knowledge, training, and experience in managing data and metadata quality [23]. Currently, the government needs to provide OGD, but this requires significant organizational change that is likely to impact back-end systems ability to quickly respond to establish information transparency, which in turn will affect transparency on a front-end portal [23]. Interaction designers and HCI researchers will therefore need to consider the role of both back- and front-end system contexts when determining what elements can promote transparency in the public sector for OGD.

6.2 Promoting Citizen Engagement in the Use of OGD

Increased participation offers the potential to promote transparency and accountability, resulting in synergistic benefits [12] because it gives citizens more opportunities to voice their opinions on policies that impact them. Additionally, transparency strengthens democracy by motivating government officers ensure that they are acting for the benefit of citizens [4]. Motivation is the first thing that HCI designers should focus on because it is the major driver of OGD to encourage more citizens to participate in OGD [15, 34]. In the context of SSS, HCI designers need to be aware of the users' extrinsic motivation, not just their transactional progression, and presenting information that insured persons are currently interested in or need to know, such as conditions for receiving benefits, required documents, and significant policy changes, because these directly affect their outcomes.

Encouraging insured persons to interact with SSO e-services can provide users with the benefits of their rights on SSB and open data portals through public relations and training. In particular, they should give open data FAQs for basic users to educate themselves [13]. In our study, workplace colleagues were the people who most influenced participants to learn about government data due to their availability and ease of seeking consultation. Other important persons included family members, HR staff, and friends in internet communities. Commonly, participants used digital media (YouTube, Facebook, Line app, etc.) to search for information, and sometimes received posts from friends through these channels. Related to this, Zuiderwijk et al. [39] recommended that convincing influential members close to the user that open data should be utilized can increase the use and acceptability of open data technologies. Consequently, HCI designers need to emphasize not only the enhancement of open data technologies but also the significance of social relationships. The government could, for instance, share success stories or important information with citizens and allow them to reshare this with their online communities. Nonetheless, the SSO website and SSO connect mobile were the primary open data portals used by participants in our research. Our participants described how the two services operate, remarking that their functions were uncomplicated. They did not discuss user interface design issues in considerable detail. However, some participants mentioned the dissemination of information using legal language, making it difficult to understand and enact. To remove this problem, government sources need to avoid jargon or technical terms [23].

Another barrier to access was around pending claims. In this situation, the SSO e-services on the website and mobile phone could not display progress, making it impossible for the insured person to track any status. To be transparent, the government must allow citizens to track the ongoing progress of their government services, ensure efficiency, and provide reasonable processing times for various services, documents, and resources [7]. In the context of the SSS, insured persons can access OGD via the SSO e-services, online communities, and direct phone calls to public officers. However, there is a concern here about data protection for privacy. Without knowing whether the data contains sensitive, personal information, it is dangerous to access it [23]. Accordingly, disclosing government data must comply with privacy rules such as covered by the GDPR [23, 27]. Similarly, public officers by [iPO01, iPO02 and, iPO05] are wary of disclosing information such as amount or disclose personal information over the phone to comply with Sect. 100 of the Thai Social Security Act B.E. 1990 on protection against disclosure of personal information (insured person).

Developing a mechanism for digital transparency in government means making data open to the public via government portals [22, 23] that allow dissemination of information or provide an electronic system to generate and follow special requests [22]. If citizens increasingly engage with OGD, government agencies will become more transparent. For example, the information transparency supplied via systems administration activities at the back-end system will influence the ability of the system to provide transparency information at the front-end via the citizens' interface, who can then utilize this transparency data to create suggestions for improvement. Citizen engagement with the

OGD will also impact the back-end system, as organizations will need to be more open and responsive to this user feedback and prompt in responding to complaints and suggestions [23] because if they do not, this too will become visible. Government transparency is therefore often seen as a means of enhancing openness through the disclosure of information regarding government processes and actions [19, 20]. Therefore, our study identifies three focal areas to improve citizen participation with OGD: motivation, social influence, and technical factors. Technological factors cannot be seen in isolation in the design of OGD portals and must also account for how government systems interact with human demands and their social context.

7 Conclusion

HCI designers and system developers should consider creating information transparency for both the back-end system and the front-end portal. In this study, we investigated the Sapiens system as a back-end system in the benefit payment process. We explored how public officers collaborated in the benefit payment process, how they interacted with one another, and how the computer system enabled their collaboration to create data, process, and policy transparency. From the perspectives of public officers, our findings show the factors that influence the transparency of payment systems, including segregation of duties, check and control mechanisms, legal and regulation support, training public officers, data sharing, accessibility, and data protection.

To encourage citizens to engage with OGD as a front-end system, we need to know who utilized data and what elements pushed them to keep interacting with OGD. We examined the SSO e-services as a portal for opening government data. Our study identified motivation, social influence, and technological factors as the key factors of citizens' increased use of OGD. Based on our literature reviews, our research employed only three technical factors: readability, timeliness, and accessibility. Our findings show the factors that influence increased public usage of OGD portals from the insured persons' perspectives consisting of discrimination of SSB information, training of citizens, social relationship, online communities, content, status updates, ease of access, and data security.

However, the design and implementation of front-end systems as open data portals is crucial that may be relevant not only to citizens but also businesses and other government agencies that can collaborate work together. For developing OGD service, we should conduct additional research on how to effectively employ emerging technologies such as cloud computing and big data for the purpose of greater transparency, as well as investigate the influence of collaboration between people and systems to increase the likelihood of their implementation in real practice.

Acknowledgment. We would like to thank the Ministry of Higher Education, Science, Research, and Innovation for funding this study. We also thank the Social Security Office in Thailand for the assistance with this project and all our interviewees for their time and participation in our research.

References

1. Dee, A.Y., Talib, S.A., Azmi, A.: Trust and justice in the adoption of a welfare e-payment system. Transform. Gov. People Process Policy **7**(2), 240–255 (2016)
2. Appelbaum, S.H.: Socio-technical systems theory: an intervention strategy for organizational development. Manag. Decis. **35**(6), 452–463 (1997). https://doi.org/10.1108/00251749710173823
3. Babbie, E.: The basics of social research: International edition (6th ed.) Wadsworth Cengage Learning (2014)
4. Frank Bannister and Regina Connolly. The trouble with transparency: a critical review of openness in e-government. Policy Internet **3**(1), 1–30 (2011)
5. Baxter, G., Sommerville, I.: Socio-technical systems: From design methods to systems engineering. Interact. Comput. **23**(1), 4–17 (2011). https://doi.org/10.1016/j.intcom.2010.07.003
6. Cai, L., Zhu, Y.: The challenges of data quality and data quality assessment in the big data era. Data Sci. J. **14**(2015), 1 (2015). https://doi.org/10.5334/dsj-2015-002
7. Bertot, J.C., Jaeger, P.T., Grimes, J.M.: Promoting transparency and accountability through ICTs, social media, and collaborative e-government. Transform. Gov. people Process policy **6**(1), 78–91 (2012)
8. Cozza, M.: Interoperability and convergence for welfare technology. In: Zhou, J., Salvendy, G. (eds.) Human Aspects of IT for the Aged Population. Applications in Health, Assistance, and Entertainment. Lecture Notes in Computer Science, vol. 10927, pp. 13–24. Springer, Cham (2018). https://doi.org/10.1007/978-3-319-92037-5_2
9. Csáki, C., O'Brien, L., Giller, K., McCarthy, J.B., Tan, K.-T., Adam, F.: The use of E-Payment in the distribution of social welfare in Ireland. Transform. Gov. People Process Policy **7**, 6–26 (2013)
10. Fereday, J., Muir-Cochrane, E.: Demonstrating rigor using thematic analysis: a hybrid approach of inductive and deductive coding and theme development. Int. J. Qual. Methods **5**(1), 80–92 (2006)
11. Fleetwood, D.: Judgmental sampling: definition, examples and advantages. QuestionPro. Accessed 27 June 2020. https://www.questionpro.com/blog/judgmental-sampling/
12. Hilgers, D., Ihl, C.: Citizensourcing: applying the concept of open innovation to the public sector. Int. J. Public Particip. **4**, 1 (2010)
13. Hogan, M., et al.: Governance, transparency and the collaborative design of open data collaboration platforms: understanding barriers, options, and needs. In: Ojo, A., Millard, J. (eds.) Government 3.0 – Next Generation Government Technology Infrastructure and Services. Public Administration and Information Technology, vol. 32, pp. 299–332. Springer, Cham (2017). https://doi.org/10.1007/978-3-319-63743-3_12
14. Hosseini, M., Shahri, A., Phalp, K., Ali, R.: Four reference models for transparency requirements in information systems. Requirements Eng. **23**(2), 251–275 (2017). https://doi.org/10.1007/s00766-017-0265-y
15. Hutter, K., Füller, J., Koch, G.: Why citizens engage in open government platforms? GI-Jahrestagung **2011**, 223 (2011)
16. ISSA. Ten global challenges for social security-2016. International Social Security Association. http://www.actuaries.org/CTTEES_PIWG/Documents/chicago_2017/7_PIWG_Tenglobalchallengesforsocialsecurity.pdf
17. ISSA. Ten global challenges for social security – 2019. International Social Security Association. https://ww1.issa.int/sites/default/files/documents/publications/2-10-challenges-Global-2019-WEB-263629.pdf

18. Klapper, L., Singer, D.: The opportunities and challenges of digitizing government-to-person payments. World Bank Res. Obs. **32**(2), 211–226 (2017)
19. Lindstedt, C., Naurin, D.: Transparency is not enough: making transparency effective in reducing corruption. Int. Polit. Sci. Rev. **31**(3), 301–322 (2010). https://doi.org/10.1177/019 2512110377602
20. Lněnička, M., Machova, M., Volejníková, J., Linhartová, V., Knezackova, R., Hub, M.: Enhancing transparency through open government data: The case of data portals and their features and capabilities. Online Information Review (2021)
21. Lochan, N., Mas, I., Radcliffe, D., Sinha, S., Tahilyani, N.: The benefits to government of connecting low-income households to an e-payment system: an analysis in India. Lydian Payments J. **2**, 38–44 (2010)
22. Luna-reyes, L.F., Bertot, J.C., Mellouli, S.: Open Government , Open Data and Digital Government. Gov. Inf. Q. **31**(1), 4–5 (2014). https://doi.org/10.1016/j.giq.2013.09.001
23. Matheus, R., Janssen, M., Janowski, T.: Design principles for creating digital transparency in government. Gov. Inf. Q. **38**(1), 101550 (2021). https://doi.org/10.1016/j.giq.2020.101550
24. Meijer, A.: Understanding modern transparency. Int. Rev. Adm. Sci. **75**(2), 255–269 (2009). https://doi.org/10.1177/0020852309104175
25. Purwanto, A., Zuiderwijk, A., Janssen, M.: Citizen engagement in an open election data initiative: a case study of Indonesian's "Kawal Pemilu." ACM Int. Conf. Proceeding Ser. (2018). https://doi.org/10.1145/3209281.3209305
26. Purwanto, A., Zuiderwijk, A., Janssen, M.: Citizen engagement with open government data: Lessons learned from Indonesia's presidential election. Transform. Gov. People Process Policy **14**(1), 1–30 (2020). https://doi.org/10.1108/TG-06-2019-0051
27. Batubara, F.R., Ubacht, J., Janssen, M.: Unraveling transparency and accountability in blockchain. In: Proceedings of the 20th Annual International Conference on Digital Government Research (2019)
28. Saunders, M., Lewis, P., Thornhill, A.: Research methods for business students (5th ed.) Pearson education (2009)
29. Shin, D., Ibahrine, M.: The socio-technical assemblages of blockchain system: how blockchains are framed and how the framing reflects societal contexts. Digit. Policy, Regul. Gov. **22**(3), 245–263 (2020). https://doi.org/10.1108/DPRG-11-2019-0095
30. Social Security Office. Annual Report 2015 (Thai & Eng Version) 2015. https://www.sso.go.th/wpr/assets/upload/files_storage/sso_th/7338c09c580d0c60d5305d8d08e8db10.pdf
31. Social Security Office. Annual Report 2018 (Thai & Eng Version) (2018). https://www.sso.go.th/wpr/assets/upload/files_storage/sso_th/f4dabf6d2e90ebe9015c5c72a50a6ff5.pdf
32. Social Security Office. Annual Report 2020 (Thai & Eng Version) (2020). https://www.sso.go.th/wpr/assets/upload/files_storage/sso_th/b5ccdc72a0f03b95b8e3ac1df3f80025.pdf
33. Sørum, H., Andersen, K.N., Vatrapu, R.: Public websites and human-computer interaction: An empirical study of measurement of website quality and user satisfaction. Behav. Inf. Technol. **31**(7), 697–706 (2012). https://doi.org/10.1080/0144929X.2011.577191
34. de Souza, A.A.C., d'Angelo, M.J., Filho, R.N.L.: Effects of Predictors of Citizens' Attitudes and Intention to Use Open Government Data and Government 2.0. Gov. Inf. Q. **39**(2), 101663 (2022)
35. Teo, T.S.H., Srivastava, S.C., Jiang, L.: Trust and electronic government success: an empirical study. J. Manag. Inf. Syst. **25**(3), 99–132 (2008). https://doi.org/10.2753/MIS0742-1222250303
36. Treiblmaier, H., Pinterits, A., Floh, A.: The adoption of public e-payment services. J. E-Government **3**(2), 33–51 (2006). https://doi.org/10.1300/J399v03n02_03
37. Wang, F.: Explaining the low utilization of government websites: using a grounded theory approach. Gov. Inf. Q. **31**(4), 610–621 (2014). https://doi.org/10.1016/j.giq.2014.04.004

38. Williams, A.: A global index of information transparency and accountability. J. Comp. Econ. **43**(3), 804–824 (2015). https://doi.org/10.1016/j.jce.2014.10.004
39. Zuiderwijk, A., Janssen, M., Dwivedi, Y.K.: Acceptance and use predictors of open data technologies: Drawing upon the unified theory of acceptance and use of technology. Gov. Inf. Q. **32**(4), 429–440 (2015). https://doi.org/10.1016/j.giq.2015.09.005

An Assessment of the Green Innovation, Environmental Regulation, Energy Consumption, and CO$_2$ Emissions Dynamic Nexus in China

Taipeng Sun, Hang Jiang⬛, and Xijie Zhang$^{(\boxtimes)}$

School of Business Administration, Jimei University, Xiamen 361021, China
17863086771@163.com

Abstract. Green innovation is a critical support to combat climate change arising from greenhouse gas emissions. Based on an environmental framework defined as the Stochastic Impacts by Regression on Population, Affluence, and Technology (STIRPAT) model, this study aimed to examine the impact of green innovation (GI), per capita GDP (PGDP), population density (PD), environmental regulations (ER), energy consumption (EC), and industrial structure upgrading (ISU) on CO$_2$ emissions (CO$_{2e}$). For this purpose, a sample dataset covering the 30 provincial regions in mainland China from 2005 to 2019 was analyzed using the Fixed Effects and System Generalized Method of Moment (SYS-GMM) Methodology. The data analysis indicated that CO$_{2e}$ in the current period were further exacerbated by the agglomeration effect of CO$_{2e}$ from the previous period. The empirical results showed that GI, ER, and ISU all exert a significant inhibitory effect on CO$_{2e}$, whereas PGDP, PD, and EC had a positive effect on carbon emissions when dynamic relationships were analyzed. It is suggested that policymakers in China should focus on the decisive role of green technology application, environmental protection, and green transformation of industrial structure in curbing CO$_{2e}$.

Keywords: Green innovation · environmental regulations · energy consumption · CO$_2$ emissions · GMM model · china

1 Introduction

Global warming is an enormous challenge facing humanity. The massive increase in CO$_2$ and other greenhouse gas (GHG) emissions has aggravated environmental degradation and brought about a series of negative impacts on human activities and on economic and social development. With the process of socio-technical progress and sustainable development constantly threatened by CO$_2$ emissions, the low-carbon economic development vision induces us to further investigate the factors that increase/reduce CO$_2$ emissions [1]. At present, many developing countries and emerging economies still have a high dependence on fossil energy consumption and limited capacity to use resources and energy efficiently. As an emerging economy with 9.89 billion tons of CO$_2$ emissions in 2020,

© Springer Nature Switzerland AG 2023
F. Fui-Hoon Nah and K. Siau (Eds.): HCII 2023, LNCS 14038, pp. 74–86, 2023.
https://doi.org/10.1007/978-3-031-35969-9_6

China accounts for 30.7% of global carbon emissions and is under enormous pressure to reduce CO_2 emissions. Confronted with the thorny challenge, the Chinese government pledged to build comprehensively green innovation-driven development system and to adopt more efficient schemes and pathways to reach peak carbon emissions by 2030 and achieve carbon neutrality by 2060. On this point, various carbon policies are widely used, including carbon cap-and-trade systems, carbon taxes, and carbon emissions offset policies [2]. However, compared to the past when China was limited by the economic development process and emphasized mainly research into conventional technologies, the Chinese government is now committed to the breakthrough and development of green technologies. Equally, during the critical period of economic growth shifting towards high-quality development in China, green innovation, environmental regulation, and industrial structure upgrading are critical forces to facilitate the implementation of the carbon neutrality goal [3, 4]. In addition, green technology will be a vital component for the transformation of the economy to a green and low-carbon direction, not only providing practical approach to achieve low, zero, and negative carbon at the technical level, but also producing continuous impetus for green development in heavy polluting industries at the practical level.

It is noteworthy that relevant research and conclusions are mostly concentrated at the global level, without any specific discussion of the Chinese provincial level. To the best of our knowledge, empirical evidence on the trade-off role of green innovation and environmental regulations in carbon reduction and green sustainable development in the Chinese context is still relatively scarce. In particular, the following issues that need to be proved by empirical tests. On the one hand, can green innovation and environmental regulations effectively curb CO_2 emissions in China? Mixed findings from existing studies by researchers have mainly focused on the impact of traditional technology innovation and R&D investment on carbon emission performance [5, 6]. On the other hand, given that the agglomeration effect existing in the lagged-period CO_2 emissions will have an impact on the current period, are there dynamic mechanisms of green innovation and other factors (e.g., EC and ISU) that affect CO_2 emissions? It is hoped that the results of an empirical study on these issues will not only provide a reference for the Chinese government to optimize the green innovation process, implement environmental regulations, and make reasonable use of energy, but also will serve as a basis for related emerging economies to pursue a low-carbon development path.

The remainder of the paper is arranged as follows: Sect. 2 reviews and summarizes the previous literature regarding the subject of the study; Sect. 3 details the data, models, and methodology used; Sect. 4 lists the results of the empirical model estimates; and Sect. 5 elaborates conclusions and economic implications.

2 Literature Review

2.1 Green Innovation and CO_2 Emissions

As a means of adapting to the challenges of climate change, green technology (such as cleaner production and pollution control technologies) has attracted the attention of researchers, and many research conclusions state that green technology can effectively capture non-desired pollutant emissions and achieve sustainable development goals [7,

8]. Compared with technological progress and development in the traditional sense, green innovation aims to reduce environmental pollutant and CO_2 emissions and achieve green economic growth. The relationship between green innovation and CO_2 emissions in China has been investigated from various perspectives, with mixed conclusions. Li et al. [8] used the NARDL model to examine the asymmetric relationship between green innovation and CO_2 emissions in China and concluded that implementing green innovation contributed to the continuous promotion of CO_2 abatement in China, and vice versa. Yuan et al. [9] found that green innovation has significantly reduced CO_2 emissions in China, but that institutional quality has a negative regulatory effect in this process. In other words, when institutional quality is consistently kept at a high level, green innovation a more vigorous presence on continuously reducing CO_2 emissions, and this effect is more significant in the eastern and western regions. It cannot be ignored that the potential role of green innovation in reducing natural resource consumption and ecological damage is spatially heterogeneous between regions. Xu et al. [10] empirically examined the spatial interaction and spillover effects of green innovation and CO_2 emissions in the 30 Chinese provinces using a spatial econometric model. The study revealed that interregional joint improvements in green innovation technologies were more effective in curbing CO_2 emissions, and green innovation transformation and technology-oriented industrial upgrading were highlighted. In addition, some researchers believe that to achieve simultaneous economic development and green transformation, the work of achieving carbon emissions reduction must include enterprises as a breakthrough generator [11, 12]. As a core component of green innovation, cleaner production technology can indirectly reduce CO_2 emissions by optimizing production processes. The extensive application of green technology in enterprise production can boost green energy consumption and reduce pollution emissions from the production side. At the same time, green innovation can help transform the energy-intensive industries into a cleaner, lower-carbon direction by empowering the coupled coordination and transformation of the industrial structure, thus restricting CO_2 emissions from production processes in enterprise. Using a mediating effects model for an empirical analysis, Zhu et al. [12] revealed that green technology innovation by enterprises could decrease regional CO_2 emissions intensity under the influence of two-way foreign direct investment. Equally important, carbon-related technologies belonging to green innovation are being widely used by heavily polluting sectors and enterprises. For example, CO_2 utilization, capture, and storage technologies can effectively control the cost of decarbonization and empower end-of-pipe treatment, thereby reducing pollutant emissions and environmental governance costs simultaneously [13].

2.2 Environmental Regulation and CO_2 Emissions

What we know about environmental regulations are largely based on empirical studies that have investigated how environment-related legislation mitigates CO_2 emissions to emphasize the importance of ecological protection behavior. For example, the results of Zhu and Ruth [14]'s study in China revealed that energy-saving policies introduced by the government could improve the energy use efficiency of industrial enterprises and gradually decreasing CO_2 emissions. Regarding the impact of three types of environmental regulations (command-and-control, market-based, and informal regulations) on environmental performance in China, Li and Ramanathan [15] concluded that compared with market-based regulation, there was a non-linear correlation between command-and-control regulations, the legal means to regulate the behavior of enterprise pollution emissions, and environmental performance. However, the obvious inverted "U-shaped" relationship expressing the determinant factor of environmental regulations toward CO_2 emissions has received attention from researchers. In other words, with the tightening of environment-friendly policies and legislations, carbon emissions show a trend of increasing first and then decreasing. By investigating panel data from the 30 provinces in mainland China, Lan and Wang [16] found that investment in industrial pollution control had a double threshold effect on carbon emission performance and that regional carbon performance presented a "U-shaped" development trend. On the contrary, when studying the dynamic threshold effect of industrial pollutant abatement expenditure on environmental degradation in China's industrial green transition, Hou et al. [17] reached conclusions that were almost contradictory.

2.3 Energy Consumption and CO_2 Emissions

Existing research classifies energy use into two main types: renewable and non-renewable. Non-renewable energy is dominated by fossil energy, with coal as the main component. On the contrary, renewable energy consists mainly of energy derived from natural sources, and adequate, reliable, and clean energy supply is the fundamental direction for economic growth and environmental improvement. Therefore, renewable energy is considered as a fundamental requirement for building a resource-saving society and achieving sustainable development goals. According to the findings of an investigation conducted by Musah et al. [18] for North African countries, Oil-based energy consumption, a potential factor that cannot be ignored, has exacerbated further CO_2 emissions. In another study on the driving factors of economic growth in the BRICS countries, Chang et al. [19] found that the interactive relationship between coal consumption and economic development was not proven, and that its unidirectional facilitating effect was only established in China. Similarly, in the case of China, the persistent impact of clean energy consumption and green economy development, i.e., a potential pathway to mitigate pollutant emissions, on the abatement of CO_2 emissions was studied by Wan [20]. According to their results, the practical role of clean energy consumption in decreasing CO_2 emissions is verified; in this regard, this impact comes from continuous increase and structural transformation of clean energy consumption.

In summary, the existing literature has yielded fruitful results on the study of carbon emissions and environmental issues, but there are still certain room for improvement. Further investigation is needed to reach a consensus on a clear understanding of the mechanisms of generation and mitigation of CO_2 emissions. Flaws in methodology, unclear results, and insufficient evidence for individual countries create a motivation to consider the dynamic relationship among green innovation, environmental regulation, energy consumption, and CO_2 emissions in China.

3 Research Methods and Data Source

3.1 Baseline Model

The estimated model is mainly reliable in the extended STIRPAT framework, which Dietz and Rosa [21] proposed from the original IPAT version to detect the impact of human activities on the environment.

$$I = \theta \cdot P^\alpha \cdot A^\beta \cdot T^\gamma \cdot \varepsilon \tag{1}$$

Here I = impact on the environment, P = population, A = affluence, and T = technology. θ =correlation coefficient, α, β, and γ = impact index, and ε = error term. The model is rewritten in natural logarithmic form as:

$$lnI_{it} = ln\theta + \alpha lnP_{it} + \beta lnA_{it} + \gamma lnT_{it} + ln\varepsilon \tag{2}$$

In the specific analysis, total CO2 emissions are used to characterize the environmental effects, and population density, GDP per capita, and green technology patents are used as proxy variables for population, affluence, and technology. In addition, according to previous research and actual conditions, CO_2 emissions are also influenced by other variables. Therefore, the extended STIRPAT model is defined in this paper as follows.

$$lnCO_{2eit} = ln\theta + \alpha lnPD_{it} + \beta lnPGDP_{it} + \gamma lnGI_{it} + \delta lnER_{it} + \mu lnEC_{it} + \rho lnISU_{it} + ln\varepsilon_{it} \tag{3}$$

3.2 Methodology

Due to the endogeneity problem in the static long panel, the estimation results are biased, so the dynamic estimation model can be used to correct the endogeneity and bias. The estimation model, represented by the SYS-GMM model, can mitigate the effect of endogeneity by introducing instrumental variables, and therefore obtain consistent estimation results. Assuming that the impacts of GI, ER, and EC on CO_{2e} are dynamic, according to the previous concept, it is hypothesized that the CO_{2e} of the previous period will affect the CO_{2e} of the current period and that GI, ER, and EC will affect the CO_{2e} of the current period. Equally, the lag term of CO_{2e} was introduced into the model as an instrumental variable. According to Eq. (3), the specific model used in this paper can be expressed as follows:

$$lnCO_{2eit} = ln\theta + lnCO_{2eit-1} + \alpha lnPD_{it} + \beta lnPGDP_{it} + \gamma lnGI_{it} + \delta lnER_{it} + \mu lnEC_{it} + \rho lnISU_{it} + ln\varepsilon_{it} \tag{4}$$

where, i = the specific province, t = the year, CO_{2eit-1} = the CO_{2e} index of a province with a lag of one period, and ε = error term.

3.3 Variable Description

Balanced panel data covering the 30 provincial administrative regions in mainland China from 2005 to 2019 were collected for this study. The variables were constructed as follows.

The main sources of CO_2 emissions include raw carbon, crude oil, and natural gas consumption and productfion. With reference to Gao et al. [22] and Rehman et al. [23], this study selected total CO_2 emissions in millions of tons to measure the carbon emission level of each province. The data were obtained from the China Carbon Accounting Database (CEADs) [24–27].

Green innovation not only aims to minimizing the threat posed by environmental degradation, but also is an effective driving force for the harmonious coexistence of ecology and economy. Traditional invention patents cannot accurately portray "green" innovation, whereas practical green patents, including those representing the environmental friendliness and advanced nature of innovation activities, can prominently reflect green innovation in transformation and application. Therefore, based on the information provided by the International Patent Classification (IPC) regarding patenting activities for green innovation, green utility patent applications from each province were collected in the data service platform and used as the explanatory variable. The data were obtained from the Chinese Research Data Services (CNRDS) Platform.

For key variables, this study has accumulated important factors affecting regional CO_2 emissions, including environmental regulation and energy consumption. Formal environmental regulation can reflect the effectiveness of regional pollutant abatement and further reduce pollutant emissions. Therefore, this paper sets environmental regulation indicators from the aspect of investment intensity, which is measured by the share of industrial pollution control investment in secondary industry. In China's energy structure, coal makes up a high proportion of total energy consumption, accounting for nearly 57% in 2020. Therefore, this paper takes carbon consumption as a proxy variable for energy consumption and introduces it into the model.

Due to the dynamic mobility of carbon factors among different industrial sectors, pollution usually migrates from high productivity to low productivity sectors. As the main sector that produces environmental pollutants, the concentration and crowding effects of industry will have a profound impact on CO_2 emissions. Referring to Wei and Hou [28], this paper selected the ratio of the added value of the secondary industry to GDP to measure the industrial structure upgrading. In addition, regions with greater affluence are more likely to attract a concentration of low-pollution, high-efficiency enterprises and to provide financial support for green innovation. Therefore, this paper used GDP per capita to measure this indicator. Population density was measured as the population per square kilometer of land area. These raw data are available from the China Statistical Yearbook and the China Energy Statistical Yearbook. Descriptive statistics for the variables are provided in Table 1 below.

Table 1. Descriptive statistics for variables

Variable	Symbols	Obs.	Mean.	S.D.	Min.	Max.
CO_2 emissions	CO_{2e}	450	5.47	0.81	2.02	7.44
Green innovation	GI	450	6.70	1.65	1.39	10.60
Environmental regulations	ER	450	−1.31	0.80	−4.35	0.90
Energy consumption	EC	450	9.21	0.95	5.21	10.85
Industrial structure upgrading	ISU	450	3.74	0.22	2.77	4.13
Per capita GDP	PGDP	450	10.41	0.64	8.56	11.99
Population density	PD	450	7.79	0.53	5.24	8.75

Note: Data are in natural log form.

4   Empirical Results and Discussions

4.1   Panel Unit Root Test

Before estimating the final model, a panel unit root test was first performed to check the nature of the data to avoid any biased results. This paper used the Levin–Lin–Chu (LLC) unit root test to check the stationarity of the study objects in the sample [29]. The results in Table 2 (LLC test) support rejection of the null hypothesis, affirming the absence of a unit root and data stationarity.

Table 2. Panel unit root results: LLC-test

Variable	At level	At first differences	Stationarity
CO_{2e}	−4.4772***	−4.4646***	YES
GI	−2.3816***	−1.4794***	YES
ER	−8.4728***	−3.4854***	YES
EC	2.1430	−1.2793*	YES
ISU	−4.9169***	−3.9580***	YES
PGDP	−3.6467***	−6.1727***	YES
PD	−25.3485***	5.5380***	YES

Notes: * $p < 0.1$, ** $p < 0.05$, and *** $p < 0.01$.

4.2   Total Sample Estimation Results

The dynamic nexus among the selected variables, GI, ER, EC, ISU, PGDP, PD, and CO_{2e}, are further investigated in this section, as per the GMM model estimator established in the methodology section. In addition, one of the rules for breaking endogeneity in panel data is the selection and introduction of instrumental variables (the lag term of CO_{2e} as an instrumental variable in this paper), and the Hansen test in the GMM model can perform an over-identification test for the validity of instrumental variables to prevent potential estimation bias [30–32]. The dynamic SYS-GMM model in Table 3 performed the testing process, and the Arellano-Bond test results showed that the perturbation terms had only first-order autocorrelation, with no second-order autocorrelation. The result of the Hansen test for the instrumental variables failed to reject the null hypothesis (over-identifying restrictions are valid), and therefore the pre-defined testing criteria confirmed the accuracy of the model selection and estimation. Meanwhile, to contrast with the static panel estimates, combined with the Hausman test, fixed-effects regression was included in the results.

Table 3. Summary of the results

Variable	Fixed-effects	SYS-GMM	
	Dependent variable: CO_{2e}	Dependent variable: CO_{2e}	
$CO_{2e\,t-1}$		0.7103***	0.5498***
		(0.020)	(0.026)
GI	−0.0220	−0.0182***	−0.1113***
	(0.025)	(0.004)	(0.007)
ER	−0.0060	−0.0289***	−0.0515***
	(0.014)	(0.005)	(0.006)
EC	0.4448***	0.2387***	0.4074***
	(0.030)	(0.032)	(0.019)
ISU	−0.0509		−0.3221***
	(0.093)		(0.064)
PGDP	0.3768***		0.2169***
	(0.054)		(0.019)
PD	0.0029		0.1224***
	(0.023)		(0.035)
_cons	−2.2383***	−0.4862**	−2.5906***
	(0.479)	(0.199)	(0.446)
Observations	450	420	420
Arellano-Bond (1)		P = 0.063	P = 0.082
Arellano-Bond (2)		P = 0.723	P = 0.652
Hansen test		P = 0.989	P = 0.988

Notes: * $p < 0.1$, ** $p < 0.05$, and *** $p < 0.01$. Standard errors in parentheses.

As seen in Table 3, the coefficient of GI is statistically insignificant in the static panel fixed effects regression, but significant in the dynamic panel SYS-GMM model, and is negatively correlated with CO_{2e}. It is clear that the lagged one-period CO_{2e} as an instrumental variable exacerbates CO_{2e} in the current period and that GI stimulates a further decrease in carbon emissions. The achievements of GI can be obviously transformed into green productivity, promoting the transformation of energy structure to clean and green, effectively accelerating transformation and harmonization of ISU, reducing EC, and thus decreasing the level of CO_{2e}. As an essential driving force for green development in the eco-society, GI is a strategic pathway for the economy to achieve low-carbon transformational development and reduce CO_{2e} in production activities.

In the case of ER and ISU, their coefficients show a negative and statistically significant correlation with CO_{2e}. It can be concluded that ER and ISU can efficiently restrain the further exacerbation of environmental pollutant emissions from the industrial and manufacturing sectors. Along with continuous investment of government pollution control funds, enterprises will improve green production efficiency and reduce environmental governance costs through technological innovation to achieve the effect of CO_{2e} abatement. This conclusion is supported by Pei et al. [33]'s research on CO_{2e} from energy-intensive industries in different provinces in China. The main reason why ISU will be beneficial in steadily capturing CO_{2e} is that with the continuous development of modern industries, enterprises can use more advanced green technologies to improve energy utilization and green productivity and can reduce carbon emissions by adjusting and optimizing the industrial structure.

Moreover, considering the empirical findings related to EC, the coefficients of EC show positive and statistically significant effects on CO_{2e}, as shown in Table 3. The results indicate that coal consumption and excessive dependence on natural resources have brought about environmental damage, which is not conducive to long-term sustainable development. Due to China's huge total energy consumption and the low proportion of clean energy, fossil energy will continue to dominate China's energy consumption structure. In the short term, coal-based fossil energy consumption remains the primary threat to pollutant emissions and environmental degradation in China.

Furthermore, the correlation coefficient between PGDP and CO_{2e} is positive and significant at the 1% level. These findings confirm the existence of the environmental Kuznets curve (EKC), which indicates that as a consequence of economic growth, CO_{2e} causes the quality of the environment to deteriorate during the initial stages. Energy consumption, especially fossil fuels, is considered to have gradually exacerbated the deteriorating relationship between economic growth and CO_{2e}, and the predicament has been validated. Under the new climate commitments, declines in energy intensity and energy carbon density will create more development space for economic growth in China.

Finally, PD also has an undesirable consequence on the increase in carbon emissions, which can be explained by the increase in resource crowding and energy consumption costs. This finding is similar to the outcomes obtained by Rahman [34], who found evidences that over-concentration of population contributed to the consequence of excessive energy consumption, and that the associated production of greenhouse gases, wastewater, and other pollutants makes the ecosystem unbalanced.

5 Conclusions and Economic Implications

Green innovation is the solution to the real dilemma of total CO_2 emissions control and economic and social development. It is foreseeable that the Chinese government accelerates the use of green technologies, environmental regulation tools, and other technical paradigms related to carbon abatement, which will be conducive to guiding sustainable development and improving the global competitiveness of domestic industry and economy. The econometric empirical results with panel data of 30 provinces in mainland China from 2005 to 2019 show that, first, under the STIRPAT framework, synergistically evolving green innovation and industrial models can significantly inhibit CO_2 emissions. The application of eco-friendly technologies can promote the transformation of traditionally high-pollution industries to green and energy-saving industries and lay the foundation for promoting deep integration of digitalization, greening, and industrialization. The restrictive impact of the investment in pollutant treatment funding on CO_2 emissions confirms the "Porter hypothesis" [35], which states that high-intensity environmental regulation can induce more enterprises to develop a willingness to reduce emissions under cost pressure and to seek technological innovation toward cleaner production, thus providing all-round support for China to achieve the "dual carbon" goal. Second, according to the results captured by the SYS-GMM model, coal-dominated energy consumption remains a potential factor in exacerbating CO_2 emissions. Due to its limited resource endowment, China will still mainly consume coal in the short and medium term, which will lead to a further increase in the intensity of atmospheric pollutant emissions. The positive relationship between per capita income, population density, and CO_2 emissions was also confirmed. At this stage, the per capita income curve of each province in China has not yet moved past the inflection point, and therefore the inverted "U-shaped" curve between the continuous economic development, with its consequent problems of over-exploitation and consumption of non-renewable resources and pollutant emissions, and eco-environmental protection has not been manifested. Finally, the impact of population density on environmental pollutant emissions mainly originates from the aspects of infrastructure and resource consumption concentration, and overcrowding and congestion have an impact on green and sustainable development.

According to the above conclusions, necessary policy implications should be recognized and acted upon so that carbon emissions reduction and climate change mitigation can be better implemented.

First of all, fostering the innovation and application of green technologies in China and other emerging economies is imperative to combat the global challenges of carbon emissions and climate change. To truly leverage the potential of green innovation in alleviating GHG emissions and environmental degradation, it is necessary to improve adaptation to climate change by enhancing green innovation capacity. Policymakers should strengthen the continuous integration of green technologies with the development of the real economy to provide new impetus for the coordinated construction of an environmental-friendly society. On the one hand, relevant governments and departments should introduce feasible policies and funds to support the development of disruptive technological innovations, build market-oriented green technological innovation systems, and focus on developing clean energy production and consumption industries. On the other hand, enterprises should take up green technology applications as a cost advantage to effectively control decarbonization and to provide technical support for the utilization, capture, and storage of CO_2 through green innovation.

Second, energy conservation and high-tech industries are an important conduit for transforming green technology research and development into practical scientific and technological achievements. It promotes industrial specialisation and sophistication by rational integrating production processes and transforming energy-intensive production models, thereby improving production efficiency. In view of the energy supply shortage and development difficulties, the government should use administrative and legal means to formulate a series of relevant policies and actively promote a resource-efficient economic development model. At the same time, the extension of the industrial chain can effectively improve the efficiency of resource use, thus reducing resource wastage and CO_2 emissions. The synergistic upgrading of industrial structures means that there is greater potential for inter-industry cooperation and linkages, which can further increase economic benefits. The "decarbonization" of the economy, created by deepening the existing industrial structure, is more relevant than the creation of new industries.

Finally, policymakers should focus on improving the environmental regulation and governance system and should formulate targeted environmental penalties and taxation to strengthen the capacity and standards of eco-environmental protection. Because environment-related regulating tools has a positive inhibitory effect on CO_2 emissions, the government should continue to incentivize enterprises to change their high consumption, high emissions production patterns, promote cleaner production, and pursue the goal of sustainable development efficiency. In addition, to reduce the uncertainty of environmental investment, enterprises need to clarify the direction of their green technology research and development to achieve a win-win situation between practical application and value enhancement. The economic, social, and ecological benefits of reducing fossil energy consumption cannot be ignored. As an inevitable choice to combat global climate change and to realize the clean energy construction and industrial upgrading in various countries, the use of green renewable energy will be a trailblazer for global carbon reduction and sustainable economic and social development.

References

1. Jiang, Q., Rahman, Z.U., Zhang, X., Islam, M.S.: An assessment of the effect of green innovation, income, and energy use on consumption-based CO2 emissions: Empirical evidence from emerging nations BRICS. J. Clean. Prod. **365**, 132636 (2022). https://doi.org/10.1016/j.jclepro.2022.132636
2. Zhang, Y., Tan, D., Liu, Z.: Leasing or selling? Durable goods manufacturer marketing model selection under a mixed carbon trading-and-tax policy scenario. IJERPH. **16**(2), 251 (2019). https://doi.org/10.3390/ijerph16020251
3. Sun, Y., Razzaq, A., Sun, H., Irfan, M.: The asymmetric influence of renewable energy and green innovation on carbon neutrality in China: Analysis from non-linear ARDL model. Renew. Energy **193**, 334–343 (2022). https://doi.org/10.1016/j.renene.2022.04.159
4. Shi, X., Xu, Y.: Evaluation of China's pilot low-carbon city program: a perspective of industrial carbon emission efficiency. Atmos. Pollut. Res. **13**(6), 101446 (2022)
5. He, Y., Fu, F., Liao, N.: An analysis of the effect of industrial R&D investment on carbon emissions based on the STIRPAT model. Sci. Technol. Manag. Res. **41**(17), 206–212 (2021). https://doi.org/10.3969/j.issn.1000-7695.2021.17.026

6. Gu, J.: Sharing economy, technological innovation and carbon emissions: evidence from Chinese cities. J. Innov. Knowl. **7**(3), 100228 (2022). https://doi.org/10.1016/j.jik.2022. 100228

7. Ali, N., et al.: FDI, Green innovation and environmental quality nexus: new insights from BRICS economies. Sustainability **14**(04), 2181 (2022). https://doi.org/10.3390/su14042181

8. Li, Y., Chuan, Z., Li, S., Usman, A.: Energy efficiency and green innovation and its asymmetric impact on CO_2 emission in China: a new perspective. Environ. Sci. Pollut. Res. **22**(31), 47810–47817 (2022). https://doi.org/10.21203/rs.3.rs-1013410/v1

9. Yuan, B., Li, C., Yin, H., Zeng, M.: Green innovation and China's CO_2 emissions – the moderating effect of institutional quality. J. Environ. Planning Manage. **65**(5), 877–906 (2021). https://doi.org/10.1080/09640568.2021.1915260

10. Xu, J., Tong, B., Wang, M.: A study on the impact of green technology innovation on CO_2 emissions from a spatial perspective. Scientific Res. **40**(11), 1–20 (2022). https://doi. org/10.16192/j.cnki.1003-2053.20220301.001

11. Desheng, L., Jiakui, C., Ning, Z.: Political connections and green technology innovations under an environmental regulation. J. Clean. Prod. **298**, 126778 (2021). https://doi.org/10. 1016/j.jclepro.2021.126778

12. Zhu, Y., Gao, H., Xu, Y.: How can coordinated development of two-way FDI reduce regional CO_2 emissions intensity? --- Based on the mediating effect of green technology innovation by enterprises and the moderating role of government quality. Soft Sci. **36**(02), 86–94 (2022). https://doi.org/10.13956/j.ss.1001-8409.2022.02.13

13. Xie, X., Zhu, Q., Wang, R.: Turning green subsidies into sustainability: How green process innovation improves firms' green image. Bus Strat. Env. **28**(7), 1416–1433 (2019). https:// doi.org/10.1002/bse.2323

14. Zhu, J., Ruth, M.: Relocation or reallocation: impacts of differentiated energy saving regulation on manufacturing industries in China. Ecol. Econ. **10**, 119–133 (2015). https://doi.org/ 10.1016/J.ECOLECON.2014.12.020

15. Li, R., Ramanathan, R.: Exploring the relationships between different types of environmental regulations and environmental performance: evidence from China. J. Clean. Prod. **196**, 1329–1340 (2018). https://doi.org/10.1016/j.jclepro.2018.06.132

16. Lan, H., Wang, L.: A study on the threshold effect of regional carbon emission performance and environmental regulation under green development - based on SE-SBM and double-threshold panel model. Soft Sci. **33**(08), 73–77 (2019). https://doi.org/10.13956/j.ss.1001-8409.2019.08.13

17. Hou, J., Teo, T., Zhou, F., Lim, M., Chen, H.: Does industrial green transformation successfully facilitate a decrease in carbon intensity in China? an environmental regulation perspective. J. Clean. Prod. **184**, 1060–1071 (2018). https://doi.org/10.1016/j.jclepro.2018.02.311

18. Musah, M., Kong, Y., Mensah, I.A., Antwi, S.K., Osei, A.A., Donkor, M.: Modelling the connection between energy consumption and carbon emissions in North Africa: Evidence from panel models robust to cross-sectional dependence and slope heterogeneity. Environ. Dev. Sustain. **23**(10), 15225–15239 (2021). https://doi.org/10.1007/s10668-021-01294-3

19. Chang, T., Deale, D., Gupta, R., Hefer, R., Inglesi-Lotz, R., Simo-Kengne, B.: The causal relationship between coal consumption and economic growth in the BRICS countries: Evidence from panel-Granger causality tests. Energy Sources Part B **12**(2), 138–146 (2017). https://doi.org/10.1080/15567249.2014.912696

20. Wan, Y.: Green economic development, clean energy consumption and CO_2 emissions. Ecolog. Econ. **38**(050), 40–46 (2022)

21. Dietz, T., Rosa, E.A.: Effects of population and affluence on CO_2 emissions. Proc. Natl. Acad. Sci. U.S.A. **94**(1), 175–179 (1997). https://doi.org/10.1073/pnas.94.1.175

22. Rehman, A., Ma, H., Ozturk, I., Radulescu, M.: Revealing the dynamic effects of fossil fuel energy, nuclear energy, renewable energy, and carbon emissions on Pakistan's economic growth. Environ. Sci. Pollut. Res. **29**(32), 48784–48794 (2022). https://doi.org/10.1007/s11356-022-19317-5

23. Gao, P., Wang, Y., Zou, Y., Su, X., Che, X., Yang, X.: Green technology innovation and carbon emissions nexus in China: does industrial structure upgrading matter? Front. Psychol. **13**, 951172 (2022)

24. Shan, Y., et al.: China CO_2 emission accounts 1997–2015. Sci. Data. **5**(1), 170201 (2018)

25. Shan, Y., Huang, Q., Guan, D., Hubacek, K.: China CO_2 emission accounts 2016–2017. Sci. Data. **7**(1), 54 (2020). https://doi.org/10.1038/s41597-020-0393-y

26. Guan, Y., Shan, Y., Huang, Q., Chen, H., Wang, D., Hubacek, K.: Assessment to China's recent emission pattern shifts. Earth's Future **9**(11), 2241 (2021). https://doi.org/10.1029/2021ef002241

27. Shan, Y., et al.: New provincial CO_2 emission inventories in China based on apparent energy consumption data and updated emission factors. Appl. Energy **184**, 742–750 (2016). https://doi.org/10.1016/j.apenergy.2016.03.073

28. Wei, L., Hou, Y.: A study on the effect of specialized and diversified industrial clusters on regional green development. Manag. Rev. **33**(10), 22–33 (2021). https://doi.org/10.14120/j.cnki.cn11-5057/f.2021.10.003

29. Levin, A., Lin, C.-F., James Chu, C.-S.: Unit root tests in panel data: asymptotic and finite-sample properties. J. Econ. **108**(1), 1–24 (2002). https://doi.org/10.1016/s0304-4076(01)00098-7

30. Arellano, M., Bond, S.: Some tests of specification for panel data: monte carlo evidence and an application to employment equations. Rev. Econ. Stud. **58**(2), 277 (1991). https://doi.org/10.2307/2297968

31. Roodman, D.: How to do Xtabond2: An Introduction to Difference and System GMM in Stata. Stand. Genomic Sci. **9**(1), 86–136 (2009). https://doi.org/10.1177/1536867x0900900106

32. Hansen, L.P.: Large sample properties of generalized method of moments estimators. Econometrica **50**(4), 1029 (1982). https://doi.org/10.2307/1912775

33. Pei, Y., Zhu, Y., Liu, S., Wang, X., Cao, J.: Environmental regulation and carbon emission: The mediation effect of technical efficiency. J. Clean. Prod. **236**, 117599 (2019). https://doi.org/10.1016/j.jclepro.2019.07.074

34. Rahman, M.M.: Exploring the effects of economic growth, population density and international trade on energy consumption and environmental quality in India. IJESM **14**(6), 1177–1203 (2020). https://doi.org/10.1108/ijesm-11-2019-0014

35. Porter, M.E., van der Linde, C.: Toward a new conception of the environment-competitiveness relationship. J. Econ. Perspect. **9**(4), 97–118 (1995). https://doi.org/10.1257/jep.9.4.97

Introduction to Ontologies for Defense Business Analytics

Bethany Taylor(✉), Christianne Izumigawa, and Jonathan Sato

Naval Information Warfare Center (NIWC) Pacific, San Diego, USA
{bethany.j.taylor3.civ,christianne.g.izumigawa.civ,
jonathan.k.sato.civ}@us.navy.mil

Abstract. In recent years, ontologies have become the leading solution for capturing corporate knowledge. Stored explicitly in documentation or tacitly in the minds of Subject Matter Experts (SMEs), enterprise knowledge, in all its forms, can be optimized into a tangible representation that gives way to more advanced business analytics. This paper seeks to review the benefits and challenges of developing ontologies through the lens of the government defense sector. To further demonstrate the learning curve in adapting to ontologies and graph-based knowledge structures in general, this paper will also provide a use-case experiment where business Subject Matter Experts (SMEs) were trained to design an ontology of the Operating Materials and Supplies (OM&S) domain at Naval Information Warfare Center (NIWC) Pacific by-hand.

Keywords: Ontologies · Knowledge Graphs · Ontology Learning · Business Analytics · Knowledge Representation

1 Introduction

Ontologies are semantic structures that represent the concepts, properties, and relationships of a domain of knowledge [19]. The most distinct advantage of using an ontology to model an area of knowledge, is that its graph framework is machine-readable [26]. Consequently, the ideas and connections stored in the ontology are now also machine-accessible, enabling advanced applications in the areas of reasoning and knowledge representation. While ontologies, and knowledge graphs, have been used to great effect in the private sector, it is still an unfamiliar technology in the Department of Defense (DoD), and even less so for its business application areas such as financial management, logistics, contracts, and corporate operations. Graph data science has the power to revolutionize the practice of business analytics, and as such organizations must not get left behind with adopting this data practice.

The goal of this paper is to introduce ontologies and knowledge graphs to the Defense sector, or any data practitioner who is unfamiliar with graph data science, as well as offer a review of the benefits and challenges that come from implementing ontologies in business practice. The secondary goal is to offer a DoD use-case of building an Operating Materials and Supplies (OM&S) Ontology for the purpose of providing an example of

© Springer Nature Switzerland AG 2023
F. Fui-Hoon Nah and K. Siau (Eds.): HCII 2023, LNCS 14038, pp. 87–101, 2023.
https://doi.org/10.1007/978-3-031-35969-9_7

what the development process can entail, and the lessons that were learned through the experience.

The following section will provide further background on ontologies and the business analytics problem created by current knowledge-storing methods. The third section will summarize the benefits organizations gain from developing ontologies, and the fourth section will provide the challenges that can be encountered in this pursuit. The final section will provide a detailed use-case in creating an ontology with business SMEs manually by-hand.

1.1 Contributions

This paper identifies a common knowledge management problem that many businesses experience and offers insight on how ontologies and knowledge graphs can increase the accessibility and use of corporate knowledge. Discussion on the benefits and challenges of ontology development also offers lessons learned to provide motivation for the adoption of ontologies and increases the chances of implementation success. In addition, this paper offers a practical ontology use-case that was implemented within the DoD and provides helpful feedback and insight gained through this experience.

2 Background

2.1 Related Works

Although ontologies are a relatively unfamiliar data structure, they are not a new method for knowledge representation. Ontologies became popular in the 1990s and early 2000s with the rise of the Semantic Web and the creation of knowledge representation languages, such as the Resource Description Framework (RDF), Web Ontology Language (OWL), and DARPA Agent Markup Language (DAML) [3, 19, 26]. Their popularity has resurged once again in recent years due to the advances in machine learning and language models that can now make use of ontologies for advanced reasoning purposes as well as assist in the automation of ontology development [1, 21].

Past ontology work has also explored the benefits and challenges of implementing ontologies, such as Noy and McGuinness's development guide which walks through a formal method for designing an ontology [26]. The guide emphasizes the ability for ontologies to establish a domain vocabulary and offer a reusable structure for transferring knowledge between people and computer applications, which are benefits that this paper will also explore in later sections [26]. While Noy and McGuinness's guide provides a helpful framework, it is aimed at a more manual development approach that could be adjusted to include more automated techniques.

In a more recent work by Al-Aswadi et al., a modern perspective is offered on ontology generation via Ontology Learning (OL), which uses a variety of machine learning, linguistic, statistical, and Natural Language Processing (NLP) techniques to help automate ontology development [1]. By transitioning away from a manual ontology development approach, automating OL reduces development time and cost, and allows for the processing of larger datasets [1]. Al-Aswadi et al. also notes the specific challenges that come from implementing OL such as trying to reduce the level of human input during development and the amount of error terms produced during automated data processing [1].

This paper extends prior analysis of benefits and challenges by analyzing ontology implementation within the Defense sector where unique opportunities and restrictions exist. As an example of prior ontology research in this field, Bowman and Tecuci document a "military course of action ontology" that was created to help capture senior military personnel knowledge in support of the tactical level of war [5]. While this work is a successful example of military-related ontology research, this paper seeks to further narrow its scope to the business domains of the DoD, where ontologies and knowledge graphs can enhance analytics and knowledge management for business operations.

2.2 The Business Problem

In the practice of business analytics, the goal is to extract insight via enterprise data and knowledge in order to support decision-making [8]. However, the quality and success of this analysis hinges on the ability to access an organization's data and information sources. One of these sources includes corporate knowledge, which for the purposes of this paper will be defined as the tacit and explicit knowledge produced by an organization. Tacit knowledge includes personal sources such as the intuition or "working knowledge" that employees learn through experience [23]. Alternatively, explicit corporate knowledge can include external sources such as company documentation in the form of PDFs, images, and PowerPoints [23]. The challenge posed by both types of corporate knowledge is they are typically stored in formats that are not understandable to machines. The accumulation of this unused, inaccessible knowledge creates an immense, missed opportunity for organizations to use the full extent of the information they generate. Ontologies offer a solution to this problem by providing the required semantic structure for representing corporate knowledge in a form that both humans and machines can comprehend.

2.3 The Ontology Structure

Ontologies are comprised of four main components: Classes, Instances, Relationships, and Properties [19]. Classes are the main concepts in a domain, which contain a set of instances, or objects that have the same characteristics and meaning [19]. Instances are granular examples of objects that make up a class. Relationships consist of the connections between classes, and between other components such as instances and properties [19]. Properties are the attributes of classes and instances in an ontology that provide further detail for these entities [26]. These elements cover the foundational pieces of an ontology that can offer a thorough description of the semantics and logic that define a field of knowledge. As noted in Sect. 2.1, ontologies can be represented using programming languages such as RDF or OWL, which makes these semantic structures machine-readable [19].

It should also be noted that this paper will often refer to knowledge graphs in addition to ontologies. The reason for this is that the two structures are closely related and are both commonly employed for knowledge representation. Knowledge Graphs are similar to ontologies in that they are network graphs with nodes and edges that represent concepts and their relationships; however, knowledge graphs store data points which attributes to their larger size [15]. To formally distinguish the relationship between these two structures, Ehrlinger and Wolfram define ontologies as the knowledge-base that initializes a knowledge graph [11]. The knowledge graph is then populated by consistently adding new data to the base ontology from external sources, which can then be used in reasoning applications [11].

3 Benefits to Implementing Ontologies

The following section reviews the advantages an organization can gain from utilizing ontologies to capture corporate knowledge. This list of benefits was compiled from lessons learned through first-hand experience in designing the OM&S Ontology, as well as insight from current literature on ontologies and graph data science.

3.1 Enabling Reasoning Applications

As stated in Sect. 1, an advantageous feature of ontologies is that they are machine-readable semantic structures and can provide new knowledge for reasoning applications. In the field of machine learning, ontologies can be used to enhance predictive modeling inputs with context and domain knowledge that were not previously accessible [21]. For example, Kulmanov et al. Writes how ontology embeddings can be used to constrain optimization search problems or measure semantic similarity [21]. Another reasoning application is practicing knowledge discovery using ontology-based search for information systems, where an ontology can be used to improve search queries through advanced inference or disambiguation of query terms [18].

In the private-sector, companies are adopting ontologies and knowledge graphs to provide reasoning features in their products and perform advanced business analytics. For example, TurboTax developed a Tax Knowledge Graph which maps the concepts and calculation logic embedded in U.S. and Canadian tax law [32]. By utilizing a knowledge graph for calculation logic and eligibility rules, as opposed to procedural programming scripts, TurboTax can now trace calculation and logic dependency, and thus offer automated explainability to customers in their "Explain Why" product [32]. As noted by Yu et al., this work is a rare example of using semantic graph structures to model complex enterprise logic [32].

These examples are just a few of the many ways ontologies can enhance an organization's predictive analytics with better reasoning capabilities, particularly through the machine-accessible domain knowledge they provide as well as their efficient graph structures.

3.2 Foundation for Enterprise Knowledge Graphs (EKGs)

EKGs are large knowledge graphs that integrate an enterprise's various information systems, data sources, and knowledge bases into one central knowledge structure [13]. As mentioned in Sect. 2.3, ontologies are the foundational knowledge base, or schema, of an EKG by providing the semantic rules and organization for the data that is added to the knowledge graph [11].

There are many well-known examples of EKGs such as the Google Knowledge Graph, Microsoft Bing Knowledge Graph, Facebook's social graph and IBM's Knowledge Graph Framework, which enable a variety of advanced reasoning applications like advanced search queries, question answering, and knowledge discovery to name a few [25]. These massive knowledge graphs can reach up to billions of nodes, highlighting the importance of automating the development of EKGs as well as designing the correct architecture for integrating knowledge sources [13, 25].

In the Defense sector, the DoD is beginning to use knowledge graphs at an enterprise-level with the introduction of its new GAMECHANGER application [17]. According to the Defense Intelligence Agency (DIA), GAMECHANGER is an unclassified "artificial intelligence and natural language processing application" designed to collect, query, and analyze policy documentation across the entire DoD [9]. The application utilizes a knowledge graph structure to create inter-document links across Department-wide policy documents [17]. While this knowledge graph application is mostly confined to the policy domain, tools such as this offer potential grassroots for further EKG development for all enterprise operations.

3.3 Creating a Domain Terminology

Developing an ontology not only provides an organizational structure for knowledge but can also help create a terminology to be used for a particular field. According to Valeontis and Mantzari, the word "terminology" refers to the unique set of words that make up the "special language of a specific subject field" [31]. As an ontology is formed, whether through automatic or manual methods, developers and domain SMEs must agree on the terms or set of rules for naming the various entities within an ontology. This practice naturally results in the formation of a domain terminology.

But why is a terminology so important? In large domains, such as the Defense sector where specialized military language is used, a terminology can help formalize the communication and comprehension of domain concepts across a wide network of sub-organizations [20]. In fact, to achieve this goal of improved communication, the Office of the Secretary of Defense (OSD) created a terminology program that aims to standardize military terminology by creating resources such as the *DoD Dictionary of Military and Associated Terms* and the *Terminology Repository of DoD (OSD/JS) Issuances* [20].

While these resources help standardize general DoD terminology, there is still more work to be done to standardize vocabulary within the DoD's business domains such as financial management, logistics, and corporate operations to name a few. These sub-domains contain unique terminology that merges their subject-specific vocabulary with military context from the mission they support. By determining the right terminology for these specialized fields, ontologies can improve communication amongst practitioners.

3.4 Acquiring Tacit Knowledge

Tacit knowledge refers to the information and intuition an employee gains through experience and hands-on training but is not externally documented [30]. It can be further described as employees' "personal knowledge which cannot always be articulated in a coded form" [7]. According to Chergui et al., capturing tacit knowledge is a key element for companies to maintain a competitive edge [7]. However, the process of "tacit knowledge acquisition", or converting tacit knowledge to a more accessible, documented form, is a very challenging procedure since this information is stored internally in employee's minds [23].

Chergui et al. describes a formal process for using ontologies in the acquisition of tacit knowledge, where employees participate in "explicitation interviews" and "self-confrontation" in order to extract and document their intuition and "know-how" [7]. This knowledge is then further processed to extract its most relevant components, which are then added to an ontology as the final knowledge model [7].

In a similar manner, this project's OM&S ontology use-case also performed tacit knowledge acquisition through SME interviews and discussion. Although the experiment was not specifically designed for this process, developing the OM&S ontology certainly uncovered pieces of this internal information as SMEs would describe their intuition when organizing or discussing domain topics. Thus, ontologies are an excellent tool for enterprises seeking to capture all sources of corporate knowledge to enhance business operations and analytics.

3.5 Human-Friendly Knowledge Representation

The field of Knowledge Visualization (KV) uses visual representations to aid in the sharing and creation of knowledge [12]. According to Eppler and Burkhard, KV is an important component in the practice of knowledge management as it improves the quality of knowledge transfers, fosters the innovation of new knowledge, and helps avoid the common problem of "information overload" [12]. Ontologies can help support the practice of KV in an organization by providing a human-friendly method for visualizing abstract, complex domain knowledge.

As Jepsen describes in their work, ontology visualization communicates relational information that RDF logic or text sources cannot express nearly as well [19]. Indeed, as was seen in the OM&S use-case, the resulting ontology presented a simplified graph that was more immediately comprehendible as compared to the text documents, tables, and diagrams that communicated the same information. Additionally, when compared to the visual quality of other graph-based representations such as large knowledge graphs, ontologies can provide a more concise visual representation. Ultimately, ontologies are not only beneficial in providing machine-usable applications, but also improve human-interaction with data and domain concepts.

4   Challenges to Ontology Adoption

The following section summarizes the difficulties that can be encountered when developing ontologies and knowledge graphs. This list particularly stresses the challenges that tend to be unique to projects within the DoD, but nonetheless hold value to any organization looking into graph data structures. As stated in the prior section on benefits, this section mostly drew upon first-hand experience from developing the OM&S ontology, as well as lessons learned through literature review.

4.1 Documentation Accessibility

Corporate knowledge is often stored across a widespread array of company documentation. The ability to locate and access the right sources of knowledge directly impacts the available content an ontology can be developed from.

In the DoD, documentation accessibility can be particularly challenging as it is well-known for the data silos that exist across its many enterprises [4]. These separate information systems only offer a "limited ability to quickly access information across classification levels or service platforms", which consequently also limits access to knowledge sources for creating ontologies [4].

Luckily, there is some progress to remedy this barrier for information sharing within the DoD. As mentioned previously, GAMECHANGER is a DoD application which uses NLP and knowledge graphs for analyzing policy documentation [17]. The application pulls text documents from "27 major sources across the Government", which is a significant step towards breaking down the silos of information [9]. Another advantage of GAMECHANGER is also that it provides a single point of access for this information, which greatly eases the search process for DoD analysts [17]. Tools such as GAMECHANGER are imperative for the DoD, or any company, to provide the information necessary for building ontologies.

4.2 Limitations with Ontology Learning

Ontology Learning (OL) uses an automated approach for developing ontologies with knowledge extracted from text sources [1, 10]. The OL process includes acquiring domain text sources, extracting keywords and relationships, mapping entity and relational terms into an ontology structure, programming the ontology using an ontology language, and then evaluating and maintaining the ontology [1]. Given how time-intensive this process is, automating OL as much as possible is essential to helping businesses create these knowledge structures faster.

However, there are many challenges that limit the performance and level of automation of OL systems. For instance, when performing term extraction to identify ontology entities, NLP techniques such as Part-of-speech (POS) Tagging or Pattern-based extraction can identify erroneous or irrelevant terms due to the nuances that exist in natural language [1]. This task is particularly difficult when performed on DoD documentation which is littered with acronyms, military-jargon, and verbose sentence structures.

Prior to implementing the OM&S ontology experiment, this project team evaluated various OL techniques on DoD business documentation to assess how much automation was possible before involving business SMEs. When tested on DoD text sources, such as manuals and policy documents, POS tagging struggled to identify the correct part of speech when acronyms were encountered—which occur frequently in government documentation. When performing synonym discovery using WordNet *synsets* or testing techniques for hierarchical clustering using WordNet *hypernyms*, these methods often failed to correctly define the input root word because the term was too domain-specific to the DoD and the business area it described [1, 28]. Furthermore, sentence dependency parsing was tested using spaCy's *en_core_web_sm* pipeline for relationship and entity extraction, which typically works well for sentences that contain simpler, repetitive syntactic structure [16]. However, lengthy text sources, like the business documentation that was tested, tend to use a legalistic, verbose style of writing that is difficult to fit to a syntactic pattern. The NLP techniques that were tested mostly follow a linguistic approach to OL, versus the statistical or machine learning techniques, however they quickly revealed general challenges involved in automating OL as well as the unique struggle that comes from working with domain-specific natural language [10].

4.3 Limited Ontology Editor Software

Ontologies have yet to be adopted as a tool that business SMEs commonly use or interact with. With this lack of demand, the availability of low-code, ontology-editing software designed for business analysts is very limited. Many ontology editors, Protégé being one of the most popularly referenced, are geared towards ontologists or programming-savvy data practitioners [24]. According to two web ontology-editor surveys, by Alatrish and Buraga, both concluded that Protégé offered one of, if not the most, user-friendly interface [2, 6]. However, when used by the project team to visualize the OM&S ontology, this software was evaluated as unsuitable for adoption by business analysts. The depth of ontology knowledge required to understand the different tabs and settings, as well as the need to enable interface plug-ins, creates a much higher learning curve compared to the interfaces of popular analytics tools such as Microsoft Excel or Tableau. When implementing ontologies in business-related domains, it is often necessary for SMEs and analysts to be involved with hands-on participation, and this process can be difficult if the ontology software is not designed for this purpose.

In addition to the scarcity of low-code ontology editors, security requirements for DoD software further restricts the list of available ontology editing tools to use. According to DoD guidance on software development, some risk factors that need to be considered when identifying available software options include avoiding "vendor lock-in", increased expenses due to licensing costs, and software associations with foreign governments [29]. These are risk factors that the OM&S ontology project had to assess when searching for a secure and affordable ontology visualization tool.

4.4 Unfamiliarity with Graph Data Structures

Business analysts in the DoD primarily interact with enterprise data via business intelligence tools, which offer spreadsheet reports or simplified dashboard views of large datasets. Over the last few years, the DoD has launched a new Department-wide data analytics platform called Advana, which is gradually becoming the main business analytics and business intelligence platform for employees in corporate domains [27]. As stated in the DoD Financial Management Regulation (DoD FMR), Advana is "a centralized data and analytics platform that provides DoD users with common business data, decision support analytics, and data tools" [27]. Through its widespread use across the DoD, Advana captures the most common data practices and data tools that business analysts employ. For instance, this includes business analytics tools such as Qlik which is used for visualizing datasets and creating reports [22].

The key point here is that most DoD business analysts understand their data in the form of tabular structures which are typically stored in relational databases. When introducing ontologies and knowledge graphs, there is a large barrier of unfamiliarity to the idea of storing data in graphical form or representing data semantically. This creates a disadvantage for the DoD and its sub-organizations, as the unawareness of graph data science and knowledge representation creates a larger training requirement when involving business SMEs in ontology projects. It also prevents ontologies and knowledge graphs from being considered as viable data solutions for business problems, as they are less commonly known about. As will be discussed in greater detail in the following section, the development team for the OM&S ontology observed some successful techniques in getting business SMEs quickly up-to-speed to participate in designing an ontology.

5 OM&S Ontology Development

This section will summarize a use-case of guiding DoD business SMEs to develop an ontology by-hand. The designated field of knowledge to model was the management of OM&S at the Naval Information Warfare Center (NIWC) Pacific, which oversees the command's inventory management and warehousing processes. This field is critical to naval operations yet remains one of the most difficult subjects to provide training on due to its complexity. The significance in being able to create alternative, simpler representations of OM&S knowledge, that are also machine comprehendible, made this a high-impact domain to experiment in.

The ontology construction process was adapted from the development methodology outlined by Noy and McGuinness, as well as some OL techniques described by Al-Aswadi et al. [1, 26]. Consequently, the final development process can be described as follows:

1. Define the ontology's domain, scope, and application purpose.
2. Extract and list keywords from domain knowledge sources.
3. Define ontology classes and their relational structure.
4. Identify class properties and property values.
5. Populate ontology with instances.
6. Evaluate and edit resulting ontology.

5.1 Selecting and Onboarding Participants

The project team worked with NIWC Pacific SMEs who operate as data stewards and systems analysts in the OM&S domain. These SMEs were ideal participants as they were from different departments and contributed unique perspectives when building an ontology of such a widespread topic.

When introducing the participants to the use-case, the ontology team prepared a training strategy for acclimating the SMEs to a semantic data perspective. Lessons learned for this step include the following:

- Begin training program by teaching participants about knowledge graphs and taxonomies before ontologies. These graph structures were easier to understand than ontologies and helped participants acclimate to graph data science.

- Provide examples of commercial products or well-known companies that utilize ontologies and knowledge graphs, such as TurboTax or Facebook [25, 32]. This step offers participants real examples that they may be familiar with and reinforces ontology theory.
- Walk through a simple ontology development example from beginning to end. This project walked through Noy and McGuinness's wine ontology example to demonstrate each development step [26].
- Review the benefits of implementing ontologies. This step is key to providing participants with motivation and enthusiasm for learning about these concepts and persevering through manually designing the ontology.

5.2 Determining Ontology Scope

The first step of the development process, in accordance with Noy and McGuinness's methodology, involves determining the scope of the ontology in order to narrow down a reasonable range of knowledge to model [26]. Given that the field of OM&S is multi-faceted, with data flowing between the logistics and finance business areas, this step was crucial.

Noy and McGuinness recommend using Competency Questions during this step to help reduce the ontology's scope by determining the purpose, level of granularity, and application of the completed ontology [26]. During the experiment, SMEs determined that one of the initial purposes of the ontology would be to correctly classify materials into their proper category: OM&S, General Equipment, Supply Management-Inventory, and Non-Accountable Property. Given this purpose, some of the relevant competency questions that the ontology needed to be able to answer are shown in Table 1.

Table 1. Sample competency questions to define OM&S Ontology scope.

Competency Questions
What is non-accountable property?
Is a laptop classified as OM&S?
Is project software classified as General Equipment?
Who is the manufacturer of a material?
What warehouse locations may material be located in?

At the end of this step, the material classification process and some key foundational OM&S concepts were designated as the scope and purpose for the ontology. Modeling this knowledge was determined to be highly valuable as it is a process that requires widespread training across the workforce, and misclassification inserts errors into the financial system which SMEs must then investigate and reconcile.

5.3 Constructing the Ontology

Once the scope of the ontology was defined, the second development task includes creating an OM&S word bank by listing keywords from the domain to include in the ontology (See Fig. 1) [26]. In order to assist SMEs with the listing process, this step was initially automated by performing term extraction on domain documentation [1, 26]. The project team selected text sources from the NIWC Pacific Financial Manual and OM&S training documents, and then automated the keyword extraction process using a series of NLP techniques.

First, PDF text extraction and sentence tokenization were performed to extract sentences from the documents. This output was then fed into a language model algorithm called keyBERT, which uses Bidirectional Encoder Representations from Transformers (BERT) embeddings to perform keyword extraction [14]. The algorithm output was a keyword list of 270 terms and phrases from the OM&S documentation. After several rounds of reconciling and cleaning this list, which involved removing irrelevant words and matching acronyms to their full names, the final OM&S word bank was generated. The final word bank count was 176 terms which provided SMEs with a reasonable amount of content to analyze and re-structure.

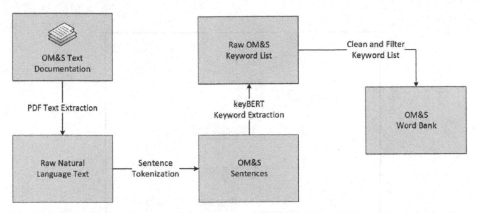

Fig. 1. Process to extract keywords and generate OM&S Word Bank for ontology construction.

Using this list, the team could then move onto steps 3–5 to identify ontology classes, their taxonomic structure and relationships, class properties, and instances. These steps often occurred simultaneously or out of order depending on whether SMEs shifted between analyzing high-level or granular concepts. This process was highly influenced by Noy and McGuinness's "combination" development process for determining class hierarchy, which uses both a "top-down" and "bottom-up" approach [26].

Initial versions of the ontology restricted possible component names to the OM&S word bank, to reduce task complexity and prevent extension of the scope. However, once SMEs gained confidence and proficiency with ontology development, new words could be added or exchanged to improve the ontology structure and ensure competency questions were addressed.

Some strategies that SMEs used for identifying ontology classes included clustering the list of keywords to identify high-level OM&S topics and discard outlier information. Additionally, SMEs engaged in active terminology discussion to designate final labeling and hierarchical structuring. This step demonstrated how ontologies help determine the vocabulary of a domain, as all SMEs had to agree on the naming of classes and identify when terminology from external domains were entering into the ontology.

During these final steps of the organization process, the project team assisted the SMEs by visualizing the ontology using Protégé software, which was chosen based on factors discussed in Sect. 4.3 regarding security restrictions and software availability options [24].

5.4 Results and Future Work

The current version of the resulting OM&S ontology consists of 73 Classes, 127 Relationships, 51 Instances, and 7 Properties, and covers high-level concepts of the OM&S domain as well as the necessary information to classify command material. Figure 2 displays a portion of the most recent OM&S ontology as visualized using WebProtégé and shows how a generic *Material*, sponsored by *Customer X*, is classified as *General Equipment* [24].

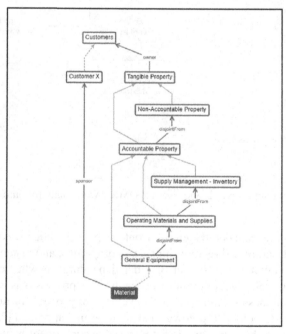

Fig. 2. Portion of final OM&S ontology demonstrating how a *Material* funded by *Customer X* is classified as *General Equipment*.

As for the final development step, evaluating and editing the ontology, this continues to be an ongoing process as the resulting ontology is initially planned to be the basis of an OM&S classification tool. The web application is intended to help make the knowledge in the OM&S ontology available to end-users to automate classification reasoning and provide additional information for common OM&S questions.

Through this experiment, the project team gained insight into the effort and considerations required to manually develop an ontology. For instance, identifying the right sources of knowledge to be modeled as well as being able to locate and access these sources can prove to be challenging in large organizations or classified domains. Training SMEs requires an investment of time and effort but is necessary to produce quality results. Selecting the best ontology and knowledge graph software also can be challenging when needing to evaluate cost, ease-of-use, and availability in restricted domains.

Despite these challenges, this experiment highlighted many of the benefits that this paper previously discussed. Through active discussion, the SMEs determined a terminology for ontology elements, which they then agreed would help formalize the vocabulary used in general OM&S operations. Additionally, the logic that participants used to add, discard, or change ontology components, helped to capture small pieces of these experts' tacit knowledge. Finally, the resulting product of this experiment sets a strong foundation for reasoning applications within this DoD business domain.

6 Conclusion

Ontologies and knowledge graphs can introduce a new path for enterprise knowledge management and enhance business analytics with advanced reasoning capabilities for deeper insight and improved predictive analytics. The development process for these structures can follow many different methods and is often determined by the use-case's data and knowledge management environments. For ontology implementation within the DoD's business domains, this can entail challenges with data access, software restrictions, and a greater learning curve for knowledge representation training.

However, as the OM&S ontology helps demonstrate, these limitations are not the end of the story. By understanding some of the domain-specific hurdles that exist, DoD practitioners can plan for these limitations and take advantage of new platforms, such as Advana and GAMECHANGER, that offer potential solutions around accessibility and software platform issues. As ontologies and knowledge graphs gain popularity, they will become a more familiar data tool to employ and will hopefully gain more use for DoD business applications.

Acknowledgements. The OM&S ontology presented in this paper, as well as many of the observed benefits and challenges that were documented, would not have been realized without the help and participation of NIWC Pacific's OM&S team: Debra Ernst, Joshua Parrish, Kristie Wood, and Tyler Renfro. This paper would also like to acknowledge the help and support of NIWC Pacific scientist Andrew Kan.

References

1. Al-Aswadi, F.N., Chan, H.Y., Gan, K.H.: Automatic ontology construction from text: a review from shallow to deep learning trend. Artif. Intell. Rev. **53**(6), 3901–3928 (2019). https://doi.org/10.1007/s10462-019-09782-9
2. Alatrish, E.: Comparison some of ontology editors. J. Manag. Inf. Syst. **8**(2), 18–24 (2013)

3. Berners-Lee, T., Hendler, J., Lassila, O.: The Semantic Web. Sci. Am. **284**(5), 34–43 (2001). http://www.jstor.org/stable/26059207
4. Booz Allen Hamilton Inc.: Overcoming Obstacles to Data Integration for Defense. Perspectives (2022). https://www.boozallen.com/insights/defense/overcoming-obstacles-to-data-int egration-for-defense.html
5. Bowman, M., Lopez, A., Tecuci, G.: Ontology development for military applications. In: Proceedings of the SouthEastern Regional ACM Conference, Atlanta, GA (2001)
6. Buraga, S., Cojocaru, L., Nichifor, O.: Survey on web ontology editing tools. Transactions Automatic Control Computer Science, pp. 1–6 (2006)
7. Chergui, W., Zidat, S., Marir, F.: An approach to the acquisition of tacit knowledge based on an ontological model. J. King Saud Univ. – Comput. Inf. Sci. **32**(7), 818–828 (2020)
8. Delen, D., Ram, S.: Research challenges and opportunities in business analytics. J. Bus. Analytics **1**(1), 2–12 (2018). https://doi.org/10.1080/2573234X.2018.1507324
9. DIA Public Affairs.: GAMECHANGER: Where policy meets AI. Defense Intelligence Agency (2022)
10. Drissi, A., Khemiri, A., Sassi, S., Chbeir, R.: A new automatic ontology construction method based on machine learning techniques: application on financial corpus. In: Proceedings of the 13th International Conference on Management of Digital EcoSystems, pp. 57–61. Association for Computing Machinery (2021). https://doi.org/10.1145/3444757.3485111
11. Ehrlinger, L., Wolfram, W.: Towards a definition of knowledge graphs. In: 12th International Conference on Semantic Systems (2016)
12. Eppler, M., Burkhard, R.: Knowledge visualization: towards a new discipline and its fields of application. Università della Svizzera italiana (2004)
13. Galkin, M., Auer, S., Kim, H., Scerri, S.: Integration strategies for enterprise knowledge graphs. In: 2016 IEEE Tenth International Conference on Semantic Computing (ICSC), pp. 242–245. Laguna Hills, CA, USA (2016). https://doi.org/10.1109/ICSC.2016.24
14. Grootendorst, M.: KeyBERT: Minimal keyword extraction with BERT. Zenodo (2020). https://doi.org/10.5281/zenodo.4461265
15. Hogan, A et al.: Knowledge Graphs. Assoc. Comput. Mach. **54**(4), 3447772 (2021). https://doi.org/10.1145/3447772
16. Honnibal, M., Montani, I., Van Landeghem, S., Boyd, A.: spaCy: Industrial-strength Natural Language Processing in Python (2020). https://doi.org/10.5281/zenodo.1212303
17. JAIC Public Affairs: Meet the Gamechanging App That Uses AI to Simplify DoD Policy Making. Chief Digital and Artificial Intelligence Office (2021)
18. Jasper, R., Uschold, M. A framework for understanding and classifying ontology applications. In: Proceedings 12th International Workshop on Knowledge Acquisition, Modelling, and Management KAW 99, pp. 16–21 (1999)
19. Jepsen, T.C.: Just what is an ontology, anyway? IT Profess. **11**(5), 22–27 (2009)
20. Katsos, G.: Department of defense terminology program. J. Force Q. **88**, 124–127 (2018)
21. Kulmanov, M., Smaili, F., Gao, X., Hoehndorf, R.: Semantic similarity and machine learning with ontologies. Brief. Bioinform. **22**(4), 1–18 (2021)
22. Lin, G.: Meet Advana: How the department of defense solved its data interoperability challenges. Government Technology Insider (2021)
23. Mohammad, A., Al-Saiyd, N.: Guidelines for tacit knowledge acquisition. J. Theor. Appl. Inf. Technol. **38**(1), 110–118 (2012)
24. Musen, M.: The Protégé project: a look back and a look forward. AI Matters. Assoc. Comput. Mach. Specific Interest Group Artif. Intell. **1**(4), 25757003 (2015). https://doi.org/10.1145/2557001.25757003
25. Noy, N., Gao, Y., Jain, A., Narayanan, A., Patterson, A., Taylor, J.: Industry-scale knowledge graphs: lessons and challenges: five diverse technology companies show how it's done. Assoc. Comput. Mach. **17**(2), 48–75 (2019). https://doi.org/10.1145/3329781.3332266

26. Noy, N., McGuinness, D.: Ontology development 101: a guide to creating your first ontology. Stanford Knowledge Systems Lab (2001)
27. Office of the Under Secretary of Defense (Comptroller) (OUSD(C)): Advana – Common Enterprise Data Repository for the Department of Defense. Department of Defense Financial Management Regulation (DoD FMR) 1(10) (2020)
28. Princeton University: About WordNet. Princeton University, WordNet (2010)
29. Sherman, J.: Guidance on Software Development and Open Source Software. U.S. Department of Defense (2022)
30. Smith, E.: The role of tacit and explicit knowledge in the workplace. J. Knowl. Manag. **5**(4), 311–321 (2001)
31. Valeontis, K., Mantzari, E.: The linguistic dimension of terminology: principles and methods of term formation. In: 1st Athens International Conference on Translation and Interpretation Translation: Between Art and Social Science, pp. 13–14 (2006)
32. Yu, J., McCluskey, K., Mukherjee, S.: Tax knowledge graph for a smarter and more personalized TurboTax. arXiv (2020)

Learning by Reasoning: An Explainable Hierarchical Association Regularized Deep Learning Method for Disease Prediction

Shuaiyong Xiao[1,2], Gang Chen[3], Zongxiang Zhang[4], Chenghong Zhang[4（✉）], and Jie Lin[1]

[1] School of Economics and Management, Tongji University, Shanghai 200092, People's Republic of China
[2] Guangdong Provincial Key Laboratory of Traditional Chinese Medicine Informatization, Guangzhou 510632, People's Republic of China
[3] School of Management, Zhejiang University, Hangzhou 310058, People's Republic of China
[4] School of Management, Fudan University, Shanghai 200433, People's Republic of China
chzhang@fudan.edu.cn

Abstract. Multimodal healthcare data provides a huge opportunity for big-data-based disease prediction, supporting the diagnosis and treatment decision-making process for doctors. Many deep learning based methods are developed to yield better performance of multimodal data based disease prediction considering their powerful representation abilities. However, the black-box nature of deep learning methods results in many serious concerns: e.g. the reliability of the prediction performance is questionable; the end-users (i.e., doctors) can not understand the reasons behind the prediction. These issues make it difficult for deep learning based disease prediction systems to apply in practice. Therefore, we aim to tackle the aforementioned challenges and propose an explainable hierarchical association regularized deep learning method to produce interpretable prediction results while maintaining prediction accuracy. The method takes the multimodal healthcare data into consideration for the disease prediction task and constructs a hierarchical association path for each sample. An ingenious loss function is designed to learn consistent features among different data modalities and the disease prediction path with a hierarchical structure. The experimental results based on a public dataset show the superiority of our proposed method. The ablation study and sensitivity analysis verify the effectiveness and necessity of the method design.

Keywords: Multimodal Healthcare Data · Expandability · Hierarchical Association Learning

1 Introduction

In the era of big data, tremendous healthcare data with different modalities, e.g., patients' geographical information, textual electronic health records, and medical images from CT, X-ray, and MRI, has been stored and facilitated to support doctors' disease diagnosis decisions. Deep learning models have largely been investigated in multimodal

© Springer Nature Switzerland AG 2023
F. Fui-Hoon Nah and K. Siau (Eds.): HCII 2023, LNCS 14038, pp. 102–113, 2023.
https://doi.org/10.1007/978-3-031-35969-9_8

healthcare data based disease prediction because of their powerful ability in representing unstructured data. Researchers developed various deep learning models including natural language processing, computer vision, and reinforcement learning to leverage the valuable information in healthcare data for medical diagnosis [1]. Existing research has witnessed the effectiveness and efficiency of deep learning models in scenarios including Alzheimer's disease detection [2, 3], diagnosis of osteoarthritis [4], breast cancer risk assessment [5, 6], and heart failure prediction [7], etc.

More and more deep learning models are designed to improve the prediction accuracy of medical diagnosis, however, since the disease diagnosis and treatments are closely related to the patient's health and life security, the consequence of misleading prediction results is irretrievable and may bring harmful effects on them. Therefore, the black-box based deep learning models have not achieved large-scale application due to the model prediction opacity issue. These models can neither give explanations upon the prediction results nor provide reliable boundaries for doctors, which heavily impedes the adoption in practice and decreases the collaboration efficiency between the deep learning models and human beings.

In face of the aforementioned challenges, some scholars try to develop explainable deep learning models to better support the human medical diagnosis decision-makings. These methods can be categorized into two folds: attention-based methods and gradient-based methods. The former kind of methods adopts the attention mechanism in the training process and allocates the most influential items by analyzing learned attention weights. The latter ones leverage the heat map or saliency map to visualize the important area of the trained models. Both two kinds of approaches lack an explanation of the model training process and can not help the end-users (i.e., doctors) understand the learning process of the deep learning model.

Thus, we delve to propose an explainable deep learning method named explainable hierarchical association learning method (XHAL), which aims to open the black-box medical diagnosis model, and make the prediction results easily understood by the patient and medical staff. Specifically, the proposed method includes three components: the feature embeddings of multimodal healthcare data based on pre-trained deep learning models, the multimodal healthcare data based learning process considering the consistency across different modalities, and the hierarchical association regularization based learning process for the model interpretability.

To verify the effectiveness of our proposed method, we tested the prediction performance on a public multimodal healthcare data. The experimental results indicate that our proposed method significantly outperformed the state-of-the-art benchmarks. The exploratory analysis shows the rationality and necessity of our proposed method.

To sum up, our work contributes to the existing research from the following three aspects:

(1) We focus on multimodal healthcare data for the disease prediction task, which is less mentioned in prior research. By coordinating different data modalities and learning consistent information, the prediction performance yields a significant improvement.
(2) We propose a novel multimodal deep learning method that can effectively capture the hierarchical associations among various diseases and leverage them into the learning objectives to train the model more reasonably.

(3) The experiment results show the superiority of our proposed XHAL in contrast with the benchmarks. The ablation study and sensitivity analysis verified the rationality of our method design. The case study shows how XHAL produces a reasonable and understandable prediction result, which can effectively attenuate the drawbacks of black-box deep learning methods.

2 Related Work

2.1 Deep Learning Based Disease Prediction

Deep learning methods are widely used in various disease prediction scenarios including heart disease [8], pancreatic cancer [9], Alzheimer's disease [10], and so on. Incorporated with deep learning methods, multimodal healthcare data has been explored for the disease prediction task. For example, Lin et al. [11] used the MRI images with deep convolutional neural network to predict Alzheimer's disease. Liu et al. [12] took both the structural patient information and the textual health records as multimodal data to predict the probability of chronic diseases such as congestive heart failure and kidney failure. Oh et al. [13] constructed a deep representation model to learn the microbiome data for disease prediction. Kumar et al. [14] predicted pulmonary disease with the patients' cough sound data.

Overall, deep learning methods can provide a fine-grained investigation of multimodal disease data and offer a better disease prediction performance. However, the black-box nature of deep learning methods has largely hindered their practical application in the real world since disease diagnoses are highly associated with patients' life security. Therefore, the interpretability of deep learning methods for disease prediction is becoming a research spotlight in healthcare scenarios [15].

2.2 Interpretability of Deep Learning Methods

Since the prediction results in healthcare scenarios have to be accountable, explainable, and responsible to the patients, the interpretability inherently becomes indispensable and has drawn scholars' and doctors' extensive attention [16, 17].

Various methods for improving the interpretability of deep learning based disease prediction methods are proposed in recent years, which can be classified into two types: attention-based methods and gradient-based methods. The attention-based methods attempt to find the most influential area to the prediction results by analyzing learned attention weights. For instance, Bai et al. [18] proposed an interpretable deep learning method Timeline to predict the risk of patients' future visits, aiming to interpret the prediction results by analyzing the attention weights of previous visits. Kwak et al. adopted a self-attentive mechanism to fuse different sources of patients' records and explored the attention weights of a specific patient case to illustrate their interpretability [19]. Choi et al. proposed a reversed time attention model to explain the prediction results by finding the most influential patients' visits and variables [20]. Different from the attention-based method, gradient-based methods interpret the prediction results through gradient backward-propagation and visualization of the input factors [21]. For example, Tang et al. [22] designed a convolutional neural network to train the plaque classification for Alzheimer's disease and they interpret**ed** the results with a saliency map generated by Grad-CAM. Jha et al. [23] proposed an enhanced integrated gradient method to identify the significant features related to the prediction task.

In summary, although previous studies endeavor to make the deep-learning-based disease prediction results understandable, they merely focus on the correlations between the data input and output and lack explanation in terms of the model training process.

3 Proposed Method

Figure 1 depicts the framework of our proposed XHAL method, which contains three components: (1) the feature embedding construction of healthcare data based on large scale pre-trained deep learning models, (2) the prediction-oriented multimodal deep learning process considering the consistency of different data modalities, and (3) the hierarchical association regularized learning process for the model explanation.

3.1 Embedding Feature Extraction

We first obtain the embeddings of multimodal healthcare data. The modalities of the data could be text (e.g., the health records), image (e.g., the X-ray images), and structured indicators (e.g., patients' age, gender, and so on). For the textual data, we apply the BioBERT [24] model, a pre-trained biomedical language representation model specialized for biomedical text mining, to extract the embeddings. For the medical images, we adapt the DenseNet121 as the feature embedding extraction tool, which is widely used by scholars for its powerful representation ability of images. For the structured data, we use the multilayer perceptron (MLP) to acquire the hidden embeddings.

3.2 Multimodal Consistency Learning

After that, we introduce a multimodal consistency loss to coordinately leverage the multimodal healthcare data to predict whether the patients are diseased or not. Specifically, we concatenate the embedding features from different modalities and regularized them to represent consistent information that indicating the patients' health condition with Frobenius regularization loss, which can effectively measure the similarity of two matrices [25]. The multimodal consistency loss is shown as Eq. (1)

$$L_1 = \lambda_F \sum_{k<q}^{M} \left\| \mathbf{X}^{(k)} - \mathbf{X}^{(q)} \right\|_F^2, \tag{1}$$

where M is the number of modalities, and λ_F is a hyperparameter.

3.3 Hierarchical Association Learning

Finally, we construct the hierarchical association regularized learning process to make the prediction process more reliable and understandable. We construct the hierarchical association matrix, which can indicate the true hierarchical path of each sample. To

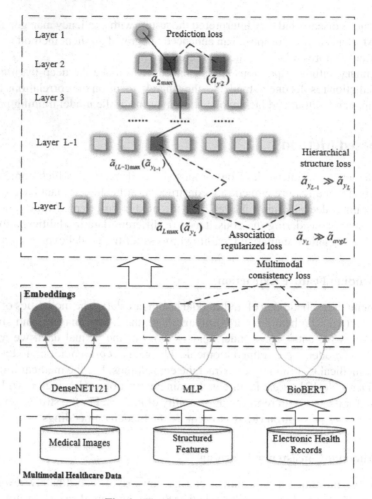

Fig. 1. The framework of XHAL.

achieve this, we build the indicator matrix $\mathbf{C} \in \mathbb{R}^{a \times a}$ to denote the possible connection between diseases of parent node and diseases of child node. The item in $\mathbf{C}$ is shown in (2):

$$c_{ij} = \begin{cases} 1, & \text{if disease } i \text{ at parent node connects to disease } j \text{ at child node} \\ 0, & \text{others} \end{cases} \qquad (2)$$

For each sample, the hierarchical association matrix $\mathbf{A} \in \mathbb{R}^{a \times a}$, predicted by multimodal embedding features, is one of the paths of the indicator matrix $\mathbf{C}$. For the predicted hierarchical association matrix $\tilde{\mathbf{A}} \in \mathbb{R}^{a \times a}$, the submatrix of each layer is denoted as $\{\tilde{\mathbf{A}}_1, \tilde{\mathbf{A}}_2, \cdots, \tilde{\mathbf{A}}_L\}$, the maximum values of each layer are the predicted diseases, which is denoted as $\tilde{\mathbf{a}}_{max}^{diag} = [\tilde{a}_{1\,max}, \tilde{a}_{2\,max}, \cdots, \tilde{a}_{L\,max}] \in \mathbb{R}^L$. For sample i, the true path label is $\mathbf{y}^i = [y_1^i, y_2^i, \cdots, y_L^i] \in \mathbb{R}^L$, Thus the disease classification loss for a batch (batch size is B) of samples is calculated as

$$L_2 = \frac{1}{B} \sum_{i=1}^{B} \left(\frac{1}{L} \sum_{j=1}^{L} y_j^i \log \tilde{a}_{j\,\text{max}}^i \right). \tag{3}$$

The association regularized loss in the same layer is

$$L_3 = \lambda_p \frac{1}{B} \sum_{i=1}^{B} \left[\frac{1}{L} \sum_{j=1}^{L} \left(\tilde{a}_{y_j}^i - \tilde{a}_{avg_j}^i \right) \right], \tag{4}$$

where $\tilde{a}_{y_j}^i$ is the predicted probability of sample i at layer j, $\tilde{a}_{avg_j}^i$ is the average probability of sample i at layer j. L_3 ensures the model to focus on the true disease class.

The hierarchical structure regularized loss across different layers is

$$L_4 = \lambda_L \frac{1}{B} \sum_{i=1}^{B} \left[\frac{1}{\frac{1}{2}L(L-1)} \sum_{j>k}^{L} (\tilde{a}_{y_k}^i - \tilde{a}_{y_j}^i) \right]. \tag{5}$$

The L_4 ensures the training process to follow the hierarchical association relations, that is, the prediction probability of disease at parent level is larger than the disease at child level. The final loss of our proposed XHAL is

$$Loss_{XHAL} = L_1 + L_2 + L_3 + L_4. \tag{6}$$

4 Experiments

4.1 Data

We used a large chest radiograph dataset released by Stanford University named CheXpert [26] to evaluate the effectiveness of our proposed method. The CheXpert dataset contains 224,316 chest radiographs of 65,240 patients. The authors designed an auto labeling tool and 14 labels are determined for each radiograph sample. The 14 labels are inherently shown hierarchical structure dependencies which is depicted in Fig. 2. The "No Finding" label denotes the absence of all pathologies, which can determine whether the sample is diseased (label 0) or not (label 1). When "No Finding" equals to 1, the rest 13 labels are 0. Thus, we take the "No Finding" label as the root node. The hierarchical structure of CheXpert follows [26]. Each sample is assigned to either positive (labelled as 1), negative (labelled as 0) or uncertain (labelled as -1) for each label. For simplicity, we removed the uncertain samples. The relation between the labels help us to capture the hierarchical association in XHAL.

In addition to the medical image data in the CheXpert dataset, we also took the quantitative features (i.e., age, gender) into the prediction, which are largely overlooked by previous studies. Thus, we obtained two modalities (i.e., medical images and structured features) in the CheXpert case.

The final prediction result is the probabilities of the 14 labels in hierarchical structure, which can not only tell whether the sample is diseased or not (assessed by the "No Finding" label), but also depict which the most possible diseases are and how the prediction path is formed.

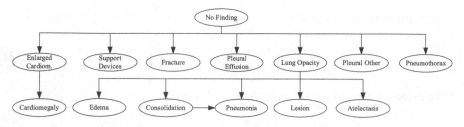

Fig. 2. Illustration of the hierarchical association structure among labels in CheXpert dataset.

4.2 Experiment Setup

We conducted four experiments to validate the effectiveness of our proposed XHAL method. First, we compared the prediction performance of XHAL with six state-of-the-art benchmarks. The details of the six benchmarks are shown in Table 1.

Table 1. The benchmark methods for disease prediction.

Study	Method	Method description
[27]	*Multimodal Ensemble*	It integrates the prediction outcomes of different feature modalities with averaging fusion strategies to form the final prediction result
[28]	*Latent Space*	It projects multimodal representation features into a shared latent space to model their connections for prediction
[29]	*Distance Loss*	It jointly trains multiple deep neural networks via relative distance (L2-norm in the experiments) maximization to produce discriminative feature representations
[26]	*Standard*	It takes the embeddings from pre-trained DenseNet121 to predict
[30]	*HDCNN*	It constructs a disease tree, then train on the conditional data where all the nodes at parent level are positive, after that, train on the full data and make predictions
[31]	*CBM*	It takes the diseases as concepts, first train a model to predict the concepts then uses the estimated concept to predict

In the second experiment, we generated the ablated XHAL model to verify the effectiveness of the key components of the proposed method. The ablated models are XHAL without modality consistency loss, XHAL without association loss, and XHAL without hierarchical structure loss. The third experiment focuses on the model sensitivity to the hyperparameters in XHAL.

All the experiments are evaluated with AUC, a metric that can well balance the trade-off between model sensitivity and specificity [32]. Moreover, to further verify the effectiveness of experimental results, we also reported another two commonly used evaluation metrics Kolmogorov-Smirnov (KS) statistic [33] and H measure [34].

The experiment results of each method are generated through five independent ten-fold cross validations. The proposed method and benchmarks are deployed on a server with an NVIDIA RTX 2080Ti GPU.

5   Results

5.1   Prediction Performance of XHAL

The prediction performance of our method and the benchmarks is shown in Table 2. Our proposed XHAL method can achieve an improvement about 3% to 17% in terms of the AUC metric compared with the benchmarks. The results on the other two evaluation metrics also indicate the superiority of our proposed method. The results validated that the consideration of multimodal data and hierarchical association structures can effectively improve disease prediction performance.

Table 2. Prediction Performance of XHAL vs the Benchmark.

Method	AUC	KS	H measure
Multimodal Ensemble	73.36	59.84	46.54
Latent Space	76.19	60.48	43.73
Distance Loss	66.97	55.46	41.19
Standard	62.61	32.64	22.03
CBM	70.44	54.69	37.44
HDCNN	77.18	61.50	47.16
XHAL	80.03	69.46	49.13

5.2   Ablation Study

Table 3 summarizes the prediction performance of XHAL in contrast with its ablated variants.

Table 3. Prediction Performance of XHAL vs the ablated variants.

Method	AUC	KS	H measure
without modality consistency loss (L_1)	78.73	68.01	31.18
without association loss (L_3)	77.02	69.88	36.64
without hierarchical structure loss (L_4)	75.83	68.69	28.84
XHAL	80.03	69.46	39.13

As shown in the results, the ablated components of XHAL led to a decrease in the prediction performance, which means the necessity of modality consistency and hierarchical association regularization. Moreover, the ablation of hierarchical association regularization loss, especially the hierarchical structure loss would lead to 4.2% decrease in terms of AUC. This indicates that hierarchical association regularization term plays an indispensable role in the disease prediction process.

5.3 Sensitivity Analysis

We further investigated the influence of the hyperparameters on the prediction performance of XHAL. The prediction performance (AUC %) of the hyperparameters λ_F, λ_p, and λ_L is summarized in Fig. 3. As shown in the results, the variations of these hyperparameters show a significant influence on the model performance. The best prediction performance is achieved when $\lambda_F = 0.001$, $\lambda_p = 0.0001$, and $\lambda_L = 0.01$, which indicates that the consideration of multimodal consistency and hierarchical association regularization is of great importance to disease prediction.

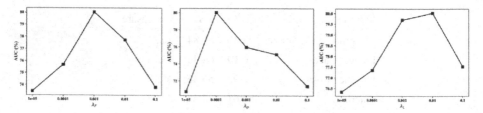

Fig. 3. The Sensitivity Analysis of XHAL.

5.4 Case Study

We randomly selected a sample to illustrate the prediction process of XHAL. The predicted result is shown Fig. 4. The grey ellipses are the positive labels and the white ellipses are the negative label. The values near the ellipses are the percentage prediction probabilities. As we can see, the root node indicates the sample is diseased, and the corresponding probability is 70.54% (for the "No Finding" indicator, 0 denotes diseased), which is consistent with the true condition. The prediction results of the second layer reveal that the association regularization is efficacious since the positive labels are predicted as the largest probabilities. Moreover, XHAL also precisely predicted the hierarchical path from "Lung Opacity" to "Atelectasis", indicating the effectiveness of hierarchical structure regularization.

The doctors and other medical staff can easily understand the logic that how XHAL makes such a result and justify whether the result is reasonable or not given Fig. 4. This can facilitate the collaboration of human-computer systems (i.e., doctors and deep learning models) to a large extent.

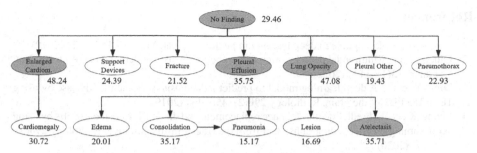

Fig. 4. An Illustrative Case of the Prediction Result Base on XHAL.

6 Conclusion

The explosive growth of the healthcare data and the rapid iteration of software and hardware infrastructure for artificial intelligence applications make it possible than ever before to support the decision-making process for the medical professions. Various deep learning method based disease diagnosis systems are developed for better human-computer collaborated diagnosis and treatment decisions. However, the black-box nature of deep learning methods makes the prediction performance hardly trust by doctors and other medical staff, largely impeding their applications and the efficiency of human-computer collaboration. To this end, we propose an explainable hierarchical association learning method aiming to investigate the prediction path of the deep learning model and provide a human-understandable, reliable, and trustworthy prediction result for the medical professions. Experimental results based on a large-scale dataset verified the effectiveness of the prediction performance and interpretable prediction path of our proposed XHAL method.

The XHAL method can be further improved by incorporating the following. In the CheXpert dataset, we removed the samples with uncertain labels, other approaches such as U-zero and U-one mentioned in [26] can be applied to further test the effectiveness of XHAL. More kinds of multimodal healthcare data can be collected to further enhance the prediction performance. Embedding features based on different large scale pre-trained deep learning model can be tested as alternatives. For example, VGG-16, ResNet, Xception and so on are potential alternatives to extract the visual embeddings from medical images. The hierarchical association rules are manually coded based on expert knowledge, which can be automatically mined from the healthcare corpus in the future.

Acknowledgments. This work was supported by the National Natural Science Foundation of China (grant # 71971067, 72271059), the National Social Science Foundation of China (grant # 22AZD136), the Research Fund Program of Guangdong Provincial Key Laboratory of Traditional Chinese Medicine Informatization (grant # 2021501), the Shanghai "Science and Technology Innovation Action Plan" Soft Science Research Project (grant # 22692108300), and the China Postdoctoral Science Foundation, (grant # 2022M722394).

References

1. Esteva, A., et al.: A guide to deep learning in healthcare. Nat. Med. **25**(1), 24–29 (2019)
2. Qiu, S., et al.: Multimodal deep learning for Alzheimer's disease dementia assessment. Nat. Commun. **13**(1), 3404 (2022)
3. Ding, Y., et al.: A deep learning model to predict a diagnosis of Alzheimer disease by using 18F-FDG PET of the brain. Radiology **290**(2), 456–464 (2019)
4. Leung, K., et al.: Prediction of total knee replacement and diagnosis of osteoarthritis by using deep learning on knee radiographs: data from the osteoarthritis initiative. Radiology **296**(3), 584–593 (2020)
5. Qian, X., et al.: Prospective assessment of breast cancer risk from multimodal multiview ultrasound images via clinically applicable deep learning. Nature Biomed. Eng. **5**(6), 522–532 (2021)
6. Han, S., et al.: A deep learning framework for supporting the classification of breast lesions in ultrasound images. Phys. Med. Biol. **62**(19), 7714 (2017)
7. Maragatham, G., Devi, S.: LSTM model for prediction of heart failure in big data. J. Med. Syst. **43**(5), 1–13 (2019). https://doi.org/10.1007/s10916-019-1243-3
8. Ali, F., et al.: A smart healthcare monitoring system for heart disease prediction based on ensemble deep learning and feature fusion, Information Fusion, vol. 63, 208–222 (2020)
9. Sekaran, K., Chandana, P., Krishna, N.M., Kadry, S.: Deep learning convolutional neural network (CNN) With Gaussian mixture model for predicting pancreatic cancer. Multimedia Tools Appl. **79**(15–16), 10233–10247 (2019). https://doi.org/10.1007/s11042-019-7419-5
10. Jo, T., Nho, K., Saykin, A.J.: Deep learning in Alzheimer's disease: diagnostic classification and prognostic prediction using neuroimaging data, Front. Aging Neurosci. **11**, 220 (2019)
11. Lin, W., et al.: Convolutional neural networks-based MRI image analysis for the Alzheimer's disease prediction from mild cognitive impairment. Front. Neurosci. **12**,777 (2018)
12. Liu, J., Zhang, Z., Razavian, N.: 'Deep ehr: Chronic disease prediction using medical notes. PMLR, vol. 85, 440–464 (2018)
13. Oh, M., Zhang, L.: DeepMicro: deep representation learning for disease prediction based on microbiome data. Sci. Rep. **10**(1), 6026 (2020)
14. Kumar, A., et al.: Towards cough sound analysis using the internet of things and deep learning for pulmonary disease prediction. Trans. Emerg. Telecommun. Technol. **33**(10), 4184 (2022)
15. Xiao, C., Choi, E., Sun, J.: Opportunities and challenges in developing deep learning models using electronic health records data: a systematic review. J. Am. Med. Inform. Assoc. **25**(10), 1419–1428 (2018)
16. Ahmad, M.A., Eckert, C., Teredesai, A.: Interpretable machine learning in healthcare. In: Proceedings of the 2018 ACM International Conference on Bioinformatics, Computational Biology, and Health Informatics, pp. 559–560 (August 2018)
17. Vellido, A.: The importance of interpretability and visualization in machine learning for applications in medicine and health care. Neural Comput. Appl. **32**(24), 18069–18083 (2019). https://doi.org/10.1007/s00521-019-04051-w
18. Bai, T., Zhang, S., Egleston, B.L., Vucetic, S.: Interpretable representation learning for healthcare via capturing disease progression through time. In: Proceedings of the 24th ACM SIGKDD International Conference on Knowledge Discovery & Data Mining, pp. 43–51 (July 2018)
19. Kwak, H., Chang, J., Choe, B., Park, S., Jung, K.: Interpretable disease prediction using heterogeneous patient records with self-attentive fusion encoder. J. Am. Med. Inform. Assoc. **28**(10), 2155–2164 (2021)
20. Choi, E., Bahadori, M.T., Sun, J., Kulas, J., Schuetz, A., Stewart, W.: RETAIN: an interpretable predictive model for healthcare using reverse time attention mechanism. Adv. Neural Inf. Proc. Syst. 29 (2016)

21. Selvaraju, R.R., Cogswell, M., Das, A., Vedantam, R., Parikh, D., Batra, D.: Grad-cam: visual explanations from deep networks via gradient-based localization. In: Proceedings of the IEEE International Conference on Computer Vision, pp. 618–626 (2017)

22. Tang, Z., et al.: Interpretable classification of Alzheimer's disease pathologies with a convolutional neural network pipeline. Nat. Commun. **10**(1), 1–14 (2019)

23. Jha, A., K Aicher, J., R Gazzara, M., Singh, D., Barash, Y.: 'Enhanced integrated gradients: improving interpretability of deep learning models using splicing codes as a case study'. Genome. Biol., **21**(1), 1–22 (2020)

24. Lee, J., et al.: BioBERT: a pre-trained biomedical language representation model for biomedical text mining. Bioinformatics **36**(4), 1234–1240 (2020)

25. Peng, X., Lu, C.Y., Yi, Z., Tang, H.J.: Connections between nuclear-norm and frobenius-norm-based representations. IEEE Trans. Neural Networks Learn. Syst. **29**(1), 218–224 (2018)

26. Irvin, J., et al.: Chexpert: A large chest radiograph dataset with uncertainty labels and expert comparison. Procced. AAAI Conf. Artif. Intell. **33**(01), 590–597 (2019)

27. Wen, G., Hou, Z., Li, H., Li, D., Jiang, L., Xun, E.: Ensemble of deep neural networks with probability-based fusion for facial expression recognition. Cogn. Comput. **9**(5), 597–610 (2017)

28. Baltrusaitis, T., Ahuja, C., Morency, L.P.: Multimodal machine learning: a survey and taxonomy. IEEE Trans. Pattern Anal. Mach. Intell. **41**(2), 423–443 (2019)

29. Ding, S., Lin, L., Wang, G., Chao, H.: Deep feature learning with relative distance comparison for person re-identification. Pattern Recogn. **48**(10), 2993–3003 (2015)

30. Koh, P.W., et al.: Concept bottleneck models'. PMLR, vol.119, 5338–5348 (2020)

31. Pham, H.H., Le, T.T., Tran, D.Q., Ngo, D.T., Nguyen, H.Q.: Interpreting chest X-rays via CNNs that exploit hierarchical disease dependencies and uncertainty labels. Neurocomputing **437**, 186–194 (2021)

32. Lin, Y.K., Chen, H., Brown, R.A., Li, S.H., Yang, H.J.: Healthcare predictive analytics for risk profiling in chronic care. MIS Q. **41**(2), 473–496 (2017)

33. Massey, F.J., Jr.: The Kolmogorov-smirnov test for goodness of fit. J. Amer. Statistical Assoc. **46**(253), 68–78 (1951)

34. Hand, D.J.: Measuring classifier performance: a coherent alternative to the area under the ROC curve. Mach. Learn. **77**(1), 103–123 (2009)

Government Initiative to Reduce the Failed or Unsuccessful Delivery Orders Attempts in the Last Mile Logistics Operation

Muhammad Younus[1]([✉]), Achmad Nurmandi[1], Misran[1], and Abdul Rehman[2]

[1] Department of Government Affairs and Administration, University of Muhammadiyah
Yogyakarta, Yogyakarta, Indonesia
mohammedyounusghazni@gmail.com, m.younus.psc22@mail.umy.ac.id
[2] Department of English, Pakistan Air Force College, Sargodha, Pakistan

Abstract. This study aims to determine the details related to the Web or Mobile-based application for the accurate prediction against the Order or Parcel ordered Online through Shipper E-commerce Site, which will be the Probability for the Successful Delivery Attempt. Researchers are doing research work in this specific use case using 'Qualitative Research.' This research data was taken from the Delivery Mobile Application, LinkedIn, Online Media, E-commerce Website, and Play Store and then represented in descriptive format. This study indicates that the creation of the Application will surely improve the quality of customer services in the logistics sector. Several milestones need to be fulfilled for this initiative's development and successful implementation—first, the analysis of the data and determining the critical indicators required for the prediction. Second, the selection of the medium for the solution and then start development of the solution based on the analysis. Finally, making it easy for accessible and straightforward to implement this solution to whoever needed. This paper is limited to the extent to which the problem is clearly described, then proposing the solution and how it should work, and lastly, what benefits the solution will be given to fight the challenge the logistics industry faces.

Keywords: Last mile delivery · Ecommerce · Merchant · Consignee · COD (Cash on Delivery)

1 Introduction

In the Logistics Operation, the most critical part is the Last Mile Delivery Operation which means When the Customer or Consignee receives the Parcel or Order or Shipment that he has Ordered Online Through the Merchant or Shipper Ecommerce Website or social media Page or Mobile Application, etc. Channel. The Ideal Scenario is that When Delivery Rider reaches the Customer's Location, The Customer is Available and has the Required COD Amount Available at the time, which the Customer has to pay against the Order he is Receiving. So, it seems straight Forward that Rider has to Deliver the Order and Receive the COD Amount, and Finally, Order Journey Will Successfully End. But in

© Springer Nature Switzerland AG 2023
F. Fui-Hoon Nah and K. Siau (Eds.): HCII 2023, LNCS 14038, pp. 114–138, 2023.
https://doi.org/10.1007/978-3-031-35969-9_9

reality, the Order Journey does not always remain as simple as it seems. The most critical challenge faced in the last-mile delivery operation is when Delivery Order Attempt gets Failed or is unsuccessful. The reasons for failure or unsuccess are also not that easy, and there is a long list that contains why a Delivery Order can Fail (Straight, B. (2022, July 8)). But Overall, Failed or Unsuccessful Orders can be of Two Major Types. One is a 'Re-Attempt' Fail Order which means the Order is Failed, and we have to Attempt that Order Again on Next Day for Successful Delivery. And the Second one is a 'Return' Fail Order which means the Order Failed, and we don't have to attempt that Order Again on Next Day because Consignee is not needed or Require that Order. He has Rejected it. To Understand better, take the Example of 'Re-Attempt' Category Failure Reasons, Like Consignee not Available, Reschedule Order on Consignee Request, COD Amount is not Available, Delivery Location is Closed, Customer wants to Open Parcel, etc., are a few of the many Failure Reasons of it. Also, To Understand better, note Examples of 'Return' Category Failure Reasons, such Like Consignee Refusing to Accept Order, Consignee Mistakenly Made Order, Customer already purchased the Item, Order Arrived Late, Non-Service Area, etc., are a few of the many Failure Reasons of it. So, Both Failed Attempt Cases are the biggest challenge for Last Mile Operation. According to the stats (N.d.). xxxx), almost 10% to 20% of Orders are Failed on the First Attempt, which causes an immense amount of pressure on Logistics because they have to face the consequence of it in terms of, Increase in Cost to Deliver Orders, Handling of the Inventory, Extra Load of work for Warehouse Staff and Delivery Riders, Time to Deliver Increases because Time wasted on Fail Attempt, Fuel Cost Increases, etc. are a Few of the many additional problems which might be faced on Logistics Operation (Abdul Rahman, M., Aamir Basheer, M., Khalid, Z., Tahir, M., & Uppal, M. 2022).

Like, understand through the below Table, which is very simply defined that a Logistics Company handles '20' Orders Daily. According to the Orders they receive daily, they need 2 Riders and Warehouse Staff to manage them based on the Capacity of One Staff or Rider, which here defined as '10', so we have divided (/) the '**No. of New Orders Daily**' with the '**The capacity of One Rider.**' In that way, we get the Required Staff or Rider Count. This calculation of operation planning will remain simple if no Failed or Unsuccessful Order exists (Table 1).

Table 1. Logistics Operation Plan without Fail Orders

No. of New Orders Daily	No. of Orders Failed Daily	Overall Count of Orders Daily	The capacity of One Rider	No. of Rider Required Daily
20	0	20	10	2

But what happens if there is a Failed Order? Then see the below Table. So, now it is visible what impact a Fail Order is making, so with '20%' average Fail Orders now, you are handling '24' Orders Daily because these Fail Orders will return to the Warehouse from Route. So Now the Order load is increased, which means Warehouse Staff and Riders also need more, and the Cost for handling per Order and Cost per Staff and Riders will also Increase. Most importantly, the overall efficiency or performance is impacted by the operation (Escudero-Santana, A., Muñuzuri, J., Lorenzo-Espejo, A., & Muñoz-Díaz, M.-L. 2022) (Table 2).

Table 2. Logistics Operation Plan with Fail Orders

No. of New Orders Daily	No. of Orders Failed Daily	Overall Count of Orders Daily	The capacity of One Rider	No. of Rider Required Daily
20	4	24	10	3

This example will help you now to imagine what happens when No. of Orders is in the Thousands or Millions. The challenge will become bigger and bigger with the Increase in Orders Load, and it is not only about the increase in Cost in terms of Staff, Riders, Fuel, Workspace, etc. It's about Performance and efficiency also, because when the Orders fail, it decreases the Success rate of the operation. Success Rate is one of the essential metrics through which the operation performance is evaluated. So how Failed Orders impact the operation performance? Understand it through the below Table. Which is very simply defined that there is a Logistics Company that enroute '20' Orders. Out of '20' orders, '16' orders were Successfully delivered, and '4' orders were Failed Attempts. So, we have divided (/) the '**No. of Orders Delivered**' with the '**No. of Orders Enroute**' and lastly converted it into Percentage (%). That way, we get the '**Operation Success Rate%**.' This is how the performance goes up and down, with Failed Orders increasing and decreasing. It directly impacts the operation, one of the most critical Key Performance Indicators (KPIs) which represent Logistics companies' standards overall (Publow, P. F. 2007) (Table 3).

Table 3. Logistics Operation Performance View

No. of Orders Enroute	No. of Orders Delivered	No. of Orders Failed	Operation Success Rate %
20	16	4	80%

So, now we are fully aware of the issues and how it increases the time, resources, and efforts required to solve them and further complicate the problem. That is why it is becoming harder for individual companies and businesses to solve this issue independently because it is the challenge of the overall industry, not of a specific company. That is why a more potent force must be involved to handle this problem. In this case, I believe the government can invest time and once and all to solve this issue. Because the time and effort required to solve are beyond any logistics company's capacity, so without government help, it cannot be done. So, when the government solution is ready, any logistics company will access it and use it for the benefit of the operation process betterment. This solution will give them information in advance so the operation would know about the coming challenge and proactively align the staff and processes according to it (Kolasińska-Morawska, K., Sułkowski, Ł., Buła, P., Brzozowska, M., & Morawski, P. 2022).

2 Literature Review

2.1 E-Commerce

The purchasing, selling, and trading of products and services over the Internet are known as e-commerce. It is often referred to as Internet commerce or electronic commerce (Xu, J., Yang, Z., Wang, Z., Li, J., & Zhang, X. 2023). Here are a few other ways to spell "e-commerce," including e-commerce, e-commerce, E-commerce, eCommerce, etc. Some businesses sell things online, but for many, e-commerce serves as a conduit of distribution as part of a larger business plan that includes physical stores and other revenue sources (Almtiri, Z., Miah, S. J., & Noman, N. 2023). In either case, e-commerce enables startups, established enterprises, and multinational corporations to sell their goods globally (Yin, L., Zhong, R. R., & Wang, J. 2023). The most typical e-commerce model types and illustrations of what they mean are as follows (Peng, Y., & Yi, J. 2023):

1. Business to Consumer (B2C): The most well-liked e-commerce model is business to consumer (B2C). Business to consumer refers to a transaction when a company and a consumer are involved, such as when you purchase something from an online retailer.
2. Business to Business (B2B): B2B e-commerce describes selling a product or service by one company to another company, such as a manufacturer, a wholesaler, or a retailer. Business-to-business e-commerce typically involves goods like raw materials, software, or combination goods and isn't targeted at consumers. Through B2B e-commerce, manufacturers can also sell directly to retailers.
3. Direct to Consumer (D2C): The newest type of e-commerce is direct consumer, and trends in this field are constantly evolving. Direct-to-consumer (D2C) refers to a brand's sales to the final consumer without using a retailer, distributor, or wholesaler. Direct-to-consumer sales frequently involve subscriptions, and social selling on websites like Instagram, Pinterest, TikTok, Facebook, and Snap Chat is also joint (Jaller, M., & Dennis, S. 2023).
4. Consumer to Consumer (C2C): The sale of a good or service to another consumer is called C2C e-commerce. Consumer-to-consumer transactions occur on websites like eBay, Etsy, Fivver, etc.
5. Consumer to Business (C2B): When a person offers services or goods to a business entity, this is known as a consumer to business. Influencers delivering exposure are included in C2B.

These are a few illustrations of several e-commerce delivery models (Pérez-Morón, J. M. 2023).

1. Retail: Direct sales to customers without the use of a middleman.
2. Dropshipping: Selling goods produced and delivered to customers by a third party.
3. Digital products: Downloadable materials that must be bought for use, such as templates, courses, e-books, software, or media. Buying software, tools, cloud-based goods, or digital assets makes up a sizable portion of e-commerce transactions.
4. Wholesale: Bulk sales of goods Typically, wholesale goods are sold to a retailer, who then sells them to customers.
5. Services: These are online-purchased professions like counseling, writing, influencer marketing, etc.
6. Subscription services: A well-liked D2C business model, subscription services involve regular, recurring purchases of goods or services.
7. Crowdfunding: Crowdfunding enables entrepreneurs to raise seed money to launch their products. The product is then made and shipped after a sufficient number of customers have purchased it (Bighrissen, B. 2023).

2.2 Merchant/Shipper

A merchant stands for an individual or business transacting in goods or services. A party that only sells products or services online is referred to as an eCommerce merchant (Hasan, F., Mondal, S. K., Kabir, M. R., Al Mamun, M. A., Rahman, N. S., & Hossen, M. S. 2022). Due to his knowledge of the products he sells, a merchant will sell the goods to the consumer for a profit, and by law, he owes the buyer a duty of care. A merchant may be a wholesaler or a retailer, and the goods may be distributed among various sources. Anybody who sells anything is referred to as a merchant, with the sole qualification being that it is being done so to make a profit (Etumnu, C. E. 2022).

2.3 Customer/Consignee

The bill of lading lists a consignee or customer in shipping (BOL). This individual or group is the recipient of the shipment and, in most cases, the owner of the transported items (Ashrafpour, N., Niky Esfahlan, H., Aali, S., & Taghizadeh, H. 2022). The company or person legally required to take the package is the consignee unless there are specific instructions to the contrary (Stefko, R., Bacik, R., Fedorko, R., & Olearova, M. 2022).

2.4 Rider/Courier

Transporting parcels and other items from a mail facility to a residential or commercial address is the responsibility of a delivery driver, often known as a carrier (Oviedo-Trespalacios, O., Rubie, E., & Haworth, N. 2022). They are responsible for loading cargo, using navigational aids to reach the correct address, and delivering packages to it (McKinlay, A., Mitchell, G., & Bertenshaw, C. 2022).

2.5 Logistics

ECommerce logistics are the procedures for storing and shipping inventory for an online store or marketplace. E-commerce logistics begins with the movement of stock from the manufacturer and continues until it reaches the final client (Li, Y., Shi, J., Cao, F., & Cui, A. 2021). A supply chain is created by coordinating several activities, including inventory management, warehousing, packaging, labeling, billing, shipping, payment collection, return, and exchange. Together, they create a challenging task that needs to be completed using a foolproof plan. In addition, eCommerce logistics requires in-depth familiarity with geographic areas, roads, conditions, and transportation legislation. The main goal of a logistics unit is to deliver packages considerably more quickly, safely, and precisely (Wu, G. 2021). The operations of an eCommerce Logistics Company are Bidirectional. Distribution and delivery of commodities to customers is in the forward direction. The Reverse Direction is exchanging or replacing incorrect, damaged, or defective shipments. There are different types of logistics that you need to be aware of. The five major types of logistics are (Muñoz-Villamizar, A., Velázquez-Martínez, J. C., Haro, P., Ferrer, A., & Mariño, R. 2021):

1. Inbound Logistics: Inbound logistics is employed to carry out strategic organizational duties for working upstream. Under this inbound logistics, different information and products from the suppliers are moved, transported, and stored in the warehouse before being transferred to the production facilities for additional processing & production. The primary focus of inbound logistics is on the movement of goods between businesses and their suppliers (Dasgupta, S., Kanchan, S., & Kundu, T. 2019).
2. Outbound Logistics: Outbound logistics moving items from producing facilities to the following supply chain link is known as outbound logistics. These goods are transported from the warehouse to the site of consumption or the clients. So, order fulfillment is another name for outbound logistics.
3. Reverse Logistics: As the name implies, reverse logistics is the movement of goods or products from end consumers to the supply chain. Reverse logistics is required when products need to be replaced or returned for refurbishing, repairing, swapping, discarding, or recycling (Das, D., Kumar, R., & Rajak, M. K. 2020).
4. Third-Party Logistics (3PL Logistics): Third-party logistics, often known as 3PL, is the practice of contracting out operational or eCommerce logistics to a third-party logistics business, which subsequently takes care of everything from inventory control to product delivery. This allows business owners to concentrate on their core competencies while a 3PL provider efficiently manages the order fulfillment processes (Cosmi, M., Nicosia, G., & Pacifici, A. 2019).
5. Fourth-Party Logistics (4PL): Companies employ 4PL, or fourth-party logistics, to delegate all of their logistics tasks to a single logistics partner. The logistics provider oversees every client's supply chain aspect, including assessment, design, construction, operation, and tracking. For this reason, a 4PL logistics partner signifies to the client a greater level of supply chain management.

2.6 Last Mile Delivery

The last mile delivery is the link in the e-commerce supply chain that links customers and brands by delivering the customer's order (Nagpal, G., Bishnoi, G. K., Dhami, H. S., &

Vijayvargia, A. 2020). Products are provided from a warehouse or distribution center to a customer's residence, place of business, or parcel locker (Kahr, M. 2022). The last mile of the product's trip is the most challenging and costly for the shipper. To ensure that every delivery reaches its destination consistently, on time, accurately, efficiently, and sustainably, excellent last-mile delivery aims (Wang, M., Zhang, C., Bell, M. G. H., & Miao, L. 2022).

2.7 Cash on Delivery/Pre-paid

With a cash-on-delivery (COD) transaction, the recipient pays for the item in cash rather than utilizing credit at the delivery time. The conditions and allowed methods of payment change depending on the purchase agreement's payment clauses. Since delivery may accept cash, cheque, or electronic payment, cash on delivery is also known as collect on delivery (Ha, X. S., Le, T. H., Phan, T. T., Nguyen, H. H. D., Vo, H. K., & Duong-Trung, N. 2021). A consumer places an order and designates a delivery address on the internet. The buyer chooses to pay when the items are delivered rather than paying for them at the time of purchase. An invoice from the seller will be included with the confirmed order (Purwandari, B., Suriazdin, S. A., Hidayanto, A. N., Setiawan, S., Phusavat, K., & Maulida, M. 2022).

2.8 Application Programming Interface (API)

An application Programming Interface, or API, is a software bridge that enables communication between two applications. You utilize an API every time you use a mobile app like Facebook, send an instant message, or check the weather. A mobile application connects to the Internet and sends information to a server (Verma, R., Dhanda, N., & Nagar, V. 2023). The server retrieves, interprets, puts to use, and sends the data back to your phone. The application then analyzes the data and displays the information you requested comprehensibly. All of this occurs through an API, which is what it is. There are four main types of web API; You can use the HTTP protocol to connect to open APIs and open-source application programming interfaces. They have established API endpoints, request, and response forms, also known as public APIs. APIs for partners are programming interfaces made available to or by strategic business partners (González-Mora, C., Barros, C., Garrigós, I., Zubcoff, J., Lloret, E., & Mazón, J.-N. 2023). Typically, a public API developer site is where developers can self-serve access these APIs. Still, they must undergo an onboarding procedure and obtain login information to access partner APIs. Application programming interfaces (APIs) that are internal are not visible to outside users. These internal development teams' efficiency and communication can be enhanced by using these secret APIs, which aren't accessible to people outside the business. Multiple data or service APIs are combined in composite APIs. These services enable developers to make a single call to many endpoints. Composite APIs are helpful in a microservices design, where a single job may require information from numerous sources.

2.9 Machine Learning

Machine learning (ML) is a topic of study focused on comprehending and developing "learning" techniques or techniques that use data to enhance performance on a particular set of tasks. It is considered to be a component of artificial intelligence (Bopage, G., Nanayakkara, J., Vidanagamachchi, K. 2019). Without being taught to do so, machine learning algorithms create a model using sample data, also called training data, to make predictions or judgments (Luo, X., Sun, Q., Yang, T., He, K., & Tang, X. 2023). Machine learning algorithms are utilized in many applications, including speech recognition, email filtering, computer vision, and medicine, when it is challenging or impractical to create traditional algorithms to carry out the required functions (Armstrong, C. E. J., Gilmore, A. M., Boss, P. K., Pagay, V., & Jeffery, D. W. 2023). Simple Linear Regression models the response variable using a single Independent/Predictor(X) variable (Y). However, there may be several situations when more than one predictor variable impacts the response variable; the Multiple Linear Regression technique is applied in these situations. As more than one predictor variable is required to predict the response variable, Multiple Linear Regression is a development of Simple Linear Regression (Yun, K. K., Yoon, S. W., & Won, D. 2023).

3 Research Method

Qualitative research methods were applied in this study, with the help of data gleaned from online stores and logistics operations. This strategy describes and further assesses the data to obtain more thorough, valid, and trustworthy information. A Business Intelligence tool was used to collect data and was carefully investigated before the data was analyzed. They used the below Assessment framework when applying the Machine learning algorithm (Fig. 1).

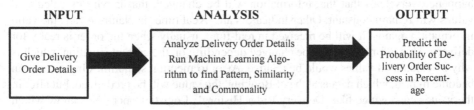

Fig. 1. Framework assessment of the working of the Model

The Following key indicators mentioned in Table 4 are prioritized based on their impact, which will help in conducting prediction.

So, in Table 4, all Possible indicators that will impact the success of the Order Delivery are mentioned with their Weightage or Thickness and their nature. Only 'Delivery Address' has the higher Weightage Value. In contrast, Other has the same Weightage, and the Reason for it is that Clarity on the Delivery Address Plays a very vital part in the success of a Delivery Order; without its Correctness, the Delivery will not be possible. That is why all other variables can have the highest Score of '6%' and Lowest score of '0%'. But in the case of the 'Delivery Address' variable, it can have the highest Score of '10%' and the Lowest score of '0%'. Also, in Nature, the data finalization against

Table 4. Key Indicators with Weightage and Nature

Indicator	Weightage	Nature
Delivery City	6%	Predefined
Delivery Area	6%	Predefined
Delivery Address	10%	Predefined
Delivery Location Type	6%	Predefined
Delivery Attempt Time	6%	Real-Time
Customer History	6%	Predefined
Merchant History	6%	Predefined
Delivery Rider History	6%	Predefined
COD Amount	6%	Predefined
Weather	6%	Real-Time
Traffic	6%	Real-Time
Public Holiday	6%	Predefined
Road Blocks	6%	Real-Time
Emergency Situation	6%	Real-Time
Customer Respond to Call/Message	6%	Real-Time
Valid Phone Number	6%	Predefined

each indicator is defined. Few are 'Predefined' in Nature means the information against it will be received at the initial point, usually when the Order is Booked, and it has happened very least that that information will be changed, that is why its calculated values will remain constant. Other Indicators are 'Real time' in Nature means the valid information against it will be received in real-time, usually when the order is ready for delivery. This information will be changed depending on the current situation, which is why its calculated values would be relative. Also, to mention sometimes Indicators can become hybrid, which means at the of Booking, the value will be predefined, but after it, the values can change like 'Delivery Rider History'; Logistics operations know which Rider works on which route and what Order is present for its Route, so operation can preselect Rider Name who will take Order enroute but in some cases, if Rider is absent because of any reason then Order will be accepted by Backup Rider, therefore in that Case Indicator value can change.

Same here, on above Table 5. The Categories of the Score on which the Predicted Success Rate against the Order will be measured and based on which we can differentiate between a good and bad score. So overall, the Predicted Success Rate% can have the highest Score of '100%' and the Lowest score of '0%'.

The assessment will be accumulated using the Calculation Formula, in which values against each indicator mentioned in Table 4. Will be Calculated. As defined and discussed, The Indicator values will be of two types one is predefined Values like Delivery City, Delivery Area, Delivery Address, etc., and the second one will be the Real Time

Table 5. Success Rate % Value Score-Category

Category	Score
Excellent	90% to 100%
Very Good	80% to 89%
Good	70% to 79%
Fair	50% to 69%
Bad	25% to 49%
Very Bad	0% to 24%

values like Weather, Traffic, Delivery Attempt Time, etc. The predefined values will be taken from the Database of the Logistics Company or Ecommerce Website, and The Real-Time values will be Taken from the Weather Forecast Sites, Logistics Company Database and Google Map Traffic Update, etc.

Predicted Attempt Success Rate % = [Delivery City + Delivery Area + Delivery Address + Delivery Location Type + Delivery Attempt Time + Customer History + Merchant History + Delivery Rider History + COD Amount + Weather + Traffic + Public Holiday + Road Blocks + Emergency Situation + Customer Response to Call/Message + Valid Phone Number]

So, individual values will be obtained against each Variable through the Machine learning algorithm. It will be converted into the criteria of weightage, and lastly, it will sum all individual indicator values. Finally, we will get the resultant value of the Predicted Success rate against any Order. To convert the value on Percentage (%), multiply it by Hundred '100'.

We suggest the 'Multiple Linear Regression' Algorithm (Bopage, G., Nanayakkara, J., Vidanagamachchi, K. 2019) make predictions for output. To analyze the data, that can provide ease to explore it properly, specifically when information is not in a small amount. Also, that way can give the Option to display analysis results in a presentable format in diverse ways. That is why we choose **Google Data Studio** (Ashrafpour, N., Niky esfahlan, H., Aali, S., & Taghizadeh, H. 2022). It is an online business intelligence platform that helps visualize data and create interactive reports and dashboards. Using this Platform doesn't require any additional setup or installation. You can use it by visiting the link; only you should have Gmail Account. Therefore, connect the information database as data sources and do the same with Real-time data. Lastly, after joining the Data, we can visualize the Data in any way possible, like Table, Graph, and Pivot Table. Also, with its Feature of adding a Filter, we can see specific Cities, Customers or merchants, etc. Records. Also, for the Literature review, we have used the in-direct citation. We have used the grounded theory Approach of the Research method.

4 Results and Discussion

Based on the detailed analysis of the problem in different ways, The solution being proposed with the potential of solving the problem is the Application or System that can Predict Success Rate against Delivery Orders Based on provided Order Details and some External Factors. So, the details of Delivery Orders will be given as Input. Then it will analyze Details against the Orders and provide the output as the Predicted Attempt Success Rate in Percentage (%).

4.1 'Probability of Delivery Attempt Success' Indicators Details

We will mention each indicator used for the calculation of the 'Predicted Attempt Success Rate %' below with the rationale behind the addition of it for the calculation (Fig. 2).

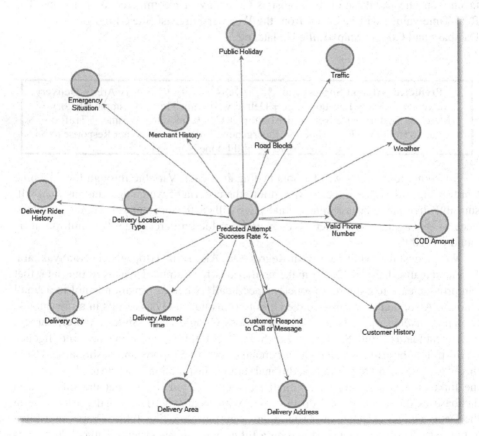

Fig. 2. Shows the Main indicators of 'Predicted Attempt Success Rate %'

'Delivery City' Indicator. The 'Delivery City' indicator will check details regarding Delivery Order City. The customer already gives this indicator's data at the time of

Delivery Order Booking. So, we will utilize that information soon. Order is booked to calculate the 'Delivery City' indicator value. The weightage of this indicator will be '6%'. The highest value against the 'Delivery City' indicator can be attained as '6%,' and the lowest value against it will be '0%'. The reason for adding 'Delivery City' as an indicator for the calculation of 'Predicted Attempt Success Rate %' is that The City where the Delivery Order is going to be attempted makes a difference in the success of the Delivery attempt because each City has its own dynamics, they have their own disadvantage and advantages based on their geographic location, Development infrastructure (Roads, Internet connectivity), Type of city (Urban or rural), etc., these all are the factors which can able to make delivery attempt process smooth or can add an additional challenge for the Last mile delivery operation. That is why we will be taking the 'Delivery City' indicator into account due to its possible impact, and we know that we can't consider the challenges of the city itself as constant it varies from city to city like we can't say the way delivery of an order in 'London' is done exactly the same way it is done in 'New York.'

'Delivery Area' Indicator. The indicator of the 'Delivery Area' will track details regarding the Area of Delivery Order. The information for this indicator is sometimes given by the Customer at the time of Delivery Order Booking separately in the different input fields, and sometimes it is mentioned within the Delivery Address input, so in case second, we have fetched the delivery area data from it before starting the calculation. We will utilize that data soon. Order is booked to calculate the 'Delivery Area' indicator value. The weightage of this indicator will be '6%'. The maximum value against the 'Delivery Area' indicator can be got as '6%,' and the minimum value against it will be '0%'. The rationale behind adding 'Delivery Area' as an indicator for the calculation of 'Predicted Attempt Success Rate %' is that The Area within the City where the Delivery Order is going to be attempted also creates it's a role in the success of the Delivery attempt because each different Area within the City has its own subtleties, they have their own weakness and strength based on their Reach (it's in the center of a city or the boundary), Development infrastructure (Roads, Internet connectivity), Type of Area (Posh or Slum), etc., these all are the aspects which can able to make delivery attempt process easy or can add further difficulty for the Last mile delivery operation. That is why we will be taking 'Delivery Area' as an indicator due to its probable effect. We know that we can't consider the area's challenges as persistent; it varies from area to area; each Area in the city has its problems.

'Delivery Address' Indicator. The indicator of the 'Delivery Address' will record details regarding the Customer Address of the Delivery Order. The Customer always gives the info for this indicator at the time of Delivery Order Booking, and it's mandatory to share so it can't be missed until there is some technical issue. We will utilize that information soon. Order is booked to calculate the 'Delivery Address' indicator value. The weightage of this indicator will be '10%'. The largest value against the 'Delivery Address' indicator can be got as '10%,' and the smallest value against it will be '0%'. The justification behind adding 'Delivery Address' as an indicator for the calculation of 'Predicted Attempt Success Rate %' and also giving the higher weightage to it compared to other Indicators is that, The Address where the Customer lives and on which Delivery Order is going to be attempted plays the most significant tasks in the Success of Delivery attempt because without having Delivery Address which is also Clean and clear can able to make delivery attempt process simple and if not its create problems for making additional efforts for getting clean delivery address and complicates the work for the Last mile delivery operation. Due to this, we have to take 'Delivery Address' as an indicator due to its immense impact on the success of the delivery attempt. The Better and clearer 'Delivery Address' against the Delivery Order will increase the value against this particular indicator.

'Delivery Location Type' Indicator. The indicator of the 'Delivery Location Type' will maintain details related to the Location type of the Delivery Order. The details for this indicator are occasionally mentioned within the Delivery Address input field. Still, in the majority, we must find the delivery location type by reading the delivery address before starting the calculation. After getting the information, we will utilize it soon. Order is booked to calculate the 'Delivery Location Type' indicator value. The weightage of this indicator will be '6%'. The uppermost value against the 'Delivery Location Type' indicator be achieved as '6%,' and the lowermost value against it will be '0%'. The logic behind adding 'Delivery Location Type' as an indicator for the calculation of 'Predicted Attempt Success Rate %' is that The Location Type, which means the Delivery Address is a Residential Address where the Customer lives or the Delivery Address is a Commercial Address where Customer works will affect the attempt success rate. The location type where the Delivery Order is going to be attempted has its intricacies, like if the Delivery Location if the Location type is Office address, then the time when the delivery rider will attempt will be important due to Office hours; delivery attempts made before or after Office Hours will have the possibility of getting Failed and similarly if Location type is Home Address then the time when the delivery rider will be important due to Customer not Available because a customer is in Office for work. These scenarios can make the delivery attempt process challenging for the Last mile delivery operation, it depends on the location type, and each location type has its complications. That is the cause for which we will be taking 'Delivery Location Type' as an indicator.

'Delivery Attempt Time' Indicator. The 'Delivery Attempt Time' indicator will hold details regarding the Time when Delivery Rider will attempt the Delivery Order. The Stats for this indicator come from real-time based on the Delivery Rider attempt time, and we can only estimate it through the Delivery Rider working sequence. We will utilize that information soon. Order is Taken en route by Delivery Rider to calculate the 'Delivery Attempt Time' indicator value. The weightage of this indicator will be '6%'. The biggest value against the 'Delivery Attempt Time' indicator can be taken as '6%,' and the least value against it will be '0%'. The explanation behind adding 'Delivery Attempt Time' as an indicator for the calculation of 'Predicted Attempt Success Rate %' is that the time when the Delivery Order will be attempted has its importance in the Success of the Delivery attempt. We cannot assume that at any time, the Delivery rider will attempt to the customer will always be available and waiting for the delivery order. That is not the case. Every customer is busy with their day-to-day tasks, and we must ensure that delivery reaches the right location at the right time for the attempt's success. Not being cautious about it makes delivery attempts fail, which is not good for the Last mile delivery operation. That is why we have to take 'Delivery Attempt Time' as an indicator due to its massive influence on the success of the delivery attempt.

'Customer History' Indicator. The 'Customer History' indicator will analyze details about the Customer of the Delivery Order. The specifics for this indicator are always stated with the Delivery Order at the time of Booking in the Customer Name input field. After attaining the data, we will utilize it soon. Order is booked for calculating the 'Customer History' indicator value. The weightage of this indicator will be '6%'. The topmost value against the 'Customer History' indicator can be achieved as '6%,' and the bottommost value against it will be '0%'. The sense behind adding 'Customer History' as an indicator for the calculation of 'Predicted Attempt Success Rate %' is that The Customer who has booked the Order has its influence over the Success of the Delivery attempt based on its previous track history like in past when Delivery Rider attempted the Delivery Order to the same customer what was the outcome of it if the order was delivered on the first attempt successfully without any issue and challenges or it was failed and even failed multiple times or customer rejected and refused delivery order from accepting. What was the outcome in history based on which we can able to predict the success of the current delivery order. Through this indicator, we can also identify the customer who unnecessarily books an order and always reject it so we can connect with the customer and understand the problem. If it is a genuine reason for rejection, then Operation can solve it with merchants. Still, if it is not justifiable, we can blacklist that customer so that customer can't bother further. That is the root for which we will take 'Customer History' as the indicator.

'Merchant History' Indicator. The indicator of the 'Merchant History' will trail info regarding the Merchant or Shipper of the Delivery Order. The information for this indicator is always cited with the Delivery Order at the time when Order is Booked in the Merchant Name input field. After getting details, we will utilize them for calculating the 'Merchant History' indicator value. The weightage of this indicator will be '6%'. The highest value against the 'Merchant History' indicator can be achieved as '6%,' and the lowest value against it will be '0%'. The motive behind adding 'Merchant History' as an indicator for the calculation of 'Predicted Attempt Success Rate %' is that The Merchant to which the Customer given an Order booking has an effect on the Success of the Delivery attempt based on the past trial history like previously when Delivery Rider has attempted the Delivery Order of the same Merchant then at that time what was the overall result of it, We can find out by checking the successful attempt count and then comparing with a failed count of specific merchant Order and finally we can able to judge the Success rate of individual merchant and then use it for this indicator value calculation. With the help of this indicator we can also recognize the merchants has lowest attempt success rate so we can able to connect with merchant about it and also can take customer point of view and understand their problem if it is a honest reason of rejection then Operation can solve it with merchants but if it is due to Merchant product quality like merchant doing fake items business and fooling the customers then it is not justifiable and we can blacklist that merchant so that merchant can't trouble more, it is also necessary because sometime last mile logistics companies don't care about merchant quality of work and say we just have to deliver order its customer and merchant issue but they should know our Delivery rider will be going to face customers and in many cases customer creates trouble for delivery rider by saying you are also involve in this fraud because they don't know the difference between last mile operation and merchant itself this why last mile operation needs to take this issue seriously to from save from any problem. That is why we will take 'Merchant History' as an indicator.

'Delivery Rider History' Indicator. The 'Delivery Rider History' indicator will examine specifics related to the Delivery Rider who will attempt the Delivery Order. The material for this indicator will be getting real-time. Once we get its information, we can get delivery order details when the last time Order was Successfully attempted or failed attempted by the same Delivery Rider. We will utilize the same information for calculating the 'Delivery Rider History' indicator value. The weightage of this indicator will be '6%'. The maximum value against the 'Delivery Rider History' indicator can be accomplished as '6%,' and the minimum value against it will be '0%'. The aim behind adding 'Delivery Rider History' as an indicator for the calculation of 'Predicted Attempt Success Rate %' is that The Delivery Rider who will attempt the given Order has its consequence on the Success of the Delivery attempt based on the Delivery Rider historical working experience like before when same Delivery Rider has made attempts of the Delivery Order then at that time what was the general outcome of it, We can easily find out by read-through the successful attempt count of delivery rider and then comparing with delivery rider failure count. Lastly, we can able to get the Success rate of a specific delivery rider and then use it for the indicator value calculation. With the support of this indicator, we can also spot the delivery rider who has the minimum attempt success rate so we can able to link with the delivery rider about his performance. If a delivery rider needs assistance, we can train and help him. Still, if it is due to a Delivery rider's low-quality work, like a delivery rider making fake delivery attempts and having complaints from a customer against it, then it is not reasonable. We can give a warning, and after no improvement, can change the delivery rider; it is necessary to do because last mile logistics companies' image depends on it; the Delivery rider will be going to face customers and will represent them. That is the purpose for which we will be taking 'Delivery Rider History' as the indicator.

'COD Amount' Indicator. The 'COD Amount' indicator will note details about the Cash on a delivery amount that the customer needs to pay the delivery rider against receiving the Delivery Order. The essentials for this indicator are every time stated with the Delivery Order at the time of Delivery Order Booking. After fetching the information, we will utilize it soon; the Order is booked for calculating the 'COD Amount' indicator value. The weightage of this indicator will be '6%'. The uppermost value against the 'COD Amount' indicator can be reached as '6%,' and the bottommost value against it will be '0%'. The purpose behind adding 'COD Amount' as an indicator for calculating the 'Predicted Attempt Success Rate %' is that The COD Amount of the Order which is booked has its impact on the Success of the Delivery attempt because the COD Amount value with its increase or decrease can make order delivered or failed, like Delivery Order can be COD which means Amount against the delivery order will be paid at the time of delivery or Non-COD which means Amount against the delivery order has already been made by the customer it is prepaid so now delivery rider just have to deliver the order. Also, even with the Delivery order being COD, how much the COD Amount also has an impact. Suppose COD Amount is lesser, like a few hundred. In that case, chances are high that the customer will not reject the order due to the amount of unavailability because the amount so less, but if the COD amount is greater in thousands, then it can happen customer can make an attempt to fail due to unavailability of cash with him at the time. That is the basis for which we will take 'COD Amount' as an indicator.

'Weather' Indicator. The indicator of the 'Weather' will work with the details about the Weather Conditions at Time and Day when Delivery Rider will attempt to deliver the Order. The information for this indicator will derive from real-time weather reporting websites. Then we will use that data based on the current date to calculate the 'Weather' indicator value. The weightage of this indicator will be '6%'. The maximum value against the 'Weather' indicator can be taken as '6%,' and the minimum value against it will be '0%'. The clarification behind adding 'Weather' as an indicator for the calculation of 'Predicted Attempt Success Rate %' is that the time and day when the Delivery Order will be expected to be attempted has its role in the success of the delivery attempt. Because the delivery rider's whole work is done outside the enroute and based on weather conditions, the delivery rider decides to attempt delivery, especially when the weather is rainy and the delivery rider is on a bike. Then in many cases, delivery riders save themselves and the order they had; the Delivery rider decides to come back from the route without attempting to order due to the weather not supporting route work. And then last mile companies usually send apologetic SMS or alerts on the websites about delivery attempts affected due to bad weather conditions. Therefore, we have to take 'Weather' as the indicator.

'Traffic' Indicator. The indicator of the 'Traffic' will check the info associated with the Traffic Condition at the Time when the Delivery Rider will be making an attempt to Delivery the Order. The data for this indicator will originate from real-time traffic reporting sites. After it, we will be going to use that information based on the current time to calculate the 'Traffic' indicator value. The weightage of this indicator will be '6%'. The most significant value against the 'Traffic' indicator can be taken as '6%,' and the smallest value against it will be '0%'. The goal behind adding 'Traffic' as an indicator for the calculation of 'Predicted Attempt Success Rate %' is that the time when the Delivery Order will likely be attempted has its effect on the success of the Delivery attempt. Because delivery riders' whole day using the roads from going one customer location to another to do work on time and efficiently, the clearing of the road from traffic will be vital for reaching the on-time to customer location for delivery. But if it does not arrive on time or cannot even get near the customer's location due to being stuck in traffic, then it will definitely make the delivery attempt fail. Hence due to the mentioned reasons, we have to take 'Traffic' as an indicator.

'Public Holiday' Indicator. The indicator of the 'Public holiday' will check the data about the possible public holiday on the Date when Delivery Rider will make an attempt to Delivery the Order. The material for this indicator will arise from sites related to calendars with all public holidays mentioned in them. After getting information on the data based on the current date, the calculation of the 'Public holiday' indicator value will be done. The weightage of this indicator will be '6%'. The largest value against the 'Public holiday' indicator can be taken as '6%,' and the smallest value against it will be '0%'. The reason behind adding 'Public holiday' as an indicator for the calculation of 'Predicted Attempt Success Rate %' is that the date when the Delivery Order will be predictable to get attempted will play its role in the Success of the Delivery attempt. Because if a public holiday is coming, which means it is likely that the last mile delivery operation will also close its working due to the public holiday, an attempt is not possible on that date. Even if the last mile company decides to work on a holiday, then on the attempt date, it is likely that the customer will be closed in case the delivery address is of Work or Office. It will be closed due to a public holiday, and even if the address is a Home, then it is also possible Customer will not be available at home because he is going out with family for picnics or visits, etc. So last mile companies needed to inform customers about delivery attempts affected due to public holidays. That is why we must take 'Public holiday' as the indicator.

'Road Blocks' Indicator. The 'Road Blocks' indicator will examine the details related to Road Block situations at the Time and Day when Delivery Rider will be attempting to deliver the Order. The Specifics for this indicator will initiate from different maps or news sites. Then we are going to be using that data based on the current date to calculate the 'Road Blocks' indicator value. The weightage of this indicator will be '6%'. The uppermost value against the 'Road Blocks' indicator can be taken as '6%,' and the bottommost value against it will be '0%'. The purpose behind adding 'Road Blocks' as an indicator for the calculation of 'Predicted Attempt Success Rate %' means that delivery riders' needed clear road routes to reach customer locations easily and effectively perform work on time. Suppose a specific road that reaches to customer's location is blocked. In that case, it means the delivery rider can't get to the customer, so either the delivery rider makes failed attempt and comes back, or the customer should come to the rider to receive the order, which is very unlikely. Thus, due to the discussed reason, we have to take 'Road Blocks' as the indicator.

'Emergency Situation' Indicator. The indicator of the 'Emergency Situation' will handle the info about any Emergency situation created at the time when the Delivery Rider will be making an attempt to Delivery the Order. The Specifics for this indicator will be attained from real-time news websites. Then we will use that information based on the current situation to calculate the 'Emergency Situation' indicator value. The weightage of this indicator will be '6%'. The highest value against the 'Emergency Situation' indicator can be taken as '6%,' and the lowest value against it will be '0%'. The purpose behind adding 'Emergency Situation' as an indicator for the calculation of 'Predicted Attempt Success Rate %' is that, Because the overall situation of the city is important for any type of business to be run, but if any instability in peace due strikes, protest or sit-ins, etc., then like any business last mile companies' operation will also be affected. The delivery rider has to come back from the route without a delivery order attempt. It is important because delivery riders are always on the roads and are more vulnerable. So, for their safety, it's the right thing to do. Just last mile companies should send SMS or alerts to customers about delivery attempts affected due to the Emergency situation. This is the reason; we have to take 'Emergency Situation' as the indicator.

'Customer Respond to Call/Message' Indicator. The indicator of the 'Customer Respond to Call/Message' will deal with the information interrelated with Customer response towards the Call or Message being done by the Delivery Rider or Customer Support department. The data for this indicator will be taken from the real-time response of the customer. After this, we will be able to use that data to calculate the 'Customer Respond to Call/Message' indicator value. The weightage of this indicator will be '6%'. The highest value against the 'Customer Respond to Call/Message' indicator can be attained at '6%,' and the least value against it will be '0%'. The motive behind adding 'Customer Respond to Call/Message' as an indicator for the calculation of 'Predicted Attempt Success Rate %' is that, through customer response, we will be able to understand better if a delivery attempt is going to be successful or failed based on customer response, so for removing the unclarity on delivery attempt success the timely and positive response from a customer will be playing a valuable role in the success of delivery order attempt. This is why we have added the 'Customer Respond to Call/Message' as the indicator.

'Valid Phone Number' Indicator. The 'Valid Phone Number' indicator will acquire details regarding the Customer's Phone Number against the Delivery Order. The Customer always gives the data for this indicator at the time of Delivery Order Booking, and it's mandatory to share so it can't be missed in any case until there is some technical issue. We will use that data soon as the Order booking is made to calculate the 'Valid Phone Number' indicator value. The weightage of this indicator will be '6%'. The maximum value against the 'Valid Phone Number' indicator can be got as '6%,' and the minimum value against it will be '0%'. The objective behind adding 'Valid Phone Number' as an indicator for the calculation of 'Predicted Attempt Success Rate %' is that when the delivery address where the Delivery Order is going to be attempted has any issue which causes unclarity in successfully reaching the location then the only way to clear confusion is to call the customer on his provided phone number and get clarity about delivery address and also get know about the availability of customer on the address at the time of delivery attempt so this indicator plays an essential role in the Success of Delivery attempt even with the clear delivery address. Without having it, the delivery rider always needs to take a chance in an attempt. Maybe it will be successful, and perhaps it can be failed depends on customer availability, and no way to confirm it proactively without the 'Valid Phone number' indicator. This is why we have added it as an indicator.

4.2 Working of 'Probability of Delivery Attempt Success'

So, through the above-mentioned indicators and an application utilizing their values can Predict the Success Rate, which will give a probability against each delivery order, and with the help of this, we will be able to judge better and use advanced learning techniques to create the most optimal routes and put only those Orders en route which have high chances of Success. The Prediction will be at the time of Order booking, which will give Logistics Operation time to do pre-planning against Orders, which will provide them with a chance to get ready and try their best to make the Order Successful on the First Attempt. The Logic will be simple If the Delivery Order details meet specific acceptable targets like More Than 90% etc. The Delivery Order will be fulfilled and dispatched for Delivery to the Customer. If it does not meet the Criteria, then the shipper or merchant can reject the order or process for additional verifications. Customer Service can Contact Customers via Call to get Clarity or Can-do SMS or Can-do Email or Auto IVR Calls, etc., in short, any way possible to reach the customer and ensure the success of Order Delivery (Fig. 3).

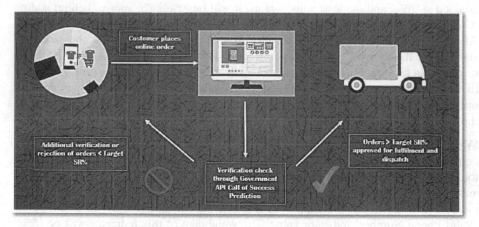

Fig. 3. Explains the working of the Prediction Model

With the Development of this solution, Government can make it accessible for Logistics Companies to use it Through Application Programming Interface (API) Calls; it will be easy for the Logistics Companies to Just Embed it into their current system and use it for the betterment of their operations, and in return, they can pay service charges to the Government.

4.3 Limitation of 'Probability of Delivery Attempt Success'

We must be clear that Delivery Failure will always remain an issue in Logistics operations. This solution will not wipe this problem out, and the reason is not technical. It relates to the Customers because, in genuine cases, failure is always possible, and we can't do much about it. But we can minimize the problem through these Digital solutions, which can wipe out unnecessary, fake, or deliberately done failure cases. So now, through this solution, we can not only minimize failure cases as much as possible but can also differentiate between fake or genuine cases and proactively deal with problems that will increase the efficiency of the Country's Logistics Sector.

4.4 Making Accessible 'Probability of Delivery Attempt Success'

In the Future, after the successful implementation of this initiative, then in the next step, more visibility of this solution will be given through Mobile-Based and Web-Based applications for the Customers and also to Merchants so they can able to view the predicted value based on their order details. They can be also able to see complete information on each indicator. Then this application will suggest to Customers and Merchants the ways through which they can make better their Prediction of success attempt percentage. By following those instructions, they proactively make efforts to make sure the successful delivery of their Order.

5 Conclusion

In summary, the broad implication of the research in this paper is that we have discussed the problem of the poor 1st attempt delivery success rate and how it challenges the Last mile delivery operation. It was essential to understand and get context about the problem before we could do an analysis on it and then move towards suggesting a possible solution. That deep understanding of the problem sharpened our analysis techniques. It made us able to get good knowledge about the recommended solution, also how this solution is suitable for solving the issue. Furthermore, the proposed solution provided the necessary foundation for building a sustainable working model and a first step towards minimizing the impact of the problem on the Last mile delivery operations performance. Regarding the solution, it is mentioned in detail that it has the potential to decrease the effect of the problem, but it can't take out the problem entirely, and the reason for that is also defined to give clarity on it. Finally, the solution has the capacity and direction to make an impactful reflection on the logistics industry. Still, its success is highly dependent on its implementation of it and the skill of the resources who are going to use and make it better through quality feedback. To be mentioned it utilizes the machine learning algorithm that means so much training data. Through many iterations, it has to pass first to be able to make accurate predictions. Also, the smooth transition from Government to the private sector will be a key factor for its acceptance. Therefore, further work and research are certainly required to minimize the complexities of it but right now, it provides a good starting point which is, based on the current situation, a novel way to face the challenge which they face daily.

Acknowledgement. The research work of this article was supported by Dr. Achmad Nurmandi from the University of Muhammadiyah Yogyakarta. The researcher would like to express his immense gratitude to his supporters who have provided all necessary insight and expertise for assistance in this research.

Funding. The authors of the paper reported that there is no funding associated with the work featured in the article.

References

(N.d.). Operations and supply chain management: an international journal. Https://www.journal. oscm-forum.org/journal/proceeding/download_paper/20191207215658_oscm_2019_paper_ 107.pdf

Abdul Rahman, M., Aamir Basheer, M., Khalid, Z., Tahir, M., Uppal, M.: Last mile logistics: impact of unstructured addresses on delivery times. In: w. C.c. s. S. (eds.), 7th international conference on smart data and smart cities, dsc 2022 (vol. 48, issues 4/w5–2022, pp. 3–8). International society for photogrammetry and remote sensing (2022). https://doi.org/10.5194/ isprs-archives-xlviii-4-w5-2022-3-2022

Alfarizi, M., Sari, R. K.: Analysis of factors affecting customer behavior of marketplace applications: a case study of cash on delivery (cod) payment systems. In: 13th International Conference on Advanced Computer Science and Information Systems, ICACSIS 2021 (2021). https://doi. org/10.1109/icacsis53237.2021.9631307

Alkis, A., kose, T.: Privacy concerns in consumer e-commerce activities and response to social media advertising: empirical evidence from Europe. Comput. Hum. Behav., 137 (2022). https:// doi.org/10.1016/j.chb.2022.107412

Almtiri, Z., Miah, S.J., Noman, N.: Application of e-commerce technologies in accelerating the success of some operation. In: Y. X.-S., S. S., D. N., J. A. (eds.), 7th International Congress on Information and Communication Technology, ICICT 2022 (vol. 448, pp. 463–470). Springer science and business media deutschland gmbh (2023). https://doi.org/10.1007/978-981-19- 1610-6_40

Anjum, S., Chai, J.: Drivers of cash-on-delivery method of payment in e-commerce shopping: evidence from Pakistan. Sage open, 10(3) (2020). https://doi.org/10.1177/2158244020917392

Armstrong, C.E.J., Gilmore, A.M., Boss, P.K., Pagay, V., Jeffery, D.W.: Machine learning for classifying and predicting grape maturity indices using absorbance and fluorescence spectra. Food chemistry, 403 (2023). https://doi.org/10.1016/j.foodchem.2022.134321

Ashrafpour, N., N, H., Aali, S., Taghizadeh, H.: The prerequisites and consequences of customers' online experience regarding the moderating role of brand congruity: evidence from an Iranian bank. J. Islamic Market. **13**(10), 2144–2172 (2022). https://doi.org/10.1108/jima-09-2020- 0277

Bighrissen, B.: A study of barriers to e-commerce adoption among cooperatives in morocco. In: A. B., H. A. (eds.), International Conference on Business and Technology, ICBT 2021 (vol. 485, pp. 557–570). Springer Science and Business Media Deutschland Gmbh (2023). https:// doi.org/10.1007/978-3-031-08093-7_37

Bopage, G., Nanayakkara, J., Vidanagamachchi, K.: A strategic model to improve the last mile delivery performance in ecommerce parcel delivery. In: 9th International Conference on Industrial Engineering and Operations Management, IEOM 2019, 2019(mar), 2018– 2019 (2019). Https://www.scopus.com/inward/record.uri?eid=2-s2.0-85067252593&partne rid=40&md5=bb0db685e3bc6ec27a89fb9bb97f853f

Chang, S., L, A., Wang, X., Wang, X.: Joint optimization of e-commerce supply chain financing strategy and channel contract. Eur. J. Oper. Res. **303**(2), 908–927 (2022). https://doi.org/10. 1016/j.ejor.2022.03.013

Cosmi, M., Nicosia, G., Pacifici, A.: Lower bounds for a meal pickup-and-delivery scheduling problem. In: H. J., K. S., M. B.R., U. V., U. M.S. (eds.), 17th Cologne-Twente Workshop on Graphs and Combinatorial Optimization, CTW 2019 (pp. 33–36). University of Twente (2019). Https://www.scopus.com/inward/record.uri?eid=2-s2.0-85084376442&partne rid=40&md5=c72ecd6e056b6120b1096351d181b069

Das, D., Kumar, R., Rajak, M.K.: Designing a reverse logistics network for an e-commerce firm a case study. Oper. Supply Chain Manage. **13**(1), 48–63 (2020). https://doi.org/10.31387/osc m0400252

Dasgupta, S., Kanchan, S., Kundu, T.: Creating a KPI tree for monitoring and controlling key business objectives of first mile logistics services. In: 9th International Conference on Industrial Engineering and Operations Management, IEOM 2019, 2019(Mar), 716–727 (2019). Https://www.scopus.com/inward/record.uri?eid=2-s2.0-85067240367&par tnerid=40&md5=6f22293a5e6e215fb702c50a7d0e22a4

Datta, S., Naruka, K.S., Sajidha, S.A., Nisha, V.M., Ragala, R.: Sanskriti—a distributed e-commerce site implementation using blockchain. In: Lecture Notes on Data Engineering and Communications Technologies (vol. 139, pp. 329–346). Springer Science and Business Media Deutschland Gmbh (2023). https://doi.org/10.1007/978-981-19-3015-7_24

Delivery failures hit nearly 75% of consumers. Multichannel merchant (2022). Https://multichan nelmerchant.com/blog/delivery-failures-hit-nearly-75-percent-of-consumers/

Escudero-santana, A., Muñuzuri, J., Lorenzo-espejo, A., Muñoz-díaz, M.-L.: Improving e-commerce distribution through last-mile logistics with multiple possibilities of deliveries based on time and location. J. Theor. Appl. Electron. Commer. Res. **17**(2), 507–521 (2022). https:// doi.org/10.3390/jtaer17020027

Etumnu, C. E.: A competitive marketplace or an unfair competitor? An analysis of Amazon and its best sellers ranks. J. Agric. Econ. **73**(3), 924–937 (2022). Https://doi.org/10.1111/1477-9552.12495

González-mora, C., Barros, C., Garrigós, I., Zubcoff, J., Lloret, E., Mazón, J.-N.: Improving open data web API documentation through interactivity and natural language generation. Computer Standards and Interfaces, 83 (2023). https://doi.org/10.1016/j.csi.2022.103657

Google data studio overview. (n.d.). Google data studio overview. Https://datastudio.google.com/ u/0/

Google sheets: sign-in. (n.d.). Https://docs.google.com/spreadsheets/u/0/

Ha, X. S., Le, T. H., Phan, T. T., Nguyen, H. H. D., Vo, H. K., Duong-Trung, N.: Scrutinizing trust and transparency in cash on delivery systems. In: W. G., C. B., L. W., D. Pietro R., Y. X., H. H. (eds.), 13th International Conference on Security, Privacy, and Anonymity in Computation, Communication, and Storage, SPACCS 2020: vol. 12382 LNCS (pp. 214–227). Springer Science and Business Media Deutschland Gmbh (2021). https://doi.org/10.1007/978-3-030-68851-6_15

Hasan, F., Mondal, S. K., Kabir, M. R., Al Mamun, M. A., Rahman, N. S., Hossen, M. S.: E-commerce merchant fraud detection using machine learning approach. In: 7th International Conference on Communication and Electronics Systems, ICCES 2022, 1123–1127 (2022). https://doi.org/10.1109/icces54183.2022.9835868

https://www.sciencedirect.com/science/article/pii/s1877042812028625

Jaller, M., Dennis, S.: E-commerce and mobility trends during covid-19. In: Springer Tracts on Transportation and Traffic (vol. 20, pp. 79–93). Springer Science and Business Media Deutschland Gmbh (2023). https://doi.org/10.1007/978-3-031-00148-2_6

Javatpoint: (n.d.). "Multiple linear regression". Https://www.javatpoint.com/multiple-linear-reg ression-in-machine-learning

Kahr, M.: Determining locations and layouts for parcel lockers to support supply chain viability at the last mile. Omega (United Kingdom), 113 (2022). https://doi.org/10.1016/j.omega.2022. 102721

Kolasińska-morawska, K., Sułkowski, Ł., Buła, P., Brzozowska, M., Morawski, P.: Smart logistics—sustainable technological innovations in customer service at the last-mile stage: the polish perspective. Energies 15(17) (2022). https://doi.org/10.3390/en15176395

Boysen, N., Fedtke, S., Schwerdfeger, S.: Last-mile delivery concepts: a survey from an operational research perspective. OR Spectrum 43(1), 1–58 (2020). https://doi.org/10.1007/s00291-020-00607-8

Li, Y., Shi, J., Cao, F., Cui, A.: Product reviews analysis of e-commerce platform based on logistic-arma model. In: 2021 IEEE International Conference on Power, Intelligent Computing and Systems, ICPICS 2021, pp. 714–717 (2021). https://doi.org/10.1109/icpics52425.2021.952 4238

Lu, S.-H., Kuo, R. J., Ho, Y.-T., Nguyen, A.-T.: Improving the efficiency of last-mile delivery with the flexible drones traveling salesman problem. Expert Syst. Appl. 209 (2022). https://doi.org/10.1016/j.eswa.2022.118351

Luo, X., Sun, Q., Yang, T., He, K., Tang, X.: Nondestructive determination of common indicators of beef for freshness assessment using airflow-three-dimensional (3D) machine vision technique and machine learning. J. Food Eng. 340 (2023). https://doi.org/10.1016/j.jfoodeng.2022. 111305

Luo, Z., Gu, R., Poon, M., Liu, Z., Lim, A.: A last-mile drone-assisted one-to-one pickup and delivery problem with multi-visit drone trips. Comput. Oper. Res 148 (2022). https://doi.org/10.1016/j.cor.2022.106015

Mckinlay, A., Mitchell, G., Bertenshaw, C.: Review article: dined (delivery-related injuries in the emergency department) part 1: a scoping review of risk factors and injuries affecting food delivery riders. EMA - Emergency Medicine Australasia 34(2), 150–156 (2022). https://doi.org/10.1111/1742-6723.13927

Muniasamy, A., Bhatnagar, R.: Analyzing online reviews of customers using machine learning techniques. In: Rathore, V.S., Sharma, S.C., Joao Manuel, R.S., Tavares, C.M., Surendiran, B. (eds.) Rising Threats in Expert Applications and Solutions: Proceedings of FICR-TEAS 2022, pp. 485–493. Springer Nature Singapore, Singapore (2022). https://doi.org/10.1007/978-981-19-1122-4_51

Muñoz-Villamizar, A., Velázquez-Martínez, J.C., Haro, P., Ferrer, A., Mariño, R.: The environmental impact of fast shipping ecommerce in inbound logistics operations: a case study in Mexico. J. Cleaner Prod. 283, 125400 (2021). https://doi.org/10.1016/j.jclepro.2020.125400

Nagpal, G., Bishnoi, G.K., Dhami, H.S., Vijayvargia, A.: Use data analytics to increase the efficiency of last mile logistics for e-commerce deliveries. In hand-book of research on engineering, business, and healthcare applications of data science and analytics (pp. 167–180). IGI global (2020). https://doi.org/10.4018/978-1-7998-3053-5.ch009

Oviedo-trespalacios, O., Rubie, E., Haworth, N.: Risky business: comparing the riding behaviours of food delivery and private bicycle riders. Accident Anal. Prev. 177 (2022). https://doi.org/10.1016/j.aap.2022.106820

Peng, Y., Yi, J.: Research on the application of big data technology in the process of cross-border e-commerce product selection. In: W. T., P. S., H. J. W.C., R. V. M.L. (eds.), 4th International Conference on Decision Science and Management, ICDSM 2022 (vol. 260, pp. 29–37). Springer Science and Business Media Deutschland Gmbh (2023). https://doi.org/10.1007/978-981-19-2768-3_3

Pereira Marcilio Nogueira, G., José de Assis Rangel, J., Rossi Croce, P., Almeida Peixoto, T.: The environmental impact of fast delivery B2C e-commerce in outbound logistics operations: a simulation approach. Cleaner Logistics Supply Chain, 5 (2022). https://doi.org/10.1016/j.clscn.2022.100070

Pérez-morón, J.M.: E-commerce in china and latin America: a review and future research agenda. In: A. B. H. A. (eds.), International Conference on Business and Technology, ICBT 2021 (vol. 485, pp. 571–587). Springer Science and Business Media Deutschland Gmbh (2023). https://doi.org/10.1007/978-3-031-08093-7_38

Publow, P.F.: Consider third-party logistics to address your company's challenges. Canadian apparel, 31(3), 8–9+37 (2007). Https://www.scopus.com/inward/record.uri?eid=2-s2.0-343 47389742&partnerid=40&md5=e67e6298fb406f713299d15c40f90af1

Purwandari, B., Suriazdin, S.A., Hidayanto, A.N., Setiawan, S., Phusavat, K., Maulida, M.: Factors affecting switching intention from cash on delivery to e-payment services in c2c e-commerce transactions: covid-19, transaction, and technology perspectives. Emerg. Sci. J., 6(special issue), 136–150 (2022). https://doi.org/10.28991/esj-2022-sper-010

Raj, N.V, Saini, J.R.: Loyalty score generation for customers using sentimental analysis of reviews in e-commerce. In d. P., c. S., b. A., d. S., S.C. (eds.), 3rd International Conference on Emerging Technologies in Data Mining and Information Security, IEMIS 2022 (vol. 490, pp. 461–473). Springer Science and Business Media Deutschland Gmbh (2023). https://doi.org/10.1007/978-981-19-4052-1_46

Rider (yc wc22) on linkedin: #rider #withrider #ecommerce #logistics #ecommercelogistics #pakistan.... (2022, june 7). Linkedin. Https://www.linkedin.com/posts/withrider_rider-wit hrider-ecommerce-activity-6939876673869746176-e9ji?utm_source=share&utm_medium= member_desktop

Sandoval, M.G., Álvarez-miranda, E., Pereira, J., Ríos-mercado, R. Z., Díaz, J.A.: A novel districting design approach for on-time last-mile delivery: an application on an express postal company. Omega (United Kingdom), 113 (2022). https://doi.org/10.1016/j.omega.2022.102687

Stefko, R., Bacik, R., Fedorko, R., Olearova, M.: Gender-generation characteristic in relation to the customer behavior and purchasing process in terms of mobile marketing. Oeconomia copernicana 13(1), 181–223 (2022). https://doi.org/10.24136/oc.2022.006

Straight, B.: Failure is not a winning strategy in the last mile. Freightwaves (2022). Https://www.freightwaves.com/news/solving-for-last-mile-delivery-failures

Tpl logistics launches "rider" | carspiritpk. Carspiritpk- pakistan's most trusted automotive blog (2019). Https://carspiritpk.com/tpl-logistics-launches-rider/

TPL logistics launches Pakistan's first live order tracking – customsnews.pk daily. Customsnews.pk daily – Pakistan customs, shipping & business news resource (2019). Https://customsnews.pk/2019/12/03/tpl-logistics-launches-pakistans-first-live-order-tracking/

Tran, N.A.T., et al.: Health and safety risks faced by delivery riders during the covid-19 pandemic. J. Transp. Health, 25 (2022). https://doi.org/10.1016/j.jth.2022.101343

Verma, R., Dhanda, N., Nagar, V.: Towards a secured IoT communication: a blockchain implementation through APIs. In: S. P.K., W. S.T., T.S., R. J.J.P.C., R. J.J.P.C., G. M. (eds.), 3rd International Conference on Computing, Communications, and Cyber-Security, IC4S 2021 (vol. 421, pp. 681–692). Springer Science and Business Media Deutschland Gmbh (2023). https://doi.org/10.1007/978-981-19-1142-2_53

Wang, M., zhang, C., Bell, M.G.H., Miao, L.: A branch-and-price algorithm for location-routing problems with pick-up stations in the last-mile distribution system. Eur. J. Oper. Res. 303(3), 1258–1276 (2022). https://doi.org/10.1016/j.ejor.2022.03.058

Wu, G.: Research on the development path of logistics management innovation in e-commerce environment. In: 2020 6th International Conference on Environmental Science and Material Application, ESMA 2020, 714(4) (2021). https://doi.org/10.1088/1755-1315/714/4/042022

Xu, J., Yang, Z., Wang, Z., Li, J., Zhang, X.: Flexible sensing enabled packaging performance optimization system (fs-ppos) for lamb loss reduction control in e-commerce supply chain. Food Control, 145 (2023). https://doi.org/10.1016/j.foodcont.2022.109394

Yin, L., Zhong, R. R., Wang, J.: Ontology based package design in fresh e-commerce logistics. Expert systems with applications, 212 (2023). https://doi.org/10.1016/j.eswa.2022.118783

Yun, K.K., Yoon, S.W., Won, D.: Interpretable stock price forecasting model using genetic algorithm-machine learning regressions and best feature subset selection. Expert Syst. Appl. 213 (2023). https://doi.org/10.1016/j.eswa.2022.118803

Mobile Commerce and e-Commerce: User Experience and Business Perspectives

The Impact of Country-of-Origin Images on Online Customer Reviews: A Case Study of a Cross-Border E-Commerce Platform

Wen-Hsin Chen[1] and Yi-Cheng Ku[2]($\boxtimes$)

[1] Department of Business Administration, Fu Jen Catholic University,
New Taipei City 242062, Taiwan
410316013@mail.fju.edu.tw
[2] Department of Business Administration and Service Design Research Center,
Fu Jen Catholic University, New Taipei City 242062, Taiwan
ycku@mail.fju.edu.tw

Abstract. In response to the global trend and the development of internet technology, many companies have entered the field of cross-border e-commerce in recent years. However, in the process of developing cross-border e-commerce, the products sold by the companies will be influenced by the country-of-origin effect and online customer reviews. And previous research literature has seldom combined the country-of-origin effect and online customer reviews for exploration, so this study quotes the world's largest cross-border e-commerce platform, Amazon, as the example, to analyze the impact of the country-of-origin effect on online customer reviews in the cross-border e-commerce environment. The study uses secondary data for analysis, and the results show that the country-of-origin effect and customer rating have a significant impact on the popularity of products (number of reviews), while the country-of-origin effect has no significant impact on customer rating. The results of this study can provide reference for cross-border e-commerce administrators in managing cross-border platforms.

Keywords: Cross-Border E-Commerce · Country-of-Origin · Customer Reviews

1 Introduction

According to an investigative agency Marketer, global e-commerce sales are expected to reach 6.169 trillion US dollars by 2023, demonstrating that e-commerce has become a significant form of trade, with the development of cross-border e-commerce attracting the attention of businesses. Cross-border e-commerce refers to an international business activity in which the trade and retail industries of various countries complete transactions and payment behaviors through e-commerce platforms and deliver goods through cross-border logistics. Under the Free Trade Agreement (FAT), the impact of cross-border e-commerce on enterprises is as follows [17]:

- Financial innovation can promote the standardization of cross-border e-commerce export operations and enhance enterprise financing ability.

F. Fui-Hoon Nah and K. Siau (Eds.): HCII 2023, LNCS 14038, pp. 141–149, 2023.
https://doi.org/10.1007/978-3-031-35969-9_10

- Supervision innovation can promote the localization of consumption in cross-border e-commerce, forcing enterprises to upgrade domestic brands while stabilizing and expanding the domestic consumption market.
- The cross-border e-commerce industry has mixed business of imports and exports. To improve the utilization rate of funds, storage facilities, and human resources, some cross-border e-commerce enterprises will develop import and export business simultaneously.

Cross-border e-commerce can be conducted in any region or country as long as there is a network connection, without being limited by time or location. Thus, cross-border e-commerce is likely to be one of the major global development trends in the future. However, the main difference between cross-border e-commerce and domestic e-commerce is that consumers may have different stereotypes of different countries. According to a survey, when customers shop online and are unfamiliar with a product, more than 90% choose to view online customer reviews to understand it. Furthermore, there have been a few discussions in the literature about the source country effect and online consumer reviews. Given these factors, the purpose of this study is to use the cross-border e-commerce platform Amazon as an example to explore the impact of the country-of-origin effect on online customer reviews and provide recommendations based on the research results for businesses operating in cross-border e-commerce or intending to enter this market.

2 Literature Review

2.1 The Country-Of-Origin Effect

The country-of-origin effect refers to the phenomenon where consumers base their purchasing decisions on the image associated with the country of origin of products, even when the products possess the same attributes and functions. Consumers hold different perceptions towards different countries, which results in higher reviews for products from some countries and lower reviews for products from others, affecting their purchasing intention. The country-of-origin effect occurs when consumers make purchasing decisions on products with similar attributes and functions. Scholars classified a country of origin (coo) into two types [5] : a country of brand and a country of manufacture. In this study, a country of origin is defined as a "country of brand" with the following feature: *The country of brand is the location of the headquarters of the company that owns the brand. Although the product may not be manufactured in that country, it is still identified as a product from that country* [9]. For example, although the smartphone iPhone is not manufactured in the US, consumers still view the iPhone as a product from the US.

The country-of-origin effect may be generated by two constructs—the halo effect and the summary effect [6]. The halo effect refers to when consumers are unfamiliar with a product from a certain country and cannot judge its quality or obtain enough product information. They will use the country's image for reference when purchasing the product. On the other hand, the summary effect refers to when consumers are very familiar with a product from a certain country. They will recall their previous purchasing experiences or exposure to the product and summarize them into an image of the country, which will affect the consumer's attitude towards the brand. In other words, the image of the country of origin is transformed into a summary concept, reflected in the consumer's overall judgment of the product. In summary, the country-of-origin effect influences consumers' evaluation of product quality and ultimately affects their purchasing intention [2]. When consumers are unfamiliar with a product, they will use the image of the country of origin for reference. When they are familiar, they will rely on their own impression of the country of origin to form a general impression of products from that country, thus affecting their attitude towards the product.

2.2 Online Customer Reviews

Online customer reviews are a form of word-of-mouth communication. Online customer reviews including numerical ratings and textual reviews allow consumers to write or read product reviews through different online platforms. Online customer reviews can reduce product uncertainty by inferring the quality of a product and help consumers make their final purchasing decisions [8]. However, just like traditional shopping, online shopping is also a social activity and consumers' purchasing decisions can be influenced by interaction with others [11]. Online customer reviews can reduce consumers' perceived risk, but not all online reviews are accurate, and consumers may face additional risks or incomplete and distorted information provided by online sellers [12]. Additionally, customer reviews can also be manipulated intentionally. For example, businesses may post positive anonymous reviews or increase public visibility in possible scenarios [13].

From the perspective of businesses, online consumer reviews can provide information on consumer demands and services for improvement, thus enhancing consumer satisfaction. On the other hand, from the perspective of consumers, online consumer reviews provide information for customers to refer to when making consumption decisions. Therefore, businesses can utilize online customer reviews for marketing planning and building a positive image for their brand. There is a positive correlation between the number of online reviews and sales of popular products [15]. Popular products tend to receive more online reviews, and a large volume of reviews make online products appear more trustworthy. Additionally, the large volume of reviews received by popular products allows consumers to easily find online reviews of popular products. As the number of online reviews from consumers increases, the overall rating tends to converge towards true quality. Therefore, the reviews of popular products can more accurately reflect product quality and thus have more influence [18].

3   Research Method

3.1   Research Model and Hypotheses

This study proposes a research model based on a literature review to examine the impact of countries of origin on product popularity (number of reviews) and online review rating (overall evaluation), as well as the impact of online reviews on product popularity, as shown in Fig. 1

The number of online reviews represents the popularity of a product, and the number of reviews is related to the number of products purchased by consumers [15]. Thus, as

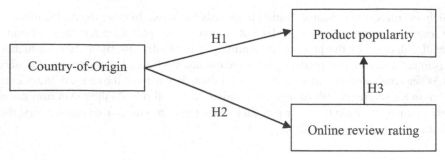

Fig. 1. Research model

the number of reviews increases, consumers' purchasing intention is also stronger. In addition, the results of the study indicate that country-of-origin image has a positive impact on purchasing intention [10]. Therefore, this study uses the number of reviews as a substitute indicator for consumer purchasing intention, and it is inferred that the country-of-origin image will affect the consumer purchasing intention. That is, product popularity (number of reviews) will be different when similar products are provided by different brands. In summary, this study proposes the hypothesis:

H1: The country of origin will affect product popularity.

The study has shown that the country-of-origin image will have a different impact on product evaluation, and products from countries with a better image will have higher evaluations than those from countries with a negative image [7]. Furthermore, consumers typically have higher evaluations of products from highly developed countries [1]. Some scholars' studies pointed out that most reviewers on Amazon give more positive evaluations [3]. In summary, this study proposes the hypothesis:

H2: The country of origin will affect online ratings.

The number of online reviews for a product represents the accumulated number of personal reviews for the product, indicating its popularity. The more reviews a product has, the more it is loved [15]. Some scholars have pointed out that online consumer reviews have two functions, one of which is to recommend and provide information to consumers on whether the product is popular on the market [16]. Therefore, based on the above, the study proposes that the higher the online review rating of a product, the greater its sales, leading to a higher number of reviews and higher product popularity. In summary, this study proposes the hypothesis:

H3: Online review ratings will affect product popularity.

3.2 Measurement

In this section, definitions and explanations of the variables in the research framework will be provided respectively. First, product popularity refers to the number of product reviews displayed on the website page. The more reviews a product has, the more consumers have used or purchased the product. Therefore, the number of reviews is considered to be related to the product's popularity [15]. Thus, in this study, a product with a higher level of popularity is considered to have a larger number of reviews. Second, the platform Amazon provides a rating scale of one to five stars for consumers, with one star being the worst and five stars being the best. The research will assign a score of 5 to a five-star rating, 1 to a one-star rating, and so on. The research will calculate the average rating for each product from different countries using a rounding method.

3.3 Empirical Data

This study utilizes the Amazon data set collected by scholar Julian McAuley, encompassing product reviews from May 1996 to October 2018. The data schema includes Asin, Type, Brand, Headquarters, Reviewer ID, Reviewer Name, Review text, review time, Unix review time, Summary, Overall (rating of the product), Verified, Num of review, and Average of overall. The data set was filtered through criteria such as file size, completeness of the file information, and whether it meets the requirements of the study. Finally, a product type, all beauty, was selected and went through data filtering. The data acquisition and analysis process are as follows: first, after preliminary data cleaning, the remaining review data was 71,231 out of 279,227 in the file. This data set included 9 countries with more than 1,000 review data, which are the United States (US), United Kingdom (UK), France, Indonesia, Germany, Japan, India, South Korea, and the Philippines.

To reduce the disparity in sample size among countries and to ensure the comparability of research samples, this study utilized pivot analysis to select the brands with a market share of 5% or higher within each country as the final sample size for that country. Finally, the average rating and the average number of reviews for each product were calculated for the review data from the 9 countries, resulting in 461 product items and 34,112 reviews. However, Indonesia and the Philippines had fewer than 10 product items, and therefore were not included in the analysis. The remaining 7 countries had 456 product items and 28,329 review data, and their product items (p) and the number of reviews (r) are as follows, US (p = 126, r = 10,191), UK (p = 179, r = 5,967), France (p = 37, r = 3,625), Germany (p = 45, r = 3,412), Japan (p = 15, r = 2,692), India (p = 16, r = 1,286), and South Korea (p = 38, r = 1,156).

4 Data Analysis

This study conducted a one-way ANOVA analysis and regression analysis to test H1-H3. The results showed that the country of origin has no significant impact on product popularity. Therefore, Hypothesis H1 was not supported.

The results show that the country of origin has a significant impact ($p < 0.01$) on online review rating, meaning that the image of the country of origin affects the online reviews. Thus, the results support hypothesis H2. In order to further understand the impact of different countries on the online review rating, this study also conducted a comparison using Scheffe multiple tests, and the results are shown in Table 1. It can be seen that France, Germany, and Japan have a significant impact ($p < 0.05$) on online review rating. Among them, France has the greatest impact on online reviews.

Table 1. Scheffe's Multiple Comparison of the Countries of Origin

Country (I) \ Country (J)	UK	France	Germany	Japan	India	Korea
US	0.1586 (0.1039)	-0.2946 (0.1671)	-0.2256 (0.1552)	0.7290 (0.2441)	0.0500 (0.2372)	-0.0226 (0.1654)
UK		-0.4532 (0.1614)	-0.3842 (0.1490)	0.5705 (0.2402)	-0.1085 (0.2332)	-0.1812 (0.1596)
France			0.0690 (0.1983)	1.0236* (0.2735)	0.3446 (0.2674)	0.2720 (0.2064)
Germany				0.9547* (0.2664)	0.2757 (0.2601)	0.2030 (0.1969)
Japan					-0.6790 (0.3212)	-0.7516 (0.2725)
India						-0.0726 (0.2663)

Note: Mean difference: I-J (Std. Error); * p<0.05

In order to further analyze the impact of brand images associated with the country of origin on online review ratings and the popularity of products, this study conducted advanced analysis. In terms of the impact of brand image on product popularity, the results showed that there were significant differences in the products from the US ($p < 0.001$) and United Kingdom ($p < 0.01$). In terms of the impact of brand image on online review rating, brand image from the United Kingdom ($p < 0.001$), France ($p < 0.01$), Germany ($p < 0.05$), and Korea ($p < 0.01$) has a significant effect on online review ratings. The brand image from other countries does not have a significant impact on online review ratings.

In order to test H3, this study performed regression analysis after taking the log of online review ratings and product popularity (number of reviews). The results showed that the effect of online review ratings on product popularity is significant ($p < 0.01$). This shows that online reviews have a significant impact on product popularity (number of reviews), meaning that the online review will affect the product's popularity (number of reviews). Therefore, this result supports hypothesis H3 (Table. 2).

Table 2. Regression analysis

	Unstandardized Coefficients		Standardized Coefficients	t	Sig.
	B	Std.Error	Beta		
(constant)	0.393	0.168		2.343	0.020
online review rating	0.816	0.273	0.139	2.991	**0.003**

Note: Dependent variable: product popularity

5 Conclusions

This study shows that the country-of-origin image does not have a significant impact on the popularity of the product, indicating that the level of the country-of-origin image does not affect consumers' comments on that country.

But this study also shows that the country-of-origin image have significantly impact online review ratings, indicating that the level of the country-of-origin image affects consumers' online reviews of that country. Therefore, consumers tend to give higher ratings to countries with a higher country of origin image, and lower ratings to countries with a lower country of origin image. Moreover, this study used multiple testing to determine that France, Germany, and Japan have a significant impact on online reviews, and among all the countries in this study, France has the greatest impact on consumers' online reviews.

Furthermore, this study conducted advanced analyses on the impact of brand image on product popularity and online review ratings. It was found that different brand images associated with different countries have different impacts on both product popularity and online reviews. Specifically, for the impact of brand image on product popularity, the US and United Kingdom have a significant impact, while other countries such as France, Germany, Japan, India, and Korea had no significant effect. On the other hand, for the impact of brand image on online reviews, the United Kingdom, France, Germany, and Korea have a significant impact, while other countries such as the US, Japan and India had no significant effect.

The study shows that online reviews have a significant impact on the popularity of a product, meaning that the quality of online reviews affects the popularity of a product. Higher-rated products have more reviews, indicating that more consumers have used the product. Conversely, lower-rated products have fewer reviews, indicating fewer consumer interactions with the product.

5.1 Implications

This research investigates the effect of a country of origin on product popularity and online review ratings. The results show that products from different countries do have different impacts on online review ratings. The results indicate that products from different countries do indeed have varying effects on online reviews. Consumers tend to rate products from countries with a higher perceived image more positively. For example, in this research, products from France have the greatest impact on online review among all the countries studied.

Therefore, it is suggested that Amazon's cross-border e-commerce operators can encourage and guide consumers to provide positive reviews of their products. Operators can add feedback mechanism for consumer reviews to encourage consumers to leave positive feedback after purchasing. For example, the shopping website Taobao gives coupons to customers after they leave positive reviews, and coupons can encourage customers to shop again, increasing product sales. In addition, previous studies about online reviews have all explored customers' final consumption intention through online reviews. This research attempts to explore the country-of-origin effect along with customer online reviews and hopes to help cross-border e-commerce companies understand the impact of the country-of-origin effect on online reviews on Amazon's website and fill the gap in the literature.

5.2 Limitations and Suggestions

This study uses secondary data for analysis. Due to limitations of the research resources, the scope of the study is fully dependent on the data source and is limited to only the online shopping platform Amazon. As a result, the research findings may not apply to other similar cross-border e-commerce websites and await further study verification.

Due to time constraints in the study, the research project is only focused on a single product category, but according to the research [14], the product category can affect consumers' reviews of the product. For example, consumers are less likely to give neutral reviews for experiential products. It is therefore suggested that future research compare and explore more product categories to make research more diverse.

Research variables in this study are represented by quantitative numbers, and the content of consumer reviews is not considered. However, the content may help to more accurately measure the quality of reviews and consumers' opinions of the evaluated item [4]. Hence, a semantic analysis will be adopted for qualitative research in the future.

Acknowledgment. This work was supported by the National Science and Technology Council (NSTC) under the grants NSTC 111–2622-H-030–001.

References

1. Bilkey, W.J., Nes, E.: Country-of-origin effects on product evaluations. J. Int. Bus. Stud. **13**(1), 89–100 (1982)
2. Bhamjee, Z.: Global Branding and Country-Of-Origin: Creativity and Passion. Routledge (2019)
3. Chevalier, J.A., Mayzlin, D.: The effect of word of mouth on sales: online book reviews. J. Mark. Res. **43**(3), 345–354 (2006)
4. Forman, C., Ghose, A., Wiesenfeld, B.: Examining the relationship between reviews and sales: the role of reviewer identity disclosure in electronic markets. Inf. Syst. Res. **19**(3), 291–313 (2008)
5. Han, C.M., Terpstra, V.: Country-of-origin effects for uni-national and bi-national products. J. Int. Bus. Stud. **19**(2), 235–255 (1988)
6. Han, C.M.: Country image: halo or summary construct? J. Mark. Res. **26**(2), 222–229 (1989)
7. Hamzaoui Essoussi, L., Merunka, D.: Consumers' product evaluations in emerging markets: does country of design, country of manufacture, or brand image matter? Int. Mark. Rev. **24**(4), 409–426 (2007)
8. Hu, N., Liu, L., Zhang, J.J.: Do online reviews affect product sales? The role of reviewer characteristics and temporal effects. Inf. Technol. Manage. **9**(3), 201–214 (2008)
9. Johansson, J.K., Douglas, S.P., Nonaka, I.: Assessing the impact of country-of-origin on product evaluations: a new methodological perspective. J. Mark. Res. **22**(4), 388–396 (1985)
10. Lin, L.Y., Chen, C.S.: The influence of the country-of-origin image, product knowledge and product involvement on consumer purchase decisions: an empirical study of insurance and catering services in Taiwan. J. Consum. Mark. **23**(5), 248–265 (2006)

11. Lu, B., Fan, W., Zhou, M.: Social presence, trust, and social commerce purchase intention: an empirical research. Comput. Hum. Behav. **56**, 225–237 (2016)
12. Liu, Z., Park, S.: What makes a useful online review? Implication for travel product websites. Tour. Manage. **47**, 140–151 (2015)
13. Mayzlin, D.: Promotional chat on the Internet. Mark. Sci. **25**(2), 155–163 (2006)
14. Mudambi, S. M., Schuff, D.: Research note: What makes a helpful online review? A study of customer reviews on Amazon. com. MIS quarterly, 185–200 (2010)
15. Park, D.H., Lee, J., Han, I.: The effect of on-line consumer reviews on consumer purchasing intention: the moderating role of involvement. Int. J. Electron. Commer. **11**(4), 125–148 (2007)
16. Tsao, W.C., Hsieh, M.T., Shih, L.W., Lin, T.M.: Compliance with eWOM: the influence of hotel reviews on booking intention from the perspective of consumer conformity. Int. J. Hosp. Manag. **46**, 99–111 (2015)
17. Xue, W., Li, D., Pei, Y.: The Development and Current of Cross-border E-commerce (2016)
18. Zhu, F., Zhang, X.: Impact of online consumer reviews on sales: the moderating role of product and consumer characteristics. J. Mark. **74**(2), 133–148 (2010)

Key Successful Factors of E-commerce Platform Operations

Yu-Jing Chiu, Hsi-His Chen, and Chin-Yi Chen[✉]

Department of Business Administration, Chung Yuan Christian University,
Chung Li District, Taoyuan, Taiwan
{yujing,iris}@cycu.edu.tw

Abstract. With the advent of the Internet era, virtual channels have gradually replaced the original traditional physical channels, and then developed the online shopping platform that is currently the most relied on by consumers, also known as e-commerce platforms. However, among the thousands of e-commerce platforms, there are still e-commerce platforms that stand tall and are thriving without being affected by the Internet bubble, and successfully occupy the market by accumulating the trust of users for a long time. Therefore, this study aims to explore why most e-commerce operators choose the top three e-commerce platforms of Shopee, momo Shopping and PChome24h as their choice for opening. This study determines the preliminary levels and factors through literature collection, and then uses the Delphi method expert questionnaire for questionnaire analysis, gathers the opinions of digital e-commerce operators and summarizes the key factors of platform use. The key criteria are "overall service quality of the platform", "high customer convenience", "complete product service". How to make consumers have a sense of trust in the e-commerce platform, and then enhance the willingness to buy, and whether the website designs the browsing route from the customer's perspective so that consumers do not need to spend too much time to complete the transaction, and sets customer service staff to solve the problems encountered online for consumers at any time, providing complete online order information, which is the focus of *attention of each platform. Our research results can provide reference for the* operation and management of other e-commerce platforms.

Keywords: Electronic Commerce · Platform · Decision Making Trial and Evaluation Laboratory (DEMATEL) · Analytic Network Process (ANP) · DANP

1 Introduction

In today's era of extremely advanced information technology and rapid changes, people's thinking mode has gradually shifted to speed, convenience and time-saving, which has led to the continuous growth of the population of Internet users, and the Internet has formed a new market channel (Kiang et al., 2000). As a result, consumers using mobile networks are becoming more and more bulky, and this new consumption model has largely replaced the traditional consumption habits of our in-store selection. Due to the

© Springer Nature Switzerland AG 2023
F. Fui-Hoon Nah and K. Siau (Eds.): HCII 2023, LNCS 14038, pp. 150–169, 2023.
https://doi.org/10.1007/978-3-031-35969-9_11

popularity of the Internet and the widespread use of smartphones, people are gradually shifting from traditional modes of shopping such as physical stores to the Internet.

Segev et al. (1995) define the above consumer behaviour on the Internet as e-commerce, which is a form of transaction that can engage in the purchase, sale, service and payment mode of products through public or private networks, it is called "electronic" means that it takes place in a virtual environment, that is, the Internet we use, but it is more convenient and faster than the traditional way of consumption.

According to the Taiwan Network Information Center (TWNIC) (2018), the overall network usage status in Taiwan (Fig. 1) shows that Taiwan's Internet access rate ranks third among Asian countries in international development, after Japan and South Korea. From the above information, it can be seen that mobile phone usage in Taiwan is extremely high, which has facilitated the vigorous development of e-commerce platforms in recent years.

With the gradual popularity of e-commerce platforms, consumers' feedback and willingness to repurchase have also become issues that everyone pays attention to, through the Taiwan Netizen Online Shopping Consumption Survey of the Information Policy Council, it can be clearly observed from the data in Fig. 2 that the three e-commerce platforms with high usage in Taiwan are Shopee shopping, momo shopping network and PChome24h.

Why do these three stand out from most similar e-commerce platforms, what are the keys to their high degree of consumer dependence, and what are their key success factors or marketing strategies?

This research aims to explore and analyse the key success factors of e-commerce platforms, and to provide reference for the operation and management of other e-commerce platforms, which is the motivation of this research.

In light of the above research background and motivation, we hope to discuss the following issues through this research: 1. To discuss the factors that influence the operation of e-commerce platforms. 2. Identify the key elements of business success from the perspective of e-commerce operators.

The research process of this study is divided into five stages. The first stage is to establish the background, motivation and purpose of the research. The second stage consists of literature review and discussion to construct the prototype structure of the research. The third stage is to use the convergence criteria of the Delphi method to form a formal research framework through experts. The fourth stage is to design the questionnaire according to the formal research structure and test the experts, and use the Decision Making Trial and Evaluation Laboratory -based Analytic Network Process (DANP) to analyze the questionnaire data and quantify the weight of the criteria. The fifth step is to summaries and formulate research conclusions and recommendations based on the analysis results.

After the literature discussion, the prototype structure was established and the five evaluation aspects and 30 evaluation criteria were summarized, and the prototype structure was used to interview the expert group of this study. The final research framework was determined by the results of the expert interviews, including 4 evaluation aspects and 25 criteria.

As e-commerce platforms become more popular, consumer feedback and repurchase intentions are on everyone's radar. According to 2018 data, Taiwan's current top three e-commerce platforms include Shopee, momo, and PChome, which have high usage rates. Therefore, this study will discuss the key factors for the success of Shopee, momo and PChome. Through the literature review, we can see that most studies are conducted from the consumer's perspective. Therefore, this study will summarise consumers' views on e-commerce platforms from secondary data. In addition, we will conduct interviews with experts. The experts are mainly operators of various e-commerce platforms and academics in industry and university related research.

2 E-commerce and Factors Influencing the Operation of e-commerce Platforms

The original meaning of the term e-commerce comes from Electronic Commerce (EC). Segev et al. (1995) pointed out that e-commerce is a type of transaction that can be used to buy, sell, service and pay for products over public or private networks. In a broad sense, e-commerce is a modern business model that integrates business activities such as buying and selling, products and services through computer networks, so that it can meet the needs of organisations, products and consumers, and then achieve cost savings. Decreasing thoughts increase their willingness to buy (Kalakota & Whinston, 1997).

Based on the definitions of the above scholars, e-commerce as defined in this study is a new type of transaction based on the Internet industry and information and communication technology, which can provide consumers with more convenience and savings through the Internet. Kalakota & Whinston (1996) proposed that the types of e-commerce platforms can be roughly divided into four categories according to the different transaction objects: B2B (business to business), B2C (business to consumer), C2C (consumer to consumer), C2B (consumer to business), etc.

(1) B2B (business-to-business): Mainly business owners negotiate with suppliers, place orders, sign contracts, etc. As a large number of product transactions form a value chain, this transaction model is derived. Laudon and Traver (2002) also point out that government is also a purchaser of goods and services and is therefore included in the B2B category.
(2) B2C (Business-to-Consumer): Refers to the transaction between the business and the consumer, changing the traditional consumption pattern into a shopping transaction without a physical store on the Internet, i.e. an online store that provides consumers with a complete pre-sales and after-sales service.
(3) C2C (Consumer-to-Consumer): Refers to all transactions between consumers. The website operator is not responsible for logistics and transport services, but assists in collecting data in the market and establishing a good credit system.
(4) C2B (Consumer-to-Business): Consumers form a purchasing group to conduct transactions with businesses. Consumers use the community to strengthen their bargaining power and transfer product leadership and first-mover rights to themselves.

Therefore, C2B sites of this type usually play the role of community initiators or operators.

With the development of the Internet, online auctions are gradually being widely used by consumers, thus expanding a new type of e-commerce model that integrates the traditional e-commerce model and connects businesses and self-employed consumers, which is B2B2C (Business to Business to Consumer), so that self-employed users can contact other businesses more quickly, and it also makes it easier for consumers to search for transaction targets.

According to Torkzadeh & Dhillon (2002), DeLone and McLean (2003), Ming (2005), Jian (2005), Huang (2012), Yang et al. (2018) and compiled a total of 30 criteria in 5 dimensions, as a prototype of the research on the key factors affecting the success of e-commerce platforms.

3   Methodology

3.1   Establishment of Research Framework

Based on the prototype framework established after the literature review, five evaluation dimensions and 30 evaluation criteria were summarized, and the expert group of this research was interviewed with this prototype framework. Four assessment dimensions and twenty-five criteria were obtained through expert interviews. Then, according to the Delphi method, the final confirmed formal research framework includes 4 dimensions and 16 criteria, as shown in Table 1 and Table 2.

Table 1. Formal Research Dimensions and Definitions

Dimensions	Definitions
Website design(A)	In the online shopping market, websites provide functions such as business flow, cash flow, logistics, and information flow for users or consumers. The website platform must consider whether it can meet the needs of consumers and increase consumers' willingness to purchase, whether the information provided is correct and complete, easy to read, and easy to operate, as much as possible to shorten the time wasted by consumers in purchasing products, and update information in real time. Provide transaction-related information after placing an order to assist in the completion of the transaction
Customer response(B)	The website provides interactive functions to support question feedback and community interaction, and the website content can provide website interface design and information that may be required based on the consumer's past visit behavior or other user information of similar attributes, and after the customer places an order Provide good after-sales service and convenient return and exchange service

(continued)

Table 1. (*continued*)

Dimensions	Definitions
Product value(C)	The internal and external characteristics of the product itself, such as product risk, consumers' willingness to choose products, and brand image, will all affect consumers' willingness to purchase. The platform should screen out products from good suppliers before they can be put on the shelves for sale
Logistics Handling (D)	Pre-sale and after-sale service process of goods, goods logistics and necessary assistance provided, etc

Table 2. Formal Research Criteria and Definitions

Criteria	Definitions
Perfect real-time information mechanism (A1)	Provide a website navigation function to assist consumers to complete browsing actions with the smallest path, such as a site map. With an easy-to-navigate and use website interface, the ability to attract consumers to browse and place orders and update the latest product information and special promotions in real time
Valid link (A2)	The link selected by the consumer is valid, and there will be no link errors or blank pages
Easy access (A3)	Means that the process of linking to the site is simple and easy
Reactivity (A4)	The response speed of the website is mainly related to the maximum load of the website platform bandwidth and the utilization efficiency of the host
High customer interaction (B1)	Pay attention to consumers' opinions and difficulties encountered at any time, and the website can provide them with the assistance and services they need in the first time
Stable customer relationship (B2)	Keep in touch with consumers at any time, and continue to maintain follow-up transaction relationships with consumers based on trust
Complete product service (B3)	Provide consumers with good quality and diversified products at reasonable prices, and provide complete, usable and easy-to-read product information, and do our best to meet consumers' needs throughout the entire transaction process from purchase to follow-up services, so that customers increase the willingness to buy

(*continued*)

Table 2. (*continued*)

Criteria	Definitions
Customized service (B4)	According to the personal needs of consumers, we provide detailed introductions of seasonal or currently most popular limited products at any time, and create exclusive intimate messages for consumers. And provide customized information services according to the personal consumption habits of different users during the consumption transaction process
High customer convenience (B5)	Does the website design browsing routes and information classifications from the perspective of customers, increase the speed of consumers' quick search, so that consumers can complete transactions without spending too much energy and steps, and set up customer service personnel to solve problems for consumers online at any time? Arrived questions, and provide complete online order information
Company Information (B6)	Provide detailed information about the company to which the business website belongs, such as company profile, brand description, etc
Trading Rules (B7)	Whether the website provides account privacy functions or protects consumers' identities and personal information to avoid the risk of personal data leakage and card theft, and provides diversified payment modes
Pricing (C1)	Consideration factors such as product price positioning and price differentiation from similar products
Overall service quality of the platform (C2)	The e-commerce platform establishes a good brand image, so that consumers have a sense of trust in it, and then increase the willingness to buy
Delivery service (D1)	Refers to the provision of logistics services, such as home delivery, fixed-point pick-up, etc
Customer Support (D2)	Refers to the mechanisms that assist consumers in completing transactions during the shopping process, such as shopping carts, checkout budgets, after-sales service or return mechanisms, etc
Shipping and return service (D3)	Online shopping products are easily or may be damaged by external forces during the delivery process. Consumers receive defective products and want to return them. The manufacturer's processing procedures will affect consumers' repurchase and have a significant impact on website loyalty

3.2 DANP

In the study by Tzeng and Huang (2011), it was pointed out that the DANP method can be exempted from Saaty's (1996) ANP (Analytic Network Process; ANP) theory of paired comparison questionnaires when the respondent is a relevant member of the company, and the total impact matrix of DEMATEL can be directly applied as an unweighted supermatrix for ANP (Chiu, Tzeng, and Li, 2013). It can be seen from the literature that DANP is widely used in various fields, such as the analysis of major investment decisions (Lee, Huang, Chang, and Cheng, 2011) and product development decisions (Chen Yiming, Xiang Jiajun, and Yang Haobo, 2010). It is confirmed that DANP is helpful for researchers to analyse the complex structural relationship between different dimensions and criteria, and it can also help this study to explore the key success factors of e-commerce platforms. Therefore, this study uses DANP as a research tool to find "key factors".

How do enterprises respond and manage these risk factors? The concept framework of this research was created through a literature review and expert interviews. The Delphi method was subsequently employed to achieve consensus and identify a study framework comprising 4 dimensions and 16 criteria. Finally, the Decision-Making Trial and Evaluation Laboratory-based Analytic Network Process method (DEMATEL-based ANP, DANP) was used weight and rank the dimensions and criteria. The feasibility and applicability of the created system was verified through a risk assessment case involving actual suppliers.

These two methods have been combined for solving multiple criteria decision-making problems in different fields. To determine the level of interdependence and significance of the indicators regarding the accomplishment of our objective, DEMA-TEL and ANP were combined in this study. The operation model of DANP involved

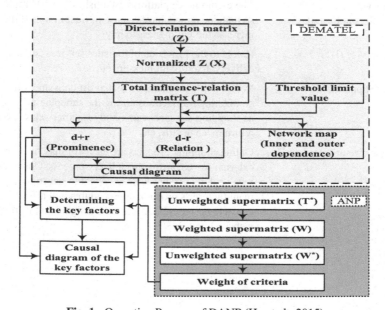

Fig. 1. Operation Process of DANP (Hu et al., 2015)

operation steps, the selection of key factors, and the design of a cause-and-effect di ram (Fig. 1).

4 Results and Analysis

4.1 Importance Analysis of Criterion

According to the assessment results of the impact of the questionnaire on the criteria, a direct relationship matrix (Z) is made, as shown in Table 3; the direct relationship matrix (Z) is normalized to obtain a normalized relationship matrix (X), as shown in Table 4; through the formula $T = X (1-X)^{-1}$ calculation (I represents the unit matrix) to obtain the total influence relationship matrix (T), as shown in Table 5; compile the total influence relationship degree table, as shown in Table 6.

From the calculation results and the table of the relationship between the total influence (as shown in Table 6), it can be seen that the top five items in the order of importance are "the overall service quality of the platform" (importance $=$ 8.4559), followed by "high customer convenience" (importance $=$ 8.2254), followed by "complete product service" (importance $=$ 8.1840), "perfect real-time information mechanism" (importance $=$ 7.9378) and "high customer interaction" (importance = 7.8213). Indicates that "the overall service quality of the platform" has the strongest overall influence relationship among the overall criteria, and is the most important influencing criterion for the key factors affecting the success of the e-commerce platform, and so on. On the contrary, "company information" (importance $=$ 4.0915) has the least degree of total influence.

Through the total impact relationship degree table (as shown in Table 6), it is known that 7 of the 16 criteria have a positive "cause degree", in order from large to small It is "customized service" (reason degree $=$ 0.4487), followed by "perfect real-time information mechanism" (reason degree $=$ 0.4194), "complete product service" (reason degree $=$ 0.4031), "company information" (reason degree $=$ 0.2419), "delivery service" (reason degree $=$ 0.1320), "pricing" (reason degree $=$ 0.0866), and finally "transaction rules" (cause degree $=$ 0.0574), the causality is biased towards "initial cause Class", which is an active influencing factor. The cause degree of "customized service" is 0.4487, which means that the direct impact on other criteria is greater than the influence of other criteria on this criterion. The analysis results of the criteria show that when it comes to the key factors affecting the success of e-commerce platforms, the top five items in terms of importance are "overall service quality of the platform", "high customer convenience", "complete product services", and "perfect real-time information mechanism" and "high degree of customer interaction" are the most critical and should be prioritized for management.

1. Build Direct Relationship Matrix (Z)

A total of 67 valid questionnaires were recovered in this study. After entering the data of each questionnaire into the computer, the values in the same column of each questionnaire were summed up and the average value was calculated to obtain the direct relationship matrix (Z), as shown in Table 3.

Table 3. Direct relationship matrix (Z)

	A1	A2	A3	A4	B1	B2	B3	B4	B5	B6	B7	C1	C2	D1	D2	D3
A1	0	1.8666	1.7333	1.3333	1.7333	1.3333	1.8	1.3333	1.9333	1.3333	1.4666	1.3333	1.8666	0.9333	1.6666	1.3333
A2	1.5333	0	1.8666	1.8666	1.2666	0.8	0.8666	0.2666	1.6	0.7333	1	0.5333	1.5333	0.5333	1	0.8666
A3	1.4666	1.7333	0	1.5333	1.5333	1.2666	0.9333	0.6	1.8666	0.7333	0.8	0.5333	1.5333	0.3333	0.8666	0.8
A4	1.3333	1.4	1.4666	0	1.4	0.9333	1.0666	0.7333	1.8666	0.5333	0.4666	0.2666	1.5333	0.8	1.1333	0.9333
B1	1.5333	1.2	1.1333	1.6	0	1.8	1.6666	1.4	1.7333	0.6666	0.8	0.6666	1.9333	1	1.6666	1.3333
B2	1.4	0.9333	0.9333	1	1.7333	0	1.7333	1.6	1.6666	0.7333	0.8666	1.0666	1.9333	1.0666	1.5333	1.5333
B3	1.6666	1.2666	1.1333	1.4666	1.8666	1.8	0	1.5333	1.8666	1.2	1.2	1.0666	2	1.7333	1.8	1.8666
B4	1.1333	0.7333	0.6666	0.8	1.7333	1.6666	1.7333	0	1.7333	0.5333	0.6666	1	1.6	1.1333	1.4	1.0666
B5	1.7333	1.4666	1.2666	1.5333	1.8	1.6666	1.4666	1.2666	0	0.6666	1.3333	0.8666	1.6666	1.6	1.7333	1.7333
B6	1.4	0.9333	0.8666	0.6666	1.0666	0.8666	0.7333	0.5333	0.6666	0	0.6	0.7333	0.8666	0.4666	0.6	0.5333
B7	1.1333	0.8666	0.8666	0.3	0.7333	1.1333	1.1333	0.5333	1.0666	0.6666	0	1.1333	1.2666	0.8666	0.8666	1.2666
C1	1.0666	0.5333	0.4	0.4	1	1.1333	1.1333	1.2	0.6666	0.4666	1.1333	0	1.0666	0.8	0.6	0.8
C2	1.8	1.8	1.8	1.6666	1.6666	1.6666	1.8666	1.2	1.5333	0.6	1.0666	0.8	0	1.5333	1.6666	1.3333
D1	1.1333	0.6	0.6	1	1.3333	1.4666	1.7333	0.9333	1.2666	0.4	0.6	0.8666	1.6	0	1.4666	1.7333
D2	1.3333	0.8666	0.9333	0.8	1.6	1.6	1.6666	1.2666	1.6666	0.4666	0.6666	0.7333	1.8	1.3333	0	1.6666
D3	0.8	0.9333	0.9333	0.8	1.4666	1.6666	1.4666	0.8	1.6666	0.4666	0.8	0.5333	1.7333	1.8666	1.5333	0

2. Normalized direct relationship matrix (X)

Select the maximum value from the sum of each column and the sum of each row of the direct relationship matrix (Z), and substitute it into the operation formula (2) to obtain the normalized direct relationship matrix (X), as shown in Table 4.

3. Establish the total influence relationship matrix (T)

Referring to the operation formula (4), the total influence relationship matrix (T) can be obtained after operation, as shown in Table 5.

4. Calculate importance and cause

The d value can be obtained by summing each column of the total influence relationship matrix (T); the r value can be obtained by summing each row. And calculate d + r and d−r, the results are shown in Table 6.

4.2 DANP Weight Value and Ranking

In this section, the DANP method is used to calculate the weight of the criteria and make a ranking. According to the operation steps of DANP, according to the total influence relationship matrix (T) of DEMATEL, it is directly used as the unweighted supermatrix (T^*) of ANP, and the unweighted supermatrix (T^*) is normalized to obtain the weighted supermatrix of ANP. Matrix (W), after multiplying the weighted supermatrix (W) by itself several times, the dependency relationship will converge to a fixed value, and the limit supermatrix (W^*) will be obtained. Read the content of the limit super matrix (W^*) to know the global weight (Global Weight), that is, "DANP weight". Evaluation of key factors for business platform success (Table 7).

The total influence matrix (T) of DEMATEL is directly used as the unweighted super matrix of ANP, and the unweighted super matrix (T) is normalized so that the elements of each row are combined into 1 to generate the weighted super matrix of ANP (W), as shown in Table 8.

In this study, after multiplying the weighted supermatrix (W) by itself three times, the overall weight of the evaluation criteria has converged, and the limit supermatrix (W^*) is obtained, as shown in Table 9 below.

It is summarized into the DANP criterion weight and sequence table, as shown in Table 10.

Table 4. Normalized direct relationship matrix (X)

	A1	A2	A3	A4	B1	B2	B3	B4	B5	B6	B7	C1	C2	D1	D2	D3
A1	0.0000	0.0780	0.0724	0.0557	0.0724	0.0557	0.0752	0.0557	0.0808	0.0557	0.0613	0.0557	0.0780	0.0390	0.0696	0.0557
A2	0.0641	0.0000	0.0780	0.0780	0.0529	0.0334	0.0362	0.0111	0.0669	0.0306	0.0418	0.0223	0.0641	0.0223	0.0418	0.0362
A3	0.0613	0.0724	0.0000	0.0641	0.0641	0.0529	0.0390	0.0251	0.0780	0.0306	0.0334	0.0223	0.0641	0.0139	0.0362	0.0334
A4	0.0557	0.0585	0.0613	0.0000	0.0585	0.0390	0.0446	0.0306	0.0780	0.0223	0.0195	0.0111	0.0641	0.0334	0.0474	0.0390
B1	0.0641	0.0501	0.0474	0.0669	0.0000	0.0752	0.0696	0.0585	0.0724	0.0279	0.0334	0.0279	0.0808	0.0418	0.0696	0.0557
B2	0.0585	0.0390	0.0390	0.0418	0.0724	0.0000	0.0724	0.0669	0.0696	0.0306	0.0362	0.0446	0.0808	0.0446	0.0641	0.0641
B3	0.0696	0.0529	0.0474	0.0613	0.0780	0.0752	0.0000	0.0641	0.0780	0.0501	0.0501	0.0446	0.0836	0.0724	0.0752	0.0780
B4	0.0474	0.0306	0.0279	0.0334	0.0724	0.0696	0.0724	0.0000	0.0724	0.0223	0.0279	0.0418	0.0669	0.0474	0.0585	0.0446
B5	0.0724	0.0613	0.0529	0.0641	0.0752	0.0696	0.0613	0.0529	0.0000	0.0279	0.0557	0.0362	0.0696	0.0669	0.0724	0.0724
B6	0.0585	0.0390	0.0362	0.0279	0.0446	0.0362	0.0306	0.0223	0.0279	0.0000	0.0251	0.0306	0.0362	0.0195	0.0251	0.0223
B7	0.0474	0.0362	0.0362	0.0125	0.0306	0.0474	0.0474	0.0223	0.0446	0.0279	0.0000	0.0474	0.0529	0.0362	0.0362	0.0529
C1	0.0446	0.0223	0.0167	0.0167	0.0418	0.0474	0.0474	0.0501	0.0279	0.0195	0.0474	0.0000	0.0446	0.0334	0.0251	0.0334
C2	0.0752	0.0752	0.0752	0.0696	0.0696	0.0696	0.0780	0.0501	0.0641	0.0251	0.0446	0.0334	0.0000	0.0641	0.0696	0.0557
D1	0.0474	0.0251	0.0251	0.0418	0.0557	0.0613	0.0724	0.0390	0.0529	0.0167	0.0251	0.0362	0.0669	0.0000	0.0613	0.0724
D2	0.0557	0.0362	0.0390	0.0334	0.0669	0.0669	0.0696	0.0529	0.0696	0.0195	0.0279	0.0306	0.0752	0.0557	0.0000	0.0696
D3	0.0334	0.0390	0.0390	0.0334	0.0613	0.0696	0.0613	0.0334	0.0696	0.0195	0.0334	0.0223	0.0724	0.0780	0.0641	0.0000

Table 5. Total influence Relationship Matrix (T)

	A1	A2	A3	A4	B1	B2	B3	B4	B5	B6	B7	C1	C2	D1	D2	D3	d
A1	0.2304	0.2709	0.2603	0.2496	0.3152	0.2878	0.3071	0.2289	0.3327	0.1705	0.2127	0.1897	0.3409	0.2236	0.2905	0.2678	4.1786
A2	0.2282	0.1468	0.2153	0.2178	0.2301	0.2023	0.2065	0.1391	0.2512	0.1156	0.1525	0.1205	0.2552	0.1561	0.2037	0.1906	3.0314
A3	0.2302	0.2176	0.1459	0.2093	0.2451	0.2244	0.2139	0.1558	0.2658	0.1179	0.1481	0.1234	0.2605	0.1525	0.2036	0.1924	3.1065
A4	0.2202	0.2007	0.1991	0.1451	0.2357	0.2082	0.2149	0.1577	0.2611	0.1074	0.1318	0.1103	0.2555	0.1675	0.2099	0.1940	3.0192
B1	0.2683	0.2271	0.2195	0.2411	0.2254	0.2837	0.2814	0.2163	0.3023	0.1336	0.1720	0.1513	0.3191	0.2102	0.2712	0.2489	3.7712
B2	0.2584	0.2120	0.2068	0.2137	0.2877	0.2098	0.2797	0.2209	0.2937	0.1336	0.1718	0.1643	0.3135	0.2098	0.2618	0.2522	3.6894
B3	0.3010	0.2527	0.2419	0.2590	0.3277	0.3130	0.2459	0.2428	0.3378	0.1679	0.2060	0.1837	0.3542	0.2609	0.3036	0.2954	4.2936
B4	0.2293	0.1873	0.1800	0.1897	0.2671	0.2554	0.2600	0.1441	0.2743	0.1160	0.1512	0.1505	0.2788	0.1967	0.2382	0.2173	3.3359
B5	0.2880	0.2474	0.2348	0.2489	0.3085	0.2919	0.2874	0.2204	0.2488	0.1400	0.2008	0.1667	0.3244	0.2428	0.2859	0.2760	4.0127
B6	0.1748	0.1409	0.1353	0.1295	0.1708	0.1562	0.1528	0.1136	0.1610	0.0618	0.1052	0.1012	0.1741	0.1148	0.1414	0.1331	2.1667
B7	0.1869	0.1568	0.1531	0.1339	0.1830	0.1910	0.1926	0.1319	0.2013	0.1001	0.0965	0.1310	0.2162	0.1512	0.1748	0.1841	2.5845
C1	0.1707	0.1316	0.1232	0.1254	0.1786	0.1776	0.1795	0.1479	0.1713	0.0854	0.1322	0.0783	0.1932	0.1377	0.1519	0.1539	2.3383
C2	0.2939	0.2633	0.2579	0.2577	0.3069	0.2941	0.3042	0.2197	0.3129	0.1395	0.1928	0.1656	0.2626	0.2414	0.2858	0.2633	4.0617
D1	0.2197	0.1747	0.1702	0.1892	0.2425	0.2389	0.2509	0.1744	0.2473	0.1061	0.1424	0.1396	0.2687	0.1452	0.2321	0.2343	3.1763
D2	0.2446	0.2000	0.1975	0.1972	0.2711	0.2614	0.2661	0.1999	0.2816	0.1176	0.1567	0.1449	0.2961	0.2113	0.1913	0.2476	3.4851
D3	0.2145	0.1930	0.1886	0.1883	0.2545	0.2530	0.2477	0.1738	0.2695	0.1117	0.1542	0.1307	0.2813	0.2228	0.2411	0.1732	3.2977
r	3.7592	3.2228	3.1293	3.1954	4.0501	3.8485	3.8905	2.8873	4.2127	1.9248	2.5270	2.2517	4.3943	3.0443	3.6870	3.5242	

Table 6. Total influence Relationship Scale

	d	r	d + r	d − r	Rank
A1	4.1786	3.7592	7.9378	0.4194	4
A2	3.0314	3.2228	6.2542	−0.1913	9
A3	3.1065	3.1293	6.2358	−0.0228	10
A4	3.0192	3.1954	6.2146	−0.1762	13
B1	3.7712	4.0501	7.8213	−0.2789	5
B2	3.6894	3.8485	7.5379	−0.1591	6
B3	4.2936	3.8905	8.1840	0.4031	3
B4	3.3359	2.8873	6.2232	0.4487	11
B5	4.0127	4.2127	8.2254	−0.1999	2
B6	2.1667	1.9248	4.0915	0.2419	16
B7	2.5845	2.5270	5.1115	0.0574	14
C1	2.3383	2.2517	4.5900	0.0866	15
C2	4.0617	4.3943	8.4559	−0.3326	1
D1	3.1763	3.0443	6.2205	0.1320	12
D2	3.4851	3.6870	7.1721	−0.2018	7
D3	3.2977	3.5242	6.8219	−0.2264	8

Table 7. Unweighted supermatrix (T^*)

	A1	A2	A3	A4	B1	B2	B3	B4	B5	B6	B7	C1	C2	D1	D2	D3
A1	0.2304	0.2282	0.2302	0.2202	0.2683	0.2584	0.3010	0.2293	0.2880	0.1748	0.1869	0.1707	0.2939	0.2197	0.2446	0.2145
A2	0.2709	0.1468	0.2176	0.2007	0.2271	0.2120	0.2527	0.1873	0.2474	0.1409	0.1568	0.1346	0.2633	0.1747	0.2000	0.1930
A3	0.2603	0.2153	0.1459	0.1991	0.2195	0.2068	0.2419	0.1800	0.2348	0.1353	0.1531	0.1222	0.2579	0.1702	0.1975	0.1886
A4	0.2496	0.2178	0.2093	0.1451	0.2411	0.2137	0.2590	0.1897	0.2489	0.1295	0.1339	0.1254	0.2577	0.1892	0.1972	0.1883
B1	0.3152	0.2301	0.2451	0.2357	0.2254	0.2877	0.3277	0.2671	0.3085	0.1708	0.1830	0.1786	0.3069	0.2425	0.2711	0.2545
B2	0.2878	0.2023	0.2244	0.2082	0.2837	0.2098	0.3130	0.2554	0.2919	0.1562	0.1910	0.1776	0.2941	0.2389	0.2614	0.2530
B3	0.3071	0.2065	0.2139	0.2149	0.2814	0.2797	0.2459	0.2600	0.2874	0.1528	0.1926	0.1755	0.3042	0.2509	0.2661	0.2477
B4	0.2289	0.1391	0.1558	0.1577	0.2163	0.2209	0.2428	0.1441	0.2204	0.1136	0.1319	0.1479	0.2197	0.1744	0.1999	0.1738
B5	0.3327	0.2512	0.2658	0.2611	0.3023	0.2937	0.3378	0.2743	0.2488	0.1610	0.2013	0.1713	0.3129	0.2473	0.2816	0.2695
B6	0.1705	0.1156	0.1179	0.1074	0.1336	0.1336	0.1679	0.1160	0.1400	0.0618	0.1001	0.0854	0.1395	0.1061	0.1176	0.1117
B7	0.2127	0.1525	0.1481	0.1318	0.1720	0.1718	0.2060	0.1512	0.2008	0.1052	0.0965	0.1322	0.1928	0.1424	0.1567	0.1542
C1	0.1897	0.1205	0.1234	0.1103	0.1513	0.1643	0.1837	0.1505	0.1667	0.1012	0.1310	0.0783	0.1656	0.1396	0.1449	0.1307
C2	0.3409	0.2552	0.2605	0.2555	0.3191	0.3135	0.3542	0.2788	0.3244	0.1741	0.2162	0.1932	0.2626	0.2687	0.2961	0.2813
D1	0.2236	0.1561	0.1525	0.1675	0.2102	0.2098	0.2609	0.1967	0.2428	0.1148	0.1512	0.1377	0.2414	0.1452	0.2113	0.2228
D2	0.2905	0.2037	0.2036	0.2099	0.2712	0.2618	0.3036	0.2382	0.2859	0.1414	0.1748	0.1589	0.2858	0.2321	0.1913	0.2411
D3	0.2678	0.1906	0.1924	0.1940	0.2489	0.2522	0.2954	0.2173	0.2760	0.1331	0.1841	0.1589	0.2633	0.2343	0.2476	0.1732

Table 8. Weighted supermatrix (W)

	A1	A2	A3	A4	B1	B2	B3	B4	B5	B6	B7	C1	C2	D1	D2	D3
A1	0.0551	0.0753	0.0741	0.0729	0.0712	0.0700	0.0701	0.0687	0.0718	0.0807	0.0723	0.0730	0.0724	0.0692	0.0702	0.0651
A2	0.0648	0.0484	0.0701	0.0665	0.0602	0.0575	0.0588	0.0562	0.0617	0.0650	0.0607	0.0563	0.0648	0.0550	0.0574	0.0585
A3	0.0623	0.0710	0.0470	0.0660	0.0582	0.0560	0.0563	0.0540	0.0585	0.0625	0.0592	0.0527	0.0635	0.0536	0.0567	0.0572
A4	0.0597	0.0718	0.0674	0.0481	0.0639	0.0579	0.0603	0.0569	0.0620	0.0598	0.0518	0.0536	0.0634	0.0596	0.0566	0.0571
B1	0.0754	0.0759	0.0789	0.0781	0.0598	0.0780	0.0763	0.0801	0.0769	0.0788	0.0708	0.0764	0.0756	0.0764	0.0778	0.0772
B2	0.0689	0.0667	0.0722	0.0690	0.0752	0.0569	0.0729	0.0766	0.0727	0.0721	0.0739	0.0760	0.0724	0.0752	0.0750	0.0767
B3	0.0735	0.0681	0.0689	0.0712	0.0746	0.0758	0.0573	0.0779	0.0716	0.0705	0.0745	0.0768	0.0749	0.0790	0.0764	0.0751
B4	0.0548	0.0459	0.0502	0.0522	0.0574	0.0599	0.0565	0.0432	0.0549	0.0524	0.0511	0.0633	0.0541	0.0549	0.0574	0.0527
B5	0.0796	0.0829	0.0856	0.0865	0.0802	0.0796	0.0787	0.0822	0.0620	0.0743	0.0779	0.0732	0.0770	0.0779	0.0808	0.0817
B6	0.0408	0.0381	0.0380	0.0356	0.0354	0.0362	0.0391	0.0348	0.0349	0.0285	0.0387	0.0365	0.0343	0.0334	0.0337	0.0339
B7	0.0509	0.0503	0.0477	0.0437	0.0456	0.0466	0.0480	0.0453	0.0500	0.0486	0.0373	0.0565	0.0475	0.0448	0.0450	0.0467
C1	0.0454	0.0397	0.0397	0.0365	0.0401	0.0445	0.0428	0.0451	0.0416	0.0467	0.0507	0.0335	0.0408	0.0440	0.0416	0.0396
C2	0.0816	0.0842	0.0839	0.0846	0.0846	0.0850	0.0825	0.0836	0.0808	0.0804	0.0837	0.0826	0.0647	0.0846	0.0850	0.0853
D1	0.0535	0.0515	0.0491	0.0555	0.0557	0.0569	0.0608	0.0590	0.0605	0.0530	0.0585	0.0589	0.0594	0.0457	0.0606	0.0675
D2	0.0695	0.0672	0.0656	0.0695	0.0719	0.0709	0.0707	0.0714	0.0713	0.0653	0.0676	0.0649	0.0704	0.0731	0.0549	0.0731
D3	0.0641	0.0629	0.0619	0.0643	0.0660	0.0684	0.0688	0.0651	0.0688	0.0614	0.0713	0.0658	0.0648	0.0738	0.0710	0.0525

Table 9. Limit super matrix (W*)

	A1	A2	A3	A4	B1	B2	B3	B4	B5	B6	B7	C1	C2	D1	D2	D3
A1	0.0703	0.0703	0.0703	0.0703	0.0703	0.0703	0.0703	0.0703	0.0703	0.0703	0.0703	0.0703	0.0703	0.0703	0.0703	0.0703
A2	0.0602	0.0602	0.0602	0.0602	0.0602	0.0602	0.0602	0.0602	0.0602	0.0602	0.0602	0.0602	0.0602	0.0602	0.0602	0.0602
A3	0.0585	0.0585	0.0585	0.0585	0.0585	0.0585	0.0585	0.0585	0.0585	0.0585	0.0585	0.0585	0.0585	0.0585	0.0585	0.0585
A4	0.0597	0.0597	0.0597	0.0597	0.0597	0.0597	0.0597	0.0597	0.0597	0.0597	0.0597	0.0597	0.0597	0.0597	0.0597	0.0597
B1	0.0755	0.0755	0.0755	0.0755	0.0755	0.0755	0.0755	0.0755	0.0755	0.0755	0.0755	0.0755	0.0755	0.0755	0.0755	0.0755
B2	0.0718	0.0718	0.0718	0.0718	0.0718	0.0718	0.0718	0.0718	0.0718	0.0718	0.0718	0.0713	0.0718	0.0718	0.0718	0.0718
B3	0.0727	0.0727	0.0727	0.0727	0.0727	0.0727	0.0727	0.0727	0.0727	0.0727	0.0727	0.0727	0.0727	0.0727	0.0727	0.0727
B4	0.0540	0.0540	0.0540	0.0540	0.0540	0.0540	0.0540	0.0540	0.0540	0.0540	0.0540	0.0540	0.0540	0.0540	0.0540	0.0540
B5	0.0787	0.0787	0.0787	0.0787	0.0787	0.0787	0.0787	0.0787	0.0787	0.0787	0.0787	0.0787	0.0787	0.0787	0.0787	0.0787
B6	0.0359	0.0359	0.0359	0.0359	0.0359	0.0359	0.0359	0.0359	0.0359	0.0359	0.0359	0.0359	0.0359	0.0359	0.0359	0.0359
B7	0.0472	0.0472	0.0472	0.0472	0.0472	0.0472	0.0472	0.0472	0.0472	0.0472	0.0472	0.0472	0.0472	0.0472	0.0472	0.0472
C1	0.0419	0.0419	0.0419	0.0419	0.0419	0.0419	0.0419	0.0419	0.0419	0.0419	0.0419	0.0419	0.0419	0.0419	0.0419	0.0419
C2	0.0820	0.0820	0.0820	0.0820	0.0820	0.0820	0.0820	0.0820	0.0820	0.0820	0.0820	0.0820	0.0820	0.0820	0.0820	0.0820
D1	0.0569	0.0569	0.0569	0.0569	0.0569	0.0569	0.0569	0.0569	0.0569	0.0569	0.0569	0.0569	0.0569	0.0569	0.0569	0.0569
D2	0.0688	0.0688	0.0688	0.0688	0.0688	0.0688	0.0688	0.0688	0.0688	0.0688	0.0688	0.0688	0.0688	0.0688	0.0688	0.0688
D3	0.0658	0.0658	0.0658	0.0658	0.0658	0.0658	0.0658	0.0658	0.0658	0.0658	0.0658	0.0658	0.0658	0.0658	0.0658	0.0658

Table 10. DANP Criteria Weight and Order

Criteria	Weight	Rank
A1	0.0703	6
A2	0.0602	9
A3	0.0585	11
A4	0.0597	10
B1	0.0755	3
B2	0.0718	5
B3	0.0727	4
B4	0.0540	13
B5	0.0787	2
B6	0.0359	16
B7	0.0472	14
C1	0.0419	15
C2	0.0820	1
D1	0.0569	12
D2	0.0688	7
D3	0.0658	8

4.3 The Key Factors

This section is mainly to sort the Ranking weight and DEMATEL importance of the DANP calculation results into a comprehensive DANP importance table, as shown in Table 11.

From the DANP importance summary table shown in Table 36, it can be known that the standard platform's overall service quality (C2), high customer convenience (B5), complete product service (B3), high customer interaction (B1) and perfect The real-time information mechanism (A1) is more important. Then review the total influence relationship matrix (T) of the criteria shown in Table 5 and the total influence relationship level table shown in Table 9, and create a causal relationship network map (NRM) for key criteria, as shown in Fig. 2.

Table 11. DANP importance summary

Dimensions /Criteria	DANP Global Weight	DANP Ranking (I)	DEMATEL Importance Ranking (II)	BORDA SCORE Value of (I) + (II)	Overall Sorting
Perfect real-time information mechanism (A1)	0.0703	6	4	10	5
Valid link (A2)	0.0602	9	9	18	9
Easy access (A3)	0.0585	11	10	21	10
Reactivity (A4)	0.0597	10	13	23	11
High customer interaction (B1)	0.0755	3	5	8	4
Stable customer relationship(B2)	0.0718	5	6	11	6
Complete product service (B3)	0.0727	4	3	7	3
Customized service (B4)	0.0540	13	11	24	12
High customer convenience (B5)	0.0787	2	2	4	2
Company Information (B6)	0.0359	16	16	32	16
Trading Rules (B7)	0.0472	14	14	28	14
Pricing (C1)	0.0419	15	15	30	15
Overall service quality of the platform (C2)	0.0820	1	1	2	1
Delivery service (D1)	0.0569	12	12	24	12
Customer Support (D2)	0.0688	7	7	14	7
Shipping and return service (D3)	0.0658	8	8	16	8

From the causality network diagram shown in Fig. 2, it can be seen that the perfect real-time information mechanism (A1) and the complete product service (B3) influence each other, and the complete product service (B3) directly influences the criteria of high customer relationship (B1), stable customer relationship (B2), high customer convenience (B5), overall platform service quality (C2), customer support (D2), and delivery and return service (D3).

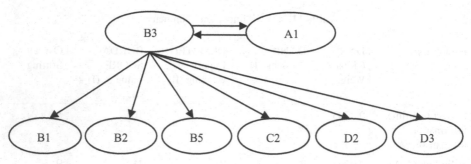

Fig. 2. Key Criteria Causality Network Diagram

5   Conclusions

The research results show that the key dimensions are 'customer response', 'website design' and 'product value'. How to make consumers trust the e-commerce platform, thereby increasing their willingness to buy, as well as ensuring that the content is complete and easy to read, minimizing the time consumers waste in purchasing products, and providing good after-sales service after customer's place orders, so as to achieve the company's brand image. The key criteria are "overall service quality of the platform", "high level of customer convenience", and "complete product services". How to make consumers trust the e-commerce platform, thereby increasing their willingness to buy, and whether the website is designed from the customer's perspective The browsing route allows consumers to complete transactions without spending too much time, and setting up customer service personnel to solve problems encountered by consumers online at any time, and providing complete online order information is the focus of each platform.

From the causal network diagram of the key criteria in Fig. 2 and the DANP importance summary table in Table 11, we see that the key criteria are complete product performance(B3) and perfect real-time information mechanism(A1), both of which are considered the most important key criteria. The two suggestions are provided as follows.

1. Establish a smooth shopping process

This means that e-commerce platforms should consider whether their websites have been designed from a customer perspective, so that consumers can complete transactions in a short time, reducing time costs for consumers and enabling consumers to experience a smooth sales service and complete online order information on their websites.

2. Enhance and focus on customer-oriented services

This means that e-commerce platforms should think about how to make consumers feel trustworthy in e-commerce platforms. This means that e-commerce platforms should think about how to make consumers feel trustworthy on e-commerce platforms. When e-commerce platforms can instantly solve customers' consumption needs and instantly respond to customers' problems, it can also be an effective way for consumers to shop online. For this reason, e-commerce platforms should continue improving their "client-focused" service processes and education and training, explore consumer needs through various channels, and incorporate consumer feedback into the system.

References

1. Chiu, W.Y., Tzeng, G.H., Li, H.L.: A new hybrid MCDM model combining DANP with VIKOR to improve e-store business. Knowl.-Based Syst. **37**, 48–61 (2013)
2. Chiu, Y.J., Chiu, S.H.: Key factors of supply chain risk for the procurement of automobile manufacturer from China. NTU Manage. Rev. **2015**(28), 61–96 (2015)
3. DeLone, W.H., McLean, E.R.: The DeLone and McLean model of information systems success: a ten-year update. J. Manag. Inf. Syst. **19**(4), 9–30 (2003)
4. Yi-Chung, H., Chiu, Y.-J., Hsu, C.-S., Chang, Y.-Y.: Identifying key factors for introducing GPS-based fleet management systems to the logistics industry. Math. Prob. Eng. **2015**, 1–14 (2015). https://doi.org/10.1155/2015/413203
5. Huang, S.C.: Key success factor on women's wear online market from the perspective of consumers. Master Thesis of the Department of Business Administration, National Cheng Kung University (2012)
6. Isakowitz, T., Bieber, M., Vitali, F.: Web information systems. Commun. ACM **41**(7), 78–80 (1998)
7. Jian, Y.S: Constructing the evaluation model for business-to-consumers electronic commerce – A consumer's perception perspective. Master Thesis of the Information Management Department, Fu Jen Catholic University (2005)
8. Kalakota, R., Whinston, A. B.: Frontiers of electronic commerce. Addison Wesley Longman Publishing Co., Inc (1996)
9. Kiang, M.Y., Raghu, T.S., Shang, K.H.M.: Marketing on the Internet—who can benefit from an online marketing approach? Decis. Support Syst. **27**(4), 383–393 (2000)
10. Konczal, E.F.: Models are for managers, not mathematicians. J. Syst. Manag. **26**(165), 12–15 (1975)
11. Lee, W.S., Huang, A.Y., Chang, Y.Y., Cheng, C.M.: Analysis of decision making factors for equity investment by DEMATEL and Analytic Network Process. Expert Syst. Appl. **38**(7), 8375–8383 (2011)
12. Mahadevan, B.: Business models for Internet-based e-commerce: an anatomy. Calif. Manage. Rev. **42**(4), 55–69 (2000)
13. Ming, P.C.: Key success factors of virtual channel industries marketing strategies. The Case Study of PC Home. Master Thesis of the Master Class of the Institute of Communication Management, Ming Chuan University (2005)
14. Rayport, J.F., Jaworski, B.J.: Cases in e-Commerce. McGraw-Hill Higher Education (2001)
15. Saaty, T.L.: Decision Making with Dependence and Feedback: The Analytic Network Process, vol. 4922. RWS Publications, Pittsburgh (1996)
16. Segev, A., Wan, D., Beam, C.: Electronic catalogs: a technology overview and survey results. In: Proceedings of the Fourth International Conference on Information and Knowledge Management, pp. 11–18 (1995)
17. Torkzadeh, G., Dhillon, G.: Measuring factors that influence the success of Internet commerce. Inf. Syst. Res. **13**(2), 187–204 (2002)
18. Tzeng, G.H., Huang, J.J.: Multiple attribute decision making: methods and applications. CRC press (2011)
19. Vladimir, Z.: Electronic commerce: structures and issues. Int. J. Electron. Commer. **1**(1), 3–23 (1996)
20. Yang, C.C., Lu, C.H., Chang, C.H.: Identifying key success factors for international logistics center operators entry into the cross-border e-commerce business. Maritime Q. **27**(2), 73–97 (2018)

On the Role of User Interface Elements in the Hotel Booking Intention: Analyzing a Gap in State-of-The-Art Research

Stefan Eibl[1](✉) and Andreas Auinger[2]

[1] University of Applied Sciences Wiener Neustadt, Zeiselgraben 4, 3250 Wieselburg, Austria
stefan.eibl@fhwn.ac.at
[2] University of Applied Sciences Upper Austria, Wehrgrabengasse 1, 4400 Steyr, Austria
andreas.auinger@fh-steyr.at

Abstract. Hotel customers nowadays predominantly conduct their bookings via digital channels of hotels and online booking platforms such as booking.com. To gain a better understanding of influencing and decisive factors for hotel booking intention, a systematic literature review as well as a semi-structured content analysis was conducted. The review reveals that user-generated reviews and price of hotel rooms have been investigated most dominantly in the last years. In comparison to that, other factors such as hotel pictures, room descriptions, scarcity cues, brand image and other User Interface (UI) elements were not in the central focus of the research. It is found that in current research several factors that influence online hotel booking intention were investigated separately and there is a considerable lack of interconnected research into these factors. The authors identify research gaps regarding the different UI categories and propose research questions and a research agenda for future research. It can be analyzed that there are blind spots in current literature which can be covered by conducting further interconnected research in the field of online hotel booking intention or online hotel choice.

Keywords: Influencing factors hotel booking intention · User interface elements · Hotel choice

1 Introduction

With the world becoming more digital, it is natural that an increasing number of customers are conducting hotel booking transactions via digital channels [1]. Online hotel booking websites have become increasingly popular over the years, as they are convenient, cost-effective, and offer a wide range of features and services. Thus, a majority of all travel bookings worldwide with a total volume of over $660bn were conducted via a booking platform or direct traffic of the hotels [3]. However, due to the abundance of options available, it can be difficult for customers to decide which hotel is the best fit for their needs in terms of price, service, and convenience [4, 5]. As such, research into online hotel booking intention is essential to understand the factors that influence consumers' decisions in hotel booking. An understanding of the motivations and preferences of customers allows companies to optimize their services and products accordingly [6, 7].

F. Fui-Hoon Nah and K. Siau (Eds.): HCII 2023, LNCS 14038, pp. 170–189, 2023.
https://doi.org/10.1007/978-3-031-35969-9_12

Furthermore, it can help hotels make informed decisions regarding their pricing and marketing strategies. When thinking about booking a hotel, there are many different User Interface (UI) elements and criteria on online hotel booking websites that influence the booking decision of a user [8, 9]. Examples for such elements are scarcity indicators like "only one room left", the photo quality of the hotel facilities, or customer reviews. All of these elements can positively or negatively affect the purchase decision during the booking process [10–12]. Hence, this research paper will explore how UI elements affect the intention to book a hotel, and how the overall user experience impacts the decision-making process. It will also provide an overview of existing research on the subject, as well as an evaluation of how UI elements interact with each other in recent research.

We briefly describe our review methodology in Sect. 2 which is followed by an outline of the outcomes in online hotel booking intention research in Sect. 3. In Sect. 4 we conclude the major outcomes based on the identified research and furthermore we propose research questions and deliver a research agenda for future research activities in this topic.

2   Review Methodology

2.1   Systematic Literature Review

In order to analyze the research conducted in the field of online hotel booking intention, a systematic literature review was conducted in first place. The literature review process began by researching the relevant existing peer-reviewed literature from the scholarly databases of Scopus and Web of Science. Our keywordset built the following search-string: (hotel* OR resort*s) AND (online OR internet OR web) AND (determinants OR factors OR influencing factors) AND (booking OR consideration OR reservation OR choice)). After collecting 158 articles from the last five years (2017–2022) to gain insights in most current research by using the respective search-string, each article was evaluated and analyzed for its relevance to the study. By removing duplicates and excluding research articles that were not relevant to our specific research, we explored a total of 30 research articles. Lastly, we conducted a forward and backward snowballing-search and so, we resulted in a set of 39 research papers which were subsequently analyzed in detail [13]. In a first process step we identified the influencing factors in online hotel booking which were then, in a second step, analyzed in more detail to find out to which extent these factors influence the consideration or booking intention. Key findings from the literature were identified, compared, and synthesized to build a comprehensive understanding of the research conducted in the field. By examining the research articles, we analyzed a set of UI Elements on online hotel booking websites that have an influence on the booking intention of users. To identify influencing factors in online hotel booking intention and respectively to form a set of UI elements, the literature was analyzed in a structured way. Therefore, it was examined, which influencing factors were identified by the authors of the selected studies regarding the UI of a hotel booking website. Furthermore, out of the data gathered and analyzed in the respective studies, it could be determined, to which extent these factors have an influence in online hotel booking intention. The set of UI elements is displayed in Table 1.

2.2 Semi-automated Content Analysis

In addition to the systematic literature review, we then conducted a semi-automated quantitative, conceptual content analysis of the analyzed research articles by using the content analysis tool MaxQDA Pro 2022. We defined a code-set which consists of the analyzed UI categories that are: Reward Points, Hotel Star Rating, Hotel Description, Hotel Pictures, Popularity, Scarcity, Price and Review. The code-set was derived through the systematic literature review based on the identified influencing factors on online hotel booking intention and furthermore was documented in a codebook. We then searched for and coded all text passages in the 39 papers which included the defined terms of our code-set, including similar semantic terms shown in Table 1. In total, we identified 2478 relevant text passages (cleaned by keywords and references) which were then screened manually to avoid coding sections which were not topic in the body of research.

We furthermore utilized the method of Cohen's Kappa to quantify interrater agreement between the two coders during the content analysis. This allowed us to establish how reliable our codes were agreeing on the coding of the research data. The use of Cohen's Kappa is considered an important step in establishing the validity of a research study and is an invaluable tool for any researcher seeking to demonstrate reliability in their work. Due to a coefficient of about 84% the results can be considered almost perfect, which stabilizes the reliability of the gained data [14].

In Table 1 the condensed UI categories based on the analyzed influencing factors on online hotel booking intention, identified via the systematic literature review, are displayed. We furthermore enriched the categories by synonyms and related terms based on the systematic literature review, which subsequently were utilized for coding during the semi-automated content analysis. The total count of text passages that were assigned to each category give an overview about the intensity of research that was conducted per category in recent years. We clearly see that there is an abundance of research regarding the review and price category. On the other hand, we see a lack of research in the field of reward points, picture, hotel-star-rating and description category. Table 2 shows example statements for each UI category, which were also used by the coders to ensure consistency.

Table 1. UI categories, code system including synonyms and total count of aggregated codes.

UI category	Code System	Total Count	Relative Count
Review Category	Review, Rating, eWOM	1242	50,12%
Price Category	Price, Room Rates, Promotion, Deal, Discount	653	26,35%
Popularity Category	Popularity, Brand	257	10,37%
Scarcity Category	Scarcity, Room Availability, Sold Out Factor	127	5,13%
Description Category	Description, Cancellation, Amenities, Facility	105	4,24%

(continued)

Table 1. (*continued*)

UI category	Code System	Total Count	Relative Count
Hotel-Star-Rating Category	Hotel Class, Star Rating, Hotel Category, Star Category	64	2,58%
Picture Category	Picture, Image, Photo	27	1,09%
Reward Points Category	Reward Points	3	0,12%

Table 2. UI categories and example statements.

UI category	Example statement
Review Category	"Comparing the effect size of each information clue, we found that review rating has the strongest influence on customers' booking intentions. This implies that maintaining a good review rating is critical to retain existing customers and attract new customers." [15]
Price Category	"The probability that consumers would consider a hotel rose as the listed price declined[...]" [9]
Popularity Category	"Another finding is that the cue types interact such as that providing a popularity cue does not increase scarcity perceptions when a supply cue is already present (see Fig. 4) – while in contrast, adding a supply-based cue always increased scarcity perceptions. This may be due to notion that when only a very limited number of rooms is available, it does simply not require other people for the rooms to sell out[...]"[16]
Scarcity Category	"[...]scarcity renders products more valuable and hence more desirable through social proof (i.e., through a must-be-good effect), it is also associated with a sense of urgency (i.e., the get-it-before-it's-gone effect). However, as shown in Table 3, urgency translates into booking intentions for B2C contexts only, while it has no effect in the C2C setting." [16]
Description Category	"Reservation policies contributed the most (about 18.1% relatively) to the warranting value of accommodation information, followed by hotel description (17.7%), hotel and room pictures (16%), hotel facilities (15.4%), and customer review ratings (15%)." [17]
Hotel-Star-Rating Category	"Star ratings of hotels are determined according to the quality of the hotel. It is one kind of guideline that hotel guests can use to choose the desired hotel and reserve the room." [18]

(*continued*)

Table 2. (*continued*)

UI category	Example statement
Picture Category	"Tourists choose hotel offers for further consideration based [...] information such as a picture of the hotel, the hotel name, the location, in-formation and map pictograms, customer rating, and sustainability certification." [19]
Reward Points Category	"[...]consumers' preferences for reward points and price discounts exerted significant impacts on their booking intentions." [9]

Table 3 summarizes (i) the 8 UI categories, (ii) an example image of each category on the hotel booking website booking.com, (iii) and the context in which the respective category is embedded into on booking.com as an example for its practical implementation. The categories are displayed in descending order regarding the total number of text passages per category (Table 4).

Table 3. UI categories, example images and its context on the platform booking.com

UI Category	Example Image	Context booking.com
Review Category	**Excellent** 8.9 1,000 reviews modern, incredible pool, spa and Sanctuary grounds, amazing food Ⓚ Kate 🇺🇸 United States of America	User-generated reviews on booking.com include several characteristics: Number of reviews, review rating (e.g., 8.9), textual description, reviewers name and location and a textual caption of the review valence
Price Category	**Today's Price** **€ 382** Includes taxes and fees	The hotel room rates are presented in regard to the length of the stay. In addition included services are displayed
Popularity Category	⏲ Someone just booked this	Popularity cues are highlighted
Scarcity Category	**Two-Bedroom Apartment** ● Only 1 left on our site	Scarcity cues can be presented in a red font color

(*continued*)

Table 3. (*continued*)

UI Category	Example Image	Context booking.com
Description Category		On booking.com there are covered several aspects of the hotel specification within the description section: A description of the hotel itself and its location, of its amenities (e.g., free WIFI), the cancellation policy and the respective hotels facility (e.g., Bars, Pools, etc.)
Hotel-Star-Rating Category	**Hotel Hammerwirt - Forellenhof** ★★★	The hotel-star rating is displayed next to the hotels name
Picture Category		Images of a hotel are presented in tiles on the hotel detail page so that the users can get a quick overview of the respective hotel
Reward Points Category	◇ Genius discount available	It is displayed if there is a "Genius" discount available

Table 4. Outcomes and research gaps in online hotel booking intention

UI Category	Total Count	Identified Research Gaps
Reward Points Category	3	The UI Element of reward points was researched least among all influencing factors of online hotel booking intention. Nevertheless, it is mentioned, that reward points (e.g., "Genius") have a similar effect on booking intention than a price discount [1]. Hence further research regarding this aspect should be carried out to get a deeper understanding of this aspect in connection to booking intention

(*continued*)

Table 4. (*continued*)

UI Category	Total Count	Identified Research Gaps
Picture Category	27	Even though it is described that pictures of hotels – whether they are about hotel rooms, breakfast, or hotel facilities – have a significant effect on hotel booking intention, there is a lack of research about this category [15]. It is not described what kind of picture influences the booking intention in what way, speaking of image content, daytime, seasonality, or other factors
Hotel Star Rating Category	64	Hotel-star-rating is a direct indicator for quality and regarding value for money [32]. Hence it is explainable that the majority of connected research has been done with the category of price. Nevertheless, as seen in Fig. 6, it is noticeable that hotel-star-category and popularity or hotel-star-category and scarcity has been found little in the context of interconnected research. Hence, we identify a research gap concerning the connection of hotel-star-rating and popularity or scarcity cues and its influence on online hotel booking intention. In general, it can be identified that hotel-star-rating and its influence on online hotel booking intention was investigated rather little in current research
Description Category	105	The hotel amenities and it's description is one of the most viewed cue of information while booking a hotel online [1]. It can be stated, that in a first step of online hotel booking, descriptions are being read to get an overview about offered services, hotel amenities or cancellation policies. In this specific stage of forming a consideration set of potential hotels to stay at, hotel pictures are being viewed even more often. As we can see in Fig. 5, description and pictures are combined in research very little (count of 7). Hence, we can identify another possible blind spot regarding interconnected research of hotel description and hotel pictures

(*continued*)

Table 4. (*continued*)

UI Category	Total Count	Identified Research Gaps
Scarcity Category	127	We have found that incorporating scarcity cues into the online hotel booking experience has a significant effect on intention to book. Hotel pictures and descriptions have been found to be effective in influencing this intention, too. However, there has been very little research into their combined effects. As such, further research needs to be conducted in order to fully understand the role these elements have on online hotel booking intention
Popularity Category	257	Popularity cues are perceived as having a moderate influence on online hotel booking intention, especially when being shown in combination with scarcity cues. Nevertheless, it can be analyzed, that scarcity and popularity cues have been researched in combination very little, although these two factors have the potential to affect each other significantly. Therefore, we propose to analyze the combination of these two categories in more detail
Price Category	653	It could be found out that some customers base their online hotel booking intention on price as the most important factor, however, a research gap has been identified indicating that there are surrounding factors that also contribute to a customers' perception of price. Research suggests that other factors such as type of traveler (business or leisure), location and reviews can play a role in influencing price sensitivity and price perception in online hotel booking. This indicates that there should be a more comprehensive approach to examining online hotel booking intentions that includes both, price and other surrounding factors

(*continued*)

Table 4. (*continued*)

UI Category	Total Count	Identified Research Gaps
Review Category	1242	The category of review is already analyzed broadly in current research. Besides researching user generated ratings in an isolated way also connected research has been undertaken, above all with the category of price. Hence, we already identified several outcomes regarding these two aspects. Nevertheless, we see a lack of interconnected research regarding user generated reviews and scarcity cues, or reviews and hotel pictures offered from hoteliers on online hotel booking websites. Since all of these factors have an influence on hotel booking intention it may be of great interest to evaluate the mutual influence of these factors

3   Outcomes in Online Hotel Booking Intention Research

Online hotel booking intention research has emerged as a key research area in recent years. A great deal of research has been conducted to understand the effects of online reviews and hotel room rates on the decision-making process of customers. It can be stated that there is a noticeable focus in research on the UI-Elements price and user generated reviews. These articles have yielded valuable insights into the way people make decisions when it comes to booking a hotel online. Nevertheless, there may be alternative research approached and methodologies to gain insights into the field of online hotel booking intention, which were not utilized so far.

Based on our systematic literature review and the semi-automated content analysis, we discuss the outcomes regarding the 8 UI-categories identified. To provide a holistic view of the topic we divided the analysis into several steps. We therefore analyzed these categories on three different layers: (i) the influence of identified UI categories on online hotel booking intention, (ii) the extent of research that was conducted in the specific field of UI category and (iii) the extent of research that has been conducted in the context of two or more UI influencing factors. Layer (iii) is of special interest since we also focus on the interaction effects between several influencing factors on online hotel booking intention. These insights will be discussed in the following.

3.1   Online Review Category

Online reviews or customer rating of hotels help users to form a choice set of potential resorts to stay at during the vacation. It is one of the most researched topics regarding online hotel booking intention [1, 20, 21]. Research results show, that the more and better ratings a hotel can assemble, the more it will be considered in the hotel booking process of users. Hence, online reviews have a significant influence on online hotel booking intention [9, 18, 20, 22]. There are several different characteristics regarding online reviews which influence the booking decision significantly: the relevance of reviews, the valence and quality of the ratings, the quantity and subsequently the recency of online reviews. Review recency and review valence have the strongest positive effect on user's online hotel booking decision [19, 20, 23–26]. To identify the relevance and thus the credibility of a review, users often look for similar reviews regarding one hotel and so determine the credibility of a specific user generated rating. The perceived credibility of online reviews (PCOR) has a significant influence on online hotel booking intention. In detail it is found that the review quality, the review consistency, and the type of review – whether it's a positive-sided or a two-sided review – have a significant effect on the perceived credibility of online reviews. In turn, a negative-sided review has no significant effect on PCOR and furthermore on booking intention. Also, the review-reader's pre-conceived knowledge influences PCOR significantly [27].

Research also shows that there is a significant effect influencing the valence and the quality of ratings in a positive way when there is a closer reviewer-reader similarity in demographics and geographics [22, 28]. Not only the characteristics of online reviews have an influence on online hotel booking intention but obviously also the content of the respective ratings. It is found that particularly the topics "quality of sleep", "value for money", "location", "service quality", "room facilities" and "cleanliness" have an effect on the booking decision [29]. Negative reviews can also outweigh the positive aspect of having a great location as a hotel, otherwise it was found that more positive reviews do not outweigh a poor location. It is described that this phenomenon is caused due to risk avoidance strategies of users [26, 30]. Research shows that the category of online review is the most sought information cue while booking a hotel online [15, 31].

Furthermore, it was found that concerning the different steps in the hotel booking process, the influence of different review characteristics seem to differ. In the consideration stage, by forming a first set of possible hotels to stay at, the valence and the volume of hotel reviews or ratings are positively associated with the likelihood that a hotel will be chosen throughput the consideration process [9, 15, 19, 31].

Figure 1 shows the number of Codes which occurred within the same scientific article. On the X-axis we see the sub-categories regarding the category "Review" while we see all other categories on the Y-axis. We summarized the number of Review-Code with each other Category-Code and hence we got an overview of which category was researched the most in combination with "Review". It can be analyzed that user generated reviews and price elements are researched the most (145) regarding the combination of two categories (always including "Review") followed by "Review" and "Popularity" (55) and "Review" and "Description" (47). We can therefore also conclude, that "Review" in combination with "Hotel Pictures" (7) or in combination with "Scarcity" (15) or "Reward-Points" (4) was little considered.

Code System	Rating (Element)	Review (Element)	EWOM (Element)	Review Category (SUM)
Reward Points Category	2	1	1	4
Hotel-Star-Rating Category	11	9	1	21
Description Category	22	14	11	47
Picture Category	5	2	0	7
Popularity Category	22	22	11	55
Scarcity Category	8	6	1	15
Price Category	61	57	27	145
Σ SUM	131	111	52	294

Fig. 1. Code matrix review category

3.2 Price Category

It is often thought that cheaper hotel room prices will lead to more bookings, but research shows that this is not always the case. A discounted price does not guarantee a higher booking intention. In fact, bad reviews of the hotel and a lower price seem to influence the decision to (not) book much more significantly than a normal price level in combination with great customer reviews. Also star rating of a hotel has a significant effect on the willingness to pay more for a stay at a higher rated hotel [9, 32]. The likelihood of booking a certain hotel even diminishes if a hotel offers a deep price discount in combination with a better rating than other comparable hotels. Potential travelers are therefore more likely to question the credibility of the hotel rating. Conversely, if seen independently, it was found that a lower room price increases the likelihood that a hotel will be considered and that a promotional discount influences the likelihood that a hotel will be considered in a positive way [9].

Research furthermore shows that there is a profound connection of price and online reviews regarding booking intention. It is often described as value for money [26, 29, 33]. Even when there is a strong connection between price and reviews, it is stated, that price is the most important factor in online hotel booking intention, regardless the booking stage or situation, even regardless which kind of vacation users tend to book [1]. Hence, other studies identify a difference between leisure and business travelers regarding the perception of the hotel room rates. Business travelers seem to be less price sensitive than leisure travelers [34, 35].

It can also be determined that certain personal circumstances influence the user's perception of high or low prices. In the area of luxury travel, hotel room rates seem to have much less influence on the hotel booking decision [36, 37]. In addition, it was found that different user groups focus on different aspects regarding online hotel booking. A user group focusing on price is willing to accept a decline of quality factors if the price is regarded as low [5, 21, 35, 38].

Booking a hotel is typically not a single-step process but consists of several process instances. First, users form a consideration set of potential hotels which will be filtered in more detail in the upcoming steps in the hotel booking process. Therefore, price and location information are perceived as the most important information cue in the process step of forming a consideration set, while consumer reviews are recognized as the most influential elements in the final booking decision [15].

Code System	Promotion (Element)	Deal (Element)	Discount (Element)	Price (Element)	Room Rates (Element)	Price Category (SUM)
> Reward Points Category	2	1	2	2	2	9
> Hotel-Star-Rating Category	4	1	2	14	8	29
> Description Category	12	6	11	26	11	66
Picture Category	1	1	1	5	4	12
> Popularity Category	12	5	7	25	6	55
> Scarcity Category	6	4	3	9	6	28
> Review Category	30	10	23	63	19	145
Σ SUM	67	28	49	144	56	344

Fig. 2. Code matrix price category

Regarding the category "Price" including its synonyms and related terms "Room Rates", "Discount", "Deal" and "Promotion" we can see, as displayed in Fig. 1 and Fig. 2 that research focused on price in combination with user generated reviews (145), followed by "Price" &"Description" (66) and "Price" &"Popularity" (55). It can be stated, that the area of price was least combined with "Reward-Points" (9), "Picture" (12) and "Scarcity" (28) followed by "Hotel-Star-Rating" (29).

3.3 Popularity Category

Besides "Review" and "Price", the "Popularity" of a hotel is another factor which influences online hotel booking intention significantly [9, 20, 24, 25]. "Popularity" of a hotel is described either as the amount of bookings in a specific time that is communicated towards users or the popularity of a hotel itself regarding the "Brand" or image [27]. Even though it was found that a hotel "Brand" or the "Popularity" of a hotel has a significant effect on hotel booking decision, compared to user generated reviews or room rates the effect size is quite low [15, 34]. "Popularity" based cues also have an interoperable effect in combination with scarcity cues. Hence, the perceived scarcity improves when there are popularity-based cues indicated. These "Popularity" based cues can for example indicate the amount of people who are currently looking at a specific hotel entry [16].

Code System	Popularity (Element)	Brand (Element)	Popularity Category (SUM)
> Reward Points Category	1	2	3
> Hotel-Star-Rating Category	2	7	9
> Description Category	1	16	17
Picture Category		4	4
> Scarcity Category	4	5	9
> Price Category	14	41	55
> Review Category	16	39	55
Σ SUM	38	114	152

Fig. 3. Code matrix popularity category

Since online reviews as well as the aspect of "Price" are the most researched fields in online hotel booking intention, we also see, that these two aspects were researched most in combination with other aspects (Fig. 3). We can identify that "Popularity" with its related term "Brand" was investigated mostly combined with "Price" (55) and "Review" (55), followed by the category "Description" (17). Besides that, we can see, that there was only little research related popularity in interrelation with other constructs. It was combined least with "Reward-Points" (3), "Picture" (4) and "Hotel-Star-Rating" (9) and "Scarcity" (9).

3.4 Scarcity Category

By presenting a "Scarcity" attribute at a hotel listing the probability of an increase in sales rises, hence it influences the booking intention of potential travelers [32, 38]. It is observed that there is a significant interaction effect between "Scarcity" and "Popularity" cues and the booking intention of users. Notably, concerning the interaction effect of scarcity and popularity, it was analyzed that if the "Scarcity" of rooms was high ("only one room left") the booking intention was at a low level, regardless of the appearance of "Popularity" cues. On the other side, when there was a low "Scarcity" (e.g., 30 rooms left) in combination with "Popularity" cues (e.g., 40 times booked in the last 12 h), the booking intention increased [25]. Outcomes of literature seem to differ since also other implications could be identified. Supply-based "Scarcity" cues (only one room left) have a significant effect on booking intention [16].

Code System	Room Availability (Element)	Scarcity (Element)	Sold Out Factor (Element)	Scarcity Category (SUM)
Reward Points Category	1	2	0	3
Hotel-Star-Rating Category	3	4	1	8
Description Category	2	2	0	4
Picture Category	1	1	0	2
Popularity Category	3	6	0	9
Price Category	10	16	2	28
Review Category	5	10	0	15
Σ SUM	25	41	3	69

Fig. 4. Code matrix scarcity category

As scarcity is discussed very controversy in recent literature regarding the influence on online hotel booking intention, it stands out, that this topic does not have a lot of inter-connecting factors in current research (Fig. 4). It is most viewed at in combination with "Price" (28), followed by "Review" (15) and least viewed at in combination with "Picture" (2), "Reward-Points" (3) and "Description" (4).

3.5 Description Category

"Descriptions" of hotel "Amenities", "Cancellation" policies, "Facilities", or the hotel itself have a significant effect on the perceived value of warranty. Especially the topic of reservation policies and the hotel description can be identified as important regarding the perceived quality and so, description also has an effect on booking intention [17, 39, 40]. The hotel amenities and it's description is one of the most viewed cue of information while booking a hotel online [1]. By forming a consideration set of potential hotels to stay at, which takes place in a very early process step in the hotel booking process, 34% of users have a look at the hotel "Description". In comparison, in this first steps, 69% of users compare "Prices" and 79% screen hotel "Pictures" [15].

Code System	Description (Element)	Cancellation (Element)	Amenities (Element)	Facility (Element)	Description Category (SUM)
> Reward Points Category	0	1	1	0	2
> Hotel-Star-Rating Category	2	3	3	1	9
Picture Category	2	2	2	1	7
> Popularity Category	4	3	5	5	17
> Scarcity Category	0	2	2	0	4
> Price Category	11	14	28	13	66
> Review Category	13	3	21	10	47
Σ SUM	32	28	62	30	152

Fig. 5. Code matrix description category

Again we observe a high interrelation of research regarding "Description" and"Price" (66) followed by "Description" and"Review" (47) (Fig. 5). The aspect of a description with its related terms "Facility", "Amenities" and "Cancellation" was least combined with "Reward-Points" (2), followed by "Scarcity" (4) and "Picture" (7).

3.6 Hotel-Star-Rating Category

The "Star rating" of a hotel itself gives a standardized overview of the quality and diversity of amenities and the service level of a resort. Alongside user generated "Reviews", the "Star Rating" is one guideline that potential travelers can use as a hint for quality [18]. The "Star Rating" of a hotel has a direct effect on price of a hotel room and the perceived value for money [17, 32]. Displaying the "Star rating" of a hotel is a cue for credibility and hence has an influence on booking intention [24]. The "Star Rating" of a hotel is often used as a filter attribute in the first steps of the hotel booking process, to form a set of potential resorts [15].

Code System	Hotel Class (Element)	Star Rating (Element)	Hotel Category (Element)	Star Category (Element)	Hotel-Star-Rating Category (SUM)
> Reward Points Category	1	0	0	0	1
> Description Category	2	6	1	0	9
Picture Category	2	3	0	0	5
> Popularity Category	2	5	1	1	9
> Scarcity Category	2	4	2	0	8
> Price Category	6	16	5	2	29
> Review Category	2	10	7	2	21
Σ SUM	17	44	16	5	82

Fig. 6. Code matrix hotel-star-rating category

As we see in Fig. 6, the aspect of hotel-star-rating with its related terms "Star Category", "Hotel Category", "Star Rating" and "Hotel Class", was most combined with the category "Price" (29) and "Review" (21). It is to be stated, that there was special attention of the authors on the term star rating during coding and cleaning process, since star rating was also used as a synonym for online reviews. We again see that the "Star ratings" of hotels were investigated least in combination with "Reward-Points" (1), "Picture" (5), followed by "Scarcity" (8), "Description" (9) and "Popularity" (9).

3.7 Picture Category

Although the UI category of "Pictures" and images is explored very little regarding the online hotel booking intention, it was found that "Pictures" are one of the most used sources of information during the hotel booking process [17, 18]. Especially, in the first steps of the hotel booking process, hotel "Pictures" and "Review" score are the most sought cue of information to form a first consideration set, this result can be seen among all types of travelers [15, 19] (Fig. 7).

Code System	Picture (Element)	Picture Category (SUM)
> 🔲 Reward Points Category	1	1
> 🔲 Hotel-Star-Rating Category	5	5
> 🔲 Description Category	7	7
> 🔲 Popularity Category	4	4
> 🔲 Scarcity Category	2	2
> 🔲 Price Category	12	12
> 🔲 Review Category	7	7
Σ SUM	38	38

Fig. 7. Code matrix picture category

It can be seen that the UI category "Picture" was researched in a more or less isolated way. Mostly "Picture" was combined with the category "Price" (12). It was least combined with the categories "Reward Points" (1), "Scarcity" (2) and "Popularity" (4).

3.8 Reward Points Category

"Reward Points" have a comparable effect on booking intention like promotion discounts in the "Price Category". Nevertheless it can be analyzed, that the topic of "Reward Points" or reward system is currently not focused regarding research in the area of hotel booking intention [1, 9].

As illustrated in Table 1, it can be stated that the category "Reward Points" was researched least regarding influencing factors on online hotel booking intention in recent years. This is also reflected in Fig. 8 "Reward Points" was mostly combined with "Price" (9) and "Review" (4). Opposite this, the category "Reward Points" was least combined in research with "Hotel-Star-Rating" (1), "Picture" (1) and "Description" (2).

Code System	Reward Points (Element)	Reward Points Category (SUM)
> 🔢 Hotel-Star-Rating Category	1	1
> 🔢 Description Category	2	2
🔢 Picture Category	1	1
> 🔢 Popularity Category	3	3
> 🔢 Scarcity Category	3	3
> 🔢 Price Category	9	9
> 🔢 Review Category	4	4
∑ SUM	23	23

Fig. 8. Code matrix reward points category

4 Conclusion and Research Agenda

Recent research shows that when it comes to online hotel booking intention, there are a variety of influencing factors. Most notably, online reviews and hotel room rates play an important role in the decision-making process [41]. Other elements such as scarcity, popularity, hotel-star-rating, a hotel description, reward-point-system, and photos of the hotel are also identified as factors that have an influence on online hotel booking intention [42]. However, most of these factors are researched and analyzed separately, without considering their interconnected influence. Collectively, these factors intertwine and can strongly influence the overall decision-making process of the customer.

4.1 Research Gaps

Although the current literature about online hotel booking intention is extensive, yet there are still blind spots in the research which fail to address the interconnectedness of these factors and the effect they have on intent to book. In order to better understand the landscape, it is necessary to conduct further research into these gaps, as well as explore potential solutions to bridge them. As such, this conclusion aims to identify these blind spots, discuss how they can be addressed, and provide recommendations for future research.

We therefore divide the outcomes of this respective research into three areas: Influencing factors of online hotel booking intention analyzed via a systematic literature review, User-Interface categories on online hotel booking websites that have been little researched in recent years and the lack of interconnected research regarding specific User-Interface elements, both identified by using a semi-structured content-analysis approach via MAXQDA. We give an overview of these outcomes in the following table.

4.2 Research Questions

Finally, we propose several research questions and a potential research agenda to take the current lack of interrelated research of the identified influencing factors into account. We therefore suggest building up on the respective literature review and semi-structured content analysis to develop a holistic methodology in terms of influencing factors on online hotel booking intention.

We propose a research agenda as follows: By using a design science research process, several aspects of online hotel booking intention can be considered and can be researched in an interconnected and holistic way, since this process can be used to carry out an in-depth analysis of the topic in several iterations. Subsequently, a theoretical decision tree model can be created as a design science artifact, which then should be tested and optimized through several iteration steps by using methods such as conjoint analysis. These results can finally be evaluated by conducting field studies in cooperation with a booking platform as a proof of concept. Based on above results, the following research questions can be derived:

RQ1: How do scarcity cues in combination with user generated reviews influence online hotel booking intention?

RQ2: How do hotel pictures in combination with user generated reviews influence online hotel booking intention?

RQ3: Which surrounding factors influence price sensitivity of users in the hotel booking process?

RQ4: Which surrounding factors influence price perception of users in the hotel booking process?

RQ5: How do scarcity cues shown in combination with popularity cues influence online hotel booking intention?

RQ6: How do scarcity cues, hotel description and hotel pictures interact with each other in the matter of online hotel booking intention?

RQ7: How do hotel pictures influence the perception of a hotel description in the matter of online hotel booking intention?

RQ8: How does hotel-star-rating shown in combination with scarcity cues influence online hotel booking intention?

RQ9: Hoe does hotel-star-rating shown in combination with popularity cues influence online hotel booking intention?

RQ10: Which content of hotel pictures on online hotel booking platforms influence online hotel booking intention?

RQ11: How do Reward-Points influence online hotel booking intention?

RQ12: How are the identified categories causally related to the booking intention and to each other?

5   Concluding Remarks

The research performed in this paper concluded that user interface elements play an important role in online hotel booking intention. It is important to consider the impact of user interface elements when designing hotel entries on hotel booking platforms, as they can affect how likely customers are to make a booking. This study adds to the growing body of literature on important factors in online hotel booking intention.

The current study is limited in that it does not consider moderating variables of online hotel booking intent such as age, gender, income or other factors. These variables could have a significant impact on online hotel booking intention and should be analyzed in future research.

Online hotel booking intention is a growing area of research and has become increasingly relevant in the modern times. Possible future research should take into account interconnected factors and moderating variables like the type of travel, the travel budget, age or gender. Such research would provide valuable insight into the decision-making process, allowing for a more comprehensive understanding of online hotel booking intention.

References

1. Jang, Y., Chen, C.C., Miao, L.: Last-minute hotel-booking behavior: the impact of time on decision-making. J. Hosp. Tour. Manag. **38**, 49–57 (2019). https://doi.org/10.1016/j.jhtm.2018.11.006
2. Chang, C.M., Liu, L.W., Huang, H.C., Hsieh, H.H.: Factors influencing online hotel booking: extending UTAUT2 with age, gender, and experience as moderators. Information (Switzerland) **10**(9), 281 (2019). https://doi.org/10.3390/info10090281
3. Statista, "Prognose zum Umsatz im Markt für Reisen & Tourismus weltweit* für die Jahre 2017 bis 2025 (in Milliarden Euro) [Graph]." Statista, 13. November, 2020 (2020). Accessed: 31. Januar 2023. https://de.statista.com/prognosen/890117/umsatz-im-markt-fuer-reisen-und-tourismus-weltweit
4. Confente, I., Vigolo, V.: Online travel behaviour across cohorts: the impact of social influences and attitude on hotel booking intention. Int. J. Tour. Res. **20**(5), 660–670 (2018)
5. Petricek, M., Chalupa, S., Chadt, K.: Identification of consumer behavior based on price elasticity: a case study of the prague market of accommodation services. Sustainability **12**(22), 9452 (2020). https://doi.org/10.3390/su12229452
6. Tseng, T.H., Wang, Y.S., Tsai, Y.C.: Applying an AHP technique for developing a website model of third-party booking system. J. Hospitality Tour. Res. (2021). https://doi.org/10.1177/1096348020986986
7. Vinhas, A.S., Bowman, D.: Online/offline information search patterns and outcomes for services. J. Serv. Market. **33**(7), 753–770 (2019). https://doi.org/10.1108/JSM-07-2017-0222
8. Talwar, S., Dhir, A., Kaur, P., Mäntymäki, M.: Why do people purchase from online travel agencies (OTAs)? A consumption values perspective. Int. J. Hospitality Manage **88**, 102534 (2020). https://doi.org/10.1016/j.ijhm.2020.102534
9. Hu, X., Yang, Y.: Determinants of consumers' choices in hotel online searches: a comparison of consideration and booking stages. Int. J. Hospitality Manage. **86**, 102370 (2020). https://doi.org/10.1016/j.ijhm.2019.102370
10. Hermanus, J., Indradewa, R.: Perceived value and attitude with trust as mediating variable toward intention to booking hotel online. Am. Int. J. Bus. Manage. **5**(3), 76–83 (2022)
11. Leite-Pereira, F., Brandão, F., Costa, R.: Is breakfast an important dimension in hotel selection? An analysis of online reviews. J. Tourism Dev. **2020**(34), 9–20 (2020). https://www.scopus.com/inward/record.uri?eid=2-s2.0-85098066353&partnerID=40&md5=3a07a9dcfc9af552b80c566c3f524248
12. Masiero, L., Yang, Y., Qiu, R.T.R.: Understanding hotel location preference of customers: comparing random utility and random regret decision rules. Tourism Manage. Article **73**, 83–93 (2019). https://doi.org/10.1016/j.tourman.2018.12.002
13. Brocke, J.V., Simons, A., Niehaves, B., Riemer, K., Plattfaut, R., Cleven, A.: Reconstructing the giant: on the importance of rigour in documenting the literature search process. In: Proceedings of the European Conference on Information Systems, 01/01 (2009)
14. Cohen, J.: A coefficient of agreement for nominal scales. Educ. Psychol. Measur. **20**(1), 37–46 (1960). https://doi.org/10.1177/001316446002000

15. Park, S., Yin, Y., Son, B.G.: Understanding of online hotel booking process: a multiple method approach. J. Vacation Market. **25**(3), 334–348 (2019). https://doi.org/10.1177/1356766671877 8879
16. Teubner, T., Graul, A.: Only one room left! How scarcity cues affect booking intentions on hospitality platforms. Electron. Commerce Res. Appl. **39**, 100910 (2020). https://doi.org/10. 1016/j.elerap.2019.100910
17. Chaw, L.Y., Tang, C.M.: Online accommodation booking: what information matters the most to users? Inf. Technol. Tourism **21**(3), 369–390 (2019). https://doi.org/10.1007/s40558-019-00146-1
18. Aeknarajindawat, N.: The factors influencing tourists' online hotel reservations in Thailand: an empirical study. Int. J. Innov. Creativity Change **10**(1), 121–136 (2019). https://www.scopus.com/inward/record.uri?eid=2-s2.0-85078894821&partne rID=40&md5=f147c89aa65706c9edf5637d47519724
19. Vinzenz, F.: The added value of rating pictograms for sustainable hotels in classified ads. Tourism Manage. Perspect. **29**, 56–65 (2019). https://doi.org/10.1016/j.tmp.2018.10.006
20. Alabdullatif, A.A., Akram, M.S.: Exploring the impact of electronic word of mouth and property characteristics on customers' online booking decision. TEM J. Article **7**(2), 411–420 (2018). https://doi.org/10.18421/TEM72-24
21. El-Said, O.A.: Impact of online reviews on hotel booking intention: the moderating role of brand image, star category, and price. Tourism Manage. Perspect. **33**, 100604 (2020). https://doi.org/10.1016/j.tmp.2019.100604
22. Chan, I.C.C., Lam, L.W., Chow, C.W.C., Fong, L.H.N., Law, R.: The effect of online reviews on hotel booking intention: the role of reader-reviewer similarity. Int. J. Hospitality Manage **66**, 54–65 (2017). https://doi.org/10.1016/j.ijhm.2017.06.007
23. Chong, A.Y.L., Khong, K.W., Ma, T., McCabe, S., Wang, Y.: Analyzing key influences of tourists' acceptance of online reviews in travel decisions. Internet Res. **28**(3), 564–586 (2018). https://doi.org/10.1108/IntR-05-2017-0212
24. Lee, Y., Kim, D.-Y.: The decision tree for longer-stay hotel guest: the relationship between hotel booking determinants and geographical distance. Int. J. Contemp. Hospitality Manage **33**(6), 2264–2282 (2020). https://doi.org/10.1108/IJCHM-06-2020-0594
25. Park, K., Ha, J., Park, J.Y.: An experimental investigation on the determinants of online hotel booking intention. J. Hosp. Market. Manag. **26**(6), 627–643 (2017). https://doi.org/10.1080/19368623.2017.1284631
26. Tanford, S., Kim, E.L.: Risk versus reward: when will travelers go the distance? J. Travel Res. **58**(5), 745–759 (2019). https://doi.org/10.1177/0047287518773910
27. Chakraborty, U.: Perceived credibility of online hotel reviews and its impact on hotel booking intentions. Int. J. Contemp. Hospitality Manage. **31**(9), 3465–3483 (2019). https://doi.org/10.1108/IJCHM-11-2018-0928
28. Yang, S.B., Shin, S.H., Joun, Y., Koo, C.: Exploring the comparative importance of online hotel reviews' heuristic attributes in review helpfulness: a conjoint analysis approach. J. Travel Tour. Mark. **34**(7), 963–985 (2017). https://doi.org/10.1080/10548408.2016.1251872
29. Ho, R.C., Withanage, M.S., Khong, K.W.: Sentiment drivers of hotel customers: a hybrid approach using unstructured data from online reviews. Asia-Pacific J. Bus. Adm. **12**(3–4), 237–250 (2020). https://doi.org/10.1108/APJBA-09-2019-0192
30. Yuan, Y.-H., Tsao, S.-H., Chyou, J.-T., Tsai, S.-B.: An empirical study on effects of electronic word-of-mouth and Internet risk avoidance on purchase intention: from the perspective of big data. Soft. Comput. **24**(8), 5713–5728 (2019). https://doi.org/10.1007/s00500-019-04300-z
31. Gavilan, D., Avello, M., Martinez-Navarro, G.: The influence of online ratings and reviews on hotel booking consideration. Tour. Manage. **66**, 53–61 (2018)
32. Castro, C., Ferreira, F.A.: Online hotel ratings and its influence on hotel room rates: the case of Lisbon, Portugal. Tourism Manage. Stud. **14**, 63–72 (2018)

33. Raja Omar, R.N., et al.: Factors that influence online behaviour in purchasing hotel room via website among tourists. Eur. J. Molecular Clin. Med. **7**(7), 219–229 (2020). https://www.scopus.com/inward/record.uri?eid=2-s2.0-85096967137&partnerID=40&md5=dd0e7fb8817a205ebd1fe7a3940cda78

34. Kim, D., Park, B.-J.: The moderating role of context in the effects of choice attributes on hotel choice: a discrete choice experiment. Tourism Manage. **63**, 439–451 (2017). https://doi.org/10.1016/j.tourman.2017.07.014

35. Nessel, K., Kościółek, S., Wszendybył-Skulska, E., Kopera, S.: Benefit segmentation in the tourist accommodation market based on eWOM attribute ratings. Inf. Technol. Tourism **23**(2), 265–290 (2021). https://doi.org/10.1007/s40558-021-00200-x

36. Xun, X., Lee, C.: Utilizing the platform economy effect through EWOM: does the platform matter? Int. J. Prod. Econ. **227**, 107663 (2020). https://doi.org/10.1016/j.ijpe.2020.107663

37. Yadegaridehkordi, E., et al.: Customers segmentation in eco-friendly hotels using multi-criteria and machine learning techniques. Technol. Soc. **65**, 101528 (2021). https://doi.org/10.1016/j.techsoc.2021.101528

38. Park, J.-Y., Jang, S.: The impact of sold-out information on tourist choice decisions. J. Travel Tourism Market. **35**(5), 622–632 (2018). https://doi.org/10.1080/10548408.2017.1401030

39. Chowdhury, R.R., Deshpande, A.: An analysis of the impact of reviews on the hotel industry. Ann. Tropical Med. Public Health **23**(17),(2020). https://doi.org/10.36295/ASRO.2020.231742

40. Xu, X.: Does traveler satisfaction differ in various travel group compositions?: evidence from online reviews. Int. J. Contemp. Hospitality Manage. **30**(3), 1663–1685 (2018). https://doi.org/10.1108/IJCHM-03-2017-0171

41. Yang, X., Yelin, F., Lai, K.K.: A social choice analysis of online hotel rating. Asia-Pacific J. Oper. Res. **38**(01), 2050039 (2021). https://doi.org/10.1142/S0217595920500396

42. Yadav, M.L., Roychoudhury, B.: Effect of trip mode on opinion about hotel aspects: a social media analysis approach. Int. J. Hosp. Manag. **80**, 155–165 (2019). https://doi.org/10.1016/j.ijhm.2019.02.002

The Mediation Role of Compatible Advantage in Mobile Wallet Usage

Brenda Eschenbrenner[1]([✉]) and Norman Shaw[2]

[1] University of Nebraska at Kearney, Kearney, NE, USA
eschenbrenbl@unk.edu
[2] Toronto Metropolitan University, Toronto, Canada
norman.shaw@torontomu.ca

Abstract. Mobile wallet adoption continues to grow but has not become a mainstream method of payment. Although the benefits of mobile wallets such as convenience, contactless features, and ease of use provide the impetus for some to adopt, others continue to hesitate. Some may be content with their existing payment methods and perceive that the benefits are not enough to change. Identifying factors that contribute to mobile wallet adoption are needed in order to expand its diffusion. Although some factors have been identified in previous research, a more in-depth inquiry is needed. Hence, we study the role of compatible advantage as a mediator in the relationship between attitudes towards mobile wallet usage and behavioral intentions. In the context of mobile wallets, compatible advantage represents one's perception of the benefits that can be derived from and the compatibility with existing methods of payment. We create a novel model by extending the Theory of Planned Behavior with compatible advantage and investigate its mediation effect. Our results show that compatible advantage mediates the relationship of attitude and behavioral intention. Potential users can be encouraged to adopt a mobile wallet if they are made aware of its benefits and compatibility with current payment methods. Our research provides guidance to practitioners and presents a model for future research.

Keywords: Mobile wallet · Theory of Planned Behavior · Compatible advantage · Mobile payment

1 Introduction

Mobile devices, especially smartphones, are an essential technology in many people's personal and professional lives. One of their many capabilities is a mobile wallet, which enables the smartphone to make payments at retailers that are able to accept them. Also, the ability to send payments directly to family and friends has generated interest in mobile wallets [1]. Although many mobile wallets rely on near field communication (NFC) (allowing a smartphone to be waved near an enabled device), others have evolved (e.g., Samsung Pay) such that payment will be accepted at traditional card terminals that are not NFC-enabled [2, 3].

© Springer Nature Switzerland AG 2023
F. Fui-Hoon Nah and K. Siau (Eds.): HCII 2023, LNCS 14038, pp. 190–201, 2023.
https://doi.org/10.1007/978-3-031-35969-9_13

Some mobile wallets provide enticements such as offering cash-back rewards [2]. For instance, consumers who patronized certain restaurants and met minimum purchase requirements could receive cash back if they used Google Pay. These incentives as well as factors such as convenience, ease of use, ability to transfer money immediately, and making carrying cash unnecessary have contributed to mobile wallet adoption [4]. However, adoption has been slow and is far from being universal. For example, although mobile wallet usage has grown in the United States from 2017 to 2021 as a percentage of total point-of-sale payment methods, the growth was 3 to 11%. Also, mobile wallet payments (i.e., 11%) is minimal compared to physical plastic cards (both credit and debit) accounting for 70% of total payment methods in 2021 [5].

Some consumers are content with their existing methods of payment and do not feel compelled to adopt a mobile wallet [6, 7]. Others may not be fully aware of the advantages of adoption because members of their social network are not using smartphones for payment. Concerns that personal data may be stolen and financial data compromised are also factors that have inhibited adoption [8]. Therefore, it is important to understand how certain factors can play a vital role in mobile wallet adoption in order for it to be more widely accepted and used.

In the Theory of Diffusion of Innovations, Rogers [9] proposed that innovations were more readily adopted if they were compatible with the current way of performing tasks and if there was a demonstrable advantage. In the context of mobile wallets, Shaw et al. [8] combined compatibility and relative advantage into the construct 'compatible advantage', defined as "the perception that the innovation is more beneficial [10] than its predecessor and that its use is compatible with current practices [11]" [8, Sect. 3.6.], and found it to be influential in mobile wallet usage intentions [8]. Considering the need for understanding the benefits of a new technology as well as its compatible nature, studying compatible advantage's role in mobile wallet usage intentions is warranted. Therefore, our research question is: Does compatible advantage affect one's attitude towards intentions to adopt mobile wallet?

Previous research has studied mobile wallet adoption intentions [12–16], continued usage intentions (e.g., [17]), as well as commitment/loyalty and recommendation intentions [18, 19]. Also, previous studies have leveraged the Theory of Planned Behavior (TPB) in the context of mobile wallets (e.g., [12]). However, we are not aware of any that have evaluated the construct of compatible advantage and its role in attitudes' influence on intention to adopt mobile wallets. Hence, a valuable contribution can be made by focusing on compatible advantage's role in mobile wallet adoption that can then be incorporated in future research as well as utilized by practitioners who wish to increase mobile wallet adoption.

This paper is organized as follows. First, we provide our theoretical foundation, proposed hypotheses, and research model. Next, we present our research methodology followed by the results and discussion of the results. Finally, we conclude with contributions, implications, limitations, and future research potential.

2 Theoretical Foundation and Hypotheses

2.1 Theory of Planned Behavior and Compatible Advantage

Although an individual's personality traits, attitude, and disposition can be influential in behavioral intentions, the Theory of Planned Behavior (TPB) proposes that context-specific factors can be used to predict behavioral intentions [20]. More specifically, the factors are: 1) one's attitude regarding the behavior under consideration (i.e., attitude toward the behavior), 2) the influence of others' perceptions of the behavior (i.e., subjective norm), and 3) the potential to perform the behavior without issues (i.e., perceived behavioral control). These three factors (i.e., attitude, subjective norm, and perceived behavioral control) predict intentions to engage in the behavior, with intentions influencing the actual behavior. Other factors, such as personality traits, are more distal to these predictions in comparison to contextual factors. Some have suggested that factors, such as general attitudes, may influence the contextual factors that are relevant to the particular behavior [20]. Hence, we focus on the three contextual factors (i.e., attitude, subjective norm, and perceived behavioral control) for this study.

The Theory of Planned Behavior (TPB) has been utilized in previous studies to understand behavioral intentions such as adoption of e-books [21], student ethical IT use [22], and information technology usage [23]. Regarding other potential theories, previous IS research has compared the predictive abilities of the Technology Acceptance Model (TAM) and TPB [24]. TPB was found to offer "more specific information that can better guide development" [24, p.173] in the context of IS usage intentions. Considering our research objective is to identify facilitators of usage intentions specific to mobile wallets that can be fostered by practitioners to enhance adoption, we utilize TPB for this study.

Attitude. Attitude has been referred to "as the user's evaluation of the desirability of his or her using the system" [24, p.175]. Attitude, as utilized in TPB, is behavior specific. In the context of this study, attitude is the desirability of mobile wallet usage, versus general attitudes towards technology usage. Previous studies have found attitude to significantly influence behavioral intentions in contexts such as telemedicine technology [25], spreadsheet usage [24], information technology usage [23], as well as intentions to use mobile wallets [14].

An individual desires payment methods that are efficient, simple to use, and provide greater convenience than other methods. Potential users may evaluate mobile wallets favorably because mobile wallets can satisfy the aforementioned criteria. They may develop a positive attitude because mobile wallets provide an alternative means of payment. Also, the utilization of their smartphone has just been extended in more diverse ways. Hence, they no longer need to be concerned with carrying the correct amount of cash or having a credit or debit card available. Instead, they can use their smartphone which they are likely to have with them wherever they go. Hence, we hypothesize,

H1: Attitude positively influences intention to use a mobile wallet.

Subjective Norm. Subjective norm has been defined as an "individual's perception of social pressure to perform the behavior" [24, p.175]. Some contexts, such as spreadsheet [24] and telemedicine technology [25] usage, did not find subjective norm to significantly influence behavioral intentions. However, in a study of information technology usage [23] and mobile wallet adoption intentions, social influence was found to be significant [14].

In the context of mobile wallets, individuals may feel pressured to use mobile wallets after seeing others do so. They may see the speed and ease with which others in their social network are using the most recent technologies and feel they may be missing out. More specifically, they may feel pressure to utilize more contemporary payment methods to benefit from the efficiency and convenience. They may also feel the need to use a mobile wallet to maintain their social status or acceptance. Hence, we hypothesize:

H2: Subjective norm positively influences intention to use a mobile wallet.

Perceived Behavioral Control. Perceived behavioral control refers to "the individual's perception of his or her control over performance of the behavior" [24, p. 175]. Previous research has identified the importance of perceived behavioral control influencing behavioral intentions (e.g., 25, 24). Perceived behavioral control was also found to positively influence behavioral intentions in the context of mobile wallets [12] and information technology usage [23].

In the context of the mobile wallet, individuals have discretion over their adoption and usage decisions. If they observe others using it with no issues, they may be less likely to identify any potential issues to its adoption. Also, if retailers provide the appropriate technology such that mobile wallets are accepted as a form of payment, they may see few, if any, impediments to its usage. Therefore, we propose the following hypothesis.

H3: Perceived behavioral control positively influences intention to use a mobile wallet.

Compatible Advantage. As noted previously, compatible advantage has been defined as "the perception that the innovation is more beneficial [10] than its predecessor and that its use is compatible with current practices [11]" [8, Sect. 3.6.]. Compatible advantage has been found to positively influence mobile wallet usage intentions [8]. In the context of mobile wallets, individuals can see the convenience of using a smartphone to pay for purchases versus having to search for other payment methods, such as acquiring the amount of cash needed. Also, using a mobile wallet is compatible with existing payment methods, such as credit cards, in that similar steps are needed to complete the transaction and obtain requisite transaction approvals.

Considering attitude reflects one's assessment of the favorability (or lack thereof) of a specific behavior, being able to identify positive factors associated with the behavior will enhance the probability of behavioral intentions. In the context of mobile wallet usage, having a favorable assessment of mobile wallets will lead to identifying the benefits as well as compatibility with familiar payment practices (i.e., compatible advantage) which will then lead to intentions to adopt the mobile wallet. Hence, we propose the following hypothesis:

H4: The relationship between attitudes and intention to use a mobile wallet is mediated by compatible advantage.

Intention to Use. One of the tenets of the Theory of Planned Behavior is that behavioral intention reflects the level of motivation and effort an individual is willing to put forth, and higher levels of intention are more likely to result in the actual behavior occurring

[20]. Also, the behavior under scrutiny must be under the individual's control, i.e., the individual can decide if they will engage in the behavior or not [20]. It also encompasses the necessary resources or situational factors that make control possible. In contexts in which the individual has complete control, behavioral intention should suffice in predicting the actual occurrence of the behavior.

In the context of the mobile wallet, individuals would need to perceive that they have the ability to adopt and use the mobile wallet at their discretion. Also, the resources necessary (e.g., instructions on how to adopt, merchants with compatible technology) need to be available as well. If the individual has the requisite situational factors as well as intention to adopt, then actual adoption should occur. Hence, we propose the following hypothesis:

H5: Intention to use a mobile wallet positively influences mobile wallet usage.

2.2 Research Model

The proposed research model is provided in Fig. 1.

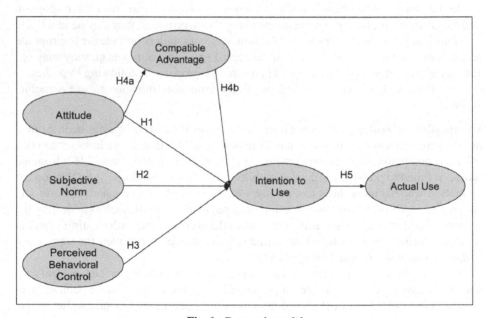

Fig. 1. Research model

3 Methodology

3.1 Participants

The methodology was quantitative. A survey was conducted and the results were analyzed to assess our research model. Participants were US residents, 19 years of age or older, and in possession of a smartphone. They were recruited via MTurk in which they were offered a small reward for completing the survey. Total responses returned were 605. Those that were incomplete or had too many answers with the same value (i.e. straight-lining) were eliminated, resulting in 310 valid surveys for a response rate of 51.2%. Table 1 shows the distribution of age and gender of participants.

Table 1. Age and gender of sample

Age	Male	Female	Totals	%
19 to 29	89	73	162	53.5%
30 to 39	45	26	71	23.4%
40 plus	34	36	70	23.1%
Totals	168	135	303	
%	55.4%	44.6%		

Note: 7 participants did not specify age or gender

Common uses of financial transactions with a smartphone by participants include paying at Starbucks, sending money to friends, and paying in a store. See Table 2 for the distribution of usage examples.

Table 2. Examples of smartphone usage

Action	Often	Sometimes	Never
Paying at Starbucks	30%	46%	24%
Sending money to friends	24%	50%	26%
Paying in a store	31%	36%	33%

The majority of the participants owned a smartwatch (59%), but only 7% used it frequently for payments.

3.2 Survey Questionnaire

The first part of the survey collected descriptive statistics, such as age and gender, as well as current ownership and usage of a smartphone for financial transactions. In the second part, constructs for attitude, compatible advantage, subjective norm, perceived behavioral control, and intention to use were measured by adapting established indicators from extant literature [24–27]. Actual usage was measured with the following item - *My use of a Mobile Wallet is best summarized as: (1) I have no plans to use a mobile wallet in the next two years, (2) I plan to use a mobile wallet within the next 12 months, (3) I use my mobile wallet but not very frequently, (4) I use a mobile wallet whenever I can, (5) I prefer not to answer this question.* All indicators were measured on a 5-point Likert scale.

The questionnaire was piloted with a small group of post-graduate students who provided feedback about its clarity. After making a few minor changes, the questionnaire was distributed through Amazon MTurk.

3.3 Data Analysis

Data was analysed with SmartPLS [28] using a two-stage approach. The first stage is the validation of the outer model to ensure that the indicators for the constructs converge. Indicators with a loading less than 0.708 were excluded [29].

Two tests were used to establish discriminant validity: the Fornell-Larcker criterion [30] and the heterotrait-monotrait (HTMT) ratio of correlations [31]. For the Fornell Larcker criterion, the square root of the AVE values are compared with the latent variable correlations. Our results satisfied the criteria in that the square root of each construct's AVE was greater than its highest correlation with any other construct. For the HTMT ratio, the HTMT value should be less than 0.9 [32]. Our results satisfied the criteria with the exception of intention to use which was slightly above 0.9. Considering intention to use is conceptually a distinct construct from the other constructs in our model and the results of the Fornell-Larcker test discussed previously, we consider discriminant validity to be adequately supported.

The second stage was the path analysis conducted by bootstrapping with replacement samples, which calculated the path coefficients and the path significance. SmartPLS also calculates the indirect effects, which were input to the 'variance accounted for' to determine if compatible advantage mediates the effect of attitude on intention to use.

In addition to confirming convergent and discriminant validity, the data was tested for common method bias using Harman's single factor test [33]. With the aid of SPSS, all indicators were input to an exploratory factor analysis. The results produced one factor whose variance was less than 50%, meaning that there was no common method bias.

4 Results

4.1 Outer Model

All indicators were greater than 0.708, with the exception of one for Compatible Advantage, which was 0.695. Hair et al. [29] suggests that indicators greater than 0.6 can be retained if their retention does not have a negative impact. As this was the case, we included this indicator. Model consistency and reliability were above the recommended values: 0.6 for Cronbach's Alpha, 0.7 for Composite Reliability, and 0.5 for Average Variance Extracted [29]. See Table 3 for results.

Table 3. Construct reliability and validity

	Cronbach's alpha	Composite reliability (rho_a)	Average variance extracted (AVE)
Att	0.873	0.875	0.663
CA	0.75	0.754	0.572
ITU	0.839	0.842	0.756
PBC	0.823	0.857	0.651
SN	0.854	0.861	0.698

NOTE: Att = Attitude; CA = Compatible Advantage; ITU = Intention to use; PBC = Perceived Behavioral Control; SN = Subjective Norm

4.2 Inner Model

For intention to use, the dependent variable in our study, 66.6% of the variance was explained by the exogenous variables (i.e., R^2, the coefficient of determination, was 0.666). To determine the path significance, bootstrapping with 10,000 subsamples was performed [34]. All paths were significant, with the exception of perceived behavioral control to intention to use. See Table 4 for results.

Table 4. Path significance

Hypothesis #		Path Coeff	T statistics	P values	
1	Att - > ITU	0.454	6.859	0	***
2	SN - > ITU	0.308	5.407	0	***
3	PBC - > ITU	0.055	0.807	0.419	
4a	Att - > CA	0.704	19.186	0	***
4b	CA - > ITU	0.131	2.288	0.022	*
5	ITU - > AU	−0.303	5.553	0	***

NOTE: Att = Attitude; ITU = Intention to use; SN = Subjective Norm; PBC = Perceived Behavioral Control; CA = Compatible Advantage. *** $p < 0.001$. * $p < 0.05$

4.3 Compatible Advantage as a Mediator

We evaluated the indirect path of attitude to compatible advantage, and compatible advantage to intention to use. Both paths were significant (see Fig. 2). The Variance Accounted For (VAF) was calculated as $= a * b / (a * b + c') = 17\%$, which is considered a small effect [35].

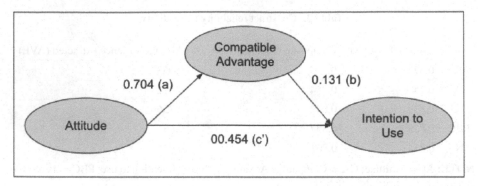

Fig. 2. Mediating effect of compatible advantage on attitude to intention to use

5 Discussion

Many studies have shown that relative advantage has a positive influence on intention to use [36]. However, for payment systems, relative advantage is not sufficient; compatibility is essential. Compatibility, from the consumer's perspective, means that paying with a plastic payment card or with their smartphone is essentially equivalent. The only difference is that the mobile wallet is presented at the payment terminal instead of the card. In the Diffusion of Innovations, Rogers [9] showed that innovations would be more readily adopted if they were compatible, meaning that there was no requirement to make major changes to the current way of achieving outcomes. Compatibility and relative advantage have been combined into the single construct, compatible advantage [8]. Our results confirm that not only does compatible advantage have a direct effect on intention to use, but it also has a mediating effect on attitude's influence on intentions. If users believe that there are no issues when paying with the mobile wallet, their attitude will be positive. However, if they believe that there will be more time reconciling payments or that merchants may not be willing to accept such payments, their attitude will be negative. The influence on behavioral intentions by users' attitudes towards the mobile wallet will be impacted by their perceptions of mobile wallet's compatible advantage.

Subjective norm had a significant influence on intention to use. Those individuals who have not yet used a mobile wallet will witness the ease and convenience when they observe others paying with their smartphones. The opinion of others, who are important to the individual, is influential in their intention to use a mobile wallet. Our results are consistent with past studies in which subjective norms had a significant effect on intention to use (e.g., [37]).

Perceived behavioral control was not significant in our study. Payments are a key service provided by financial institutions. Hence, there has been considerable investment to ensure that the underlying mechanisms work flawlessly. Networks connecting merchants with card issuers are reliable and transactions flow securely between consumer and retailer. Furthermore, the action of paying with a mobile wallet is easy to learn and requires no training. The occurrence of a problem is rare. The facilitating conditions that affect perceived behavioral control are transparent to the user and, therefore, are of no concern.

Not surprisingly, there is a strong relationship between intention to use and actual use [38]. Many merchants have installed payment terminals which are ready for tap and pay. It is no longer theoretical where the intention is thwarted by the lack of merchant acceptance.

5.1 Theoretical Contribution

Our research extends the Theory of Planned Behavior with the novel construct compatible advantage. The Theory of Planned Behavior proposed three constructs that will influence behavioral intention in a specific context (i.e., attitude regarding the behavior, subjective norm, and perceived behavioral control). Our findings support the importance of attitude and subjective norm in the context of mobile wallet adoption. However, perceived behavioral control was not found to be significant in this context. Considering the reliability of using a mobile wallet, perceived behavioral control may be of less relevance in the context of intention to use a mobile wallet.

Consistent with previous findings, the constructs of relative advantage and compatibility originally proposed in the Theory of Diffusions of Innovations should be considered as one construct, ie., compatible advantage, in the context of mobile wallets [8]. Additionally, compatible advantage provides some mediation of the relationship of attitude towards mobile wallets and behavioral intentions. Hence, future studies of mobile wallet adoption can consider applying this extended Theory of Planned Behavior.

5.2 Implication for Practitioners

To expand adoption and usage of mobile wallets, practitioners should consider implementing measures to enhance attitudes by emphasizing the benefits of the mobile wallet and stressing its compatibility with current payment methods. Practitioners can provide demonstrations of mobile wallet usage in comparison to other payment methods (e.g., credit cards) so potential users can easily recognize the similarities and at the same time witness the advantages of speed and convenience with contactless payments. Due to the significance of subjective norms, practitioners should emphasize adoption by social influencers and possibly provide testimonials from these individuals.

5.3 Limitations and Future Research

Our survey results were collected using MTurk to recruit participants. These individuals were compensated for their time and, hence, this may have influenced their objectivity when responding to the survey items. Also, all participants were citizens of the United States. Future research can replicate this study in other countries to assess the generalizability of our findings. Although the Theory of Planned Behavior was extended with the construct compatible advantage, future research can consider other potentially relevant variables, such as mobile wallet self-efficacy. Moderating factors, such as gender and age, can be added to study the adoption by different segments of the population.

6 Conclusion

Smartphones have become an essential element in many aspects of people's lives, such as entertainment, communicating with others, and shopping. As the features and functionality of smartphones proliferate, encouraging users to adopt these enhancements will require an understanding of the factors that can influence behavioral intention. Our study identified essential factors for mobile wallet adoption that future studies can utilize. Also, practitioners can leverage these factors to enhance mobile wallet adoption intentions and allow mobile wallets to pervade the mobile payment landscape.

References

1. King, W.B.: How credit unions can succeed in mobile where others have failed: By learning from others' mistakes, CUs may be in a position to make real strides forward in digital wallets, despite lackluster adoption rates seen in this space thus far. Credit Union J. **21**, 10–13 (2017)
2. Gerstner, L.: 4 ways to use mobile wallets. Kiplinger's Personal Finance **75**, 59 (2021)
3. Gustin, S.: Wired.com FAQ: near field communications big (money) moment (2011). https://www.wired.com/2011/05/wired-nfc-faq/. Retrieved, 16 Jan 2023
4. Bagla, R.K., Sancheti, V.: Gaps in customer satisfaction with digital wallets: challenge for sustainability. J. Manage. Dev. **37**, 442–451 (2018)
5. Statista: market share of cash, credit cards, and other payment methods at point of sale (POS) in the United States in 2017, 2019, 2020 and 2021 (2023). https://www.statista.com/statistics/568523/preferred-payment-methods-usa/
6. Ninia, J.: The united states lags behind china in adopting mobile payments (2019). https://business.cornell.edu/hub/2019/10/23/united-states-china-mobile-payments/
7. Why consumers aren't buying the idea of mobile wallets - yet: Knowledge at Wharton (2014). https://knowledge.wharton.upenn.edu/article/consumers-arent-buying-mobile-wallets-yet/
8. Shaw, N., Eschenbrenner, B., Brand, B.M.: Towards a mobile app diffusion of innovations model: a multinational study of mobile wallet adoption. J. Retail. Consum. Serv. **64**, 102768 (2022). https://doi.org/10.1016/j.jretconser.2021.102768
9. Rogers, E.M.: Diffusion of Innovations, 4th edn. Free Press, New York (1995)
10. Rogers, E.M.: Diffusions of Innovations, 5th edn. Free Press, New York (2003)
11. Karahanna, E., Agarwal, R., Angst, C.M.: Reconceptualizing compatibility beliefs in technology acceptance research. MIS Q. **30**, 781–804 (2006)
12. Chatterjee, D., Bolar, K.: Determinants of mobile wallet intentions to use: the mental cost perspective. Int. J. Hum. Comput. Interact. **35**, 859–869 (2019)
13. Hasan, A., Gupta, S.K.: Exploring tourists' behavioural intentions towards use of select mobile wallets for digital payments. Paradigm **24**, 177–194 (2020)
14. Mew, J., Millan, E.: Mobile wallets: key drivers and deterrents of consumers' intention to adopt. Int. Rev. Retail Distrib. Consum. Res. **31**, 182–210 (2021)
15. Sharma, G., Kulshreshtha, K.: Mobile wallet adoption in India: an analysis. IUP J. Bank Manage. **18**, 7–26 (2019)
16. Shaw, N.: The mediating influence of trust in the adoption of the mobile wallet. J. Retail. Consum. Serv. **21**, 449–459 (2014)
17. George, A., Sunny, P.: Why do people continue using mobile wallets? An empirical analysis amid COVID-19 pandemic. J. Financ. Serv. Mark. (2022). https://doi.org/10.1057/s41264-022-00174-9
18. Lubaba, H., Rohman, F.: Surachman: Leveraging experience quality to increase loyalty of digital wallet user in Indonesia. Int. J. Res. Bus. Soc. Sci. **11**, 46–56 (2022)

19. Na, N.T.L., Hien, N.N.: A study of user's m-wallet usage behavior: the role of long-term orientation and perceived value. Cogent Bus. Manage. **8**, 1899468 (2021)
20. Ajzen, I.: The theory of planned behavior. Organ. Behav. Hum. Decis. Process. **50**, 179–211 (1991)
21. Hsu, C.-L., Chen, M.-C., Lin, Y.-H.: Information technology adoption for sustainable development: green e-books as an example. Inf. Technol. Dev. **23**, 261–280 (2017)
22. Riemenschneider, C.K., Leonard, L.N.K., Manly, T.S.: Students' ethical decision-making in an information technology context: a theory of planned behavior approach. J. Inf. Syst. Educ. **22**, 203–214 (2011)
23. Taylor, S., Todd, P.A.: Understanding information technology usage: a test of competing models. Inf. Syst. Res. **6**, 144–176 (1995)
24. Mathieson, K.: Predicting user intentions: comparing the technology acceptance model with the theory of planned behavior. Inf. Syst. Res. **2**, 173–191 (1991)
25. Chau, P.Y.K., Hu, P.J-H.: Investigating healthcare professionals' decisions to accept telemedicine technology: an empirical test of competing theories. Information & Management. vol. 39, 297–311 (2002)
26. Chandra, S., Srivastava, S.C., Theng, Y.-L.: Evaluating the role of trust in consumer adoption of mobile payment systems: an empirical analysis. Commun. Assoc. Inf. Syst. **27**, 561–588 (2010)
27. Jaradat, M.-I.R.M., Moustafa, A.A., Al-Mashaqba, A.M.: Exploring perceived risk, perceived trust, perceived quality and the innovative characteristics in the adoption of smart government services in Jordan. Int. J. Mobile Commun. **16**, 399–439 (2018)
28. Ringle, C.M., Wende, S., Becker, J.-M.: Oststeinbek: SmartPLS (2022). https://www.sma rtpls.com
29. Hair, J.F., Ringle, C.M., Sarstedt, M.: PLS-SEM: indeed a silver bullet. J. Mark. Theory Pract. **19**, 139–152 (2011)
30. Fornell, C., Larcker, D.F.: Evaluating structural equation models with unobservable variables and measurement error. J. Mark. Res. **18**, 39–50 (1981)
31. Henseler, J., Ringle, C.M., Sarstedt, M.: A new criterion for assessing discriminant validity in variance-based structural equation modeling. J. Acad. Mark. Sci. **43**(1), 115–135 (2014). https://doi.org/10.1007/s11747-014-0403-8
32. Hair, J.F., Hult, G.T.M., Ringle, C.M., Sarstedt, M.: A primer on partial least squares structural equation modeling (PLS-SEM), 3rd edn. SAGE; Los Angeles, California (2022)
33. Podsakoff, P.M., MacKenzie, S.B., Podsakoff, N.P., Lee, J.Y.: The mismeasure of man(agement) and its implications for leadership research. Leadersh. Quart. **14**, 615–656 (2003). https://doi.org/10.1016/j.leaqua.2003.08.002
34. Streukens, S., Leroi-Werelds, S.: Bootstrapping and PLS-SEM: a step-by-step guide to get more out of your bootstrap results. Eur. Manag. J. **34**, 618–632 (2016)
35. Shrout, P.E., Bolger, N.: Mediation in experimental and nonexperimental studies: new procedures and recommendations. Psychol. Methods **7**, 422–445 (2002)
36. Kamis, T., Shukor, S.A., Nawai, N.: The determinants that influence mobile payment adoption: a systematic literature review. In: Editors: Nawai, N., et al (eds.) e-Proceedings of The 10th Islamic Banking, Accounting and Finance International Conference 2022 (iBAF2022), pp. 359–377 (2022). https://fem.usim.edu.my/ibaf2022-proceeding/
37. Liébana-Cabanillas, F., Ramos de Luna, I., Montoro-Ríos, F.: Intention to use new mobile payment systems: a comparative analysis of SMS and NFC payments. Economic Research-Ekonomska Istraživanja. vol. 30, 892–910 (2017)
38. Hossain, M.A., Hasan, M.I., Chan, C., Ahmed, J.U.: Predicting user acceptance and continuance behaviour towards location-based services: the moderating effect of facilitating conditions on behavioural intention and actual use. Australasian J. Inf. Syst. 21 (2017). https://doi.org/10.3127/ajis.v21i0.1454

Research on the Optimization Design of Mobile Vending Cart Service Process

Cui Hangrui[✉]

College of Art Design and Media, East China University of Science and Technology, Shanghai,
China
cui_hangrui@163.com

Abstract. Mobile vending is an important business model that facilitates residents' lives, but the current fragmented organization of the mobile vending economy conflicts with urban governance; there are information barriers between various interested parties and it is impossible for the mobile vending economy to establish a trust mechanism. The chaotic management further causes the pain point of poor working environment comfort for operators. To improve this situation, this paper uses literature analysis and interview observation to determine user needs, identifies service system design opportunities with a user journey map, uses the KANO model to determine service system functional demand attributes, analyzes and organizes the internal movement lines of vending carts and investigates the human-machine dimensions of activities inside vending carts. In the design practice scheme, the product service model combining physical object and service is proposed to optimize the mobile vending system, and an integrated platform service solution is constructed to integrate information from all parties in the vending activity; an easy-to-use mobile vending cart with functional space is designed to improve the working environment of operators. The mobile vending service system changes the status quo of disorganized business operation, low effective information dissemination, and unfriendly working environment for operators, and promotes the benign development of the mobile vending economy.

Keywords: Mobile Vending Cart · Service System Design · Ergonomics

1 Introduction

Mobile vending is a business model with no fixed place and no fixed business content, which has the characteristics of low employment threshold, low risk of failure and low commodity prices. [1] As a mobile vending business model has the quality of expressing regional culture, the industry has ushered in new opportunities with the emergence of cultural and creative industries and their popularity among consumers. However, the fragmented economic organization of mobile vending conflicts with urban governance. [2] The lack of access by operators to information on approved mobile vending locations has caused a chaotic management situation, which has further led to the pain point of a less comfortable working environment for operators. Moreover, consumers has no vending information sources and after-sales feedback entry, which makes it impossible for the mobile vending economy to establish user trust mechanism. Kotler categorizes the products and services provided by companies into five types based on the proportion of services in the product, and this paper conducts research on mobile vending service process optimization according to the combination of physical products and services among the five types. An integrated platform is applied to organize mobile vending economy business information, promote urban space governance, and provide an information platform for operators and consumers; through the design of the internal space of the physical products of vending carts, the working environment of operators is improved, and the rentable service is offered for operators to reduce the entry barrier.

2 Analysis of the User Needs of Mobile Vending Economy

2.1 Analysis of Organizational Forms

American sociologist Terry Clark proposed the place theory, which not only studies consumers as consumption activities themselves but also focuses on the social organization of consumption, using scenes to organize consumption into meaningful social forms. [3] According to place theory, the content of mobile vending operations follows the traditional customs or cultural atmosphere of the region and generally depends on the specific cultural space environment. [4] The study of mobile vending organization form is based on the cultural space perspective and revolves around three forms: traditional forms of mobile vending, community street vending, and commercial bazaars.

1. Traditional forms of mobile vending: The early American flea markets and Chinese street vending economy are spontaneous forms, not controlled by professional organizers, [5] self-organized vending in conventional high-traffic public spaces, such as the Spitalfields Market in London, England, and the Panjiayuan bazaar in Beijing. 2. Community street vending: Under the modern governance system, mobile vending is operated by the government,[6] but temporary mobile vending still gathers at community streets and alleys to facilitate residents' daily life needs, with a high degree of participation in residents' lives, [7] such as the mobile vending stalls of "assured breakfast". 3. Commercial bazaars: Under the influence of postmodernism culture, mobile vending bazaars, known as the birthplace of the newest street behavior, are operated by social organizations and cultural companies, and the creators organize the sale in public commercial places such as squares and shopping malls, [8] such as "Art Edge Day" in Yokohama Minato Park in Japan and "Fanji Common Rare" in China.

According to literature analysis, it is concluded that traditional forms and community street mobile vending have significant fragmented organization problems, which conflict with urban governance, and it is difficult to establish consumer trust because of the uneven quality of the goods sold; commercial bazaars rely more on social media to release information, and the randomness of online information release causes problems for operators and consumers.

2.2 Demand analysis of Interested Parties

Demand Analysis of Operators. Operators assume the identity of cultural value constituents in the mobile vending economy and can be divided into two categories based on different organizational forms: traditional operators with the attributes of socially disadvantaged groups, [9] and creative bazaar operators representing the nascent youth subculture phenomenon [8].

According to interviews and observations, it is concluded that the phenomenon of urban governance troubles caused by operators in the traditional model is due to the failure of information transfer, [10] and the lack of effective ways for operators to understand the planning of vending areas by managers, which leads to a cat-and-mouse game between operators and managers; [11] to make a living, operators pay less attention to the comfort of the working environment and operate for longer hours, resulting in the inability to rest even during periods of low patronage. [12] Therefore, if managers provide information on sites conducive to vending activities and assist operators in reducing operating costs, they can promote the initiative of vendor to understand urban planning policies and achieve the order of mobile vending.

The youth group participating in the creative bazaar relies on online social media platforms, and the products for sale have creative attributes, so they can generate a fan base, but they also need to obtain traffic through earned media, so the operators need the online platform to organize the whole process of the sale.

Consumer Demand Analysis. There are two main types of consumer participation in mobile vending: meeting basic needs around life through traditional and community mobile vending, or meeting interests by participating in creative bazaars. Through consumer interviews and observations, the needs are summarized as follows: 1. Mobile vending, which can provide basic services around life, leaves consumers with the impression that there is no quality assurance of goods, so it is necessary to establish a trust mechanism to ensure that the quality of goods can be traced. 2. By participating in creative bazaars, consumers are easily interested, so it is necessary to provide consumers goods with traceable information and a complete map of the bazaar to increase their sustainable consumption experience.

2.3 Summary of Demand Analysis

By summarizing the pain points of mobile vending, we can divide the demand into the following five dimensions: 1. Information dissemination efficiency: operators need to obtain information on site planning to comply with the city order and understand the cultural attributes of the site to plan the vending plan; consumers need to understand the information of mobile vending activities, and the information of continuous operators can be traced after consumption to facilitate after-sales feedback. 2. Information dissemination platform: operators need to be provided with a platform to obtain information on the site; consumers need to be provided with the vending information access platform and after-sales service entrance; 3. Services to reduce operating costs; 4. Services to provide continuous consumption; 5. Product comfort: the product should provide operators with a comfortable working environment; 6. Product order guidance: to protect the order of service movement line process.

3   Mobile Vending Service System Design Analysis

3.1   Identification of Key Touch Points with User Journey Map

User experience map is a tool to visualize the interaction between users and services and systems in an experience from the user's perspective. As a visual diagram showing individual user experiences, it is used to study user behavior in established domains and reveal the relationships between people, places, and things in service design. [13] The study uses user journey maps to identify design opportunity points for operators and consumers in mobile vending services.

User role Construction. The roles of operators and consumers were constructed, and the role model analysis was conducted based on the content of needs analysis of interested parties in Sect. 2.2, as shown in Table 1 [14].

Table 1.   User Role Model

Role models	User characteristics	User requirements	User behavior
Operators	Informal working group, involved in selling for a long time, dependent more on social media platforms	Comfortable business environment, reduced operating costs, combination with social media platforms	Look for vending locations Prepare products and carts for sale Sell products Complete a transaction Complete a sale
Consumers	Focus on efficiency and experience, as well as the sense of consumer online and offline experience	To guarantee the quality of goods, offline consumption order, online accountability platform, online re-consumption portal	Learn about the marketplace online or generate random consumer motivation offline Make a purchase Complete the transaction Product reviews

User Journey Map Plotting. By using the role model constructed in the user role model diagram, the user journey map was drawn according to the content of the organizational model analysis in Sect. 2.1, and the operator user journey map was drawn as shown in Fig. 1.

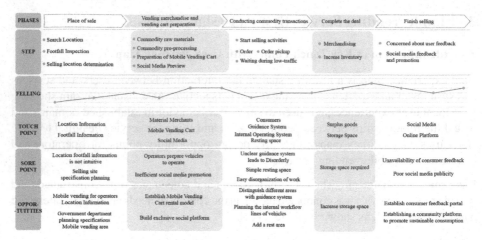

Fig. 1. Operator user journey map

The consumer user journey map is shown in Fig. 2.

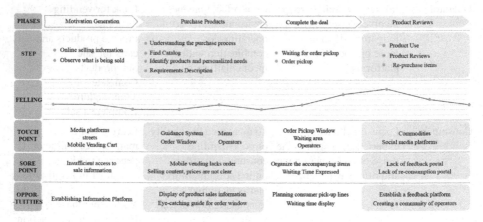

Fig. 2. Operator user journey map

Key Touch Points Identification. The satisfaction, touch points, pain points, and opportunity points of each behavior stage between operators and consumers were analyzed. Based on the frequency of contact points and the degree of influence on pain points, the following key touch points were identified.

1. Media platform. The media platform provides information on mobile vending services for buyers and sellers in the pre-sale period, and provides an entry point for regulating vending management. Later in the sale, consumers provide feedback through the media platform, and building a service platform can help interested parties communicate and build a trust mechanism for mobile vending consumers.

2. Vending cart. Providing vending cart rental services can reduce costs for operators and facilitate operator management. The internal movement line of the vending cart can guarantee the orderliness of the vending activities, and the design of the human-machine size can enhance the work comfort of operators.
3. Guidance system. The guidance system is designed for the mobile vending process, to rectify the business order, improve service efficiency and service experience.

3.2 Analysis of Opportunity Points with KANO Model

Questionnaire Design. Based on the identification of key touch points, the questionnaire was designed to investigate the three functions of media platform, vending cart and guidance system, and to address the pain points of operators, 7 functional requirements around mobile vending were proposed and each functional KANO attribute was calculated (Table 2).

Table 2. Operator questionnaire questions

Dimension	Number	Contents for operator questions
Information dissemination efficiency	Q1	Providing officially planned vending site location and foot traffic information
Information dissemination platform	Q2	Reservation of vending space in the system platform
Operating cost reduction services	Q3	Obtaining mobile vending carts through leasing
Product orderliness guidance	Q4	Mobile vending carts can distinguish between ordering and pick-up areas
Product comfort	Q5	The interior of the vending cart provides a comfortable resting space
Provision of ongoing consumer services	Q6	Continuously operating online stores through the platform after the sale
	Q7	Building consumer communities

For consumers, 5 functional requirements around mobile vending were proposed and each functional Kano attribute was calculated (Table 3).

Table 3. Consumer questionnaire questions

Dimension	Number	Contents for consumer questions
Information dissemination efficiency	Q8	Focusing on selling information release through the platform
Product order guidance	Q9	Ensuring the order through visual guidance for offline mobile vending
Information dissemination platform	Q10	Online platform allows for feedback on commodity issues
Provision of ongoing consumer services	Q11	If consumers have good offline experience, they can continue to spend online through the platform
	Q12	If consumers have better experience after multiple purchases, they can join the operator community through the platform

Based on the KANO model, forward and reverse questions were set, and the questionnaire options were shown in Table 4.

Table 4. Sample questionnaire

Question 1	For provision of officially planned vending site location and foot traffic information				
Function realization	Satisfied	Merited	No Matter	Reluctantly accepted	Disgusted
Function not implemented	Very Inconvenient	Inconvenient	No Matter	Merited	Disgusted

Data Analysis. Based on the data from the returned questionnaires, the functional KANO attribute was analyzed with Better-Worse coefficients, and the summarized data were shown in Table 5.

Table 5. KANO function data summary

Function	Better Coefficient	Worse Coefficient	Function	Better Coefficient	Worse Coefficient
Q6	60.47%	−39.53%	Q3	48.84%	−23.26%
Q7	68.29%	−21.95%	Q10	64.71%	−41.18%
Q4	61.9%	−26.19	Q8	58.82%	−17.65%
Q1	48.84%	−9.3%	Q11	58.82%	−17.65%
Q5	59.52%	−28.57%	Q12	47.06%	−5.88%
Q2	55.81%	−18.6%	Q9	29.41%	-2.3%

According to Table 5, the Better-Worse coefficient diagram was plotted in Fig. 3, [15] and the KANO attributes were divided. Q6 - online guarantee of after-sales quality of goods is an One-dimensional Quality. When the demand of this attribute is not met, users will be dissatisfied; Q7 - online continuous sales, Q4 - distinguish between ordering and pick-up areas, Q1 - understand information about the vending location through the platform, Q10 - online platform product feedback, Q8 - attention to sales information through the platform, Q11 - online platform continuous consumption, belong to Attractive Quality. When the demand of this attribute is not satisfied, the user is dissatisfied to the greatest extent; the remaining requirements belong to Indifferent Quality.

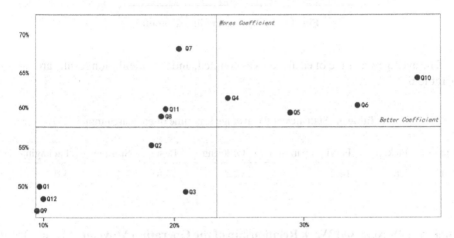

Fig. 3. Better-Worse coefficient diagram

4   Mobile Vending Cart Design Analysis

4.1   Analysis of the Internal Movement Line of the Vending Cart

The planning of the internal movement line of the vending cart is beneficial to the operator in the vending activities to ensure the work order. By using the method of user behavior research, the activity process was observed, measured and described, and the user behavior was analyzed according to the data to optimize the user behavior movement line [16].

User movement line records and data statistics. The study recorded and analyzed the mobile vending process, and recorded the user behavior with one-hour vending as event unit, and observed the behavior variables as the operation movement line and stay time. The user movement lines were plotted in Fig. 4.

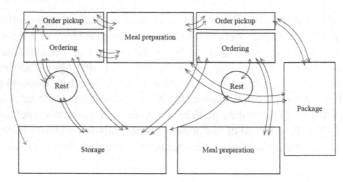

Fig. 4. User movement line diagram

The average stay time of each area was counted, and the calculation results are shown in Table 6.

Table 6. Statistics of the average stay time of each area mm

Area	Pickup	Food preparation	Ordering	Rest	Storage	Packaging
Time	7.6	14.2	12.7	13.8	7.1	4.6

Analysis of Strong and Weak Relationship of the Operation Movement Line Based on the operation movement line diagram and the average stay time statistics, the network of strong and weak relationship between each area can be organized and drawn, and the strong and weak relationship between each area is expressed in coarse and fine lines, as shown in Fig. 5.

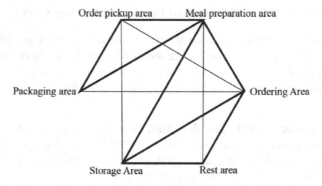

Fig. 5. User operation movement line relationship network diagram

It is concluded through the analysis of the diagram that the food preparation area is used more frequently and interacts with each area more frequently; the pick-up area, the food preparation area and the packaging area constitute a triangular work area; the food preparation area, the ordering area and the storage area constitute a triangular work area; the pick-up area, the storage area and the rest area constitute a triangular work area. By using the theory of "kitchen work triangle", the movement lines of each area are organized as shown in Fig. 6.

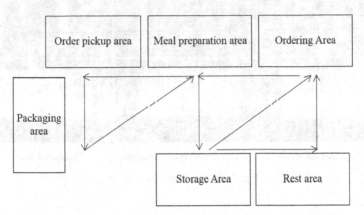

Fig. 6. Organization of the movement line in each area

4.2   Determination of Vending Cart Internal Man-Machine Size

Process action analysis. According to the movement line arrangement of each area in Fig. 5, the working actions of operators in each area were analyzed, to determine the appropriate man-machine size of the operating table. The height of the operating table needs to be considered throughout the vending activities, and the height of the operating table for arm extension is H1; in the operating process of the ordering table, the operator needs to complete the process of operating the electronic menu and consumer communication, and the range of activity of the forearm is L1; in the operating process of the food preparation area and the packaging area, the operator completes the operation on the table, and the range of activity of the big arm is L2; in the operating process of the pick-up area, the operator delivers the goods to the consumer, and the arm extension distance is L3; the sitting position of the operator should be considered in the rest area, and the sitting height is H2; the cabinet height needs to be determined in the storage area, and the arm extension height is H3. The main simulation actions are shown in Fig. 7.

Man-Machine Size Research. The man-machine sizes involved in different actions were investigated, and the average value of the human body was taken to establish the 50th percentile female mannequin by referring to the data study "GB10000-88 Chinese Adult Human Body Dimensions". [10] The main action sizes during the vending activity were researched, and the size of the vending cart was determined based on the research sizes, as shown in Table 7.

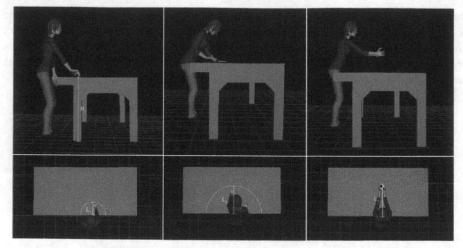

Operating electronic menus   Performing desktop operations   Product delivery

Fig. 7. Main action simulation

Table 7. Main sizes and applications of vending activities mm

Number	H1	L1	L2	L3	H2
Sizes	900	300	450	700	1650
Applications	Operating table height	Ordering screen width	Operating table width	Width of delivery area	Storage cabinet height

5 Design Practice

5.1 Design Principles

1. Principle of orderliness. Service system design should be a continuous experience. The mobile vending service system contains reservation management, providing information, after-sales service and other links, should guarantee the order of the back stage behavior and support the interaction between the front desk and the user.
2. Principle of wholeness. Service design should be holistic, including the entire service system background and application environment, the mobile vending service system should show the operator and consumer comprehensive vending place information, to ensure that the front desk information can effectively mobilize user behavior.
3. Collaboration principle. All interested parties in the service design should be involved in the design process, the front desk should mobilize operators and consumers to be active, and the back stage should actively collaborate with various departments to support the service system.
4. Convenience principle. The vending cart mainly serves the process of the operator's selling activities and meets the orderliness of the operator's work.

5. Comfort principle. The vending cart is the main workplace of the operator, and the interior comfort of the cart needs to be considered from two dimensions: man-machine size and interior decoration. The operating table and resting space need to be of suitable man-machine size to enhance the comfort of work, and the interior decoration should adopt natural materials to enhance the comfort of the environment.

5.2 Mobile Vending Service System Design

Service blueprint is one of the important tools for service design and management, which helps service providers to save service time, improve service efficiency and complete the service process with high quality and profitability. [11] The mobile vending service system design integrates information provided by various interested parties, builds an integrated platform to break the information barriers, promotes the efficiency of information dissemination, and realizes the orderliness of mobile vending service. The system blueprint visualizes elements such as the physical vending cart, the user behavior and front desk behavior of the operator and consumer, and the back stage behavior and support processes of the organizer, as shown in Fig. 8.

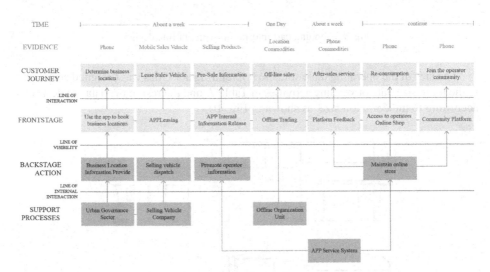

Fig. 8. Blueprint of mobile vending service system

5.3 Mobile Vending Cart Design

Interior Design of the Vending Cart. The interior of the mobile vending cart will be designed around two dimensions, namely, the movement line and the size. The movement line design will be presented based on the movement line arrangement conclusion in Fig. 5 of Sect. 4.1. The ordering and pick-up areas have the nature of strong interaction with consumers, the operating table adopts a wooden table to create a neat and intimate experience for consumers, and an electronic menu is embedded in the ordering area to provide convenience for operators; the operating table in the food preparation and packing areas adopt a quartz stone table for easy cleaning by operators; the resting area adopts wooden resting chairs, as shown in Fig. 9.

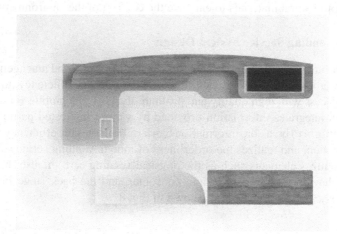

Fig. 9. Vending cart internal movement line design

The size of the operating table suitable for vending activities was designed according to the main size data of vending activities in Table 7, and the rest of the data are designed according to the common size, as shown in Fig. 10.

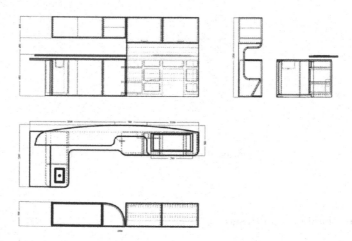

Fig. 10. Internal sizes of the vending cart

Vending Cart Appearance Design. The design of the appearance of the mobile vending cart emphasizes the guidance system in the vending activity, distinguishing between the ordering area and the pick-up area. [19] The appearance design adheres to the design principle of simplicity and fluency, presenting consumers with a clean, hygienic and trustworthy visual effect, as shown in Fig. 11.

Fig. 11. Vending cart exterior design

5.4 APP Design

The APP design serves the mobile vending service system as the front desk to interact with users. Based on the front desk behavioral services in the mobile vending blueprint in Fig. 7, key APP interaction interfaces can be derived: 1. Business location reservation interface: provide operators with planned business location information and guide reservations. 2. Vending cart rental interface: introduce operators to vending carts and guide rental of vending carts. 3. Information release interface: book vending activities for operators and deliver vending information to consumers. 4. Feedback interface of merchandise selling: provide consumers with the entrance of merchandise feedback. 5. Operator online store interface: provide consumers with the entrance of continuous consumption. 6. Operator community entrance interface: organize and maintain continuous consumption groups for operators. The logic of APP visual display and external interaction is shown in Fig. 12, and the APP information architecture is shown in Fig. 13.

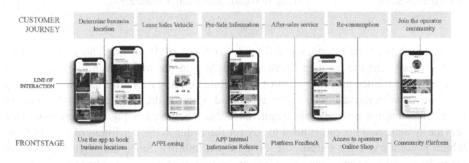

Fig. 12. App visual display and external interaction logic

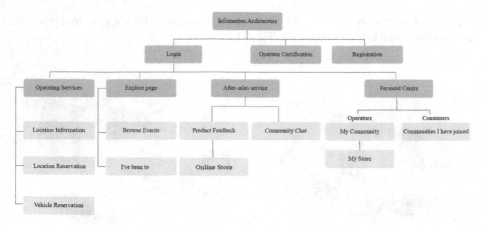

Fig. 13. APP information architecture

6 Design Evaluation

Heuristic design interview was conducted to measure the effectiveness of the design solution. A mobile vending scenario was simulated for the interviewees, simulating the existing mobile vending and the service system designed in this paper. [20] Satisfaction evaluation was conducted at the end of the interviews to measure the effectiveness of the mobile vending service system.

The interview questions are shown in Table 8.

Table 8. The interview questions

Problem Dimension	Contents of heuristic interview questions
Urban Governance	How to determine the location of mobile vending
Selling Costs	How to solve mobile vending cart tools
User trust mechanism	How to protect the goods after-sales issues
Continuous Sales	How to consolidate a stable clientele

The results of the interviews indicated that 8 out of 10 people chose the solution of this paper, and the positive feedback thought that the mobile vending system can effectively guarantee the orderliness of the mobile vending economy, establish a healthy relationship with managers and consumers, and promote the long-term development of the mobile vending economy; the negative feedback believed that it costs to maintain the service system, and operators need to pay for the order to the detriment of the interests, which is contrary to the purpose of low-cost profitability. Since the negative feedback came from older groups, age and perceptions were not excluded. Overall, the design evaluation shows that users have a positive opinion of the mobile vending system.

7 Conclusion

The study focused on the mobile vending economy, analyzed the organizational form and interested parties based on literature, identified the opportunity points of existing service models using user journey maps and KANO models, and analyzed the design of movement lines and size for vending carts. The specific research results include: first, the mobile vending service is optimized with the product service mode that combines the physical object with service; second, by integrating site information and after-sale information, an integrated service platform is built to facilitate information exchange among interested parties, alleviate conflicts between mobile vending and urban governance, and build consumer trust in mobile vending; third, the vending cart product provides a more comfortable business environment for operators, reduces operator access costs in the form of leasing, and improves the working environment of mobile vending economy. Compared with the traditional mobile vending service, the service system built in this study realizes information visualization, easy dissemination, orderly process and easy operation, which helps to improve the urban governance mechanism; compared with the traditional vending cart, the mobile vending cart designed in this study improves the ease of use of functional space and the comfort of working environment, which improves the unfriendly working environment of operators. This study also has some limitations, as the design of the evaluation content presents the problem of "who pays for the order", which needs to be studied more deeply in the field of sociology and management. It is hoped that the optimization of the economic model of mobile vending in this study will facilitate the development of the market environment for all stakeholders.

References

1. Wang, C.: Guiding the orderly development of "stall economy". J. People's Forum **2020**(20), 77–79 (2020)
2. Scott, J.C.: Perspectives on the state, pp. 69–70. translated by Wang Xiaoyi, Social Science Literature Press (2019). Terry N. CIark, Daniel Aaron Silver. Scenescapes: How Qualities of Place Shape Social Life[J],University Of Chicago Press 2016
3. Qi, S.: Building cultural scene cultivating endogenous power of urban development - A perspective of living cultural facilities. Dongyue Ser. **38**(01), 25–34 (2017). https://doi.org/10.15981/j.cnki.dongyueluncong.2017.01.003
4. Chen, Y., Wang, Y.: On the administrative regulation path of "stall economy" embedded in modern governance system--a sample of regulatory policies in 15 cities. Hun. Soc. Sci. **06**, 93–103 (2020)
5. Zhou, Z., Zhang, J., Wang, Z.: The "stall economy" and good street governance: Insights from the management strategies of the "stall economy" in the Asia-Pacific region. Int. Urban Plan. **37**(02), 74–81 (2022). https://doi.org/10.19830/j.upi.2020.372
6. Liang, Z.-T., Du, Y.: The landscape of daily life: the renewal of traditional vegetable market and the exploration of garden bazaar. Urban Dev. Res. **26**(07), 30–36 (2019)
7. Yu, H.: View the quality of the "elephant"-a micro-exploration of the phenomenon of "creative bazaar" in Jingdezhen ceramic market. Art Rev. **04**, 142–145 (2014). https://doi.org/10.16364/j.cnki.cn11-4907/j.2014.04.028
8. Qingmei, Z.: Creative bazaar: field production, landscape consumption and group revelry of youth subcultural capital. China Youth Stud. **11**, 5–11+28 (2017). https://doi.org/10.19633/j.cnki.11-2579/d.2017.0017

9. Jiang, X., Lai, X.: Spatial rights conflict and adaptation in the rule of law of municipal social governance - a sample observation of informal economy. J. Anhui Normal Univ. (Hum. Soc. Sci. Edn.) 50.02(2022):95–102. doi:https://doi.org/10.14182/j.cnki.j.anu.2022.02.012

10. Solano, A., et al.: Smart vending machines in the era of internet of things. Fut. Gen. Comput. Syst. **76** (2017). https://doi.org/10.1016/j.future.2016.10.029

11. Rakha, A., et al.: Safety and quality perspective of street vended foods in developing countries (2022) .FOOD CONTROL. https://doi.org/10.1016/j.foodcont.2022.109001

12. Turner, S., Oswin, N.: Itinerant livelihoods: Street vending-scapes and the politics of mobility in upland socialist Vietnam. Singapore J. Trop. Geograph. **36**(3) (2015). https://doi.org/10.1111/sjtg.12114

13. Wu, C., Chen, L., Li, P.: Research on user experience map model in shared product service design. Pack. Eng. **38**(18), 62–66 (2017). https://doi.org/10.19554/j.cnki.1001-3563.2017.18.018

14. Tang, C.-J., Long, Y.-L.: Research on personalized demand acquisition method based on Kano model. Soft Sci. **26**(02), 127–131 (2012)

15. Bai, T., Li, Z.: A method for calculating the importance of customer demand based on fuzzy Kano model. China Mech. Eng. **23**(08), 975–980 (2012)

16. Na, C., Wu, Z.: A study of L-shaped kitchen user behavior based on naturalistic observation. Pack. Eng. **42**(18), 165–171. https://doi.org/10.19554/j.cnki.1001-3563.2021.18.018

17. GB/T10000–1988Chinese Adult Body Size.S

18. Li, F.: Omnichannel service blueprint - a study based on the perspective of customer experience and service channel evolution. J. Beijing Univ. Technol. Bus. (Soc. Sci. Edn.) **34**(03), 1–14 (2019)

19. Design of Urban Guidance Systems. Urban Dev. Res. **01**, 64–69(2004)

20. Hu, F..: Application and design of heuristic techniques in teaching case of user interviews for older adults. New Art **38**(04), 124–131 (2017)

E-Commerce and Covid-19

An Analysis of Payment Transactions and Consumer Preferences

Sarah Krennhuber[1]([⊠]) and Martin Stabauer[2]([iD])

[1] Netcetera, Linz, Austria
sarah.krennhuber@netcetera.com
[2] Johannes Kepler University, Linz, Austria
martin.stabauer@jku.at

Abstract. This paper investigates the changes in e-commerce transactions in Germany, Switzerland and Austria in the years from 2019 to 2022, during the course of the Covid-19 pandemic. Due to temporary closures of brick-and-mortar stores, the fear of consumers to be infected with the Corona virus and shortages of goods in supermarkets, online retail in general has experienced a strong growth, even though some industries were suffering from severe declines in sales. We show the changes in transaction volumes in Germany, Switzerland and Austria and detail some of the developments. Based on categorized 3-D Secure transaction data, the results are presented on a country and industry level. The results of the transaction analysis show that during the pandemic, more food and groceries were purchased online, and also food delivery services were used more often. By contrast, the sectors of tourism and culture and events suffered immense losses. However, many categories were able to (over)compensate their losses in the following years. Differences across countries include, that for tourism in Germany it took longer to catch up than in the other two countries, and that in Austria food delivery services and groceries grew a lot more than in the other two countries. The total amount of sales over all categories remained stable from 2019 to 2020 but increased later. In order to include the consumers' points of view and to derive further insights, a supplementing quantitative research in the form of a survey study was conducted.

Keywords: E-commerce · Covid-19 · E-payment · 3-D Secure

1 Introduction

The Covid-19 pandemic has caused a severe health and economic crisis in most parts of the world. Since the outbreak of the virus in 2019 countries worldwide enforced restricting regulations including social distancing and even closing shops and restaurants to minimize the risk of the virus spreading even further [6]. These lockdowns had a high impact on wide parts of the economy, as the drop in sales in brick-and-mortar retail associated with the Covid-19 pandemic posed an unprecedented challenge for retailers and producers [7]. In addition to the

© Springer Nature Switzerland AG 2023
F. Fui-Hoon Nah and K. Siau (Eds.): HCII 2023, LNCS 14038, pp. 219–229, 2023.
https://doi.org/10.1007/978-3-031-35969-9_15

economic factors, a change in the population's consumer culture could also be observed. Goods that were primarily bought in-store before the pandemic, such as groceries, were increasingly being ordered online. In contrast, consumers were cutting back on spending on hotels, flights and leisure activities [13].

In the German-speaking countries Germany, Switzerland and Austria (GSA), the focus for many companies selling their products online, resembled a "just-in-time" approach in the sense that they were focusing on purchasing or producing goods when they are purchased and not having large quantities of their products in stock. A lot online-sellers were therefore not able to comply with the increased demand in online purchases, due to supply bottlenecks and unexpected delays in their supply chain [4]. Large online warehouses, such as Amazon and OTTO, gained a clear competitive advantage from this situation. Due to the size of the company and the many storage centers, as well as the optimized logistics and delivery processes, Amazon was able to offer better product availability than most other online shops. [8]

The aim of this paper was to create an industry and country comparison of online transactions during the Corona Pandemic for the GSA countries. The questions asked are therefore (a) How did the actual sales in various e-commerce sectors in the GSA countries develop during the course of the pandemic, and (b) How do consumers perceive their change in behavior towards e-commerce in these years.

2   Related Work

2.1   E-Commerce During the Pandemic

The Covid-19 pandemic had a massive impact on consumers' shopping behavior. People tended to avoid crowded areas because of the risk to get infected with the virus. Instead, many preferred to buy products and services online instead of visiting a store. This led to increased sales volumes in e-commerce in many areas [3]. Some products, such as groceries or hygiene items, have been commonly bought offline before the pandemic. Another important factor for the online retail increment is that many companies, which had the opportunity to do so, introduced working from home for their employees. For these people their everyday routines changed in the sense that shops that had been visited on the way to and back from work, were now replaced by online shops. Lunch, which was bought in the supermarket next to their working place, was more often ordered through an online delivery service. [9]

Especially the increase as a percentage in the area of online grocery retail has grown enormously during the pandemic in a lot of countries. A study conducted in Germany in 2021 showed that more than half of the asked participants started to buy groceries online during the pandemic and most of them also stated that they were planning to continue to do so [11]. Moreover, consumers often had no other choice than to obtain the goods they needed online due to the temporary, ordered closures of the retail trade. These temporary closures meant enormous sales losses for stationary retail. A study from 2020 stated that the stationary retail in Austria lost an average of around 110 million euros in gross sales per day due to those restrictions. [10]

2.2 Online Payment in Germany, Switzerland and Austria

Consumers in GSA countries are notoriously skeptical towards cashless forms of payment. This applies to both physical payments at a point of sale and electronic payments for online shopping. In Germany, for example, cash is still the most preferred way to pay for goods. Even in 2020, when there were hygiene measures in place at shops and people kept social distancing to not get infected with the virus, 6 out of 10 transactions were done with cash. The percentage of cashless payments is increasing in Germany but it is still far higher than any other payment method. A research from 2020 showed that after paying with cash (60%), debit cards are the second preferred payment method (23%), followed by credit card (6%). Only 2% of the examined transactions were paid with a mobile payment method, such as Apple Pay. [2]

The low level of cashless payment in the GSA countries may be caused by a lack of trust and transparency in financial service processes [12]. Surveys done in Germany before the outbreak of the Covid-19 virus showed that in Germany people tend to stick on their preferred payment methods and that it is not likely that cash payments will decrease as in other countries, unless a fundamental change would happen in the payment industry [5]. According to the European Central Bank (ECB), the percentage of physical cashless payment in German stores in 2020 has increased by 21% compared to before the pandemic has started in 2019. Also in the German online world the amount of transactions has increased, albeit slowly, especially for payment service providers like the market leader PayPal. In 2020 the transaction volume in this area has increased by over 30% compared to the previous year [14]. In Austria, the preferred types of payment in e-commerce in 2021 were credit cards with 31%, followed by bank transfer with 32%, E-wallets with 15% and E-invoice with 6%. These invoices have a larger market share in Germany with 13%, while in Switzerland credit cards represent 53% of e-commerce revenues.[1]

2.3 3-D Secure

The European Union's Revised Directive on Payment Services 2015/2366 (PSD2) aims to regulate payment services and payment service providers. Article 97(1) requires that payment service providers "apply strong customer authentication where the payer ... (b) initiates an electronic payment transaction". Strong customer authentication is defined in Article 4(30) as "an authentication based on the use of two or more elements categorized as knowledge (something only the user knows), possession (something only the user possesses) and inherence (something the user is) that are independent, in that the breach of one does

---

[1] Figures from https://www.statista.com/outlook/dmo/ecommerce/austria, last accessed 6 Feb 2023.

not compromise the reliability of the others, and is designed in such a way as to protect the confidentiality of the authentication data".

3-D Secure is the most common user authentication protocol and bank security mechanism when it comes to online payments with credit card. The protocol minimizes the risk of credit card fraud in the online world and makes online payments safer for users. When the 3-D Secure protocol is implemented, the cardholder needs to authenticate first before the payment will proceed, this procedure is ensured by the card-issuing-bank. This is not only a benefit for the user, but also for the merchant itself [1].

3-D in this regard stands for the three interacting domains (acquirer, issuer and interoperability). Version 2.0, published in 2016, aims to comply with the requirements defined in the PSD2 and introduces a distinction between high and low risk transactions and varies the steps required from consumers: Two-factor-authentication versus simplified processes. This allows for more flexibility and less consumer dropouts [15]. The protocol is implemented by major card issuers in slightly different variants.

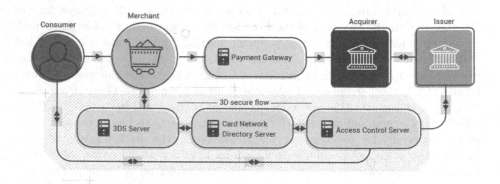

Fig. 1. Payment Process 3-D Secure (https://3dsecure2.com, last accessed 6 Feb 2023)

Figure 1 depicts the payment process when applying the 3-D Secure protocol. When a customer is purchasing something from an online merchant who is using the 3-D Secure protocol, the customer will be asked to authenticate. To do so, the customer needs to enter the 3-D Secure enrollment password (knowledge in the sense of the PSD2) and either a SMS-TAN (possession of the mobile phone) or use biometric authentication from within a mobile application (inherence). This authentication is processed on the issuer domain side. As soon as the cardholder has authenticated the payment authorization is processed along the merchant–acquirer path. The third included party, the interoperability domain describes the 3-D secure flow, including the communication path between credit card network and 3-D Secure server for verification, as well as the access control server for authentication.

3   Methodology

In order to shed light on the research questions mentioned earlier, our study's methodology is twofold: It is composed of (a) an analysis of 3-D Secure transactions and (b) an online questionnaire. These two components are addressed in detail in this section.

3.1   Transaction Analysis

The data analyzed for the first part of this work consists of 3-D Secure transactions (see Sect. 2.3 for details) from Germany, Switzerland and Austria from January 2019 to December 2022. The data was contributed by Netcetera, a provider of 3-D Secure processing services. As not all credit card transactions include the 3-D Secure protocol, our data cover around 25% of all e-commerce sales paid with credit cards in Germany and Austria, and around 90% in Switzerland. In total we accessed and analyzed payments in the amount of EUR 26,9bn. Figure 2 shows an exemplary transaction as included in our analysis.

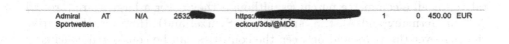

Fig. 2. Example transaction

The transactions were grouped into the categories shown in Table 1 and compared across countries and years. The study design implicates the following limitations:

- Transactions carried out by external payment providers such as PayPal or Revolut can not be categorized into one of the established groups. The reason for this is that the transaction data does not contain any hints on the categories of the product(s) or service(s) bought in these cases.
- The same goes for the large online retailers selling goods from various type (e.g., Amazon, OTTO). We can not determine the specific product category from the transaction data and therefore introduced an additional category 'Online Warehouses'.
- To make transactions in Switzerland comparable to the ones in Germany and Austria, we converted all sales from Swiss franc to Euro using yearly average exchange rates as published by the European Central Bank, which are 1 CHF = 0.89 EUR (2019), 0.92 EUR (2020), 0.93 EUR (2021), and 0.99 EUR (2022).
- While the data can be compared very well longitudinally and also across borders, it can not be seen as representative for the e-commerce sector as a whole. Not all online purchases are paid for via credit card, and not all credit card payments include the 3-D secure protocol. However, as explained above, we were able to cover a wide range of transactions, depending on the country.

3.2 Online Questionnaire

The second part of our study is an online questionnaire sent out electronically from 24 March 2021 to 1 May 2021. The participants were asked questions revolving around their changing shopping behavior due to the pandemic. In order to amend the results of the quantitative study, we used the same categories and we included questions regarding experiences and intentions.

305 questionnaires were filled out at least partially, and 284 completed questionnaires were used for the evaluation. 53% of the participants were female, 43% were male. More than half of the participants were 30 years or younger, 20% were between 31 and 40, and 19% were 41 and older. 32% were from Germany, 44% from Austria, and 20% from Switzerland. 56% had a university degree, and 26% a high-school diploma.

4 Results

As elaborated earlier, our data set consists of credit card transactions making use of the 3D-Secure protocol. Therefore, absolute numbers are not representative for all e-commerce purchases, although they cover a large share (around 25% in Germany and Austria, and 90% in Switzerland). However, the relative changes over the years and between the countries can be considered conclusive and meaningful.

Our results are mostly in line with the widespread assumptions that e-commerce sales in many categories increased significantly when comparing the years 2019 and 2020. This makes sense, as in the examined countries the impact of the pandemic had its peak starting in March 2020. This applies, e.g., to electronic devices with an increase of 58% across all 3 countries (from EUR 1,046 m to 1,652 m), food delivery (+82% from EUR 93 m to 170 m), and especially to groceries rising by 180% (from EUR 76 m to 212 m). However, some product categories saw a substantial decline. These categories are culture & events falling by 65% (from EUR 349 m to 123 m), public transport with −59% (from EUR 1,195 m to 489 m), and tourism with −45% (from EUR 1,181 m to 650 m).

All these developments from 2019 to 2020 seem logical, as tourism, public transport and events were all heavily restricted by law across all 3 countries in the course of the year 2020, and many people started ordering their groceries and other products online. Some of these restrictions were loosened in the following months and years, so we wanted to see what that meant in terms of e-commerce sales. Our findings show that most categories (over-)compensated their 2020 shrinkage in the following years 2021 and 2022. Tourism, for example, increased by 89% from 2020 to 2021 and by 16% from 2021 to 2022, meaning that in 2022 we saw EUR 1,426 m in sales in this category, compared with 1,181 m before the pandemic. Public transport sales increased by 67% from 2020 to 2021 and by 40% from 2021 to 2022, ending up at EUR 1,145 m – very close to the pre-covid level of EUR 1,195 m.

Most categories that showed rising sales from 2019 to 2020, bottomed out in the following years or even decreased slightly. For example, furniture sales increased by 47% (EUR 395 m in 2019 to 580m in 2020), this was followed by +4% (EUR 606 m in 2021) and −7% (EUR 562 m in 2022). A special case is the category of online warehouses like Amazon and other big retailers as OTTO, which had a boost of 67% from 2019 to 2020 (EUR 325 m to 544 m), an even bigger one in the year after (+119%, EUR 1,194 m in 2021) and another rise of 8% after that (EUR 1,290 m in 2022).

Interestingly, the total amount of sales over all categories remained almost stable from 2019 to 2020 (EUR 5,420 m vs 5,415 m), but then increased to EUR 7,571 m in 2021 and EUR 8,525 m in 2022. Table 1 shows the numbers from our analysis.

Table 1. Findings (numbers in million Euros)

	2019	**2020**	**2021**	**2022**
Public transport	1,195	489	819	1,145
Tourism	1,181	650	1,232	1,426
Electronic devices	1,046	1,652	1,685	1,641
Furniture	395	580	606	562
Fashion	365	521	570	674
Culture & Events	349	123	292	526
Online Warehouses	325	544	1,194	1,290
Sport Equipment	190	192	279	337
Tools & Garden Equipment	125	185	193	203
Food delivery	93	170	285	351
Drugstore	79	96	155	162
Groceries	76	212	262	208
Total	**5,420**	**5,415**	**7,571**	**8,525**

In the next step we compared the sales figures across countries, which brought up some interesting results and surprisingly considerable differences. Figures 3, 4 and 5 show the respective relative changes over years.

When comparing the tourism transactions, there were consistent decreases from 2019 to 2020, but while the sales in Switzerland (+103%) and Austria (+71%) caught up in 2021, it took longer in Germany (+48% in 2021 and +53% in 2022). Similarly, Germany constitutes an exception when it comes to online warehouses. The considerable growth in 2020 (48% in Germany, 75% in Switzerland, 48% in Austria) was followed by a comparable growth in 2021 in Switzerland (55%) and Austria (54%), but by a way stronger rise in Germany (310%).

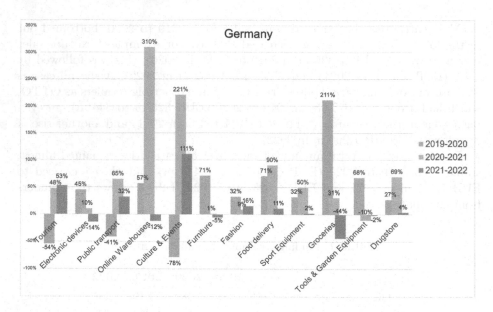

Fig. 3. Results Germany

Austria, in contrast to both other countries, saw a continuous growth over the years for both sport equipment (10%, 75%, and 88%) and also tools & garden equipment (76%, 141%, and 27%). The development of both categories was fundamentally different in Germany (32%, 50%, and 2% for sport equipment; 68%, −10%, and −2% for tools & garden equipment) and Switzerland (−19%, 30%, and 2%; 30%, −8%, and 1%, respectively). In Austria, the sales of food delivery services went up by an astonishing 596% from 2019 to 2022, whereas the increases in Germany (360%) and Switzerland (307%) were notably more moderate. A very similar case can be observed with the category of groceries (564% in Austria, 229% in Germany, 197% in Switzerland).

To complement these insights into transaction data, we conducted a survey among consumers in the three countries, details on the survey itself can be found in Sect. 3.2.

The participants were asked whether since the beginning of the Covid-19 pandemic they had generally bought more products online than usual and whether there were any specific goods or services as stated in Table 1, that they had purchased online for the first time. Overall, around 70% indicated that they generally purchased more products online during the pandemic than they did before. This behavior was more distinct among younger consumers with 87.5%. The answers to the second question varied stronger. 37% answered that they were no goods or services from the mentioned categories that they bought online for the first time. The categories chosen by the largest share of the participants were groceries (16%), tools and garden equipment (16%), furniture (15%), and food delivery services (13%). The distribution across age groups varied, younger

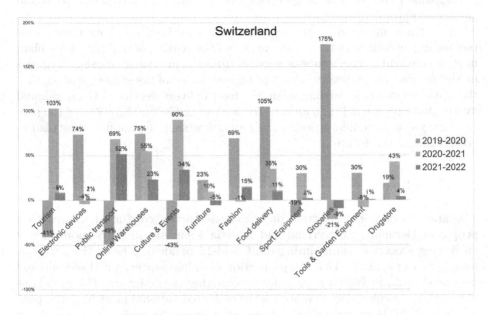

Fig. 4. Results Switzerland

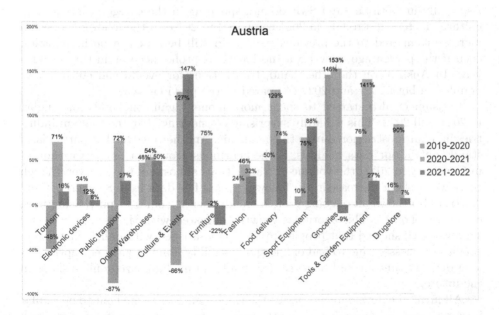

Fig. 5. Results Austria

participants most often chose groceries (28.6%), older ones preferred tools and garden equipment (20.5%).

As a follow-up question and to gain some more insights about their online purchasing intentions in the future, we asked the participants if they were planning to continue buying goods or services from the previously chosen categories in the future. The category with the highest share of consumers stating that they plan to continue buying online are food delivery services (84%), followed by drugstore products (81%) and electronic devices (75%). Only 30% of the people surveyed who bought groceries from an online shop for the first time plan to do so again in the future.

5 Conclusion

The analysis of the transaction data and the survey conducted showed that people in Germany, Switzerland and Austria spent significantly more money on buying groceries online during the Covid-19 pandemic. From the data collected, we can conclude that the proportion of online grocery purchases already decreased again in 2022 and may decrease further in the future. The industries of tourism and culture and events are the ones that suffered most from the pandemic in 2020. However, inferring from our survey, the majority of consumers plan to increase their purchases in these categories again in the future. This is confirmed by the sales data analyzed for these categories in the years 2021 and 2022, in which the tourism and culture/events industries managed to recover. Especially in Germany and Switzerland, spending in the category "events and culture" increased strongly in 2021 compared to 2020. In the following year, an increase compared to the previous year could still be noted in both countries, even if the percentage growth was no longer as pronounced as in the previous year. In Austria, on the other hand, transactions for events and culture have doubled in both 2021 and 2022 compared to the previous year.

Consumers also started to spent more money again on hotels and flights in 2021 and 2022. This is hardly surprising considering that the tourism industries had hard restriction during the pandemic. Another insight that particularly stands out in our data analysis, is the enormous growth of online warehouses in Germany. In 2021, the volume of transactions almost six-folded, even though there was a slight decrease in the following year. The data from Switzerland and Austria also shows an ongoing growth trend when it comes to purchasing from online warehouses. This enormous growth compared with the results from our survey, which showed that people are often not satisfied with online shops from smaller businesses, due to bad online shop usability, limited payment options or long delivery times, could be a risk for small and medium-sized online shops in the future.

All in all, the developments across the three countries turned out to be similar in some cases and quite diverse in others. Therefore, a future avenue of research could be to include more countries with a different attitude on cashless forms of payment. Also, longitudinal analysis over future years may be interesting as at the time of writing the pandemic is not that present in the consumers' minds anymore and their behavior may change again.

References

1. Ali, M.A., Arief, B., Emms, M., van Moorsel, A.: Does the online card payment landscape unwittingly facilitate fraud? IEEE Secur. Privacy **15**(2), 78–86 (2017). https://doi.org/10.1109/MSP.2017.27
2. Balz, B.: Zahlungsverhalten in Deutschland 2020 - Erhebung im Jahr der Corona-Pandemie. Tech. rep, Deutsche Bundesbank (2021)
3. Bhatti, A., Akram, H., Basit, H.M., Khan, A.U., Naqvi, S.M.R., Bilal, M.: E-commerce trends during COVID-19 Pandemic. Int. J. Future Gener. Commun. Netw. **13**(2), 1449–1452 (2020)
4. Bofinger, P., et al.: Wirtschaftliche Implikationen der Corona-Krise und wirtschaftspolitische Maßnahmen. Wirtschaftsdienst **100**(4), 259–265 (2020). https://doi.org/10.1007/s10273-020-2628-0
5. Brancatelli, C.: Preferences for cash vs. card payments: an analysis using german household scanner data. SSRN Electronic Journal (2019). https://doi.org/10.2139/ssrn.3400118
6. Chakraborty, I., Maity, P.: COVID-19 outbreak: Migration, effects on society, global environment and prevention. Sci. Total Environ. **728**, 138882 (2020). https://doi.org/10.1016/j.scitotenv.2020.138882
7. Coccia, M.: The relation between length of lockdown, numbers of infected people and deaths of Covid-19, and economic growth of countries. Sci. Total Environ. **775**, 145801 (2021). https://doi.org/10.1016/j.scitotenv.2021.145801
8. Demary, V.: Onlinehandel. Warum Corona Amazon weiter stärkt. IW-Kurzbericht 32 (2020)
9. Engels, B., Rusche, C.: Corona: Schub für den Onlinehandel. IW-Kurzberichte 29 (2020)
10. Gittenberger, E., Teller, C.: Einkaufsverhalten in Zeiten des Coronavirus. Teil 2: Shutdown. Tech. rep., Institut für Handel, Absatz und Marketing JKU Johannes Kepler Universität Linz (2020)
11. Gruntkowski, L., Martinez, L.: Online grocery shopping in germany: assessing the impact of COVID-19. J. Theoret. Appl. Electron. Comm. Res. **17**, 984–1002 (2022)
12. Jünger, M., Mietzner, M.: Banking goes digital: the adoption of fintech services by german households. Finan. Res. Lett. **34**, 101260 (2020). https://doi.org/10.1016/j.frl.2019.08.008
13. Kim, R.Y.: The impact of COVID-19 on consumers: preparing for digital sales. IEEE Eng. Manage. Rev. **48**, 212–218 (2020)
14. Mai, H.: Bezahlen in Europa 2020 Angleichung an der Ladenkasse, Zementierung nationaler Unterschiede online (2021). https://www.dbresearch.de/PROD/RPS_EN-PROD/PROD0000000000521274/Payment_choices_in_Europe_in_2020%3A_Convergence_at_.PDF
15. Noctor, M.: PSD2: Is the banking industry prepared? Comput. Fraud Secur. **6**, 9–11 (2018). https://doi.org/10.1016/S1361-3723(18)30053-8

Consumers' Intentions to Use Mobile Food Applications

Ralston Kwan[✉] and Norman Shaw

Toronto Metropolitan University, Toronto, ON M5B 2K3, Canada
{ralston.kwan,norman.shaw}@torontomu.ca

Abstract. The use of mobile food applications (MFAs) significantly increased in the past decade. The MFA ecosystem is evolving, and companies need to consider the changing consumer habits to stay competitive. This quantitative study proposes a comprehensive model integrating the Unified Theory of Acceptance and Use of Technology (UTAUT2) and the Theory of Consumption Values (TCV) that examines the factors that affect the behavioural intent to use MFAs. 170 participants were surveyed using the convenience sampling technique. The statistical results and discussions showed that social influence was the most significant predictor on consumers' intention to use MFAs. Performance expectancy, hedonic motivation, and habit were also positively associated with the consumers' intention to use MFAs, while effort expectancy, food safety concerns, and affordance values were not. The findings partially supported the proposed model.

Keywords: Mobile Food Applications · UTAUT2 · TCV

1 Introduction

Mobile food applications (MFAs) are intermediary platforms that offer the ability to order food from restaurants using a smartphone where the order arrives to the consumer using a contactless delivery model. Mobile food applications are convenient for consumers as they show a range of restaurants, menus, and ratings. Consumers can finalize, confirm, and pay for the order securely while tracking their order status with minimal interaction with the restaurant. Mobile food applications are also beneficial to restaurants and delivery drivers as an added channel for revenue generation and flexible earning opportunities respectively. Research on mobile food applications is varied but has mostly discussed the intention to use from the consumer's perspective.

The use of mobile food applications significantly increased in the past decade and revolutionized the way consumers and restaurants interact with each other [1]. Mobile food applications are a subset of online food delivery platforms that utilize the consumers' use of smartphones, mobile internet, and global positioning services to order food [2]. In the Canadian market, the online food delivery sector is expected to generate $6.45 billion USD, a change in 23.6% year-over-year [3]. Worldwide, revenues from MFA platforms should reach $343.80 billion USD, 15.8% change year-over-year where the market in China is expected to be the highest growth at $158.10 billion USD [3].

F. Fui-Hoon Nah and K. Siau (Eds.): HCII 2023, LNCS 14038, pp. 230–248, 2023.
https://doi.org/10.1007/978-3-031-35969-9_16

User penetration is growing and is expected to continue a positive trend for the next five years [4].

This paper's goal is to fill the gap of factors determining users' intention to use mobile food applications and provide knowledge on understanding consumer behaviours that will benefit all stakeholders. This empirical paper develops and validates a novel model that introduces additional constructs for a more holistic view of adoption intentions for mobile food applications. Specifically, we enhance a contemporary technology adoption theory - Unified Theory of Adoption and Use of Technology (UTAUT2) [5] with a consumer behavioural theory - Theory of Consumption Values (TCV) [6].

2 Literature Review

2.1 Mobile Food Applications

We reviewed the extant literature through the lens of motivations and technology adoption models on mobile food applications. Historically, restaurants directly handled every aspect of the delivery order. Fast forward to today, and it is evident that technology has enabled the creation of mobile applications that allow consumers to order food with a few clicks or swipes [7]. Lee et al. [8] extended the UTAUT2 framework to look into information quality and continued intention on mobile food applications where the results indicated that information quality, performance expectations, habit, and social influence as factors in inducing the consumer's continued usage. Alalwan [9] investigated the factor of satisfaction and continued intention to use mobile food applications using the UTAUT2 framework and the results indicated that performance expectations and hedonic motivation had the strongest influence. Meanwhile, Gunden et al. [10] analyzed impulse-buying tendencies in the UTAUT2 framework where the results indicated that performance expectations, habit, and social influence are critical antecedents to using a mobile food application. A qualitative study by Pigatto et al. [11] revealed that mobile food application content, usability, and functionality were important aspects for consumers' intention to adopt. Cho et al. [12] analyzed the differences in perceptions about mobile food applications between single-person and multi-person households and revealed that single-person households were concerned about the variety of restaurants, trustworthiness, and price whereas multi-person households were concerned about convenience, trustworthiness, and application design. Ray et al. [13] applied the U&G Theory to determine the key gratifications that led to intention to use a mobile food application and the results revealed that customer experience, ease of use, and restaurant listings were the most significant factors. Kaur et al. [14] concluded using the Theory of Consumption Values (TCV) that price, prestige, affordance, and visibility had the most significant and positive influence on intention to use. Meanwhile, Tandon et al. [2] revealed that visibility, price, and quality using the TCV framework were the main determinants of intention to use.

2.2 UTAUT2

The Unified Theory of Acceptance and Use of Technology 2 (UTAUT2) was derived to predict technology adoption for a consumer use setting from the original UTAUT which was used to predict technology adoption in an organizational use setting [5]. The original UTAUT framework did not measure voluntary use from the consumer's perspective and as such, hedonic motivation, price value, and habit were added to the framework to create UTAUT2. The original UTAUT created a unified model by synthesising eight technology adoption models: Theory of Reasoned Action (TRA) [15], Theory of Planned Behaviour (TPB) [16], Technology Acceptance Model (TAM) [17], Combined TAM and TPB, Motivational Model (MM) [18], Model of PC Utilization (MPCU) [19], Innovation Diffusion Theory (IDT) [20], and Social Cognitive Theory (SCT) [21]. UTAUT2 as shown below predicts intention to use by measuring performance expectancy, effort expectancy, social influence, facilitating conditions, hedonic motivation, price value, and habit [5].

Previous studies using UTAUT2 [5] have been successful in explaining the factors that influence behavioural intention which is the most significant predictor to system adoption. Such studies pertain to instant messaging applications [22], mobile wallets [23, 24], mobile shopping applications [25], diet applications [26], mobile banking applications [27], healthcare applications [28], and mobile food applications [1, 8, 10, 29–31]. These studies have a commonality focused on mobile applications regardless of their industry. This study continues the investigation of mobile food applications using the UTAUT2 framework.

2.3 TCV

The Theory of Consumption Values (TCV) offers a means to discern consumers' product and service choices [6]. TCV has been extended to research that focuses on components in the online context such as social media brand communities, virtual goods, and frequent flyer programs [32–34]. Previous studies have confirmed TCV is applicable to mobile apps [35, 36], food tourism [37], mobile banking services [38, 39], and restaurant satisfaction [40].

As the TCV framework only provides generic categories of value, it has been extended to the context of mobile applications. Choe and Kim [37] defined functional value in terms of price value and food safety concerns in their study of food tourism. Thomé et al. [41] extended functional value to be measured in health consciousness in their study that focused on healthy food choices. Kaur et al. [14] defined functional value in terms of price value, health consciousness, and food safety concerns in their study on mobile food applications. With regards to social value, Kaur et al. [14] defined it by prestige value. Emotional value has been defined by hedonic value [14, 33] and epistemic value was measured by visibility [14]. Gupta et al. [42] defined conditional value in terms of affordance value in their study on tourist mobile applications. Studies on mobile wallets [23] and mobile food applications have also extended conditional value to be measured by affordance value. The above studies adopted an extended TCV framework and we adopted a similar model in the context of mobile food applications.

2.4 Proposed Framework

There have been studies on mobile food applications using the UTAUT2 framework [1, 8–10, 29–31, 43]. Mobile food application studies have also used the TCV framework

[2, 14]. This study proposes a novel framework that combines both UTAUT2 and TCV-based models to examine the association between technology adoption and consumption values as the antecedents of intention to use a mobile food application. The independent UTAUT2 constructs are performance expectancy, effort expectancy, social influence, facilitation conditions, hedonic motivation, and habit. The independent TCV constructs are functional (price, health consciousness, and food safety concerns), epistemic (visibility), and conditional (affordances). The proposed research model is shown below as Fig. 1.

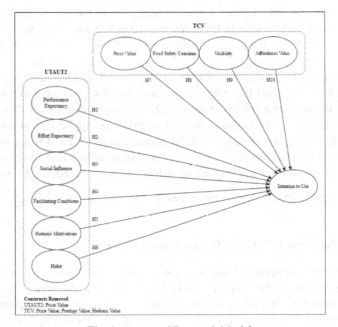

Fig. 1. Proposed Research Model.

3 Hypotheses Development

3.1 Performance Expectancy (UTAUT2)

Performance expectancy is "the degree to which an individual believes that using the system will help him or her" [44]. Mobile food applications allow users to order food and have it delivered to their location to which it needs to be measured as a utility tool. Performance expectancy in the context of MFAs is an important measurement because the application needs to be perceived as useful in completing the task of acquiring food and without spending a lot of time. Previous studies have validated that performance expectancy has a significantly positive effect on intention to use, such as mobile banking [27], mobile instant message applications [22], healthcare programs [28], and mobile wallets [23]. With regards to studies on mobile food applications, the findings show that when users perceived a higher functionality of mobile food applications, it led to a higher intention to use [1, 30, 31]. Consumers who use MFAs would find it useful because the platform allows them to view and compare the various food options and prices based on their location. This leads to:

H1. Performance expectancy positively influences the intention to use a MFA.

3.2 Effort Expectancy (UTAUT2)

Effort expectancy is "the degree of ease associated with the use of the system" [44]. Mobile food applications should be accessible such that users can utilize them with minimal difficulty. Effort expectancy in the context of MFAs is another important measurement because the learning curve of the application should not be steep and should be understandable by the general population. Studies with similar context to mobile food applications have validated that effort expectancy has a significant and positive effect on intention to use such as mobile banking [27], diet applications [26], and self-serve restaurant technology [45]. In the context of mobile food application studies, research has shown that effort expectancy is a valid measurement [1, 29, 43]. Consumers who use MFAs would expect the platform to use consistent themes and be intuitive such as having an easy-to-use interface to place an order or undo unintended actions. Therefore:

H2. Effort expectancy positively influences the intention to use a MFA.

3.3 Social Influence (UTAUT2)/Social Value (TCV)

Social influence is "the degree to which a consumer perceives that other significant people believe technology use to be important" [44]. This construct is also part of the TCV model and defined as social value where it is the image the consumer wants to portray through their decisions [6]. This construct overlaps from both models and will be referred to as social influence. In recent years, there has been a growing trend of companies partnering up with influencers to review their product or service [46]. Previous studies have validated that social influence has a significant and positive effect on intention to use in various sectors such as instant messaging [22], tourist mobile applications [42], and mobile payments [47–49]. Studies that focused on MFAs have also found social influence as a valid measurement in affecting intention to use [1, 8, 10, 43]. Consumers who place orders through MFAs would be influenced by their friends, family, and social groups and thus:

H3. Social influence positively influences the intention to use a MFA.

3.4 Facilitating Conditions (UTAUT2)

Facilitating conditions is defined as "the degree to which an individual believes that the organizational and technical infrastructure exists to support use of the system" [44]. The intent to use a mobile food application is affected by the condition the consumer is facing with regards to having the knowledge and support on using the software. Studies have concluded that facilitating conditions has a significant and positive effect on intention to use with regards to mobile shopping applications [25] and mobile wallets/payments [23, 49]. In the context of MFAs, studies have shown similar findings of the positive relationship [29, 30, 43]. Consumers who use MFAs would expect the platform to have multiple channels for instant technical support, be quick to load, have minimal broken links, and have quick ordering capabilities. Therefore:

H4. Facilitating conditions positively influences the intention to use a MFA.

3.5 Hedonic Motivation (UTAUT2)

Hedonic motivation is described as the fun or pleasure derived from using a technology [5]. This construct is also in the TCV model and will be discussed together. Mobile food applications should be enjoyable and arguably fun and entertaining. Previous studies have shown evidence to support hedonic motivation as a valid measurement of intention to use in mobile shopping applications [37] and mobile wallets/payments [24, 49]. MFA studies have similarly shown that hedonic motivation has a positive and significant effect on adoption intentions [9, 29, 30, 43]. Consumers who use MFAs would expect to see a variety of food options with vivid and vibrant colours, images, and animations that will stimulate a sense of excitement. Thus:

H5. Hedonic motivation positively influences the intention to use a MFA.

3.6 Habit (UTAUT2)

Habit is "the extent to which people tend to perform behaviours automatically because of learning" [5]. The intention to use mobile food applications is affected by the consumer's frequency in using the program. Previous studies have supported that habit has a positive and significant effect on intention to use in mobile payments [49], and mobile shopping applications [25, 50]. Mobile food application studies have also confirmed that habit has a positive effect on intention to use [8, 10, 30]. Consumers who continuously use other mobile applications such as mobile banking and mobile shopping may adopt MFAs as part of their habitual use of mobile applications. In addition, consumers who use MFAs may have had a positive experience using the platform and may have developed a habit to continually use it. Therefore:

H6. Habit positively influences the intention to use a MFA.

3.7 Price Value (UTAUT2) / Price Value (TCV)

Price value is the monetary cost incurred to use the technology [5]. A vast amount of mobile applications are generally free to download on smartphones [24]. In the context of MFAs, they are also free to use on Google Play and the App Store [51] and thus price value is not relevant from the UTAUT2 framework. However, it is applicable under the TCV framework. Price value in the context of TCV represents the functional value which is the degree to which an individual perceives the benefits offered, such as price reasonableness, and value for money [14]. Adoption of mobile food applications are influenced by the consumer's perceived belief the value they pay on the platform is acceptable. Studies have supported that price value and intention to use have a positive and significant effect on food tourism [37], and mobile banking services [38, 39]. A meta-analysis of price value validated it as a measurement of adoption intentions [52]. For studies on mobile food applications, Palau-Saumell et al. [30] identified and validated that price value is an important antecedent for intention to use. Consumers who use MFAs would expect a good price value on their food order with respect to the perception that the food order is at a reasonable price while factoring in money saved on transportation. A good price value can also be derived with no hidden costs when completing the order and without the need to pay gratuities. Hence:

H7. Price value positively influences the intention to use a MFA.

3.8 Food Safety Concern (TCV)

Food safety concern in the context of TCV represents functional value. Food safety concern is the degree to which an individual perceives the quality of food ordered through mobile food applications is safe to consume [14]. Adoption of mobile food applications are influenced by the consumer's view of food quality through the platform. Suhartanto et al. [53] have shown that food quality has a significant and positive impact on loyalty for mobile food applications. Food safety is an important aspect of food quality where it implies that the hazards of preparing the ingredients or meal have been minimized to prevent harm or death when food is consumed. Studies have shown the negative impact of improper food handling in mass media [54] and the importance of food safety throughout the food industry [55]. Consumers who have a high degree of concern for food safety have low intentions to adopt a mobile food application due to an unknown amount of people touching the food container that could spread germs easily. Consumers who use MFAs are not bothered by the possibility of cross-contamination and food-borne illnesses. Therefore:

H8. Food safety concerns negatively influences the intention to use a MFA.

3.9 Visibility (TCV)

Visibility in the context of TCV represents epistemic value, which is the consumer's desire to arouse curiosity or provide innovativeness by seeing others use a product or service [6, 14]. A study on food tourism determined epistemic value as antecedent to consumer behaviour where epistemic value is measured by the curiosity and desire to seek new foods as a learning opportunity [37]. Thomé et al. [41] also validated epistemic value has a positive effect of consumer behaviour in the context of healthy foods where consumers are willing to seek novel information and nutritional facts about healthy food. Mobile banking studies have found similar findings to support epistemic value on intentions to use where epistemic value is defined as the consumer's preference to be ahead on the latest technology trends and having a versatile life [38, 39]. Studies on wearable technologies have shown that visibility has a positive relationship with adoption intentions [56]. Visibility has also been supported as a valid measurement in the context of mobile applications [57] and mobile food application adoption [14]. Consumers who use MFAs are affected by visual stimuli such as online and television advertisements and observing colleagues use the application. Thus:

H9. Visibility positively influences the intention to use a MFA.

3.10 Affordance Value (TCV)

Affordance value in the context of TCV represents conditional value which is the impact of the available choices at a specific moment in time the consumer is facing. Affordance value is the degree to which an individual perceives the features and advantages that drive value [14]. Omigie et al. [39] validated that conditional value as an antecedent of intention to use in the context of mobile banking apps is an easier payment alternative to other services while reducing the threats of carrying physical cash. Consumers choosing to purchase from online travel agencies (OTAs) are affected by conditional values and in the context of OTAs are measured by free cancellations, promotional incentives and the amount of listing of properties [34]. In the context of mobile food applications, Kaur et al. [14] has validated that affordance has a positive and significant effect on intention to use. Consumers who use MFAs are affected by promotional offers such as coupons, delivery times and charges, and the availability of the consumers' preferred restaurants. Hence:

H10. Affordance positively influences the intention to use a MFA.

4 Methodology

4.1 Construct Measurement and Sampling

To address the research objectives and hypotheses, a quantitative methodological approach was adopted. A survey instrument was developed based on existing literature on mobile food applications with respects to UTAUT2 and TCV scales. For each indicator, a 5-point Likert scale was used ranging from 1-strongly disagree to 5-strongly agree. The survey has two sections: the first section captures behavioural and demographic information such as gender, age, income, and mobile food application frequency and the second section is the measurement items of independent and dependent variables from extant literature. A pilot survey was sent to colleagues to confirm that the respondents understood the questions. Their comments and suggestions were incorporated, and the study was launched with the finalized questionnaire.

Using convenience sampling, the questionnaire was accessed by the student research pool at the Ted Rogers School of Management. A total of 248 responses were collected and used in the data analysis. Of these responses, 220 were complete with a response rate of 88.7%. After eliminating 50 responses that were problematic (e.g., 'speeders', 'laggards', 'straightliners'), there were 170 usable survey responses.

4.2 Partial Least Squares

The statistical analysis method used in the study was Partial Least Squares (PLS). PLS is a multivariate analysis that simultaneously analyzes multiple variables and is suitable for exploratory studies using priori established theories [58]. PLS achieves high levels of statistical power with small sample sizes and does not require normal distributed data because PLS-SEM is a nonparametric method [58].

PLS analysis has two major components: the structural model and the measurement model where the structural model illustrates the relationship between the constructs whereas the measurement model illustrates the relationship between the constructs and their respective latent variables [58].

Reflective and Formative Constructs. For this study, there was a mix of reflective and formative constructs. The reflective constructs in the study are Effort Expectancy (EE), Social Influence (SI), Facilitating Conditions (FC), Hedonic Motivation (HM), Habit (HA), Price Value (PV), Visibility (VS), Affordance Value (AV), and Intention to Use (ITU). The formative constructs in the study are Performance Expectancy (PE) and Food Safety Concerns (FS). These two constructs have measurements that capture a specific aspect of the construct domain.

PLS Procedures. Upon the creation of the measurement models, the structural model was tested using the PLS algorithm that calculated the path coefficients between the constructs and the coefficient of determination (R^2) for endogenous (dependent) constructs. PLS assumes the data is not normalized and bootstrapping was required to test the statistical significance of the path coefficients, Cronbach's alpha, HTMT, and R^2 values. Hair et al. [58] suggests a bootstrapping sample of 10,000 samples and was consequently used for this study.

5 Results

5.1 Descriptive Statistics

Descriptive statistics were tabulated on the demographic questions pertaining to gender, age, income and MFA experience. As the variables were ordinal, only the frequency was calculated. Of the 170 participants (N = 170), the majority of the participants were women (68.2%), men accounted for 29.4% and the remaining participants were non-binary and preferred not to disclose at 1.2% each. Furthermore, a majority of the participants were between the ages of 18–21 (74.1%) and followed by 22–25 (15.9). The minority were ages >25 and <18 at 7.6% and 2.4% respectively.

5.2 Statistical Analysis

PLS analysis begins with the evaluation of the measurement models and then proceeds with the evaluation of the structural model. Models with reflective indicators have different procedures than models with formative indicators and are explained below.

For reflective constructs, the first step is to assess the measures' reliability (indicator reliability and internal consistency reliability) and validity (average variance extracted

(AVE) and heterotrait-monotrait (HTMT) ratio of correlations). The next and final step concludes with the evaluation of the structural model to assess collinearity (VIF), the significance of the path coefficients, and the explanatory power (R^2) [58].

For formative constructs, the first step is to assess the measures' convergent validity, collinearity issues, and the significance and relevance of the formative indicators. Like the reflective model, the final steps remain the same to assess collinearity, the significance of path coefficients and the explanatory power (R^2) [58].

5.3   Reflective Construct Results

Indicator Reliability. The size of the outer loading is also called the indicator reliability and the rule of thumb is that the outer loadings should be 0.708 or higher [59].

All measurement indicators with the exception of the following constructs: facilitating conditions (FC), price value (PV), and visibility (VS) reached the minimum reliability value of 0.708. Consequently, FC, PV, and VS were dropped from the research model and correspondingly hypotheses 4, 7, and 9 were dropped from the study. In addition, the measurement HA1 for habit did not reach the threshold and was removed.

Internal Consistency and Convergent Validity. The rule of thumb for internal consistency is that values should be greater than 0.7 [60] but for exploratory research, values between 0.60 and 0.70 are acceptable [58] such as this study that incorporated two priori established frameworks. The results in Table 1 below show that constructs have coefficients greater than 0.6 for both Cronbach's alpha and composite reliability. For convergent validity, the standard measurement is the average variance extracted (AVE) and should be 0.5 or higher and has been met for the reflective constructs in the study as shown in Table 1.

Table 1. Construct Reliability and Validity

Construct	Cronbach's Alpha	rho_A	Composite Reliability	AVE
AV	0.818	0.853	0.869	0.571
EE	0.873	0.925	0.911	0.72
HA	0.661	0.669	0.855	0.746
HM	0.876	0.91	0.923	0.801
ITU	0.829	0.856	0.878	0.59
SI	0.835	0.844	0.901	0.752

Discriminant Validity. The Fornell-Larcker scores were greater than the correlation coefficients for other constructs and indicates evidence of discriminant validity. The suggested heterotrait-monotrait (HTMT) ratio maximum threshold value is 0.90 when the constructs are conceptually similar and 0.85 when the constructs are more distinct [58]. This study's constructs are distinct, and the 0.85 threshold was used. The results revealed that the HTMT values for all pairs of constructs are lower than 0.85 and demonstrate the discriminant validity in the constructs.

5.4 Formative Construct Results

Formative Convergent Validity. The first step in assessing formative constructs was to perform a redundancy analysis that tested whether the formatively measured construct is highly correlated with a reflective measurement of the same construct [58]. A redundancy analysis utilized creating a separate model that links the respective formative construct into a reflective construct with a single indicator. The correlation between the formative and reflective construct should be a minimum of 0.7 and above [61]. The constructs PE and FS have shown convergent validity with correlations of 0.918 and 0.954 respectively.

Formative Collinearity Issues and Outer Weights. For formative constructs, the correlations should be low as each indicator is representative of the construct which should have minimal overlap with each other. VIF values should be 3 and lower and has been confirmed for all the formative indicators. The values of the outer weights provide a relative contribution to the formative construct and are typically smaller than the outer loadings of reflective constructs [59]. As formative constructs are formed by the linear combination of the indicators, the values were derived from linear regression. There were no outliers in the outer weights of the indicators for their respective formative constructs.

5.5 Structural and Measurement Model

As mentioned previously, additional constructs were included in the proposed research model composed of Facilitating Conditions (FC) with four measurement indicators, Price Value (PV) with four measurement indicators, and Visibility (VS) with three measurement indicators but were excluded from the model because the indicators did not converge.

5.6 Structural Model Evaluation

As the reflective constructs have been assessed for internal consistency, convergent validity, and discriminant validity and formative constructs assessed for convergent validity and collinearity issues, the final process was the evaluation of the structural model for collinearity issues, the significance and relevance of the path coefficients, and the model's explanatory power [58].

Collinearity. Checking for collinearity among the constructs is the first step to ensure that the constructs are not highly correlated as it skews the statistical significance and undermines the explanatory power of the model [62]. Hair et al. [58] was suggested that the coefficients for VIF be below 3 and the results revealed that the VIF scores for the predictor construct ITU are below the threshold.

Path Coefficients Significance and Relevance. The next step in the structural model evaluation was determining whether the path coefficients were statistically significant

[58]. Significance (p) values are between 0 and +1 and this study used a significance level of 5% with a bootstrap of 10,000 samples. For a 5% significance level, p-values should be 0.05 or lower or using the t-statistics of 1.96. Results showed that the statistical significance of the path coefficients where HA -> ITU, HM -> ITU, PE -> ITU, and SI -> ITU are statistically significant and AV -> ITU, EE -> ITU, FS -> ITU, PE -> FS are not statistically significant.

Explanatory Power. The next step on evaluation the structural model is to determine the coefficient of determination (R^2) [58]. In this study, there is one endogenous construct, Intention to Use (ITU) and the R^2 value was 0.486 that is statistically significant and was considered as a moderate impact.

5.7 Summary of Results

The path coefficients for hypotheses 1 to 10 were calculated with 10,000 bootstrap samples using two-tailed test at a significance level of 5%. As shown in Table 2, three constructs could not be tested in this research model: Facilitating Conditions, Price value, and Visibility. Three constructs did not have a statistical influence on Intention to Use and are Effort Expectancy, Food Safety Concerns, and Affordance Value. All other constructs reached statistical significance: Performance Expectancy, Social Influence, Hedonic Motivation, and Habit.

Table 2. H1–10 Summary of Results

Hypothesis	Path	Beta (β)	Supported
1	PE -> ITU	0.268***	✓
2	EE -> ITU	0.053	
3	SI -> ITU	0.335***	✓
4	FC -> ITU	n/a	
5	HM -> ITU	0.211**	✓
6	HA -> ITU	0.134*	✓
7	PV -> ITU	n/a	
8	FS -> ITU	−0.051	
9	VS -> ITU	n/a	
10	AV -> ITU	0.053	

Note: * p < .05, ** p < .01 *** p < .001

6 Discussion of Findings.

The results from the PLS analysis of the 170 valid survey responses that was collected through Qualtrics revealed that the constructs Performance Expectancy (PE), Social Influence (SI), Hedonic Motivation (HM), and Habit (HA) were statistically significant and positively associated with Intention to Use (ITU) a MFA while Effort Expectancy (EE), Food Safety Concerns (FS), and Affordance Value (AV) were not statistically significant and have no association with Intention to Use (ITU). The model has a moderate explanatory power where $R^2 = 0.486$ Each of the hypotheses are summarized along with the findings below.

6.1 Performance Expectancy

The p-value between performance expectancy and intention to use was <.001 and therefore supports hypothesis 1 that PE positively affects ITU a MFA. However, performance expectancy had a small effect with a path coefficient of 0.268. These results support the findings of other MFA studies [1, 8–10, 29, 31, 43] that posited performance expectancy had a positive and significant effect on intention to use a MFA.

6.2 Effort Expectancy

Based on the PLS analysis, the p-value for effort expectancy on intention to use was 0.384 and therefore does not support hypothesis 2 that EE positively affects ITU a MFA. The path coefficient for effort expectancy on intention to use was 0.053, leading to no effect on the model. It was predicted that MFAs needed to have a gentle learning curve for consumers to adopt this technology. Unfortunately, the result was inconsistent with other MFA-related studies that effort expectancy influenced the intention to use [1, 29, 43]. The researcher suspected that as consumers are becoming increasing familiar with mobile technology, effort expectancy no longer directly affects the intention to use MFAs as shown in studies relating to mobile shopping apps [25] and mobile banking apps [63].

6.3 Social Influence

The p-value between social influence and intention to use was <.001 and thus supports hypothesis 3 that SI positively affects ITU a MFA. Social influence had a medium effect with a path coefficient of 0.335. The result is consistent with other MFA-related studies [1, 8, 10, 43] that argued social influence had a positive and statistically significant effect on intention to use a MFA. Social influence was the largest effect on intention to use a MFA and has implications for practitioners. Social connections have a powerful effect on the consumers' intentions [64], and in particular a growing trend of influencers on social media platforms can grow or sustain the platform [46, 65]. Practitioners should consider establishing connections with individuals that have a distinguished status and well-known in their respective area to increase the adoption of MFAs [66, 67].

6.4 Performance Expectancy

Based on the PLS analysis, the p-value for hedonic motivation on intention to use was <.01 and therefore supports hypothesis 5 that HM positively affects ITU a MFA. The path coefficient for hedonic motivation on intention to use was 0.221. This finding was consistent with other MFA studies [9, 29, 30, 43] that hedonic motivation had a positive and statistically significant influence on intention to use. Practitioners should continue to evolve their platform by determining what consumers find most enjoyable about their technology and incorporate them into their marketing strategy.

6.5 Habit

The p-value between habit and intention to use was $<.05$ and thus supports hypothesis 6 that HA positively affects ITU a MFA. Habit had a small effect with a path coefficient of 0.134. The result is consistent with studies pertaining to MFAs [8, 10, 30] that had evidence supporting habit had a positively and statistically significant influence on behavioural intention. The researcher hypothesized that consumers who continuously use other mobile applications such as banking and shopping may adopt MFAs as part of their habitual use mobile applications. Practitioners may wish to expand the incentives and promotions to persuade the consumer into building a habit on using their food-ordering platform instead of the competitors.

6.6 Food Safety Concerns

Based on the PLS analysis, the p-value for food safety concerns on intention to use was 0.491 and therefore does not support hypothesis 8 that FS negatively affects ITU a MFA. The path coefficient for food safety concern on intention to use was -0.051 and thus has no effect overall. It was predicted that consumers who had a high degree of concern for food safety had low intentions to use a MFA. A study showed that proper food quality had a significant and positive impact on the consumers' loyalty for MFAs [53]. On the opposite end, there were also negative impacts from improper food handling shown through mass media [54]. As the results were inconsistent with other studies, this indicated that consumers are not concerned about food safety. One possible explanation that can be attributed to this finding was the sampling method used in this study. As the participant pool was composed of university students, the average age group of the participants was categorized as millennials.

6.7 Affordance Value

The p-value between habit and intention to use was 0.486 and thus does not support hypothesis 10 that AV positively affects ITU a MFA. Affordance had no effect with a path coefficient of 0.053. It was anticipated that consumers are affected by the current promotions available on the platform such as promotional incentives, free or reduced delivery charges, and the availability of preferred restaurants. Affordance value for MFAs has been validated once by Kaur et al. [14] and regrettably, this study did not reach the same conclusion. A potential reason for this could be related to the age group of the participants in this study which the majority were 18–25 years of age. Inherently, the prices for food on MFAs are higher compared to dining at the restaurant directly [68]. Their intention to use MFAs may be related to the price sensitivity of their food purchases rather than affordance value defined in this study.

7 Implications

This research contributes to the current mobile food applications literature by integrating technology acceptance and consumption value theories. A UTAUT2-TCV model was proposed for a comprehensive model predicting consumer behavior on adopting mobile food applications as part of their lifestyle. While technology acceptance models have been studied in the context of MFAs, this study is the first to propose additional constructs from the theory of consumption values consisting of price value, food safety concerns, visibility, and affordance values.

There are practical implications for mobile food platforms and the goal of increasing the conversion ratio on consumers adopting their application. As this research identified social influence, performance expectancy, hedonic motivation, and habit as significant factors in behavioural intentions to use MFAs, it is suggested that practitioners focus on promotional strategies that emphasizes the ease of use by selecting appropriate distribution channels to capture the target market. In addition, the marketing strategy should emphasize the various and positive emotions that are related to seeing a wide selection of food offerings. Lastly, practitioners can develop strategies that link the habitual use of mobile applications can also benefit the consumer if they use mobile food applications.

8 Limitations and Future Research

First, this study used convenience sampling of university students that reside in Ontario, Canada and the results may not be generalizable to different cultures, cities, and countries. Second, this study used a cross-sectional design where the collection of data was collected over a short period of time and the results may differ when conducting a longitudinal study on intentions to use a MFA. Third, three constructs: facilitating conditions (FC), price value (PV), and visibility (VS) have been removed from research model as the construct's internal consistency and measurement indicator did not meet the minimum threshold to explain the variance in the model. Consequently, the proposed UTAUT2-TCV model is less comprehensive when determining the consumers' intention to use a MFA.

The current model addressed behavioural intentions to use MFAs and can be explored to a greater extent with the continuous usage of MFAs. Researchers could examine the intention to use MFAs in group setting such as family gatherings and group are friends. Lastly, future research could explore the constructs that were dropped from the proposed model due to indicators not converging.

9 Conclusion

The growth of MFAs has increased over the years. The meal delivery market started off with pizza and Chinese food but has significantly grown and now includes all sorts of ethnic cuisines and speciality foods. Eating habits have changed with the rise of user-friendly apps, lockdowns and physical-distancing requirements, and dynamic consumer expectations. This study examined the factors that influence the consumers' intention to use mobile food applications.

This paper was a quantitative and survey-based study that examined the factors that influence the consumer's intention to use MFAs. This study contributed to the creation and evaluation of a UTAUT2-TCV model in the context of MFAs and the results revealed that based on 170 participants, consumers are affected by performance expectancy (PE), social influence (SI), hedonic motivation (HM), and habit (HA) on their intention to use (ITU) MFAs. Effort expectancy (EE), food safety concerns (FS), and affordance value (AV) had no impact on the consumer.

References

1. Muangmee, C., et al.: Factors determining the behavioral intention of using food delivery apps during COVID-19 pandemics. J. Theor. Appl. Electron. Commer. Res. **16**(5), 1297–1310 (2021)
2. Tandon, A., et al.: Why do people purchase from food delivery apps? A consumer value perspective. J. Retail. Consum. Serv. **63**, 102667 (2021)
3. Statista. Online Food Delivery - Canada | Statista Market Forecast (2022). https://www.statista.com/outlook/dmo/eservices/online-food-delivery/canada
4. Perri, J.: Which company is winning the restaurant food delivery war? (2022). https://secondmeasure.com/datapoints/food-delivery-services-grubhub-uber-eats-doordash-postmates/
5. Venkatesh, V., Thong, J.Y., Xu, X.: Consumer acceptance and use of information technology: extending the unified theory of acceptance and use of technology. MIS Q. 157–178 (2012)
6. Sheth, J.N., Newman, B.I., Gross, B.L.: Why we buy what we buy: a theory of consumption values. J. Bus. Res. **22**(2), 159–170 (1991)
7. Ahuja, K., et al.: Ordering in: The rapid evolution of food delivery (2021). https://www.mckinsey.com/industries/technology-media-and-telecommunications/our-insights/ordering-in-the-rapid-evolution-of-food-delivery
8. Lee, S.W., Sung, H.J., Jeon, H.M.: Determinants of continuous intention on food delivery apps: extending UTAUT2 with information quality. Sustainability **11**(11), 3141 (2019)
9. Alalwan, A.A.: Mobile food ordering apps: an empirical study of the factors affecting customer e-satisfaction and continued intention to reuse. Int. J. Inf. Manage. **50**, 28–44 (2020)
10. Gunden, N., Morosan, C., DeFranco, A.: Consumers' intentions to use online food delivery systems in the USA. Int. J. Contemp. Hospit. Manage. (2020)
11. Pigatto, G., et al.: Have you chosen your request? Analysis of online food delivery companies in Brazil. Br. Food J. (2017)
12. Cho, M., Bonn, M.A., Li, J.J.: Differences in perceptions about food delivery apps between single-person and multi-person households. Int. J. Hosp. Manag. **77**, 108–116 (2019)
13. Ray, A., et al.: Why do people use food delivery apps (FDA)? A uses and gratification theory perspective. J. Retail. Consum. Serv. **51**, 221–230 (2019)
14. Kaur, P., et al.: The value proposition of food delivery apps from the perspective of theory of consumption value. Int. J. Contemp. Hospit. Manage. (2021)
15. Fishbein, M., Ajzen, I.: Predicting and understanding consumer behavior: attitude-behavior correspondence. Underst. Attit. Pred. Soc. Beh. **1**(1), 148–172 (1980)
16. Ajzen, I.: The Theory of Planned Behavior-Organizational Behavior and Human Decision Processes **50** (1991). Ajzen, I.: Perceived behavioural control, self-efficacy, locus of control and the theory of planned behaviour. Journal of Applied Social Psychology, **32**(4), 665–683 (2002)
17. Davis, F.D., Bagozzi, R.P., Warshaw, P.R.: User acceptance of computer technology: a comparison of two theoretical models. Manage. Sci. **35**(8), 982–1003 (1989)

18. Davis, F.D., Bagozzi, R.P., Warshaw, P.R.: Extrinsic and intrinsic motivation to use computers in the workplace 1. J. Appl. Soc. Psychol. **22**(14), 1111–1132 (1992)
19. Thompson, R.L., Higgins, C.A., Howell, J.M.: Personal computing: Toward a conceptual model of utilization. MIS Q. 125–143 (1991)
20. Moore, G.C., Benbasat, I.: Development of an instrument to measure the perceptions of adopting an information technology innovation. Inf. Syst. Res. **2**(3), 192–222 (1991)
21. Compeau, D.R., Higgins, C.A.: Computer self-efficacy: Development of a measure and initial test. MIS Q. 189–211 (1995)
22. Lai, I.K.W., Shi, G.: The impact of privacy concerns on the intention for continued use of an integrated mobile instant messaging and social network platform. Int. J. Mob. Commun. **13**(6), 641–669 (2015)
23. Madan, K., Yadav, R.: Behavioural intention to adopt mobile wallet: a developing country perspective. J. Indian Bus. Res. (2016)
24. Shaw, N., Sergueeva, K.: The non-monetary benefits of mobile commerce: extending UTAUT2 with perceived value. Int. J. Inf. Manage. **45**, 44–55 (2019)
25. Chopdar, P.K., Sivakumar, V.: Understanding continuance usage of mobile shopping applications in India: the role of espoused cultural values and perceived risk. Beh. Inf. Technol. **38**(1), 42–64 (2019)
26. Okumus, B., et al.: Psychological factors influencing customers' acceptance of smartphone diet apps when ordering food at restaurants. Int. J. Hosp. Manag. **72**, 67–77 (2018)
27. Alalwan, A.A., Dwivedi, Y.K., Rana, N.P.: Factors influencing adoption of mobile banking by Jordanian bank customers: extending UTAUT2 with trust. Int. J. Inf. Manage. **37**(3), 99–110 (2017)
28. Kijsanayotin, B., Pannarunothai, S., Speedie, S.M.: Factors influencing health information technology adoption in Thailand's community health centers: Applying the UTAUT model. Int. J. Med. Informatics **78**(6), 404–416 (2009)
29. Agarwal, V., Sahu, R.: Predicting repeat usage intention towards O2O food delivery: extending UTAUT2 with user gratifications and bandwagoning. J. Foodservice Bus. Res. 1–41 (2021)
30. Palau-Saumell, R., et al.: User acceptance of mobile apps for restaurants: an expanded and extended UTAUT-2. Sustainability **11**(4), 1210 (2019)
31. Zhao, Y., Bacao, F.: What factors determining customer continuingly using food delivery apps during 2019 novel coronavirus pandemic period? Int. J. Hosp. Manag. **91**, 102683 (2020)
32. Kaur, P., et al.: Why people use online social media brand communities: A consumption value theory perspective. Online Inf. Rev. (2018)
33. Mäntymäki, M., Salo, J.: Why do teens spend real money in virtual worlds? A consumption values and developmental psychology perspective on virtual consumption. Int. J. Inf. Manage. **35**(1), 124–134 (2015)
34. Talwar, S., et al.: Why do people purchase from online travel agencies (OTAs)? A consumption values perspective. Int. J. Hosp. Manag. **88**, 102534 (2020)
35. Wang, H.-Y., Liao, C., Yang, L.-H.: What affects mobile application use? The roles of consumption values. Int. J. Mark. Stud. **5**(2), 11 (2013)
36. Zolkepli, I.A.: Domination of mobile apps market: the effect of apps value on apps rating and apps cost in determining adoption. In: Proceedings of Penang International Symposium on Advanced in Social Sciences and Humanities, Royale Bintang, Penang (2016)
37. Choe, J.Y.J., Kim, S.S.: Effects of tourists' local food consumption value on attitude, food destination image, and behavioral intention. Int. J. Hosp. Manag. **71**, 1–10 (2018)
38. Karjaluoto, H., et al.: Consumption values and mobile banking services: Understanding the urban–rural dichotomy in a developing economy. Int. J. Bank Mark. (2021)
39. Omigie, N.O., et al.: Customer pre-adoption choice behavior for M-PESA mobile financial services: extending the theory of consumption values. Ind. Manage. Data Syst. (2017)

40. Peng, N., Chen, A., Hung, K.-P.: Dining at luxury restaurants when traveling abroad: incorporating destination attitude into a luxury consumption value model. J. Travel Tour. Mark. **37**(5), 562–576 (2020)
41. Thomé, K.M., Pinho, G.M., Hoppe, A.: Consumption values and physical activities: consumers' healthy eating choices. Br. Food J. (2018)
42. Gupta, A., Dogra, N., George, B.: What determines tourist adoption of smartphone apps? An analysis based on the UTAUT-2 framework. J. Hospit. Tourism Technol. (2018)
43. Gârdan, D.A., et al.: Enhancing consumer experience through development of implicit attitudes using food delivery applications. J. Theor. Appl. Electron. Commer. Res. **16**(7), 2858–2882 (2021)
44. Venkatesh, V., et al.: User acceptance of information technology: Toward a unified view, 425–478. MIS Q. (2003)
45. Jeon, H.M., Sung, H.J., Kim, H.Y.: Customers' acceptance intention of self-service technology of restaurant industry: expanding UTAUT with perceived risk and innovativeness. Serv. Bus. **14**(4), 533–551 (2020). https://doi.org/10.1007/s11628-020-00425-6
46. Evans, N.J., et al.: Disclosing Instagram influencer advertising: the effects of disclosure language on advertising recognition, attitudes, and behavioral intent. J. Interact. Advert. **17**(2), 138–149 (2017)
47. Cao, Q., Niu, X.: Integrating context-awareness and UTAUT to explain alipay user adoption. Int. J. Ind. Ergon. **69**, 9–13 (2019)
48. de Sena Abrahão, R., Moriguchi, S.N., Andrade, D.F.: Intention of adoption of mobile payment: an analysis in the light of the Unified Theory of Acceptance and Use of Technology (UTAUT). RAI Revista de Administração e Inovação **13**(3), 221–230 (2016)
49. Morosan, C., DeFranco, A.: It's about time: revisiting UTAUT2 to examine consumers' intentions to use NFC mobile payments in hotels. Int. J. Hosp. Manag. **53**, 17–29 (2016)
50. Singh, M., Matsui, Y.: How long tail and trust affect online shopping behavior: an extension to UTAUT2 framework. Pacific Asia J. Assoc. Inf. Syst. **9**(4), 2 (2017)
51. digitaltrends. The best food-delivery apps for 2021 (2021). https://www.digitaltrends.com/home/best-food-delivery-apps/
52. Tamilmani, K., Rana, N.P., Dwivedi, Y.K.: Consumer acceptance and use of information technology: a meta-analytic evaluation of UTAUT2. Inf. Syst. Front. **23**(4), 987–1005 (2021)
53. Suhartanto, D., et al.: Loyalty toward online food delivery service: the role of e-service quality and food quality. J. Foodserv. Bus. Res. **22**(1), 81–97 (2019)
54. Liu, P., Ma, L.: Food scandals, media exposure, and citizens' safety concerns: a multilevel analysis across Chinese cities. Food Policy **63**, 102–111 (2016)
55. Nayak, R., Waterson, P.: The assessment of food safety culture: an investigation of current challenges, barriers and future opportunities within the food industry. Food Control **73**, 1114–1123 (2017)
56. Chuah, S.H.-W., et al.: Wearable technologies: the role of usefulness and visibility in smartwatch adoption. Comput. Hum. Behav. **65**, 276–284 (2016)
57. Yang, H.C.: Bon Appétit for apps: young American consumers' acceptance of mobile applications. J. Comput. Inf. Syst. **53**(3), 85–96 (2013)
58. Hair, J.F., et al.: A Primer on Partial Least Squares Structural Equation Modeling (PLS-SEM). 3rd ed. ed. SAGE Publications, Inc. (2021)
59. Henseler, J., Sarstedt, M.: Goodness-of-fit indices for partial least squares path modeling. Comput. Stat. **28**(2), 565–580 (2013)
60. Cronbach, L.J., Meehl, P.E.: Construct validity in psychological tests. Psychol. Bull. **52**(4), 281 (1955)
61. Chin, W.W.: The partial least squares approach to structural equation modeling. Modern Meth. Bus. Res. **295**(2), 295–336 (1998)

62. Hair Jr, J.F., et al.: Partial least squares structural equation modeling (PLS-SEM): An emerging tool in business research. Eur. Bus. Rev. (2014)
63. Yuan, S., et al.: An investigation of users' continuance intention towards mobile banking in China. Inf. Dev. 32(1), 20–34 (2016)
64. Bonfield, E.H.: Attitude, social influence, personal norm, and intention interactions as related to brand purchase behavior. J. Mark. Res. 11(4), 379–389 (1974)
65. Teo, L.X., Leng, H.K., Phua, Y.X.P.: Marketing on Instagram: Social influence and image quality on perception of quality and purchase intention. Int. J. Sports Mark. Sponsorship (2018)
66. Alalwan, A.A.: Investigating the impact of social media advertising features on customer purchase intention. Int. J. Inf. Manage. 42, 65–77 (2018)
67. Daneshvary, R., Schwer, R.K.: The association endorsement and consumers' intention to purchase. J. Consumer Mark. (2000)
68. Watsky, D.: DoorDash, Grubhub, Uber Eats: We Finally Figured Out Which One Is Cheapest (2022). https://www.cnet.com/home/kitchen-and-household/doordash-grubhub-uber-eats-we-finally-figured-out-which-one-is-cheapest/

Influence of Artificial Intelligence Recommendation on Consumers' Purchase Intention Under the Information Cocoon Effect

Siyi Liang, Nurzat Alimu, Hanchi Si, Hong Li, and Chuanmin Mi[✉]

College of Economics and Management, Nanjing University of Aeronautics and Astronautics, Nanjing, China
Liangsiyi@nuaa.edu.cn

Abstract. With the rapid development of e-commerce, artificial intelligence (AI) recommendation technology is also evolving rapidly. In the long run, the information obtained by consumers has become homogeneous, leading to the formation of "information cocoons effect". However, few researches have examined the influence of information cocoons effect on consumers' purchase intention. Drawing on the Expectation Confirmation Model (ECM) and Information Ecology Theory, we develop a research model to investigate how information factor (Information Homogeneity), human factor (Personal Innovation), environment factor (Social Influence), and technology factor (Personalized Recommendation Quality, Platform Interaction Quality) affect consumers' satisfaction and purchase intention. Data has been collected from 283 respondents and analyzed with SmartPLS 3.3.9. Findings indicate that Personalized Recommendation Quality, Platform Interaction Quality satisfaction and Perceived Time Risk significantly influences Purchase Intention. Potential theoretical and practical contributions are also discussed.

Keywords: Expectation Confirmation Model · Information Ecology Theory · Information Cocoons · Satisfaction Purchase Intention

1 Introduction

The thorough development of the information era and the growing rise of the online economy have led to the rapid increase in the scale of commodity information in an unimaginable way. Although consumers are willing to buy online, they are tired of browsing dizzy commodity information. At the same time, not all consumers have clear shopping goals and can clearly describe their needs when inputting search keywords. To solve such problems, the recommendation system has emerged, which can help people choose the goods they want according to their own wishes and needs [1]. The recommendation system is oriented by users' information needs and uses algorithms such as Item-based collaborative filtering, user social relations, and interest bias to personalize the pushed information, so that users can obtain external information basically consistent with their interests. However, the speed of information narrowing is accelerated, resulting in the phenomenon of "information cocoon". Personalized recommendation mechanism

© Springer Nature Switzerland AG 2023
F. Fui-Hoon Nah and K. Siau (Eds.): HCII 2023, LNCS 14038, pp. 249–259, 2023.
https://doi.org/10.1007/978-3-031-35969-9_17

enhances the distribution mode of preference content, which makes self-understanding of users guided by interest when users viewing the external world. Moreover, this information distribution mode is controlled by the program, which is difficult to adjust artificially [2]. The concept of information cocoons comes from American scholar Thornstein, who pointed out in his book "Information Utopia - How People Produce Knowledge" that information cocoons mean that we only listen to what we choose and delight us. [3] Nowadays, people's access to information is more diversified. Therefore, we need to comprehensively analyze the multiple ways and filtering mechanisms of people's access to information to determine what factors affect people's access to information, thereby judge the factors that lead to the information cocoon [15].

The research on information cocoon is mainly focused on news, video and other entertainment platforms. Few scholars start from the perspective of the impact of e-commerce platform consumers' purchase. The research on the impact of users' purchase decisions is mostly focused on the quality of the recommendation system. Few scholars start from the perspective of information cocoon. Based on information ecosystem theory and ECM model, this paper studies consumers' purchase intention under the influence of information cocoon from four factors, human (Personal Innovation) - environment (Social Influence) - technology (Personalized Recommendation Quality, Platform Interaction Quality) - information (Information homogeneity). Determinants are divided into three categories: satisfaction, perceived time risk, and technology factors (Personalized Recommendation Quality and Platform Interaction Quality). Among them, Perceived Time Risk is affected by Personal Innovation, Social Influence and Information Homogeneity, and Satisfaction is affected by Individual Innovation, Social Influence, Information Homogeneity and Perceived Time Risk. This research studies the purchase intention of consumers by analyzing consumers and e-commerce platforms, which can provide consumers with insights on purchase and provide a direction for e-commerce platforms to optimize recommendation algorithms and interaction quality.

2   Theoretical Background

Expected to Confirm Model (ECM) was proposed by Oliver in 1980 to study satisfaction and consumer post purchase behavior [4]. The concept of information ecosystem was put forward by Crawford. He believed that information ecosystem is an organic system composed of people, practice, value and technology in a specific environment, which can be used as a supporting theory for the coordinated development of people, information, information technology and information environment [5]. Xiwei Wang and others further pointed out that information, information people, information technology and information environment are important factors that constitute the information ecosystem [6].

To sum up, based on ECM model and information ecosystem theory, this study takes human factor, environment factor and information factor in information ecosystem theory as independent variables to indirectly affect purchase intention through satisfaction and perceived time risk in ECM model, and takes technology factors as independent variables to directly affect purchase intention.

3   Research Models and Hypotheses

The research model is shown in Fig. 1. Corresponding assumptions are discussed in detail as followings.

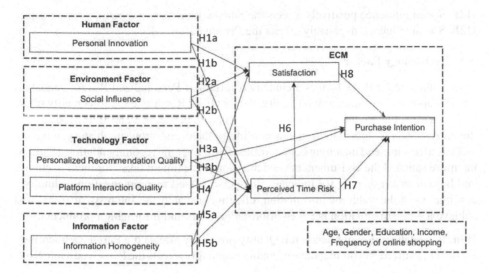

Fig. 1. Research model

3.1   Human Factor

Differences in the degree of personal innovation leads to different attitudes of consumers towards new things. Users with high personal innovation tend to accept innovative things or advanced technologies. Therefore, Personal Innovation can positively affect purchase satisfaction to a certain extent [8]. Moreover, users will also find things they are interested in according to their own needs and interests, and will not lose their way in the various recommendations of the recommendation system, which can reduce the perceived time risk. For example, if a person has the initiative to contact some novel things, his degree of personal innovation is high. Thus, we assume:

H1a. Personal Innovation positively affects the Satisfaction.
H1b. Personal Innovation negatively affects the Perceived Time Risk.

3.2   Environment Factor

When users make decisions based on the information provided, they generally pursue a balance between social relations and information needs. Therefore, the information environment in which they live can not only influence the extent to which users access information, but also have an important impact on their information attitudes and behaviors. In this study, community influence refers to the fact that users will over-consider the information from social relations that are not very close to them. As for the information released by people with strong social ties, users tend to pay attention to it and have a high degree of acceptance [7]. At this time, they are forever immersed in the values reflected in the information they have received, resulting in an information cocoon room effect. Therefore, we propose:

H2a. Social Influence positively affects the Satisfaction.
H2b. Social influence negatively affects the Perceived Time Risk.

3.3 Technology Factor

In this study, technology factors include two variables: Personalized Recommendation Quality and Platform Interaction Quality. Personalized Recommendation Quality refers to the fact that the recommendation system will build preference and style models with the corresponding algorithm system according to the users' browsing history and click records to recommend matching commodities. The higher the recommendation accuracy, the more satisfied the consumers are and the shorter the time it takes for them to choose and buy. Sean's research [10] indicates that personalized recommendation technology can maximize the value of information, effectively match the interests of users, and achieve a win-win situation between information users and information producers.

H3a. Personalized Recommendation Quality positively affects the Purchase Intention.
H3b. Personalized Recommendation Quality negatively affects the Perceived Time Risk.

Platform Interaction Quality means that the platform will provide a channel for feedback to consumers [10]. When the recommendation system recommends a product that consumers are not interested in, they can use the platform interaction function, such as clicking the words "I don't like this type of product" appearing on the screen. In this way, consumers can avoid the product that does not match their own interests from appearing again. The higher the interaction quality, the higher the accuracy of its recommendation system, and the consumers' purchase intention will naturally increase. Thus, we hypothesize:

H4. Platform Interaction Quality positively affects the Purchase Intention.

3.4 Information Factor

Information homogeneity means that when the recommendation system blindly recommends products according to consumers' interests and preferences, they will often receive the same type of products [5]. As a result, they will be limited to their own commodity values, and will not receive heterogeneous information which may arise their potential interests. Therefore, we assume:

H5a. Information Homogeneity positively affects the Satisfaction.
H5b. Information Homogeneity negatively affects the Perceived Time Risk.

3.5 Perceived Time Risk

Perceived time risk refers to the fact that if the consumers often receive the products that do not match their interests due to the inaccuracy of the recommendation system, it will take a long time for them to select the favorite products. Pires et al. [11] believe that if the quality of the products and the personalized recommendation service obtained fails to meet consumers' expectations, it will lead to unsatisfactory consequences, including returns, exchanges, and failure to select the right product, etc. Thus, we assume:

H6. Perceived Time risk negatively affects the Satisfaction.
H7. Perceived Time Risk negatively affects the Purchase Intention.

3.6 Satisfaction

For users, when they are satisfied with the personalized recommendation system, the satisfaction, including pleasure and the satisfaction with the personalized recommendation system will become the reference for the next use, and the loyalty gradually increases in the continuous satisfaction, and then forms purchase intention [7]. Thus, we hypothesize:

H8. Satisfaction positively affects the Purchase Intention.

4 Research Methodology

4.1 Data Collection

Data was collected by distributing an online survey via Sojump.com in mainland China. The questionnaire with 15 questions in total was divided into two parts, including the measurements items of the variables in the research model and the demographic information questions. All measurement items were adapted from relevant previous studies and had been modified to match our research context. Items of Personal Innovation were adapted and adjusted from Lixu Li et al. [8], while items of Social Influence were derived from Oliver [4] and Bhattacherjee [7]. Items of Personalized Recommendation Quality and Platform Interaction Quality were adapted from Sean's research [10]. And items of Information Homogeneity were adapted from Holly Crawford [5]. Furthermore, items of Satisfaction were derived from Bhattacherjee [7] and items of Perceived Time Risk were derived from Pires et al. [11].

Measurement items of Purchase Intention were adapted from Loiacono et al. [9]. All the items were measured with Likert seven-point scale, ranging from "strongly disagree" (1) to "strongly agree" (7). The survey was conducted from July 22, 2022 to July 26, 2022. 312 valid responses were received. Among them, 40.71% of the respondents were male and 59.29% were female. Majority of the respondents were between 18 and 40 years old. Besides, 67.31% respondents' highest education is a college degree. Majority of the respondents' monthly income ranged from 2000 to 10000 yuan. Approximately 70% of the respondents shop online 3–10 times a month. Detailed demographic information was shown in Table 1.

Table 1. Respondent demographics.

Measure	Item	N = 312 Frequency	Percentage (%)
Gender	Male	127	40.71%
	Female	185	59.29%
Age	Under 18	17	5.45%
	18–24	64	20.51%
	25–30	85	27.24%
	31–40	91	29.17%
	41–50	39	12.50%
	51–60	8	2.56%
	61 and above	8	2.56%
Education	Primary School	0	0.00%
	Middle School	36	11.54%
	High School	61	19.55%
	Junior College	157	50.32%
	Undergraduate College	53	16.99%
	Master and above	5	1.60%
Monthly Income	Below 2000 yuan	17	5.45%
	2000–5000 yuan	143	45.83%
	5000–10000 yuan	137	43.91%
	Above 10000 yuan	15	4.81%
Monthly Online Shopping Frequency	Less than 3 times	54	17.31%
	3–8 times	153	49.04%
	8–10 times	67	21.47%
	10–15 times	27	8.65%
	More than 15 times	11	3.53%

4.2 Data Analyses and Results

SmartPLS 3.3.9 was used to analyse the data [13]. We tested the measurement model and structural model respectively [12]. Firstly, the reliability and validity of the measurement model were verified via checking factor loadings (>0.7), composite reliability (>0.8) and Cronbach's Alpha values (>0.7). Moreover, average variance extracted (AVE) values should be greater than 0.50 [14]. As shown in Table 2, all the factor loadings are greater than 0.9. AVE value of each construct ranged between 0.867 and 0.913, indicating good convergent validity.

According to Fornell and Larcker [14], discriminant validity of the measurement model can be evaluated by comparing the square root of AVE with the correlation between the measurement items. The results in Table 3 indicate sufficient discriminant validity.

In summary, our measurement model has sufficient reliability, convergence validity and discriminant validity.

Table 2. Individual item reliability.

Measures	Item	Loading	Mean	Composite reliability	Cronbach's Alpha	AVE
Personal Innovation	PI1	0.987	5.108	0.968	0.956	0.883
	PI2	0.931				
	PI3	0.914				
	PI4	0.925				
Personalized Recommendation Quality	PRQ1	0.988	5.080	0.966	0.953	0.878
	PRQ2	0.916				
	PRQ3	0.926				
	PRQ4	0.917				
Information Homogeneity	IH1	0.987	5.083	0.966	0.953	0.878
	IH2	0.915				
	IH3	0.912				
	IH4	0.931				
Platform Interaction Quality	PIO1	0.992	5.108	0.97	0.961	0.867
	PIO2	0.910				
	PIO3	0.916				
	PIO4	0.910				
	PIO5	0.924				
Perceived Time Risk	PTR1	0.990	3.378	0.974	0.964	0.904
	PTR2	0.941				
	PTR3	0.931				
	PTR4	0.940				
Satisfaction	SAT1	0.988	5.048	0.971	0.96	0.894
	SAT2	0.930				
	SAT3	0.934				
	SAT4	0.929				
Social Influence	SI1	0.987	5.073	0.968	0.955	0.883

(continued)

Table 2. (*continued*)

Measures	Item	Loading	Mean	Composite reliability	Cronbach's Alpha	AVE
	SI2	0.929				
	SI3	0.926				
	SI4	0.914				
Purchase Intention	PB1	0.980	4.98	0.969	0.952	0.913
	PB2	0.944				
	PB3	0.941				

Table 3. Discriminant validity.

Variables	Personal Innovation	Personalized Recommendation Quality	Information Homogeneity	Platform Interaction Quality	Perceived Time Risk	Satisfaction	Social Influence	Purchase Intention
Personal Innovation	0.94							
Personalized Recommendation Quality	0.408	0.937						
Information Homogeneity	0.362	0.293	0.937					
Platform Interaction Quality	0.42	0.41	0.368	0.931				
Perceived Time Risk	-0.52	-0.382	-0.367	-0.399	0.951			
Satisfaction	0.438	0.326	0.437	0.41	-0.413	0.946		
Social Influence	0.456	0.433	0.418	0.522	-0.458	0.465	0.939	
Purchase Intention	0.559	0.513	0.439	0.553	-0.525	0.516	0.613	0.955

Notes: The diagonal elements show the square root of the AVE; the off diagonal elements show the correlations among constructs.

Results of the regression analysis of the structural model were shown in Fig. 2, in which Personal Innovation to Satisfaction ($\beta = 0.184$, $p < 0.01$), Personal Innovation to Perceived Time Risk ($\beta = -0.334$, $p < 0.001$), Social Influence to Satisfaction ($\beta = 0.227$, $p < 0.01$), Social Influence to Perceived Time Risk ($\beta = -0.199$, $p < 0.01$), Personalized Recommendation Quality to Purchase Intention ($\beta = 0.239$, $p < 0.001$), Personalized Recommendation Quality to Perceived Time Risk ($\beta = -0.122$, $p < 0.05$), Platform Interaction Quality to Purchase Intention ($\beta = = 0.267$, $p < 0.001$), Information Homogeneity to Satisfaction ($\beta = 0.228$, $p < 0.001$), Information Homogeneity to Perceived Time Risk ($\beta = -0.127$, $p < 0.05$), Perceived Time Risk to Satisfaction ($\beta = -0.129$, $p > 0.05$), Perceived Time Risk to Purchase Intention ($\beta = -0.231$, $p < 0.001$), Satisfaction to Purchase Intention ($\beta = 0.233$, $p < 0.001$). Hence, H1a, H1b H2a, H2b, H3a, H3b, H4, H5a, H5b, H7 and H8 were supported. Personal Innovation, Social Influence, and Information Homogeneity jointly explained 33.9% variance of Satisfaction.

Personal Innovation, Social Influence, Personalized Recommendation Quality and Information Homogeneity jointly explained 35.8% variance of Perceived Time Risk. In addition, Personalized Recommendation Quality, Platform Interaction Quality, Satisfaction and Perceived Time Risk jointly explained 51.2% variance of Purchase Intention.

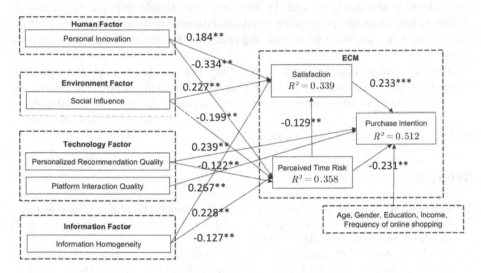

Fig. 2. Structural model

5 Conclusion

With the rapid development of e-commerce, artificial intelligence recommendation is evolving rapidly. As a rapidly developing technology, we know very little about it and what it does to consumers' willingness to buy. Thus, based on the ECM, TAM and TPB models, we built a research model and conducted a questionnaire which mimics the actual online shopping scene for empirical testing. We explored from four perspectives, including information factor (Information Homogeneity), human factor (Personal Innovation), environment factor (Social Influence), and technology factor (Personalized Recommendation Quality, Platform Interaction Quality). The results of the study identifies and verifies how five factors affect Satisfaction, Perceived Time Risk and Purchase Intention. As displayed in Fig. 2, all the proposed hypotheses are statistically supported. Findings indicate that Information Homogeneity, Personal Innovation and Social Influence positively affect the Satisfaction and negatively affect the Perceived Time Risk. Satisfaction, Platform Interaction Quality and Personalized Recommendation Quality positively affects the Purchase Intention. And Perceived Time Risk negatively affects Purchase Intention. On the one hand, consumers' shopping preferences tend to stay the same. On the other hand, in today's fast-paced life, consumers are willing to embrace new technology to achieve the purpose of fast shopping.

It is worth noting that, this study pioneeringly verifies the influence of Artificial Intelligence recommendation on consumers' purchase intention under the information cocoon effect. Practically, findings of this study can provide beneficial enlightenment for website managers who need to have a deeper understanding of how to increase consumers' purchase intentions. This means they need to capture users' needs in a more comprehensive and timely manner. Website managers should improve the quality of platform interactions and personalized recommendations, in order to make it easier for users to find what they want when browsing products, thereby reducing the time risk of consumers.

There are still several limitations to this study. For instance, the age distribution of respondents in our study was uneven and the survey was limited to Chinese mainland. We should send out more questionnaires to survey larger areas. In addition, the influencing factors we considered were limited, and it is recommended to further consider the impact of other factors on the results of the study.

References

1. Zhengxue, B., Xingxing, L., Changyong, L.: Will personalized recommendation win consumers' continuous favor?—Based on the perspective of use and satisfaction theory. J. Lanzhou Univ. Finan. Econ. **33**(06), 108–115 (2017)
2. Chunlin, P.: Analysis of "Information Cocoon" effect under personalized recommendation algorithm—taking "Headlines Today" as an example. China Media Technol. **06**, 61–63 (2022)
3. Hu, Y.: New words discussion. Echo room effect. News Commun. Res. **22**(6), 109–115 (2015)
4. Oliver, R.L.: A cognitive model of the antecedents and consequences of satisfaction decisions. J. Mark. Res. **17**(4) (1980)
5. Crawford, H.: Information ecologies: using technology with heart. Inf. Soc. **16**(3), 249–250 (2000). https://doi.org/10.1080/01972240050133706
6. Jiaxing, L., Xiwei, W., Shimeng, L., et al.: Research on the influencing factors of WeChat use behavior of elderly user groups from the perspective of information ecology. Lib. Inf. Work **61**(15), 25–33 (2017)
7. Bhattacherjee, A.: Understanding information systems continuance: an expectation-confirmation model. MIS Q. **25**(3) (2001)
8. Li, L., Wang, Z., Li, Y., Liao, A.: Impacts of consumer innovativeness on the intention to purchase sustainable products. Sustain. Prod. Consump. **27**, 774–786 (2021). ISSN 2352-5509, https://doi.org/10.1016/j.spc.2021.02.002
9. Loiacono, E.T., Watson, R.T., Goodhue, D.L.: WebQual: an instrument for consumer evaluation of web sites. Int. J. Electron. Commerce **11**(3), 51–87 (2007). http://www.jstor.org/stable/27751221
10. McNee, S. M., Riedl, J., Konstan, J. A.: Making recommendations better: an analytic model for Human-Recommender Interaction. In: CHI'06 Extended Abstracts on Human Factors in Computing Systems, CHI EA 2006, pp. 1103–1108 (2006)
11. Pires, G., Stanton, J., Eckford, A.: Infuences on the perceived risk of purchasing online. J. Consum. Behav. Int. Res. Rev. **4**(2), 118–131 (2004)
12. Chin, W.W., Marcolin, B.L., Newsted, P.R.: A partial least squares latent variable modeling approach for measuring interaction effects: results from a Monte Carlo simulation study and an electronic-mail emotion/adoption study. Inf. Syst. Res. **14**(2), 189–217 (2003)
13. Ringle, C.M., Wende, S., Will, A.: SmartPLS 2.0 (beta). Hamburg (2005)

14. Fornell, C.R. Larcker, D.F.: Evaluating structural equation models with unobservable variables and measurement error. J. Mark. Res. **18**(3), 375–381 (1981)
15. Peng, L.: Multiple factors leading to information cocoon and the path of "Breaking cocoon". J. Mass Commun. (01), 30–38+73 (2020)

The Dynamic Update of Mobile Apps: A Research Design with HMM Method

Xinhui Liu[1], Kaiwen Bao[1], Lele Kang[1(✉)], Jianjun Sun[1], and Yanqing Shi[2]

[1] Laboratory of Data Intelligence and Interdisciplinary Innovation, School of Information Management, Nanjing University, Nanjing 210023, Jiangsu, China
MF21140083@smail.nju.edu.cn, {lelekang,sjj}@nju.edu.cn
[2] Nanjing Agricultural University, Nanjing 210023, Jiangsu, China
yqs4869@njau.edu.cn

Abstract. The essential attribute of mobile apps is their dynamic update reflected by the weekly new versions. To gain a competitive advantage in the fierce competition, developers optimize their update strategy to improve app performance. However, the impact of the dynamic update on app performance is still unknown. This study proposes an approach to capture the sequential update of mobile apps in digital platform. Specifically, Hidden Markov Models (HMM) are established to estimate the influence of sequential updates on mobile app performance. As a theoretical paper, this research proposes that the update strategy determines the transmission among different levels of user satisfaction, which is the hidden states in HMM. Then, the transmission of user satisfaction influences app performance. The critical contribution of our study is that the process of mobile apps update is depicted by a dynamic approach. In this study, we propose the research design and discuss the implications.

Keywords: Mobile Apps · Product Update · Hidden Markov Model · Multi-attribute Utility Theory

1 Introduction

In recent years, the rapid development of the digital economy has led to a flood of mobile apps entering the market. As market needs and trends evolve quickly, mobile apps must be continuously updated to maintain an advantage in the fierce competition. Mobile app updates are a typical form of digital innovation, which helps them to remain current without bothering users [1, 2]. Maximizing user satisfaction is the primary objective of mobile app updates, as it is the basis for achieving a competitive advantage. App developers should constantly adjust and update their app designs to maintain or improve their market position and provide the best user experience.

Updates are an essential means of product management, typically used to correct errors and vulnerabilities, and to maintain product stability and security in the early stages [3, 4]. In recent years, updates are increasingly used to provide new functions to consumers through wireless means, which enhances consumer perception of utility

© Springer Nature Switzerland AG 2023
F. Fui-Hoon Nah and K. Siau (Eds.): HCII 2023, LNCS 14038, pp. 260–270, 2023.
https://doi.org/10.1007/978-3-031-35969-9_18

and increases their consideration of the product [5]. The update behavior of mobile apps can be characterized by frequency [6, 7], scale [8], major and minor [9], functional and non-functional [10], among other dimensions. However, not every update is going to satisfy users, and the inappropriate design of the update strategy can alienate users and generate negative feedback. For instance, adding new functions in updates creates new attractions for new users, but it increases the adaptation cost of old users, who may display emotional behaviors and a decreased sense of psychological belonging, leading to a decline in overall satisfaction [5]. Too rapid iteration of app versions may also lead to difficulties in consumer acceptance, making it challenging to imitate and improve by competitors, thus leading to a regression in app market performance [2].

From a theoretical perspective, the design of software update strategies is essential in pleasing users, and has been extensively studied in the literature. Developers have been paying more attention to user personalization, and the introduction of the recommendation system, which provides a personalized menu for users, has gradually become common [11]. To boost sales, sellers must exploit the natural segmentation of consumer tastes offered by different categories [10]. Herding behavior, in which users follow trends when using software, has also been studied in depth [12]. Despite the substantial and theoretical importance of updating strategies, there is still no theoretical consensus on the design of update strategies. Furthermore, the theoretical mechanism underlying the impact of developer update behavior on user satisfaction has not been thoroughly explored, often treated as static, although recent studies have shown that user feedback patterns for updates are significantly dynamic [13]. Our understanding of the dynamic relationship between mobile app update policy design mechanisms and user satisfaction is still limited, leaving a gap in our understanding of how user satisfaction and feedback are dynamically affected by policy design.

To address this gap, we focus on studying how to design developer strategies to motivate users when user satisfaction changes. The following research questions are discussed: (1) Which developer update strategies can effectively change mobile apps from a low user satisfaction state to a high user satisfaction state? (2) Which developer update strategies can effectively keep mobile apps in a high user satisfaction state? (3) Under the market-oriented mechanism, how do developers, mobile app platforms, and competent authorities make management decisions according to the changes in market performance brought about by mobile app updates? Understanding the dynamic effects of the above three dimensions is crucial to better design developer update strategies and effectively motivate users to maintain long-term app usage.

We adopt a dynamic approach to explore the relationship between developer strategy and user satisfaction in the machine learning field. Specifically, we propose a structural econometric model that integrates hidden Markov models (HMMs) into a consumer utility framework based on market orientation. This structured approach captures the dynamics of consumer utility across various satisfaction states and the transition mechanisms between them. Using a dataset of mobile apps, we employ Bayesian estimation to jointly evaluate the influence of different update strategies on the transition probability between user satisfaction states and the impact of market performance on their satisfaction states.

Our study has several features. Firstly, we explore how mobile app updates affect market performance changes from a dynamic perspective. Based on the consumer utility theory, we construct a dynamic relationship between mobile app updates and user satisfaction levels, which brings management insights that are not available in previous studies. Second, our structural model helps advance modeling approaches in the mobile app update literature because it explicitly describes the dynamics of user satisfaction levels at the app level. Third, our dynamic approach provides more nuanced insights into the increasingly important iterative patterns of digital products and applies to a wide range of mobile app updates, which makes further theoretical developments possible.

2   Literature Review

We draw on the literature to develop a theory of mobile app market performance and users' dynamic satisfaction. We first examine the factors that influence user satisfaction and show how the update strategy affects the market position of an app through satisfaction. We then find that the dynamism of user satisfaction and market performance is a gap in the literature, which motivates our hidden Markov model to characterize this dynamism.

In the mobile app market, consumer satisfaction with the app is the decisive factor for good market performance. It is characterized by Marshall's classical theory of supply and demand: the more satisfied consumers are with the app, the better their evaluation of the app will be, and the " demand" for downloading the app will also increase, and the "price" of the app, in other words, the market position, will also increase [14]. Consumer utility can represent consumer satisfaction. For example, adding new features may increase the perceived value of the app for consumers, while also increasing the cost for consumers to switch from a known product to an unknown product -- if the switching cost of the update exceeds the utility gained, Consumers may react negatively to updates [5]. This perceived value is manifested through a series of user feedback, such as user ratings, user reviews, downloads, etc. Therefore, the research on the relationship between consumers and app development has been a hot topic in recent years. The literature has conducted in-depth discussions on the relationship between software ranking and app updates, product improvement based on user ratings and user reviews, and how product release and update affect download changes [12, 13, 15]. Software ranking represents the best-selling degree of an app in the mobile app market, which is closely related to user needs [16]. Observable user feedback, such as ratings, reviews, and downloads, is a direct indicator of a user's psychological state, which influences both how consumers perceive an app and how developers react. For example, users will be more inclined to download software with higher ranking and better ratings, and will naturally resist software with more negative reviews [12, 17]. Exploring the reasons behind bad reviews and good reviews and how to drive more downloads has been a topic of great interest for app developers in recent years [1, 18]. At the same time, for products of the same type, lower-ranked products will actively imitate higher-ranked products to gain competitive advantages [19].

Updates are "independent modules of software that are provided free of charge to users so that they can be modified or extended after the software is rolled out and put into use" [20]. Updates are not standalone products but integrated into basic products [21, 22]. This product update links to several different concepts, including upgrades, generations, and a brochure [5]. This paper takes mobile apps as the research object to explore the dynamic relationship between different product renewal strategies and their market performance. Developers should actively adjust the update strategy and choose the update direction based on their product positioning and market performance. For example, through the impact on consumers' perceived value, mobile app developers can influence user satisfaction by adjusting different product update strategies as incentive mechanisms, to maintain a good market position of products [2]. Examples of such product update strategies include adjusting update frequency, making major or minor updates, functional or non-functional updates, etc. [6, 7, 9, 10] Among the various update strategies, we focus on three types of update strategies based on the updated content: functional, reliable, and convenient updates. At present, the research on app update types has just started and is not perfect. Our research improves and complements the relevant theories of mobile app update strategy research.

Although there is growing literature exploring user satisfaction and market performance of mobile apps, most of these studies are based on an implicit static assumption. In other words, the relationship between mobile app user satisfaction and market performance does not change over time. If we examine the relationship over time, user satisfaction in mobile apps often evolves in response to users' changing personal characteristics and their interaction with app updates, which leads to fluctuations in market performance. In this paper, we propose a dynamic theoretical framework. First, we propose the satisfaction state of a mobile app as a general structure characterizing its tendency to market performance development and model it as a mediating variable between developer update strategies and app market performance. Second, we assume that the user satisfaction status of mobile apps can change over time. Our model also allows the impact of a developer update behavior to be different when the mobile app is in different user satisfaction states. Therefore, we introduce a general model to explain the dynamic relationship between mobile app market performance and user satisfaction status. One challenge in capturing individual-level dynamics is that such dynamics are often unobservable. To capture this underlying structure, discrete state-space models are a useful approach in the literature [23]. For example, an individual's current decisions depend on his past decisions. In most of these models, the state is observable (for example, a customer's brand transition). Still, they tend to ignore other dynamic factors that may cause state changes. However, there are many other scenarios in which we cannot observe the underlying states that drive the dynamics at the individual level, for example, the motivational states in our study environment. In this case, hidden Markov models (HMMS) can be useful.

An HMM is a stochastic process consisting of three elements: a finite set of hidden states, observations conditional on the hidden states, and the probability of moving from one state to another [24]. In recent years, it has been widely used in e-commerce [25], fraud detection [26], intelligent sensing services [27], stock prediction [28], etc. With the further development of the model, it is also gradually applied in the field of product sales, such as personalized recommendation services [11]. The research of HMM in the field of mobile apps mainly focuses on network security, such as malicious app detection

[29] and encrypted Internet traffic [30]. Some scholars analyzed mobile apps from the perspective of users, such as mobile app popularity modeling based on user feedback [31], personalized goal realization based on the user participation stage [32], etc. At present, no scholars have used HMM structural model to explore the dynamic relationship between mobile app updates and user satisfaction levels from the perspective of developers. To contribute to the development of relevant theories, this study explores the dynamic effect of mobile app developers' updating behavior on their market performance changes by using HMM structural model in the framework of consumer utility.

3 Research Design

In order to answer the research questions raised above, we construct a hidden Markov model as shown in Fig. 1. It illustrates how an app's market performance states switch under the influence of various developer update strategies, and how policy choices affect the transition probabilities between these states. Specifically, our HMM has three elements:

(1) User satisfaction: from good to bad N states, the N states are theorized and operated into several hidden states, which are used as hidden variables in the HMM model. At any time t, the user's satisfaction with the app is only in one state.
(2) A certain state at time t only depends on its state at the previous time, has nothing to do with the state and observation at another time, and also has nothing to do with time t. The observed policy implementation at any time t is only related to the state of the hidden Markov chain at that time and is independent of other observations and states.
(3) From time t-1 to time t, the user's satisfaction with the app has a certain probability to switch to any state, which is affected by the app update strategy.

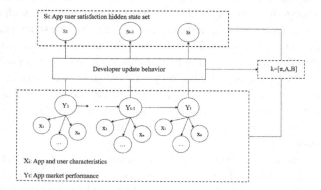

Fig. 1. Hidden Markov Model based on consumer utility framework

3.1   Model Development: Structural Modelling of User Behavior

In this section, we describe the details of our structural model with HMM, where the updated behavior of app developers determines the overall level of user satisfaction with that app.

Analyze App Market Performance in the Framework of Consumer Utility
In the app market, apps are commodities in nature, because they are labor products that meet people's needs and are exchanged. The market performance of commodities comes from the interaction between consumer demand and producer supply, which means that producers will make continuous efforts to obtain consumer preferences. Therefore, the research on consumer effect has been continuously deepened [33–36]. Under the framework of consumer utility, we adopt the Multi-Attribute Utility Theory (MAUT) to simulate the overall consumer utility of a single app in the mobile app market. Multi-attribute utility theory (MAUT) assumes that a product is a collection of attributes, and consumers evaluate products by evaluating their attributes [35]. Each attribute A that affects a consumer's decision is described by a weighted single-attribute utility (SAU) function $u(x_a)$, where x_a is the level of attribute A. The SAU linear function for a single attribute can be expressed as:

$$u(x_a) = a_a + b_a x_a \tag{1}$$

For multiple product attributes, the multi-attribute utility MAU that affects consumer decision-making can be expressed as a function [37]:

$$f(u_1, \ldots\ldots, u_n) = \sum\nolimits_{a=1}^{n} w_a \cdot u_a \tag{2}$$

where n is the number of attributes, u_a is a single-attribute utility function over attribute, w_a is the weight for attribute a, and $\sum_{a=1}^{n} w_a = 1$ ($0 \leq w_a \leq 1$ for all a). The MAU function can be determined by multiple regression analysis [38, 39].

For the convenience of analysis, we refer to the study of Chen et al. [24]. And assume that a single app's overall user satisfaction level at time t is s_{it}. Under this condition, the market performance of a single app is linearly related to the user utility at that time. And s_{it} is further determined by the potential transition tendency L_{it} (detailed later), then the market performance of the app at time t, Y_{it} can be expressed as:

$$Y_{it} = \alpha f(u_1, \ldots\ldots, u_n) \tag{3}$$

Since the representation of MAU function is linear, we can further get:

$$Y_{it} = X_{it}^{'} \beta_{s_{it}} + \varepsilon_{it}, ((\varepsilon_{it}|X_{it}, s_{it}) \sim N\left(0, \sigma^2\right)) \tag{4}$$

where X_{it} is expressed as the influence of the vector of consumer utility, which specifically t is the app and user characteristics, the error term ε_{it} follows zero mean and variance of the normal distribution of σ^2, attribute weights of utility w_i time-varying of the individual characteristics by X_{it} said. Our goal is to estimate the coefficient vector $\beta_{s_{it}}$, which captures the effect of the vector X_{it} on the app market performance Y_{it}. Note that the vector $\beta_{s_{it}}$ depends on the overall user satisfaction level s_{it} of the app at time t, which is related to the update behavior $W_{i,t-1}$ taken by the app developer at the previous time. In the following model, we describe the app's satisfaction state and its transition in detail.

Hidden States in HMM

Our proposed HMM describes the dynamic properties of app market performance as two stochastic processes: the observed app market performance and the unobserved potential user satisfaction state. We use s_{it} to represent the overall user satisfaction status of a single app i at time t. An app can have j hidden satisfaction states: $s_{it} \in S = \{1, 2, \ldots, J\}$.

The hidden state captures the time-dependent characteristics of a user's preference for an app, that is, how satisfied he is with the app. If the app is in a high satisfaction state at time t-1, its market performance will be very good at the high preference level of users. Based on his status, the users of the app, and personal characteristics (i.e., vector X_{it}). There are different responses. For example, if the app is in a state of high satisfaction, users are more likely to give a more positive sentiment analysis review. The observed market performance can be viewed as a noisy signal of the hidden state process. The hidden state and the observed market performance together form the hidden Markov chain [40].

From time t-1 to t, the app can stay in one state, or switch to another state. In our HMM, the state process $\{s_{it}\}_{t\geq0}$ is characterized as the state space $S = \{1, 2, \ldots, J\}$ of the first-order Markov chain. Along with Y_{it}, at the time t observed app i market performance, we were able to vector-valued stochastic process (Y_{it}, s_{it}) is modeled as a hidden Markov chain. The probability of its transition from one period to the next can be decomposed as follows:

$$P\big((Y_{it}, s_{it})|Y_{i,t-1}, s_{i,t-1}\big) = P(Y_{it}|s_{it}) \cdot P\big(s_{i,t-1}|s_{it}\big) \tag{5}$$

Here, $P\big(s_{i,t-1}|s_{it}\big)$ is the transition probability from state $s_{i,t-1}$ to state s_{it}, and $P(Y_{it}|s_{it})$ is the conditional probability describing the state-related market performance. We will describe these two probabilities separately in the following subsections.

Transition Probabilities of App User Satisfaction States

An app can switch between all possible states S. The following transition probability matrix $P\big(s_{i,t-1}|s_{it}\big)$ describes this transition probability:

$$P\big(s_{i,t-1}|s_{it}\big) = \begin{bmatrix} p(1, 1) & \cdots & p(1, J) \\ \vdots & \ddots & \vdots \\ p(J, 1) & \cdots & p(J, J) \end{bmatrix} \tag{6}$$

where $p(j, k)$ is the probability of moving from state j to state k, and $\sum_k p(j, k) = 1$ for all j, k $\in$ S. We assume that $p(j, k)$ is affected by the developer's update behavior, which may lead to changes in user satisfaction with the app(Hassan et al., 2020). The feedback from users will further affect the market performance of the app. We use the probability unit model to simulate the transition probability(Wooldridge, 2010), and we assume that the state is determined by the potential transition propensity L_{it}:

$$L_{it} = W_{it}' \varepsilon_{s_{i,t-1}} + \delta_{it}, \left(\delta_{it} | W_{i,t-1}, s_{i,t-1}\right) \sim N\left(0, \sigma_\delta^2\right) \tag{7}$$

such that $s_{it} = j$ if $L_{it} \in [\mu_{j-1}, \mu_j)$, where $W_{i,t-1}$ is a vector of a developer update behavior of mobile apps, $\varepsilon_{s_{i,t-1}}$ is a vector of the corresponding coefficients, and δ_{it} is a normal error term from the probit model. Then we refer to the article by Chen et al., to obtain the transition probability as follows:

$$p(j, k) = P\left(s_{it} = k | s_{i,t-1} = j, W_{i,t-1}\right)$$

$$= P\left(\mu_{k-1} \leq L_{it} \leq \mu_k | s_{i,t-1} = j, W_{i,t-1}\right)$$

$$= P\left(L_{it} < \mu_k | s_{i,t-1} = j, W_{i,t-1}\right) - P\left(L_{it} < \mu_{k-1} | s_{i,t-1} = j, W_{i,t-1}\right) \tag{8}$$

$$= \Phi\left(\frac{\mu_k - W_{it}' \varepsilon_j}{\sigma_\delta}\right) - \Phi\left(\frac{\mu_{k-1} - W_{i,t-1}' \varepsilon_j}{\sigma_\delta}\right)$$

State-Dependent Market Performance

Given the above states, we now derive conditional probabilities $(Y_{it} | s_{it})$ to describe the state-dependent market performance. Since the observed user contribution is non-negative, we adopt the standard Tobit model [42], following the Bayesian literature [43]:

$$Y_{it}^* = X_{it}' \beta_{s_{it}} + \varepsilon_{it}, \left((\varepsilon_{it} | X_{it}, s_{it}) \sim N\left(0, \sigma^2\right)\right) \tag{9}$$

For $Y_{it} > 0$, the probability density function is:

$$g(Y_{it} | X_{it}, s_{it}) = \frac{1}{\sigma} \phi\left(\frac{Y_{it} - X_{it}' \beta_{s_{it}}}{\sigma}\right) \tag{10}$$

where ϕ is the standard normal density function.

4 Discussion

Through the establishment of the above model, we will discuss which developer update strategies can effectively shift the mobile app from a poor user satisfaction state to a good user satisfaction state, and which developer update strategies can effectively keep the app in a good user satisfaction state. It also analyzes how developers, mobile app platforms, and competent authorities make management decisions based on the changes in market performance brought about by the update under the market-oriented mechanism. By discussing the above issues, we hope to provide different management insights for developers based on the dynamic relationship between mobile app updates and user satisfaction levels, which can be widely applied to mobile app updates and provide possibilities for further theoretical development.

Acknowledgement. This research is funded by the National Natural Science Foundation of China (NSFC 72072087) and Laboratory of Data Intelligence and Interdisciplinary Innovation.

References

1. Li, X., Zhang, B., Zhang, Z., et al.: A sentiment-statistical approach for identifying problematic mobile app updates based on user reviews. Information **11**(3) (2020)
2. Lin, F.Y., Zhao, J., Chi, M.M.: A study on temporal effects of different types of mobile application updates. Sustainability **14**(3) (2022)
3. Banker, R.D., Davis, G.B., Slaughter, S.A.: Software development practices, software complexity, and software maintenance performance: a field study. Manage. Sci. **44**(4), 433–450 (1998)
4. Cavusoglu, H., Cavusoglu, H., Zhang, J.: Security patch management: share the burden or share the damage? Manage. Sci. **54**(4), 657–670 (2008)
5. Foerderer, J., Heinzl, A.: Product updates: attracting new consumers versus alienating existing ones. SSRN Electron. J. (2017)
6. Hassan, S., Shang, W., Hassan, A.E.: An empirical study of emergency updates for top android mobile apps. Empir. Softw. Eng. **22**(1), 505–546 (2016). https://doi.org/10.1007/s10664-016-9435-7
7. Zhou, G., Song, P.J., Wang, Q.S.: Survival of the fittest: understanding the effectiveness of update speed in the ecosystem of software platforms. J. Organ. Comput. Electron. Commer. **28**(3), 234–251 (2018)
8. Lin, J., Sugiyama, K., Kan, M.-Y., et al.: New and improved: modeling versions to improve app recommendation. In: 37th Annual International ACM Special Interest Group on Information Retrieval Conference on Research and Development in Information Retrieval on Proceedings, Australia, pp. 647–656, ACM Digital Library (2014)
9. Tian, H., Zhao, J.: Antecedents and consequences of app update: an integrated research framework. In: Cho, W., Fan, M., Shaw, M.J., Yoo, B., Zhang, H. (eds.) WEB 2017. LNBIP, vol. 328, pp. 64–78. Springer, Cham (2018). https://doi.org/10.1007/978-3-319-99936-4_6
10. Lee, G., Raghu, T.S.: Determinants of mobile apps' success: evidence from the app store market. J. Manag. Inf. Syst. **31**(2), 133–169 (2014)
11. Danaf, M., Becker, F., Song, X., et al.: Online discrete choice models: applications in personalized recommendations. Decis. Support Syst. **119**, 35–45 (2019)
12. Zhao, X., Tian, J., Xue, L.: Herding and software adoption: a re-examination based on post-adoption software discontinuance. J. Manag. Inf. Syst. **37**(2), 484–509 (2020)
13. Gokgoz, Z.A., Ataman, M.B., Van Bruggen, G.H.: There's an app for that! understanding the drivers of mobile application downloads. J. Bus. Res. **123**, 423–437 (2021)
14. Medema, S.G.: The "subtle processes of economic reasoning": marshall, becker, and theorizing about economic man and other-regarding behavior. In: Fiorito, L., Scheall, S., Suprinyak, C.E. (eds.) Research Annual, pp. 43–73. Emerald Group Publishing Limited (2015)
15. Chen, R., Wang, Q., Xu, W.: Mining user requirements to facilitate mobile app quality upgrades with big data. Electron. Commerce Res. Appl. **38** (2019)
16. Garg, R., Telang, R.: Inferring app demand from publicly available data. MIS Q. **37**(4), 1253–1264 (2013)
17. Sallberg, H., Wang, S., Numminen, E.: The combinatory role of online ratings and reviews in mobile app downloads: an empirical investigation of gaming and productivity apps from their initial app store launch. J. Mark. Anal. (2022)

18. Hassan, S., Tantithamthavorn, C., Bezemer, C.-P., Hassan, A.E.: Studying the dialogue between users and developers of free apps in the Google Play Store. Empir. Softw. Eng. **23**(3), 1275–1312 (2017). https://doi.org/10.1007/s10664-017-9538-9
19. Liu, H., Wang, Y., Liu, Y., et al.: Supporting features updating of apps by analyzing similar products in App stores. Inf. Sci. **580**, 129–151 (2021)
20. Fleischmann, M., Amirpur, M., Grupp, T., et al.: The role of software updates in information systems continuance—an experimental study from a user perspective. Decis. Support Syst. **83**, 83–96 (2016)
21. Clements, P., Northrop, L.: Software Product Lines. Addison-Wesley, Boston (2002)
22. Ulrich, K.: The role of product architecture in the manufacturing firm. Res. Policy **24**(3), 419–440 (1995)
23. Heckman, J.J.: Statistical models for discrete panel data. Struct. Anal. Disc. Data Econ. Appl. **114**, 178 (1981)
24. Chen, W., Wei, X., Zhu, K.X.: Engaging voluntary contributions in online communities: a Hidden Markov Model. Mis Q. **42**(1), 83-+ (2018)
25. Liu, C., Ouzrout, Y., Nongaillard, A., et al.: The reputation evaluation based on optimized hidden markov model in e-commerce. Math. Probl. Eng. 1–11 (2013)
26. Srivastava, A., Kundu, A., Sural, S., et al.: Credit card fraud detection using Hidden Markov Model. IEEE Trans. Dependable Secure Comput. **5**(1), 37–48 (2008)
27. Li, X., Zhuang, Y., Lu, B., et al.: A multi-stage Hidden Markov Model of customer repurchase motivation in online shopping. Decis. Support Syst. **120**, 72–80 (2019)
28. Hassan, M.R.: A combination of Hidden Markov Model and fuzzy model for stock market forecasting. Neurocomputing **72**(16–18), 3439–3446 (2009)
29. Xu, S., Ma, X., Liu, Y., et al.: Malicious application dynamic detection in real-time API analysis. In: 2016 IEEE International Conference on Internet of Things (iThings) and IEEE Green Computing and Communications (GreenCom) and IEEE Cyber, Physical and Social Computing (CPSCom) and IEEE Smart Data (SmartData) , Chengdu, China, pp. 788–794. IEEE (2016)
30. Fu, Y., Xiong, H., Lu, X., et al.: Service usage classification with encrypted internet traffic in mobile messaging apps. IEEE Trans. Mob. Comput. **15**(11), 2851–2864 (2016)
31. Zhu, H., Liu, C., Ge, Y., et al.: Popularity modeling for mobile apps: a sequential approach. IEEE Trans. Cybernet. **45**(7), 1303–1314 (2014)
32. Zhang, Y., Li, B., Luo, X., et al.: Personalized mobile targeting with user engagement stages: combining a structural Hidden Markov Model and field experiment. Inf. Syst. Res. **30**(3), 787–804 (2019)
33. Cao, Y., Li, Y.: An intelligent fuzzy-based recommendation system for consumer electronic products. Expert Syst. Appl. **33**(1), 230–240 (2007)
34. Corner, J.L., Buchanan, J.T.: Capturing decision maker preference: experimental comparison of decision analysis and MCDM techniques. Eur. J. Oper. Res. **98**(1), 85–97 (1997)
35. Scholz, M., Dorner, V., Franz, M., et al.: Measuring consumers' willingness to pay with utility-based recommendation systems. Decis. Support Syst. **72**, 60–67 (2015)
36. Scholz, M., Dorner, V.: Estimating optimal recommendation set sizes for individual consumers **3** (2012)
37. Huang, S.-L.: Designing utility-based recommender systems for e-commerce: evaluation of preference-elicitation methods. Electron. Commer. Res. Appl. **10**(4), 398–407 (2011)
38. Laskey, K.B., Fischer, G.W.: Estimating utility functions in the presence of response error. Manage. Sci. **33**(8), 965–980 (1987)
39. Schoemaker, P.J., Waid, C.C.: An experimental comparison of different approaches to determining weights in additive utility models. Manage. Sci. **28**(2), 182–196 (1982)
40. Rabiner, L.R.: A tutorial on hidden Markov models and selected applications in speech recognition. IEEE Proc. **77**(2), 257–286 (1989)

41. Hassan, S., Bezemer, C.-P., Hassan, A.E.: Studying bad updates of top free-to-download apps in the Google Play Store. IEEE Trans. Software Eng. **46**(7), 773–793 (2020)
42. Wooldridge, J.M.: Econometric Analysis of Cross Section and Panel Data. MIT Press (2010)
43. Rossi, P.E., Allenby, G.M.: Bayesian statistics and marketing. Mark. Sci. **22**(3), 304–328 (2003)

An Analysis of Survey Results on the User Interface Experiences of E-wallet Services

Kwan Panyawanich(✉) ⓘ, Martin Maguire, and Patrick Pradel

Loughborough University, Epinal Way, LE11 3TU Loughborough, UK
k.panyawanich@lboro.ac.uk

Abstract. Previous studies on electronic wallets have primarily focused on analyzing existing services using theoretical models based on behavioral factors, including technology adoption, user experience metrics, and marketing attributes. In contrast, this study takes a different approach by employing a customized mixed-method questionnaire administered to 381 participants. The aim is to identify the pain points associated with current electronic wallet services. The research identifies building trust as the highest priority for e-wallet usage, followed by performance expectancy, effort expectancy during the verification process, and registration. Moreover, the findings suggest that features such as topping up, rewards, and advertising have an impact on consumer satisfaction levels. The main objective of this research is to raise awareness of the issues highlighted by current users in the study and provide design recommendations and insights for improving the user interface experience for service providers. Additionally, the study explores the usage of various e-wallets in both Thailand and the United Kingdom to evaluate the core issues of e-wallet interfaces in diverse circumstances.

Keywords: Technology acceptance model · Unified theory of acceptance and use of technology · UTAUT2 · User experience

1 Introduction

The adoption of e-wallets has been growing and as it has been accelerated by the Covid pandemic. The prediction for the adoption of mobile and digital wallets states that in 2025, e-wallets will account for 72% of e-commerce payment throughout the Asia Pacific region. This will be an increase from 2021, when digital wallets accounted for 69% of Asia-Pacific e-commerce payments [33]. China have over 676 million monthly active users using Alipay wallet, followed by UnionPay and Best Pay by a wide margin in May 2021 [15]. In 2021, the number of registered mobile banking accounts in Thailand reached approximately 85.3 million, an increase from the previous year. In the last four years, the number of internet banking accounts has gradually increased [4]. True Money held 52.6% of the Thai mobile wallet market in 2020. This was followed by Rabbit LINE Pay, which accounted for 24.7% of the country's market share [12]. In the United Kingdom's e-commerce landscape (2020), the percentage of users using digital and mobile wallets remains the highest (32% of all payment types). Between 2018 and 2019, the population of young millennials, senior millennials, and generation X shifts.

© Springer Nature Switzerland AG 2023
F. Fui-Hoon Nah and K. Siau (Eds.): HCII 2023, LNCS 14038, pp. 271–292, 2023.
https://doi.org/10.1007/978-3-031-35969-9_19

their spending habits to mobile payment [32]. Despite the increasing of registered users in the UK, only 8% of multichannel retailers offer Apple Pay and Google Pay services in 2019 [2].

People's perceptions of cashless payment are changing as the world places an increased emphasis on sustainability [11, 22]. It is an opportunity to learn more about this payment technology as well as the underlying issues that may lead to dissatisfaction with the service. This includes user-unfriendly interfaces and issues that directly affect how consumers use electronic wallets. The overall goal of the study is to see if understanding user pain points through a survey can aid in the development of existing e-wallet services. The findings of an e-wallet user survey on the usability of these services are described in this paper, along with insights that can be used to improve the service's interface experience.

2 Literature Review

2.1 Theoretical Implications

Marketing Qualities: This paper undertaken a scoping review of relevant studies concerning the adoption and utilization of financial technology. Before delving into mobile payment-related topics, it is fundamental to understand the definitions of electronic wallets, also known as E-wallets. These are mobile payment applications that store electronic money on mobile phone networks. They provide different systems based on their design characteristics, including semi-closed wallets, semi-open wallets, open wallets, and closed wallets [29]. Saputri and Pratama (2021) emphasize that different e-wallet services have distinct design elements tailored to specific user characteristics and demographics. Their study focused on three Indonesian e-wallet services: GoPay, OVO, and DANA. Among the 409 e-wallet users surveyed, OVO was the most commonly used, at 40%, followed by GoPay with a usage rate of over 30%. DANA ranked third, attracting approximately 11% of participants. The majority of e-wallet users were young individuals aged 17 to 25, accounting for 70.42% of the sample. Logistic regression analysis showed that the DANA service was favoured by young middle to low-income males, while GoPay and OVO were more popular among individuals with middle to high-income levels [25]. This preference can be attributed to "Gojek" and "Grab," two popular transportation services in Indonesia that also offer food and package delivery services. GoPay is the e-wallet service associated with Gojek, while OVO is the payment system used by Grab. On the other hand, DANA is linked to Kartu Prakerja, a social security program aimed at supporting unemployed individuals [25].

A study investigated the impact of security technology acceptance, service image, and marketing mix (4C's) on the decision of students or entrepreneurs to use electronic wallets, specifically True Money wallet at Kasetsart University in Chon Buri province. The study emphasized the collaboration with Generation Z consumers, highlighting convenience as the main reason for adopting the electronic wallet due to its time-saving benefits and alignment with a fast-paced lifestyle. Government support, including the amendment of a law to allow electronic wallet use in online business transactions, enhanced consumer trust. Consequently, it is important for merchants and local shops to adjust their payment options according to changing consumer behavior [5, 23]. Another study involving 400 participants in Bangkok examined the factors influencing the usage of True Money Wallet and Rabbit Line Pay. Results revealed that factors such as gender, occupation, reference group, education, occupation, and motivations were related to the choice of e-wallet services. Moreover, the marketing mix elements of price, place, promotion, and process influenced the selection of e-wallet services, while aspects like product, price, promotion, people, and physical evidence and presentation influenced the choice of a specific brand of e-wallet services [30].

Technology Acceptance Model: Forster et al. [10] utilize the Technology Acceptance Model (TAM) as a standardized questionnaire metric to measure user experience. The TAM primarily focuses on two factors that influence technology acceptance: perceived usefulness (PU) and perceived ease of use (PEU). Perceived usefulness means users believe a technology will improve their performance and productivity. It measures how valuable and beneficial users think the technology will be for them. Perceived ease of use refers to users' belief that the technology will be easy and straightforward to use [36]. However, when applying the TAM as a measurement metric, it only considers the likelihood rating rather than the agreement rating. This is because the purpose of the TAM is to predict future product and service usage, rather than evaluating the actual experience of using the product [19, 24]. The study found that using different formats for the questionnaire structure with TAM measurement did not have a significant impact. In contrast, there were more response errors observed when the magnitude sliding scale of agreement increased from right to left. However, arranging the increasing strength of agreement from left to right, similar to measures of perceived usability like the System Usability Scale (SUS), was found to be the most effective method in research [24].

Matemba and Li (2018) identified several essential factors that influence the adoption of WeChat Wallet, an e-wallet in South Africa. These factors include trust, security, privacy concerns, and the Technology Acceptance Model [22]. Pal et al. (2019) highlighted that factors influencing mobile payment usage encompass not only usefulness and ease of use but also risk, trust, and costs. Pramana (2021), focusing on developing Asian countries, concurred that the TAM model, along with risk, trust, and social influence, plays a role in driving people to adopt financial technology [20]. Trust emerges as a significant role in technology adoption, with a focus on creating a trustworthy environment for users. Security is crucial in e-payment, enhancing consumer trust. Trust can be categorized into trust in technology's risk mitigation and trust in service providers meeting customer expectations [14]. A study in Malaysia during the pandemic found that e-wallet usage intention was influenced by TAM, trust, reliability, and health-related factors [32].

Unified Theory of Acceptance and Use of Technology/UTAUT2
The UTAUT model neglects social influence, a variable that significantly affects the intention to use technology [1]. While UTAUT2, such as performance expectancy, may have an indirect effect on designing quality perceptions [1]. Widyanto, Kusumawardani, and Septyawanda (2020) emphasized their findings that the case of LinkAja mobile payment, effort expectancy was found to be insignificant with user behavioural intention, but it did have a significant relationship with performance expectancy. While social influence was

found to significantly predict behaviour intention and the significantly predicting performance expectancy. Furthermore, entertainment values a hedonic motivation was discovered to be a significant in predicting behaviour intention. While the security aspect does not substantially influence intention behavour to use the service [31]. Contrary to the findings of Chresentia and Suharto, analyses user adoption of OVO e-wallets based on the UTAUT2 model, discovered that trust an enriching existing attributes of effort expectancy and performance expectancy [9], price value and habit all have a significant impact on consumers' behavioural intention, while behavioural intention has a significant impact on consumers' actual use. The most crucial factor influencing OVO use is price value, as the service charges a reasonable price for the value of the service to use in an e-commerce site (Tokopedia). Consumers can become motivated to use OVO in daily life as a payment option in Tokopedia. In terms of effort, users can learn how to use OVO as a payment option in Tokopedia site and experience a low level of complexity [9].

2.2 Practical Implications

Ecosystem Barriers: The country's ecosystem, characterized by a lack of collaboration among stakeholders in the mobile payment service industry and an absence of a successful business model, has delayed the adoption of mobile payment platforms. This lack of collaboration makes the process of using mobile payment services complex for users, as the country's ecosystem does not fully support the installation of these systems [22]. Financial institutions are hesitant to collaborate with other institutions due to concerns about losing customer relationships. Commercial banks face challenges in striking a balance between cooperation with mobile payment services in the ecosystem and maintaining competitiveness. Additionally, public bodies have neglected to establish appropriate rules and regulations for capturing users' identities in online transactions, posing future risks to the mobile wallet landscape [18]. As evidence, merchants face barriers in adapting to mobile payment services, primarily due to severe penalties for non-compliance with tax liabilities and service obligations. In India, approximately 98% of merchants, equivalent to 40 million, continue to rely on offline transactions to avoid Value Added Tax (VAT) and Goods and Services Tax (GST). This approach impacts their revenue margins, pricing, and adds complexity to digital transactions [23, 26]. Conversely, Thailand has successfully implemented mobile financial services with strong government support. To address the economic impact of the pandemic lockdown, the government introduced a 50:50 co-payment scheme through the Pao tang e-wallet application. This scheme provides half-price subsidies and allows users to experience mobile phone payment for the first time. A study involving 506 participants, analyzed using structural equation modeling, found that economic benefits, enjoyment, health benefits, and ease of use significantly influence Thai citizens' positive attitudes toward mobile payment within the government's co-payment program. It is crucial for Thai consumers to understand the promotional benefits of using the service, as it increases their likelihood of adopting new technologies [6].

According to Leong, Hew, Ooi, and Wei (2020), the adoption of mobile payment platforms can be time-consuming due to various barriers. One significant barrier is related to the usage aspect, where limited visibility of small keypads on mobile phone devices strongly influences resistance to mobile wallets. Another influential barrier is tradition, as some people continue to prefer using cash, especially those who do not own mobile phones. However, some individuals are attracted to using e-wallets due to the services' rewards points and cashback features [8, 17, 18]. The image barrier plays a role in how users perceive mobile payment services negatively. Users may have low intentions to use these services due to a lack of trust in system quality, information quality, and service quality provided by the service provider [18]. For instance, Alipay experienced an image barrier as some users chose not to use the service because of the firm's failure to protect customer rights [35].

Technical Barriers: Apple Pay users often encounter transaction failures due to technical issues that prevent the service from functioning as intended. Users frequently face unsuccessful transactions when trying to verify their identity or when the point-of-sale (POS) reader fails to process payments through Apple Pay. Interestingly, users experience the lowest level of satisfaction when faced with identity authorization failures and often prefer shifting to card payments. The motivation behind customers using Apple Pay is not solely based on convenience but rather on creating an identity that portrays them as cool and technologically savvy in the eyes of the public. The study reveals that Apple Pay users seek perceived praise from the public for being superior in using advanced payment technology. Apple Pay is specifically designed for iPhone owners, starting from iPhone 6 and newer models, utilizing biometric technology for payment. However, if transactions through Apple Pay do not run smoothly, users may face negative criticism from the public, leading to feelings of embarrassment that can be more uncomfortable than a failed card payment. To address these issues, merchants should understand the causes of payment failures and provide detailed reports to the IT department. Employees should receive formal training on processing and debugging Apple Pay transactions within the point-of-sale (POS) system. They should be familiar with Apple Pay on iPhone to educate and assist new customers in using the service correctly. Visible instruction comics or video clips can be displayed near the registered counter or in waiting areas. In the event of a service failure with Apple Pay, frontline staff should swiftly engage in service recovery and potentially offer compensation to alleviate customers' feelings of embarrassment. Companies can also place prominent Apple Pay signs at store entrances, waiting lines, or on POS devices to notify users that the payment method is accepted. Since Apple Pay users prefer praise from others, firms can create express checkout lines specifically for Apple Pay users, creating a psychological perception of being first-class customers to enhance positive customer satisfaction [21].

Teng and Khong (2021) asserts that consumers frequently encounter technical problems when attempting to top-up their electronic wallets. These issues arise when the application fails to reload or add credit or bank cards to the consumer's account. Users also face difficulties when transferring cash or paying bills using the e-wallet service. Despite the reputation of e-wallets for providing rewards [8, 17, 18], users often report not receiving them. Many users complain about receiving a lower amount of cashback than expected. For instance, even after making five bill payments, users only receive three rewards from the e-wallet. In order to obtain the promised rewards, users must file a complaint with the service provider and wait for three working days for the cashback to appear in their e-wallet. Issues with cashback primarily stem from poor information quality, where service providers fail to deliver on their promotional campaigns, rewards, and redemption. Lack of clarity regarding the expiry date of rewards leads users to feel deceived by the service provider. Delone and Mclean further investigated information quality and introduced the information system success model, which includes accuracy, timeliness, completeness, relevance, and consistency of information systems. To prevent unclear announcements, it is important for rewards terms and conditions to be concise and easily understandable. Precise statements will enhance the information transparency of the e-wallet's promotional campaigns [29].

The user experience of e-wallet applications is often marred by the complexity of their interfaces, leading to dissatisfaction among users. A notable issue arises when regular users, who have used the application before, are required to verify their identity when withdrawing cash from their bank account to their digital wallet. Failing to complete the verification process has resulted in users abandoning the service. Additionally, users of the Boost App wallet often lack awareness of online spending transfer limits, leading to difficulties in conducting online shopping [29]. Users have also encountered unsuccessful transactions and accidental double payments while using e-wallet accounts. The maintenance of these applications poses significant inconveniences, particularly during cash transfers. Furthermore, the incompatible infrastructure of Boost wallet and Apple Pay has caused system crashes and sluggish loading the application [21]. Some users have expressed dissatisfaction with unfriendly interfaces that prompt unwanted Global Positioning System (GPS) pop-ups in the Boost wallet. E-wallet users often express dissatisfaction with the customer service support provided, as they frequently do not receive timely responses from service staff. In instances where users do receive a reply, the responses are often generic email template that fail to effectively address users' specific problems [29].

3 Methodology

The advantage of using a mixed method allows researchers to create works with richer information on the subject. Using the method will help supplement the strengths of single design studies while overcoming their weaknesses [27]. As a source of mutual validation exchange, the combined method is used to generate personas from a dataset [7]. For this research, to better understand user experience with electronic wallets, this study employs a mixed method.

Earlier researchers used standardised questionnaires to better understand a product's user experience and often used as it holds a consistent set of questions presented in the same order so that participants can self-respond at own pace [13]. The opinions of users on the pragmatic or hedonistic features of the products were gathered using a Likert scale or sigmatic differentials. UX perception evaluates product or service qualities such as ease of use, clarity, confusion, originality, and others. AttrakDiff, User Experience Questionnaire (UEQ), and meCUE are common questionnaires for measuring user experience, while the System Usability Scale (SUS) and Post-Study System Usability Questionnaire (PSSUQ) are known for measuring usability, the UTAUT and TAM for measuring technology acceptance, and the ATS for measuring trust [3, 10].

This survey implementing a standardised bipolar response option as it allows researcher to determine whether the respondents' experience was negative or positive

[13] and add finer graduation of respondents to improve data quality of feeling beyond five points. Although, there is no single number of response options for a scale that is appropriate for all circumstances according to Cox (1980). Seven plus or minus two is a reasonable range for the best number of response alternatives. Lozano et al. (2008) discovered that increasing the number of response options improved the reliability of the associated scales. Furthermore, Weijters et al. (2010) claimed that 7-point options were better for younger and more educated samples, such as university students, based on the metrics. Referring to Van Schaik and Ling (2007), most participants preferred Likert scale over VAS, and when the Single Ease Question in a 7 Likert type was compared to the Subjective Mental Effort questionnaire, a 151-point visual scale from 0–150 assessing perceived usability, both approaches produced equivalent results in terms of scale sensitivity. Funke and Reips (2012) agree that the five-item Likert-type item was not statistically significant [16].

3.1 Designing of the Survey

This survey design had several appealing features first opens with a simple question (Dillman et al.) about whether users have registered or tried to register electronic wallets. Following that, the set of questions will be divided into two branches: electronic wallet users and non-electronic wallet users. Participants will proceed to the next set of questions based on their first response to the survey. The written language used in the survey is simple and clear, with a readable level to accommodate a broader audience. Apart from the explanation part by illustrating short form of examples; using 'e.g.' meaning examples, abbreviation will not be used on the questions. Because of the technological subject, unfamiliar technical terms are used, but an explanation is provided to help participants understand. Biometric finger is finger scan, and contactless is a sensory technology by tapping on the device. To avoid recognising identity, demographic questions are placed at the end of the survey and kept to a minimum. The survey begins with a landing page that states the study's aims and clearly clarify what types of questions will be asked that reflect those objectives. The use of closed-ended questions allows participants to spend less time completing them and makes analysis easier. Include an "other" response along with the clarification "please describe/specify" to avoid missing important data from respondents. The scale is presented in a balanced form from negative to positive in a 7-Likert scale with equal intervals for multiple choice answers. Lewis (2021) said that implementing a fine-grained grading scale in the user experience study will have a significant interaction between the metric and the number of response options, so he recommended the seven response options. To avoid encouraging participants to agree on one side or the other, the survey includes a neutral response option [19]. Each section of the questions will progress from broad to in-depth context of user experience ordering by general electronic wallets questions, security and verification processes, electronic wallet application experiences, and demographic background.

Open-ended questions allow respondents to respond in their own words, reflecting their personal beliefs and insights, and are less likely to be influenced by the researcher's expectations. It allows researchers to better understand respondent responses and may lead to the development of new response options for closed-ended questions [27].

To control the survey's reliability, this research implemented an intraobserver refers to the stability of responses over time in the same individual and interrater reliability monitoring how two or more respondents answer the same questions. Along with an alternate form of the question, reword the question on the exact content and capture a similar response [27]. Receiving similar responses indicates that the survey is reliable. The survey asked participants how satisfied they were with the experience of registering an account in their electronic wallet through a scale of satisfaction; completely dissatisfied, dissatisfied, somewhat dissatisfied, neither dissatisfied satisfied, somewhat satisfied, satisfied, and completely satisfied. Because some electronic wallet designs do not include this process, never experience included as part of the choice. An alternate question in the form of an open-ended question asked participants to explain their previous answer about the satisfaction of' registering an account' by describing their experience with it. This technique is used to find the consistency and dependability of answers in which open-ended responses should be applicable to multiple choice answers. The questions were customised and adapted from existing studies from earlier researchers that proven the scope of the topic [24, 27].

The sampling population was intended to be distributed globally, with prior experience with financial applications or electronic wallets required. This survey has been approved by Loughborough University's ethics committee and assures participants that any information entered will be treated confidentially. Participants were required to sign their consent, agreement, and declaration that they are over the age of 18. Because response fatigue reduces the time spent on each question as the number of questions increases [27], the survey design intended for respondents to spend only 5–10 min on average answering 25 questions. Between July 17, 2022, and January 10, 2023, the survey was administered via social media platforms such as LinkedIn, Facebook in the research community group, university emails and direct messages from relatives.

4   Results

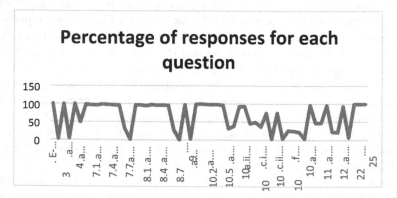

Fig. 1. A figure shows the trend of responses to each question.

The consistency of participant responses was evaluated based on data from 381 participants who confirmed their use or registration of electronic wallets (see Fig. 1). Non-users (n = 68) were excluded from the analysis. The exclusion of non-electronic wallet users from the analysis was due to the fact that their questions followed a different route and involved a separate set of questions. Open-ended questions requesting an "other" option had a low response rate (mean approx. 3%), possibly due to participants feeling no relevant reasons to mention. These questions aimed to capture additional answers that may not have been included as predefined choices in the survey. Conversely, multiple-choice questions had higher response rates (mean approx. 90%, median approx. 96%, mode approx. 96%). In regards to the open-ended questions that corresponded to the multiple-choice answers, the mean response rate was approx. 35%, median approx. 37%, and mode approx. 49%. These particular questions requested participants to provide reasons to support their claims or experiences related to the multiple-choice answers. For instance, participants were asked to explain why they were satisfied or dissatisfied with using e-wallets or to describe their experiences of payment failures. In short, irrelevant questions received a low response rate from participants, which is not causing them to drop out of the survey due to survey fatigue.

4.1 Demographics

The survey had 381 participants from Thailand (50%), the United Kingdom (30%), Japan, and the Netherlands (2% each). Convenience sampling was used, without given incentives. Participants with missing data were included because some questions may not have been relevant to the respondents (e.g., those who did not respond to all survey questions). Gender distribution: 250 female, 121 male, and 5 undisclosed. Educational background: 48% held a bachelor's degree, 35% a master's degree, and 7% have graduated from high school or secondary school. Age-wise, just about half of the respondents belonged to the millennial generation (aged 26 to 41), followed by the Z generation 32% (aged 18 to 25).

In the United Kingdom, Apple Pay, Google Pay, and PayPal are the most popular electronic wallet services (see Fig. 2). In Thailand, True Money Wallet, Pao Tang and

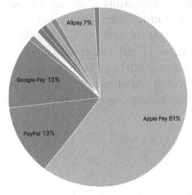

Fig. 2. A figure shows the percentage of e-wallet markets in the UK

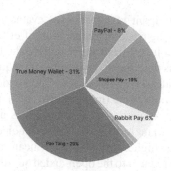

Fig. 3. A figure shows the percentage of e-wallet markets in Thailand

Shopee Pay dominate the market (see Fig. 3). Among users, females (n = 250) are the most common electronic wallet users, compared to males. Users between the ages of 18 and 25, as well as those between 26 and 41, show a preference for Apple Pay and True Money Wallet. On the other hand, users between 42 and 57 tend to favour Pao Tang. Regarding education, individuals with a doctorate or master's degree are more likely to use Apple Pay, followed by PayPal. Participants with a bachelor's degree or trade school education often use Pao Tang, or True Money Wallet.

The bar chart (see Fig. 4) depicts the majority of user satisfaction with electronic wallets. The service was either somewhat satisfied (n = 103) or satisfied (n = 171) by users. This result implies that there is room for improvement in the service. When considering different experiences provides a closer look at whether payment failure or transfer delay experiences influence user satisfaction with e-wallets.

Payment Failure Experiences: The Spearman correlation coefficient was used to examine the relationship between payment failure and customer satisfaction levels. The analysis indicated that there was no significant association between these two variables, approximate significance 0.121, (see Fig. 5).

This section explores various factors contributing to payment failures, including issues with identity verification, insecure network connectivity on the user's device, incompatible technology (such as an NFC reader) used by the merchant, malfunctioning point-of-sale (POS) devices, system errors, capacity limitations leading to inability to accommodate a large user base, insufficient funds in either the bank or e-wallet account, and daily or transaction spending limits (e.g., Google Pay). Currently, customers are required to split their payments in half and conduct two separate transactions at the service counter. Notably, if a user attempts to make two consecutive payments to the same merchant in quick succession, it often results in a denied payment. Each payment failure brings public embarrassment to consumers, and some individuals have once made a double payment.

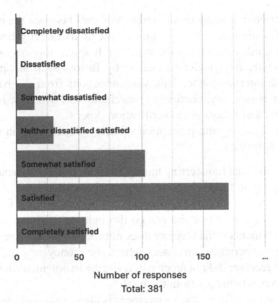

Fig. 4. Survey answers (Question 6.1.a)

| | | Fail Payment | | | | | |
		Never	Rarely	Occasionally	Frequently	Always	Total
Satisfaction	Dissatisfied	2	10	7	2	0	21
	Neither	7	6	11	3	0	27
	Satisfied	67	148	82	20	4	321
Total		76	164	100	25	4	369

Symmetric Measures

		Value	Asymptotic Standard Error[a]	Approximate T[b]	Approximate Significance	Exact Significance
Ordinal by Ordinal	Spearman Correlation	-.081	.053	-1.554	.121	.121
N of Valid Cases		369				

a. Not assuming the null hypothesis.

b. Using the asymptotic standard error assuming the null hypothesis.

Fig. 5. Correlation test (Question 6.1.a/11.1. a)

Delay Transferring Experiences: The relationship between transfer delay and service satisfaction was examined using the Spearman correlation coefficient, revealing a weak negative correlation between e-wallet satisfaction and the frequency of experiencing transfer delays (Spearman's $\rho = -0.137$, with a significant association of 0.009 (see Fig. 6). This indicates that as individuals experience transfer delays more frequently, their satisfaction with the e-wallet service decreases.

The term "delay transferring" encompasses various aspects. It includes the speed of refunding e-money from an e-wallet to the original bank account, the speed of completing payments at the receiver's end, and the speed of transferring funds from bank accounts to e-wallet accounts. In this survey, it specifically refers to the frequency of experiencing delays when transferring e-money from an e-wallet back to a bank account. Different interpretations arise from users' varying experiences with e-wallets. Some e-wallets have a top-up function, requiring users to transfer funds between the e-wallet and bank account or card. On the other hand, certain applications directly link the e-wallet to a bank account or card, eliminating the need for frequent top-ups or refunds via the bank account. Transfer delays can be caused by factors such as application system instability, unstable internet connections, disconnections from the bank's system due to banking system issues, malfunctioning merchant POS systems, limitations of the e-wallet application, and delays in the verification process.

The responses regarding the perception of delay transferring in electronic wallet experiences are as follows:

1. Consumers perceive that transferring money from an e-wallet to a bank account takes longer than topping up from a bank account to an e-wallet.
2. Payment delays can last for days, causing anxiety and uncertainty for individuals who are unsure why the transaction did not go through successfully. This increases the risk of double payments if the receiver does not receive the transaction.
3. Some users have experienced instances where the money was deducted from their e-wallet, but the receiver did not receive it, resulting in potential financial issues such as late credit payment charges from the bank.
4. Hints are provided in Alipay when a payment is about to be delayed, indicating that it should not take more than 1–2 h.
5. Delays in making payments via Shopee Pay have been reported, especially during promotional periods (Same day same month offers e.g. February 2/ April 4) when many users access the application simultaneously. This causes frustration and delays in completing payments.
6. The overall experience of topping up an e-wallet involved multiple steps that led to delay transferring.

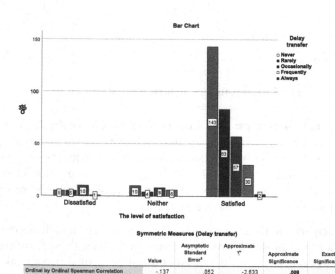

Fig. 6. Correlation test (Question 6.1.a/12.1.a)

4.2 Functions

Topping up Electronic Wallet
The findings clearly indicate that imposing minimum on top-up amounts is viewed as an unpleasant feature by the participants. One respondent shared their frustration with the waiting time involved in the top-up process. Users who only use the service occasionally find top-up to be an undesirable function. They prefer to add the minimum required amount or specify the exact amount needed for their purchase. The purpose of setting minimum top-up is to enhance security and prevent excessive spending, while also allowing others to top up the wallet. The term "top up" is subject to various interpretations, with some participants referring to linking their bank account or card details to the application. Initially, "top up" refers to the manual process of adding funds to an e-wallet by selecting an amount within the e-wallet application and initiating a transfer from a bank account to the e-wallet. Users have reported confusion and inconvenience when attempting to top up e-money via mobile banking for True Money Wallet, Pao Tang, and Shopee Pay. The multiple steps involved in logging into the mobile banking app and retrieving bank account information have proven to be cumbersome. Pao Tang wallet, for instance, necessitates switching between apps and copying the wallet number to add funds through the banking app, with a minimum top-up amount of 100 baht. True Money Wallet imposes top-up amounts in multiples of 100–300 baht, which some customers consider unreasonable, especially when buying lower-priced items at convenience stores like Seven Eleven. Topping up more money than intended can result in leftover e-money balances in users' wallets, which can be frustrating, especially considering that smaller amounts of unused e-money are non-refundable. Consequently, users often feel compelled to continue using the wallet, even if they wish to discontinue, due to these factors. Additionally, in the case of STCPay, the service does not automatically save information for future top-ups, resulting in inconvenience during subsequent use. Towards the end, topping up becomes inconvenient for users as they are required to ensure they have sufficient funds in their e-wallet accounts before making payments.

Rewards
The findings reveal that respondents find the concept of rewards fascinating and motivating when using e-wallet services. Rewards such as claiming rewards, collecting miles, saving money, or exchanging products encourage consumers to use the service more frequently. However, it is interesting to note that while some users express a desire for rewards in Apple Pay and Rabbit Pay users, another group of users is primarily focused on quick payments and not concerned with rewards. One respondent mentioned that privileged rewards may be designed to influence consumers to spend more rather than providing direct benefits.

Google Pay offers a limited selection of rewards, which require a significant amount of time and money to obtain noticeable benefits. True Money wallet provides cashback, product discounts, and True points, but some users did not claim rewards due to unclear terms and conditions, as well as uncertainty about eligibility and suitability for their lifestyle. Shopee Pay offers various discounts, cashbacks, accumulated points, and free delivery in conjunction with Shopee Food (Food delivery service) and Shopee (e-commerce). Alipay provides promotion codes to users based on their spending, which can be redeemed on their next purchase. However, one Alipay user mentioned a reduction in provided rewards after getting accustomed to using the application.

Advertisement

Many participants expressed a lack of trust in advertisements and found them annoying, particularly when they were excessive in number. Some users preferred not to have advertisements in their e-wallets as they felt they distracted from the interface and decision-making process. For example, Alipay displays ads five seconds before logging in, while Shopee Pay shows ads after completing a transaction, disrupting the user experience when attempting to initiate a new payment. Some users found e-wallet advertisements unrelated to the service's purpose, including spam emails, which led to dissatisfaction. However, there was a perspective that understood the need for platforms to generate revenue and deemed a small number of advertisements acceptable. True Money wallet displayed in-app ads for products within the same company, which users considered reasonable as a sales strategy to increase usage. Importantly, these ads were not intrusive to the payment process. Interestingly, some customers found advertisements interesting and relevant when they provided useful information. For example, ads that informed users about retail merchants accepting Alipay, Apple Pay, and PayPal for discounts, quick payment options, or other benefits, both in physical stores and online,

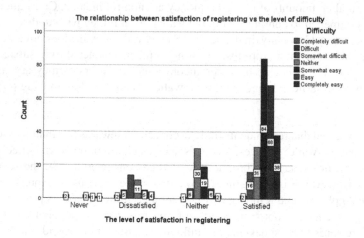

Fig. 7. Correlation test (Question 10.1. a/10.a.ii.1. a)

were appreciated. Shopee Pay also offered helpful tutorials on utilizing wallet cashbacks and discounts through their advertisements.

Registering Experiences: The Spearman correlation was used to examine the relationship between the level of difficulty in registering and the level of satisfaction with the registration process. The analysis revealed a significant positive moderate relationship between the two variables, with a correlation coefficient of .435 and a p-value of less than .001 (see Fig. 7). This indicates that when users perceive the registration process as easy, they are more likely to feel satisfied with the overall registration experience.

The requirement for personal information during registration is often perceived as excessive, personal, and frequently necessitates corrections or additional information. A user of Shopee Pay expressed dissatisfaction with the inclusion of their address, considering it unnecessary. Users find time-saving if services could offer automated filling for information like bank details. In reality, despite Apple Pay offering various registration options such as passport numbers and identification cards, not solely limited to a bank account, none of these choices proved effective, resulting in user dissatisfaction. Some users have expressed dissatisfaction with the readability of the terms and conditions during the e-wallet registration process. They believe that clearer assurances should be provided by the service. Additionally, a user suggested rearranging the registration process, starting with scanning the card before entering bank account details. This recommendation was based on their frustrating experience with Google Pay, where they had to go through multiple steps of inputting bank account details, only to find out at the end that some cards were ineligible. This led to wasted time and incomplete registration. For Pao Tang wallet, it is recommended to remove the guided frame for national ID scanning during registration. This adjustment is necessary as it currently creates difficulties for users when scanning their IDs. Respondents also mentioned the difficulties of registration, including the complexity of the process and the lack of guidance throughout the process. To enhance usability, it would be beneficial for e-wallet services to provide the option of inputting personal information by linking with other accounts during registration.

Security

Responses emphasize the importance of e-wallet providers instilling confidence and trust in the security measures to ensure the confidentiality of user information. Concerns arise regarding the requirements of laser numbers and ID card images during e-wallet registration, as unauthorized individuals may exploit this information for financial purposes. Users prefer the use of a chip on their national ID card or an open ID system like the National Digital ID (NDID) for simplified identity verification. NDID streamlines the process by requiring only one password and being compatible across multiple platforms. PayPal user's share concerns about the service's safety when encountering varying interfaces and processes on different websites, which affects both user trust and the speed of transactions. Instances where the interface appears as a pop-up on an unfamiliar window and the app frequently slows down further contribute to user anxiety.

Changing Locations

Changing the country of an Apple account can complicate future payments through the Apple Pay system. One user reported that failing to switch the account back resulted in no cards being linked in the wallet, causing embarrassment in front of other customers who queue at the back of the line for paying. Similarly, Google Pay presents challenges for users who relocate abroad, as updating the Google Play account becomes problematic. Furthermore, when using the service underground poses the risk of a dead phone when trying to tap out of the underground station without a signal.

Verification Issues

In wet weather, Apple Pay facial verification may slow down, affecting the user's ability to scan their face or interact with the touch screen for card selection. This can lead to frustration when the application prompts for a forgotten password instead of verifying face ID. Google Pay's registration process involves uploading numerous documents, but the built-in camera for verification may not function properly, causing delays and the need for multiple scanning attempts. Verifying identity through facial scan in True Money Wallet can be time-consuming, requiring users to go to an ATM machine, which is inconvenient. Responses suggest services to provide clearer instructions for identity verification during registration and improving the facial scan technology. The verification process via SMS and OTP can result in payment delays as it redirects to another pop-up page, causing users to wait without the ability to refresh. Individuals may face challenges in confirming their identity when they move to a new region or make temporary or permanent changes to their phone number, especially if their financial account is linked to the old number.

Accessibility

Certain countries face challenges in using e-wallets like PayPal, as eligibility criteria need to be met in countries such as Nepal. This restricts the availability of e-wallet services in certain regions. Similarly, WeChat Pay does not accept all bank cards.

Anticipated Features

The most interesting insight is that despite consumer demand, some existing service features were discontinued or not existed before. These include:

1. Notification feature in WeChat Pay to alert consumers about the possibility of double payments or delayed transfers.
2. The ability to receive e-receipts via email or through Facebook Messenger in PayPal, facilitating convenient storage and review of payment records.
3. Sending the terms and conditions associated with account registration to the applicant's email address for later review.
4. Providing of a physical e-wallet card to the user's address.
5. The option to physically visit different bank branches to register an e-wallet account, offering a quick and more comfortable alternative to online registration that requires less authorization.

By incorporating these features, e-wallet providers can meet the preferences and expectations of their users.

5 Discussions

5.1 Theoretical Implications

The objective of this study is to improve e-wallet experience and enhance user satisfaction levels. The survey findings indicate the potential for development in e-wallets, with customers expressing a range of satisfaction levels from somewhat satisfied to satisfied. However, there are few respondents who report being completely satisfied, indicating room for improvement and the possibility of enhancing the perceived value of the service. Additionally, some consumers rate their satisfaction as neutral, suggesting that they are still evaluating their service experience and have not encountered any issues yet. It is important to note that neutrality does not necessarily indicate a positive perception of the service.

In this study, the term "performance expectancy" refers to the expectation of swift transferring of e-money. The findings indicate a negative correlation between the expectation of fast transfers and the level of satisfaction among the respondents. This finding aligns with previous literature, which suggests that performance expectancy significantly influences consumers' behavioral intention to use technology [9, 31] The service's performance expectations encompass various aspects, such as the ability to interact with the phone's touch screen effectively, particularly when selecting the appropriate payment card in wet weather conditions. Additionally, the stability of the application's built-in camera to accurately scan documents during the registration process is also considered. Effort expectancy [9] refers to the expected ease experienced during the registration process, which is identified as a factor in this study. When using e-wallets, consumers expect a user-friendly method for inputting their personal information, including the ability to autofill information and integrate with other registered services. There is an anticipation for convenience when it comes to entering bank information into an e-wallet, with a preference for a shorter time.

These findings align with the research conducted by Matemba and Li (2018), Pal et al. (2019), Pramana (2021), Chresentia and Suharto (2020), and Hidayat et al. (2021). These studies have consistently identified trust in financial services as a significant factor influencing people's willingness to adopt financial technology. To foster trust, service providers need to prioritize the protection of users' confidential information. They should also focus on developing security measures that assure users of the safety and reliability of the service. By instilling confidence in the security of the e-wallet service, providers can enhance trust and encourage wider adoption among users. Generating that trust is gain through people-to-people constructing a holistic environment for people to have confidence in the product, includes maintain a stable mobile payment infrastructure and affordable subscription fee to accessing connection [22]. Electronic wallet providers should have the choice to register an account through bank with the help of professionals to increase comfort or registered using NDID verification as said by the sampling responses. PayPal interface is a decent illustration of how trust is essential that effect the user's usage on speed. E-wallets should maintain consistency with its interface, process when using on different websites and move away from opening a separate window to prevent dissatisfaction from the complex interface experience [29]. The country's ecosystem also accords with our earlier observations, public bodies implemented rules and regulations for not being supportive towards using electronic wallet impact the holistic adoption to the service. The case of PayPal in Nepal that it is required to meet eligible criteria [18].

5.2 Practical Implication

By addressing the challenges identified in the study and incorporating desired features [25], e-wallet providers can meet user preferences and enhance the e-wallet interface experience. There are several recommendations to achieve this:

1. Only store consumer information that is necessary for the service. This helps ensure data privacy and reduces the burden of unnecessary of data collection.
2. Provide opportunities for users to receive the terms and conditions of the service via email for later review. This allows users to have access to the terms in a more convenient and easily accessible format.
3. Adjust the sequence of inputting information during the registration process for consumer convenience. Specifically, scan the card first before entering bank details. This avoids wasting time if the card is ineligible to create the account and streamlines the registration process.
4. Improve the scanning card interface by removing the guided frame. This enhances the user experience by providing a cleaner and friendlier interface in registration.

Payment failures are caused by verification issues and the application's incompatible infrastructure [18, 21, 29], Shopee Pay and Pao Tang are unable to handle many users using the service at the same time. According to our statistical test, payment failure may not be a strong variable that impacts overall satisfaction using an e-wallet because people assume technology services occur errors on occasion. When the results of this study are compared to those of other studies, payment failure can cause great embarrassment for the cashier when using e-wallets [21]. Interestingly, it is recommended that the system should redesigned to be more flexible so that people do not experience payment failure because of the system's design. Specifically, if an Apple Wallet user has made changes to their Apple account region or if a Google Pay user has physically relocated to another country, it will not have any impact on their e-wallet account. Furthermore, it is the responsibility of merchants to ensure that their reading device (e.g., NFC reader, POS system) is updated and capable of supporting the latest system of user payment, as well as trained staff who know how to handle the situation to reduce customer embarrassment in public [21].

Over half of the respondents agreed that the restrictions of a minimum and maximum topping up function are frustrated experience in managing e-wallet balances [29]. In fact, the absence of this function in some electronic wallet services can indicate that it is not a compulsory function to develop. Even though the function's original purpose is to control spending, it actually encourages users to keep spending in order to get rid of left-over money in their e-wallet.

These findings independently support recent research that suggests people do not perceive many rewards as valuable. Some consumers have expressed disinterest in rewards, considering them worthless. They believe that to receive significant rewards, they must invest significant time in reading the terms and conditions associated with the rewards

[29]. However, it is important to note that the literature highlights the significance of understanding users' characteristics in aiding service development and designing functions that align with their values and address their challenges [25]. By optimizing user motivations and benefits, e-wallet providers can increase the value of their service. This involves offering valuable rewards, maintaining transparent terms, tailoring rewards to user preferences and lifestyle, and delivering relevant and informative advertisements.

Limitations and Recommendations
This study was unable to detect significant clusters of results due to the mean response rate for open-ended questions of being approximately 35% compared to the multiple-choice questions, which is three times less. To ensure decent accuracy, it is essential to have an equal response rate for both multiple-choice and open-ended options. Another weakness is the demographic distribution, with a majority of female participants and uneven representation across age groups and education levels. This may introduce bias into the results.

Further work needed to maintain an equal number of participants across different demographic segments, including age, education, and gender. It is also recommended to consider designing separate surveys for each electronic wallet service, such as Apple Pay, Google Pay users, etc. This approach will tailor the survey to each specific brand of e-wallet, focusing only on the existing experiences and functions of that particular wallet, thereby reducing the non-response rate. In the current study, all functions of e-wallet brands were combined into one survey, leading to some participants leaving blank or providing "unauthorized" (n/a) answers. This occurred because certain functions mentioned in the survey are not available in some services, such as Apple Pay, PayPal, Google Pay, and Rabbit Pay, which do not provide a top-up function. By designing surveys specific to each e-wallet brand, this issue can be addressed.

6   Conclusion

In conclusion, this research conducted a wide-ranging review of the literature, focusing on various barriers to the adoption of mobile payments and relevant studies evaluating the fintech services by using technology acceptance models, the Unified theory of acceptance and use of technology and its second iteration (UTAUT2), and marketing principles. The following section explored user experience research techniques and outlined the scope of this design study. The survey conducted as part of this study generated numerous insights, undercover factors influencing the adoption and usage of e-wallet services. These insights were compared and related to existing work in the field, contributing to a deeper understanding of the subject area. The findings can provide awareness to financial service providers in achieving user preferences and develop e-wallet interface experience.

References

1. Alghatrifi, I., Khalid, H.: A systematic review of UTAUT and UTAUT2 as a baseline framework of information system research in adopting new technology: a case study of IPV6 adoption. In: 2019 6th International Conference on Research and Innovation in Information Systems (ICRIIS), pp. 1–6. IEEE, December 2019

2. Ampersand. Mobile wallets: share of retailers offering payment methods on mobile such as Apple Pay or Google Pay in the United Kingdom (UK) in 2019*. Statista. Statista Inc. (2019). Accessed 02 Feb 2023. https://www.statista.com/statistics/1031364/retailers-who-offer-mobile-wallet-payments/

3. Schankin, A., Budde, M., Riedel, T., Beigl, M.: Psychometric properties of the user experience questionnaire (UEQ). In: Proceedings of the 2022 CHI Conference on Human Factors in Computing Systems (CHI 2022), Article 466, pp. 1–11. Association for Computing Machinery, New York (2022). https://doi.org/10.1145/3491102.3502098

4. Bank of Thailand. Number of registered mobile banking accounts in Thailand from 2017 to 2021 (in millions). Statista. Statista Inc. (2022). https://www.statista.com/statistics/1276722/thailand-number-of-mobile-banking-accounts/. Accessed 02 Feb 2023

5. Boonloy, P., Tangpattanakit, J.: Factors affecting decisionmaking to use electronic true money wallet of students of Kasetsart University Si Racha Campus Chon Buri Province. UBRU Int. J. 1(2), 1–10 (2021)

6. Boonsiritomachai, W., Sud-On, P.: Promoting habitual mobile payment usage via the Thai government's 50: 50 co-payment scheme. Asia Pacific Manage. Rev. (2022)

7. Boyle, R.E., Pledger, R., Brown, H.F.: Iterative mixed method approach to B2B SaaS user personas. Proc. ACM Hum.-Comput. Inter. 6(EICS), 1–44 (2022)

8. Chauhan, M., Shingari, I., Shingari, I.: Future of e-wallets: a perspective from under graduates'. Int. J. Adv. Res. Comput. Sci. Softw. Eng. 7(8), 146 (2017)

9. Chresentia, S., Suharto, Y.: Assessing consumer adoption model on e-wallet: an extended UTAUT2 approach. Int. J. Econ. Bus. Manage. Res. 4(06), 232–244 (2020)

10. Díaz-Oreiro, I., López, G., Quesada, L., Guerrero, L.A.: Standardized questionnaires for user experience evaluation: a systematic literature review. Proceedings 31(1), 14 (2019). https://doi.org/10.3390/proceedings2019031014

11. Flavián, C., Guinalíu, M., Lu, Y.: Mobile payments adoption-introducing mindfulness to better understand consumer behavior. Int. J. Bank Mark. 38(7), 1575–1599 (2020)

12. Fortumo. Market share of leading mobile wallet apps in Thailand in 2020. Statista. Statista Inc. (2021). https://www.statista.com/statistics/1322911/thailand-popular-mobile-wallet-market-share/. Accessed 02 Feb 2023

13. Geuens, M., De Pelsmacker, P.: Planning and conducting experimental advertising research and questionnaire design. J. Advert. 46(1), 83–100 (2017)

14. Hidayat, D., Pangaribuan, C.H., Putra, O.P.B., Taufiq, F.J.: Expanding the technology acceptance model with the inclusion of trust and mobility to assess e-wallet user behavior: evidence from OVO consumers in Indonesia. In: IOP Conference Series: Earth and Environmental Science, vol. 729, no. 1, p. 012050. IOP Publishing, April 2021

15. iiMedia Research: Number of monthly active users (MAU) of the leading payment apps in China as of December 2021 (in millions). Statista. Statista Inc. (2022). Accessed 02 Feb 2023. https://www.statista.com/statistics/1211923/china-leading-payment-apps-based-on-monthly-active-users/

16. James R. Lewis and Oguzhan Erdinç. 2017. User experience rating scales with 7, 11, or 101 points: does it matter? J. Usability Studies 12(2), 73–91 (February 2017)

17. Jung, J.H., Kwon, E., Kim, D.H.: Mobile payment service usage: US consumers' motivations and intentions. Comput. Hum. Behav. Rep. **1**, 100008 (2020)
18. Leong, L.Y., Hew, T.S., Ooi, K.B., Wei, J.: Predicting mobile wallet resistance: a two-staged structural equation modeling-artificial neural network approach. Int. J. Inf. Manage. **51**, 102047 (2020)
19. Lewis, J.R.: Measuring user experience with 3,5,7, or 11 points: does it matter? Hum. Factors **63**(6), 999–1011 (2021)
20. Liu, R., Li, X., Chu, J.: Promising or influencing? Theory and evidence on the acceptance of mobile payment among the elderly in China. In: Soares, M.M, Rosenzweig, E., Marcus, A. (eds.) DUXU 2022. LNCS, pp. 447–459. Springer, Cham (2022). Doi:10.1007/ 978-3-031-05897-4_31
21. Liu, S.Q., Mattila, A.S.: Apple pay: coolness and embarrassment in the service encounter. Int. J. Hosp. Manag. **78**, 268–275 (2019)
22. Matemba, E.D., Li, G.: Consumers' willingness to adopt and use WeChat wallet: an empirical study in South Africa. Technol. Soc. **53**, 55–68 (2018)
23. Ng, D., Kauffman, R.J., Griffin, P., Hedman, J.: Can we classify cashless payment solution implementations at the country level? Electron. Commer. Res. Appl. **46**, 101018 (2021)
24. Lewis, J.R.: Comparison of four TAM item formats: effect of response option labels and order. J. Usabil. Stud. **14**(4) (2019)
25. Saputri, A.D., Pratama, A.R.: Identifying user characteristics of the top three e-wallet services in indonesia. In: IOP Conference Series: Materials Science and Engineering, vol. 1077, no. 1, p. 012028. IOP Publishing, February 2021
26. Singh, N., Sinha, N.: How perceived trust mediates merchant's intention to use a mobile wallet technology. J. Retail. Consum. Serv. **52**, 101894 (2020)
27. Story, D.A., Tait, A.R.: Survey research. Anesthesiology **130**(2), 192–202 (2019)
28. Tamilmani, K., Rana, N.P., Wamba, S.F., Dwivedi, R.: The extended unified theory of acceptance and use of technology (UTAUT2): a systematic literature review and theory evaluation. Int. J. Inf. Manage. **57**, 102269 (2021)
29. Teng, S., Khong, K.W.: Examining actual consumer usage of E-wallet: a case study of big data analytics. Comput. Hum. Behav. **121**, 106778 (2021)
30. Udomsin, T., Laosuthi, T.: Factors affecting the selection of electronic payment services (e-Wallet): The case study of TrueMoney Wallet and rabbit Line Pay (2020)
31. Widyanto, H.A., Kusumawardani, K.A., Septyawanda, A.: Encouraging behavioral intention to use mobile payment: an extension of Utaut2. Jurnal Muara Ilmu Ekonomi Dan Bisnis **4**(1), 87–97 (2020)
32. Wong, H.W. and Kwok, A.O., 2022. Going Cashless? How Has COVID-19 Affected the Intention to Use E-wallets?. In: Rau, P.-L.P. (ed.) CCD 2022. LNCS, pp. 265–276. Springer, Cham (2022). https://doi.org/10.1007/978-3-031-06050-2_20
33. Worldpay. Distribution of e-commerce spending in United Kingdom (UK) in 2021, by paymentmethod. Statista. Statista Inc. (2022). https://www.statista.com/statistics/1031317/card-not-present-payments-in-the-united-kingdom-by-payment-method/. Accessed 02 Feb 2023
34. Worldpay. Forecasted share of e-commerce payment methods in the Asia-Pacific region in 2025 [Graph]. In Statista, 23 February 2022. https://www.statista.com/statistics/1039167/apac-forecasted-share-of-e-commerce-payment-methods/. Accessed 02 Feb 2023

35. Wu, J., Jiang, N., Wu, Z., Jiang, H.: Early warning of risks in cross-border mobile payments. Procedia Comput. Sci. **183**, 724-732 (2021)

36. Dewi, G.M.M., Joshua, L., Ikhsan, R.B., Yuniarty, Y., Sari, R.K. and Susilo, A., 2021, August. Perceived risk and trust in adoption E-wallet: the role of perceived usefulness and ease of use. In: 2021 International Conference on Information Management and Technology (ICIMTech), vol. 1, pp. 120–124. IEEE

Acceptance of Mobile Payment: A Cross-Cultural Examination Between Mainland China, Taiwan, and Germany

Vipin Saini[1], Julian Reckter[1], Yu-Chen Yang[1]($\boxtimes$), and Yong Jin[2]

[1] National Sun Yat-sen University, Kaohsiung, Taiwan
ycyang@mis.nsysu.edu.tw
[2] Hong Kong Polytechnic University, Kowloon, Hong Kong

Abstract. This research paper examines the influence of cross-culture on the acceptance of mobile payment by evaluating a user acceptance model in three areas: Germany, Taiwan, and Mainland China. The study uses the Hofstede framework of cultural dimensions and the Unified Theory of Acceptance and Use of Technology (UTAUT) to analyze the impact of various factors on the acceptance of mobile payment. Preliminary results of the study indicate that national culture is a significant factor in the adoption of new technologies and suggest that the performance expectation, effort expectation, social influence, and perceived security can be used to understand the impact of culture on the acceptance of mobile payment. The results confirm that cultural dimensions such as uncertainty avoidance, individualism, and power distance are related to social influence and perceived security. The study suggest that national culture should be considered when launching new technologies.

Keywords: Cross-Culture · UTAUT · Hofstede · Mobile Payment · Adoption

1 Introduction

During the last couple of years, digital transformation has affected every aspect of life, including relevant industries. The speed of the process differed significantly for the individual sectors. The digital transformation progress in the banking and finance sectors was significantly slower [1] compared to journalism or the music industry. The most significant and widespread development of digital transformation in these sectors can be seen in the online-banking [1]. Nevertheless, cash is still some European countries' most popular payment methods. An entirely accepted and used digital payment method is still not present in most countries. However, the new payment options will soon support the transition to a cashless society.

The rapid development of technology can support the notion of new payment options, wherein mobile shopping has had an enormous impact on the commercial sector for years [2]. Yet, mobile payment did not play a significant role in in-store payments until recent years. Nonetheless, the development of successful mobile payment solutions for

© Springer Nature Switzerland AG 2023
F. Fui-Hoon Nah and K. Siau (Eds.): HCII 2023, LNCS 14038, pp. 293–301, 2023.
https://doi.org/10.1007/978-3-031-35969-9_20

smartphones, like Near Field Communication (NFC), has enabled new mobile financial services applications such as bill payment and person-to-person transfers.

There can be multiple factors for the success of mobile payment. In spite of that utilization of new technology depends on the acceptance of technology by the consumers themselves [3]. The acceptance of new technology is influenced by consumer attitude and intention to use it. In the business information domain, there are a number of models, including the popular Technology Acceptance Model (TAM) and Unified Theory of Acceptance and Use of Technology (UTAUT). These models consider factors such as perceived usefulness and ease of use, as well as side factors such as gender and age. Further, the success of innovation can differ internationally, partially due to cultural differences, and national cultural aspects are believed to influence user acceptance to some degree.

Previous studies have widely used Technology Acceptance Models (TAM) to predict user behavior regarding an innovation. The focus of the paper is to study the acceptance of mobile payments and the impact of various factors by utilizing UTAUT. Though previous studies on this topic have been conducted in specific countries or regions [4]. Comparing countries can provide insights into differences in usage rates but existing studies cannot be compared as they use different models by adding additional determining or moderating variables. Even if it is assumed that the models are the same, different questions would thwart the results.

This paper analyzes the impact of national culture on user acceptance of technology by evaluating a user acceptance model in three areas. We use the Hofstede framework of cultural dimensions to formulate hypotheses for cross-cultural studies [5]. This research aims to investigate the differences in the adoption of digital technology, specifically mobile payments, across three areas Germany, Taiwan, and Mainland China based on their culture. Though these areas share some similarities but also exhibit differences in many aspects, making them an interesting group to study in terms of user acceptance of technology and how various factors influence it.

This research paper aims to develop and analyze a user acceptance model for mobile payment from the perspective of non-users in Germany, Taiwan, and Mainland China. The main focus of the research is the cultural aspect and how it affects the acceptance of mobile payment. The research question is "How do cultural aspects influence acceptance of mobile payment?" The study can also be applied to other technology acceptance cases if it shows the relationship between national culture and technology usage.

2 Literature Review

2.1 Hofstede Cultural Dimension

Culture is a pattern of feeling, thinking, and behavior that are learned and repeated throughout an individual's lifetime [5]. Culture is not uniform within a society and can vary based on the social environment they are in, such as nationality, ethnicity, or profession [5]. When discussing the impact of culture on the adoption of mobile payment, we will focus on national culture as it represents a general cross-section of the cultural norms within a country. So, based on Hofstede's theory, different dimensions can be derived from the basic core values of society in a country.

Hofstede's Cultural Dimensions theory measures cultural values across multiple dimensions: Power Distance Index (PDI), Uncertainty Avoidance (UAI), Individualism (IDV), Masculinity (MAS), and Long-term orientation (LTO). PDI measures the acceptance of unequal power distribution in culture, with high PDI in Asia and low in countries with flatter hierarchies. UAI measures tolerance for uncertainty, with high UAI in middle Europe and low in Northern European countries and Singapore. IDV values individual interests over the collective good in wealthy countries, while collectivism is strong in poorer countries. MAS measures gender roles, with rigid gender roles in masculine cultures and equal distribution in feminine cultures. LTO values future rewards over immediate gains, with high LTO in East Asian countries and low in countries like the United States.

In this paper, we are focusing on Mainland China, Germany, and Taiwan. According to Hofstede cultural dimension, both Mainland China and Taiwan have a strong collectivist cultural background influenced by Confucianism. Moreover, for Mainland China, Mao Zedong, the founding father of the People's Republic of China influenced Chinese culture significantly. He promoted collectivism and propagated individualism negatively, in accordance with his political objectives [6]. Chinese managers are typically more authoritarian and autocratic compared to Western managers, with a large power distance and subordinates assuming the leader's decision is final. The same principles can be transferred, in a mitigated extent on Taiwan. However, Taiwan has undergone modernization and Americanization, with a mix of modernizing Confucian manifestations [6].

In contrast, Germany is classified as highly individualistic with low power distance, medium masculinity, and medium uncertainty avoidance. Direct and participative communication and meetings are common, and leadership is expected to show expertise. Personal fulfillment and independence are prioritized over close relations [7]. When it comes to uncertainity avoidance Mainland China has a relatively low score in uncertainty avoidance compared to other industrialized countries, including Germany and Taiwan. This low score may be due to the fast economic growth and the optimism of the society in handling uncertain situations. Germany has a high score in uncertainty avoidance and is described as preferring deductive rather than inductive approaches and valuing details to create certainty. This is reflected in their legal system and combined with a low power distance, Germans rely heavily on expertise to compensate for their high uncertainty [8].

2.2 Mobile Payment

Mobile Payment (MP) is a payment method where monetary transactions are conducted between parties using a mobile device, such as a smartphone. The process involves connecting to a server, authentication, payment, and confirmation of the transaction [9]. MP is different from mobile banking, as the latter only involves payment handling by banks, while in MP the mobile device is a key component of the process [10].

Classification of Mobile Payment. We can classify mobile payment based on transmission technology, location, and application type. The transmission technology involves information exchange between the merchant system and the user's phone. There are five mobile payment technologies: SMS, Bluetooth, internet-based, 2D barcodes, and

NFC. NFC is seen as the most secure and convenient with a high potential for future use in proximity payments. Proximity payments involve direct interaction between the customer's phone and a merchant's device, while remote payments can be made without any additional devices [11]. This research paper focuses on proximity transactions as they are not yet widely adopted.

2.3 Unified Theory of Acceptance and Use of Technology (UTAUT)

The main aim of UTAUT was to unify the concepts of pre-existing technology acceptance models. The authors of UTAUT criticize the use of only one model to analyze specific problems, leading to inaccuracies [12]. The model includes five models: Combined TAM and TPB, Model of PC Utilization, Social Cognitive Theory, and Innovation Diffusion Theory [13–16]. In UTAUT there are three major determinants that Venkatesh et al., (2003) identified.

Performance Expectancy (PE). It is defined as the degree to which an individual believes that using technology will improve their job performance. Performance expectancy is seen as the most important predictor of intention to use technology, as it was the strongest predictor in each of the above-mentioned technology models. Effort Expectancy (EE). Refers to the perceived ease of using technology. It is based on the idea that technology is more acceptable when it is perceived to be easy to use. The determinant is more prominent in the early stages of new technology but becomes less important as instrumental concerns become more prominent. Social Influence (SI). Plays a role in the user's acceptance of technology. It is the perception of the user about how important others believe they should use the new technology and can impact the user's decision to use it.

UTAUT also has suggested that moderators can significantly impact the behavioral intention (BI) and overall user acceptance of technology. Further work should identify and test additional moderators, especially with different technologies and user groups. Our study filled this gap by introducing cultural dimensions by Hofstede as moderating variables for the acceptance of mobile payment.

3 Conceptual Framework

3.1 Hypothesis Development

Performance Expectancy, Effort Expectancy, and Social Influence. The UTAUT model by Venkatesh is used as the base to analyze the acceptance of mobile payment in different national cultures. The model is adaptable to any kind of application and is the most widespread, approved, and tested technology acceptance model. However, beyond the existing model, special aspects need to be considered and the impact of national culture needs to be taken into account. Three hypotheses are stated based on the original UTAUT model:

Hypothesis 1. Performance expectancy has a significant and positive relationship with the behavioral intention of mobile payment.

Hypothesis 2. Effort expectancy has a significant and positive relationship with the behavioral intention of mobile payment.

Hypothesis 3. Social influence has a significant and positive relationship with the behavioral intention of mobile payment.

Perceived Security. Besides the perceived benefits of mobile payment systems, potential risks are also associated with money transfers between two parties which could impact consumer acceptance [17]. Our literature review suggests that perceived security (PE) is a critical factor in consumer research about innovation and that the perceived risk of using mobile payments is high due to uncertainty and privacy concerns [18]. In this regard, we conclude that perceived risk should be considered in a user acceptance model for mobile payment and will be added to the original UTAUT model.

Hypothesis 4: Perceived security has a significant and positive relationship with the behavioral intention of mobile payment.

Uncertainty Avoidance. The influence of national culture on user acceptance has been explored in past studies but is relatively unexplored currently [19]. To fill this gap we aim to test the impact of national culture on user acceptance for private customers by linking Hofstede's three cultural dimensions to the determinants of the UTAUT model. People with high Uncertainty Avoidance are expected to attach more importance to a secure and reliable payment process, resulting in lower perceived security about this technology. We state:

Hypothesis 5: Uncertainty Avoidance has a significant and negative relationship with the perceived security regarding the acceptance of mobile payment.

Hypothesis 5a: Chinese have higher perceived security than Taiwanese and Germans regarding mobile payment.

Power Distance. People with low Power Distance are less likely to adopt the characteristics and behavior of more powerful people. The Power Distance is linked to the Social Influence determinant in the acceptance of mobile payment.

Hypothesis 6: Power Distance has a significant and positive relationship with social influence regarding the acceptance of mobile payment.

Hypothesis 6a: Taiwanese have a higher social influence than Germans regarding mobile payments, but a lower one compared to the Chinese.

Individualism. Individuals with high individualism care more about themselves and their goals rather than other people. In the context of the acceptance of new technology, these individuals are expected to be less influenced by their social group.

Hypothesis 7: Individuality has a significant and negative relationship with social influence regarding the acceptance of mobile payment.

Hypothesis 7a: Germans have a lower social influence than Taiwanese and Chinese regarding mobile payment.

The perceived security factor of risk may raise the concern of privacy, which could affect the acceptance of mobile payment, especially for individuals with a high strive for

individuality. This is because they value their control over personal information and relationships, and the potential loss of privacy through mobile payment could compromise
their autonomy.

Hypothesis 8: Individuality has a significant and negative relationship with perceived
security regarding the acceptance of mobile payment.

Hypothesis 8a: Germans have lower perceived security than Taiwanese and Chinese
regarding mobile payment.

3.2 Research Model

The research model shows that the determinants of Social Influence and Perceived Security are expected to be influenced by culture, while Performance Expectancy and Effort
Expectancy are independent of cultural impacts (Fig. 1).

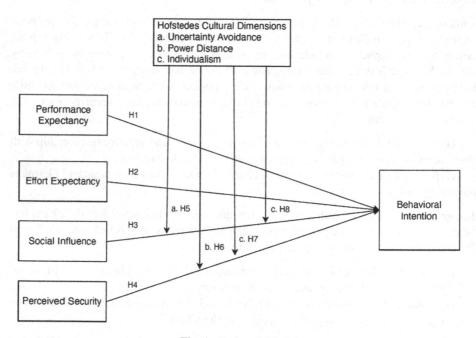

Fig. 1. Research Model

4 Preliminary Analytical Results

4.1 Survey Development

The research model for the acceptance of mobile payment includes four main determinants and four moderating effects by cultural dimensions. The cultural dimensions were
determined using Hofstede's calculations from 1991, while the main determinants were

measured through a survey in German, traditional and simplified Chinese for Taiwan and Mainland China respectively. The survey was adapted from Venkatesh's UTAUT research from 2003 and related literature to test with participants for each language, with questions being revised based on their feedback. Each item was measured on a 5-point Likert scale.

4.2 Sampling and Questionnaire Distribution

This research work focused on students in three areas as test subjects as they form a relatively homogenous group across all areas, with less fluctuation in age and experience compared to the entire population. The results can be partially projected on the entire population as the findings are a more qualitative comparison across areas. Only samples without prior experience with mobile payment were considered for data analysis to avoid overlap with actual usage experiences.

4.3 Data Analysis

The study has a sample size of 388 with 141 participants from Taiwan, 131 from Mainland China, and 116 from Germany. The sample is evenly split between male (181) and female (207) participants. The average age of participants from Taiwan and Mainland China is younger than those from Germany, but all are of a low level. The data was analyzed using SPSS and SmartPLS, and the validity and reliability of the questionnaires were reconfirmed. The results of the KMO criterion and Bartlett test showed that the sample data was suitable for component analysis, with a score significantly higher than the recommended minimum.

The factor loadings were calculated by a factor analysis and varimax rotation to classify the questionnaire into four latent variables. One item was removed due to low factor loadings. The remaining items showed satisfactory factor loadings. The reliability of the data was examined using the Spearman-Brown prediction formula and Cronbach's alpha, with a value of >0.7 considered acceptable and >0.9 considered excellent. Convergent validity was assessed using composite reliability and average variance extracted, with scores significantly higher than the recommended thresholds.

The discriminant validity of the instrument was confirmed by using the Fornell-Larcker criterion. This was done by checking if the average detected variance of a construct was higher than any squared correlation with another construct. The results showed that the instrument had adequate discriminant validity.

The sample was divided into three country groups to assess the influence of culture on mobile payment acceptance. A Multigroup analysis was performed using the Measurement Invariance of Composite Models (MICOM) [20] procedure to check for compositional invariance across the groups. The results showed that the original correlations met the requirements, and the mean and variance original differences were within the acceptable boundaries, indicating full invariance across all groups.

The mean values and standard deviations of the acceptance of mobile payment were analyzed and the results showed that there were significant differences between the areas. The UTAUT model was tested for the whole sample as well as for each subgroup and showed good fit, with R^2 explaining 0.75 of the variance in behavioral intention for the whole sample. The relationships between the UTAUT acceptance variables were verified using the structural equations procedure for all areas separately. The results are presented in a Table 1.

Table 1. The Unified Theory of Acceptance and Use of Technology model with path coefficients for the entire sample, and sub-groups. $* = p < 0.05$; $** = p < 0.01$

Path relationships	Mainland China	Taiwan	Germany	All
PE → BI	0.291 (<0.000**)	0.478 (<0.000**)	0.464 (<0.000**)	0.496 (<0.000**)
EE → BI	0.116 (0.030*)	0.086 (0.045*)	0.122 (0.043*)	0.092 (0.032*)
SI → BI	0.346 (0.009*)	0.164 (0.026*)	0.176 (0.038*)	0.182 (0.004**)
PS → BI	0.189 (0.011*)	0.360 (<0.000**)	0.334 (<0.000**)	0.293 (<0.000**)

The results shows that the latent variables perceived security and performance expectation have a significance level under 1%, while the others have a significance level under 5%. The results confirm hypotheses 1 to 4, which represent the relationship between the latent variables and the behavioral intention. The study also examines cultural dimensions and their influence on the behavioral intention. Participants from Mainland China showed the highest social influence and a high power distance, low individualism. Germans showed the lowest mean value for social influence and had low scores in power distance and high individualism. Taiwanese had low individualism like Mainland China but a moderate Uncertainty Avoidance. Germans showed less confidence in the security of mobile payment compared to Chinese, who had low scores for both Uncertainty Avoidance and Individualism. These observations support hypotheses 5 to 8a.

5 Discussion and Conclusion

This research works examines the influence of cultural aspects on the acceptance of mobile payment. The findings suggest that Hofstede's cultural dimensions of Uncertainty Avoidance, Individualism, and Power Distance are related to the technology acceptance model constructs of social influence and perceived security. The study found that individuals from Mainland China had the highest acceptance of mobile payment, while those from Germany had the lowest acceptance, with Taiwanese individuals in between. Results indicate that the proposed model (performance expectation, effort expectation, social influence, and perceived security) can be used to understand the influence of national culture on the acceptance of mobile payment. The study concludes that national culture is an important factor in the adoption of new technologies and should be considered when launching new technologies.

References

1. Linnhoff-Popien, C., Zaddach, M., Grahl, A. (eds.): Marktplätze im Umbruch. X, Springer, Heidelberg (2015). https://doi.org/10.1007/978-3-662-43782-7
2. Global Commerce Review Deutschland, Q1 2018 (2018)
3. Ginner M.: Akzeptanz von Digitalen Zahlungsdienstleistungen. Springer Fachmedien Wiesbaden (2018). https://doi.org/10.1007/978-3-658-19706-3
4. Slade, E.L., Dwivedi, Y.K., Piercy, N.C., Williams, M.D.: Modeling consumers' adoption intentions of remote mobile payments in the United Kingdom: extending UTAUT with innovativeness, risk, and trust. Psychol Mark. **32**(8), 860–873 (2015). https://doi.org/10.1002/MAR.20823
5. Hofstede, G.: Culture's Consequences: Comparing Values, Behaviors, Institutions and Organizations across Nations. Sage (2001)
6. Zhang, W.: Taiwan's Modernization: Americanization and Modernizing Confucian Manifestations. World Scientific (2003)
7. Wong, N.Y., Ahuvia, A.C.: Personal taste and family face: Luxury consumption in Confucian and Western societies. Psychol Mark. **15**(5), 423–441 (1998). https://doi.org/10.1002/(SICI)1520-6793(199808)15:5%3c423::AID-MAR2%3e3.0.CO;2-9
8. Germany - Hofstede Insights. Accessed February 6 (2023). https://www.hofstede-insights.com/country/germany/
9. Antovski, L., Gusev, M.: M-payments. In: Proceedings of the International Conference on Information Technology Interfaces, ITI. University of Zagreb, pp. 95–100 (2003). https://doi.org/10.1109/ITI.2003.1225328
10. Döpke, C.F.: Ökonomische Und Juristische Aspekte Des Mobile Payments.. Accessed February 6, 2023 (2017). www.abida.de
11. Alimi, V., Rosenberger, C., Vernois, S.: A mobile contactless point of sale enhanced by the NFC and biometric technologies. Int. J. Internet Technol. Secured Trans. **5**(1), 1–17 (2013). https://doi.org/10.1504/IJITST.2013.058291
12. Venkatesh, V., Morris, M.G., Davis, F.D.: User acceptance of information technology: toward a unified view. MIS Q. **27**(3), 425–478 (2003)
13. Taylor, S., Todd, P.A.: Understanding Information Technology Usage: A Test of Competing Models. **6**(2), 144–176. https://doi.org/10.1287/ISRE.6.2.144.101287/isre62144
14. Thompson, R.L., Higgins, C.A., Howell, J.M.: Personal computing: toward a conceptual model of utilization. MIS Q. **15**(1), 125–142 (1991). https://doi.org/10.2307/249443
15. Compeau, D.R., Higgins, C.A.: Computer self-efficacy: development of a measure and initial test. MIS Q. **19**(2), 189–210 (1995). https://doi.org/10.2307/249688
16. Rogers, E.M.: Diffusion of Innovations. The Free Press, New York, NY. Published 1995. Accessed February 6, 2023. https://catdir.loc.gov/catdir/bios/simon052/2003049022.html
17. Cho, J.: Likelihood to abort an online transaction: influences from cognitive evaluations, attitudes, and behavioral variables. Inf. Manage. **41**(7), 827–838 (2004). https://doi.org/10.1016/J.IM.2003.08.013
18. Lim, N.: Consumers' perceived risk: sources versus consequences. Electron. Commer. Res. Appl. **2**(3), 216–228 (2003). https://doi.org/10.1016/S1567-4223(03)00025-5
19. Straub, D., Keil, M., Brenner, W.: Testing the technology acceptance model across cultures: a three country study. Inf. Manage. **33**(1), 1–11 (1997). https://doi.org/10.1016/S0378-7206(97)00026-8
20. Latan, H., Noonan, R.: Partial Least Squares Path Modeling: Basic Concepts, Methodological Issues and Applications. (Latan, H., Noonan, R. (eds.). Springer (2017). https://doi.org/10.1007/978-3-319-64069-3/COVER

Types of Mobile Retail Consumers' Shopping Behaviors from the Perspective of Time

I-Chin Wu(✉) ⓘD, Hsin-Kai Yu, and Shao-I. Lien

Graduate Institute of Library and Information Studies, School of Learning Informatics,
National Taiwan Normal University, Taipei, Taiwan
icwu@ntnu.edu.tw

Abstract. The aim of this research is to explore consumers' shopping behavior on mobile apps and to mine different types of consumers. We have selected the mobile Taobao app as our research target and designed two simulated shopping tasks—goal-oriented shopping and exploratory-based shopping—based on consumers' need-states to examine consumers' shopping journeys. We have also investigated the effect of the time factor which has been regarded as an important incentive (i.e., saves time and is convenient) for consumers to shop by using mobile apps. We used a k-means clustering algorithm based on nine features of the interface of a mobile app to analyze types of consumers' online shopping behaviors. Our results have shown that there are five types of consumers which are *transitional-oriented*, *price-sensitive*, *recommendation-adopting*, *information-consuming*, and *comparing* groups. Our results reveal that the time factor is a primary means to differentiate consumers' mobile shopping behaviors instead of need-states. The most obvious difference between the five groups of consumers is that the *transitional-oriented* group spent the least time browsing information in the app whereas the *information consumption* group spent the most. Interestingly, both groups went to their shopping carts frequently and consequently made purchase decisions. While the *comparing* group spent a great deal of time on checking, comparing, and evaluating information, they seldom went to the shopping cart page. In conclusion, we have found that consumers' shopping behavior on mobile apps is deeply affected by the factor of time and exhibited different shopping journeys among groups. Our findings can provide a reference for mobile retailers to refine their app interfaces and develop marketing strategies tailored to different types of consumers.

Keywords: Clustering · Consumers' Type · Mobile Retailing · Consumer Journey · Time Factor

1 Introduction

The COVID-19 pandemic has accelerated the growth of mobile commerce. US retail mobile commerce estimates suggest that annual sales will nearly double their share of total retail sales between 2020 and 2025 (eMarketer 2021). Shopping online by using mobile apps has thus become typical in daily life. In parallel, retailers have sought to

© Springer Nature Switzerland AG 2023
F. Fui-Hoon Nah and K. Siau (Eds.): HCII 2023, LNCS 14038, pp. 302–313, 2023.
https://doi.org/10.1007/978-3-031-35969-9_21

increase awareness of the importance of the market in developing new mobile marketing strategies by improving consumers' shopping experiences.

Pantano and Priporas (2016) have investigated factors that affect consumers' movement from e-channels to mobile channels for shopping by using a qualitative approach from a cognitive viewpoint in the Italian market. Their results showed that convenience in terms of saving time is one of the drivers for using retailers' mobile apps. Moreover, saving money, supporting lifestyle, and achieving security are also important factors for increasing consumers' willingness to use mobile apps. Almarashdeh et al. (2018) conducted a survey of 143 people about why they shop using mobile apps and websites from the perspectives of access and search convenience. They found that users adopted mobile apps for shopping due to their accessibility but websites due to their searchability. However, they did not find any significant difference between shopping on websites versus mobile apps.

Previous studies have examined consumers' online shopping research from the perspective of their online shopping motivations, preferences, or cognitive perceptions. However, little research has analyzed their shopping behaviors from the viewpoint of their journeys as consumers, conceptualized in the three primary stages: pre-purchase, purchase, and post-purchase (Lemon and Verhoef 2016). Rennie et al. (2020) have pointed out that consumers want to find products with expected value by searching and comparing various sources of website-based information to make decisions. The overall shopping process can be differentiated into stages of exploration and evaluation. During the exploration stage, because consumers lack a clear picture of the products, they seek to find information. During the evaluation stage, they compare products from several alternatives and may make final decisions. The borderline between the stages is blurred. Consumers may also have loops between stages (i.e., at evaluation stage but then return to exploring information) to clarify their need-states in order to make final shopping decisions.

Recent research on consumers' journeys in studies on consumer behavior has differentiated consumers' shopping process into exploratory, evaluative, and purchase stages (Dasgupta and Grover 2019; Rudkowski et al. 2019; Tupikovskaja-Omovie and Tyler 2022). As shown, consumers' shopping behaviors can be traced and observed by following an effective consumer journey design (Kuehnl et al. 2019), and that viewpoint upholds the mentioned concepts of users' need-states. Moreover, Wu and Yu (2020) have differentiated two types of consumer need-states from the perspectives of consumers' search behaviors during the entire shopping process—namely, goal-oriented and exploratory-based types. The users' need-states or demand situations refer to their current or situational needs (Zhang et al. 2015). The goal-oriented state refers to consumers with a specific shopping goal or planned purchases, whereas the exploratory-based state refers to consumers who lack specific goals, are less deliberate, and may not consider purchasing. Their results show that consumers' search behaviors in online stores can be differentiated based on their need-states with dramatically different interactive sequences, especially in the initial stage of the online shopping process (Wu and Yu 2020).

In a recent study, Tupikovskaja-Omovie and Tyler (2022) have addressed the importance of observing consumers' online shopping behaviors by using techniques to observe their experiences as they interact with the interface throughout the shopping journey. By using an eye-tracking technique, they found that experienced and inexperienced mobile app users alike exhibited different online browsing and decision-making processes. Among several factors, time spent and frequencies of viewing product pages were quite different for the two groups of consumers. Inexperienced users took more time and twice visited more product pages than their experienced counterparts. Furthermore, the two groups had different liked, used, and desired features in relation to the apps, revealed by analyzing the qualitative interview data. Those results suggest that the developers of apps can learn more from inexperienced users than from experienced ones.

In our research, we have sought to determine whether need-states and time factors influence the decision-making process of users who have adopted mobile apps for shopping by analyzing their interaction with those apps. Furthermore, we have sought to identify different types of customers based on how they browsed pages and used functionalities provide by the apps. Regarding data collection, most similar current research relies on survey methodologies or log data analysis without information about the online search behavior of consumers by tracking their browsing, clicking, and search behaviors in real time on mobile apps or webpages (Almarashdeh et al. 2018; Kau et al. 2003; Liu et al. 2015; Moe 2003; Pantano and Priporas 2016; Zhang et al. 2015). In response, we have focused on the online shopping experiences of mobile consumers to collect consumers' behavioral data in situ. To that end, we formulated two research questions:

1. *How do need-states (i.e., goal-oriented and exploratory) and time factors (i.e., longer and shorter shopping journeys) influence consumers' search behaviors?*
2. *How can k-means clustering be used to explore whether different types of consumers browse targeted different types of elements in the interface, and how do consumers experience their shopping journeys depending on their consumer group?*

We have sought to strengthen our evaluation model explaining consumers' shopping behaviors based on the results obtained for each type of consumer and to provide implications for mobile retailers to conduct marketing in the future.

2 Methods

2.1 Participants and Tasks

We have recruited participants by purposive sampling based on our task design. We invited 37 users to participate in a preliminary evaluation using two kinds of simulated shopping tasks, for a total of 74 tasks. The participants are 23 to 42 years old, with an equal distribution of men and women. All have mobile Taobao app accounts, and all have at least two years of experience shopping on mobile apps. In fact, 82% of participants had shopped on the Taobao app at least five times a month for the past two years. Accordingly, they have similar shopping experience with shopping on mobile apps.

Laran et al. (2016) state that users may be aware (conscious) or unaware (unconscious) of an active goal when making a choice; thus, people with different levels of consciousness adopt different approaches when making choices. We preliminarily differentiate between goal-oriented and exploratory-based consumer need-states to analyze search behaviors throughout the shopping process and to understand how consumers employ web features. We designed simulated online shopping tasks based on the concepts of the experimental components for the evaluation of interactive information retrieval systems (Borlund 2000). The simulated task is a "cover story" that describes a situation that triggers the user's information needs, as shown in Table 1. During the execution of a simulated work task, we explained 1) the source of the information need, 2) the environment of the situation, and 3) the problem to be solved (Borlund 2000). We confirmed that the participants understood the objective of the search task to interpret the information needs as in real life.

Table 1. Task design.

Notes:
- You may add the product to or delete it from the shopping cart at any time
- There is no time limit and the shopping budget is based on your monthly shopping expenses

Mobile app: Mobile Taobao app

Task types	Descriptions
Goal-oriented	**Task:** Your best friend's birthday is coming. You are planning to buy a present for her/him **Requirement:** • She/he wants to buy some new clothes for the coming winter • She/he is into fashion with an interest in "athleisure" products
Exploratory-based	**Task:** For some reason, you are planning to buy products for yourself or your friend in the near future There is no requirement. It can be based on you or your friend's preferences to buy the gift

2.2 Evaluation Process

We adopted an exploratory sequential mixed-methods approach to form comprehensive answers to the research questions, as shown in Fig. 1. The approach was characterized by an initial quantitative phase of data collection and analysis data, followed by the collection and analysis of qualitative data. We aimed to explain and interpret and results of the consumers' shopping behaviors based on the mixed-methods approach. First, we conducted a questionnaire survey to collect the participants' demographic data and experiences with shopping online while using smartphones. We designed simulated online shopping tasks behind the concepts of the experimental components for the evaluation of interactive information retrieval systems (Borlund 2000). We recorded and tagged search and online shopping behaviors with the usability evaluation tool Morae. We also sought to analyze the consumers' usage behaviors also by using k-means clustering to explore the characteristics of different groups of consumers which are exhibited in their app-using behaviors. Moreover, we conducted post-experiment interviews to investigate users' perceptions of using the app Taobao to accomplish their tasks. Last, we conducted a survey addressing attitudes toward shopping online based on questionnaire questions designed by Kau et al. (2003) to gain a more in-depth understanding of each group of consumers.

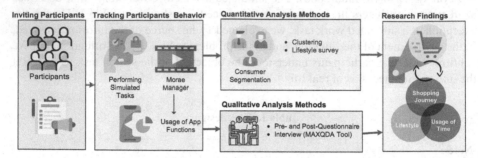

Fig. 1. The evaluation process

2.3 Variables for Customer Segmentation

We used k-means clustering algorithm to analyze how the use of features provided by the mobile app connect with the evaluation results concerning page traversal behaviors to gain a comprehensive understanding of consumers' shopping behaviors on the mobile app. We referenced variables selected from our previous research on online shopping (Wu and Yu 2020) and the features of mobile apps for the 13 variables for customer segmentation. We conducted a t test in advance to examine whether the variables have statistical significance in relation to the time factor or consumers' need-states. The results show that four variables that were not significant for the use of time. Interestingly, the functionalities of recommendations, review pages, and search were not significant, despite being important features for different types of consumers by using the Taobao website for online shopping in our previous research (Wu and Yu 2020). Furthermore, no variables were significant in relation to consumers' need-states by using the app. Thus, nine predictors with 99% significant level emerged to predict type of total time spent (i.e., longer and shorter) and are considered in our research, as shown in Table 2. The need-states will be not selected as a target variable in this research. We have set 147 s as a threshold to divide the consumers' total time into longer and shorter groups. The following four variables were removed.

- Proportion of using the recommendation function pages of all pages;
- Proportion of viewing review comments pages;
- Iterations of keyword or image searches; and
- Frequency of browsing category-level product pages.

Although the above variables did not pass significance test which will be not included as input variables during the clustering step, we will still adopt the variables to investigate each cluster.

Table 2. Indicators with associated variables used for customer segmentation.

ID	Variables	t-test (P-value) time factor
V_1	Frequency of browsing store-level pages	5.276^{***} (<0.01)
V_2	Frequency of browsing product pages	5.479^{***} (<0.01)
V_3	Frequency of browsing brand-level product pages	3.787^{***} (<0.01)
V_4	Frequency of repeatedly browsing same stores	5.506^{***} (<0.01)
V_5	Frequency of repeatedly browsing same products	3.410^{***} (<0.01)
V_6	Frequency of repeatedly browsing same categories(brands)	4.915^{***}(<0.01)
V_7	Proportion of using product pages of all pages	4.235^{***}(<0.01)
V_8	Frequency of clicking on advertisements	2.757^{***} (<0.01)
V_9	Time spent for sorting, filtering and selection products	4.411^{***} (<0.01)

Note:***: $P < 0.01$

3 Results

3.1 Clustering Process

We selected nine variables and 12 variables of the features provided by the apps (see Table 2) to cluster consumers and adopted compactness and purity metrics to evaluate the clustering results (Manning et al. 2008). We mainly focus on the intra-similarity of each cluster; thus, we compute the compactness of each cluster. For compactness, we compute the intra-cluster cohesion by measuring how near the instances in a cluster are to the cluster centroid. The smaller the compactness (i.e., closer to 0), the higher the similarity to the centroid point of all of the data instances. Compactness helped us examine the quality of the clustering results to select the best number of clusters and variables of each cluster and can be used to determine consumer segments. Furthermore, we used ground truth information to construct a confusion matrix to compute the purity value of each cluster—that is, a longer or shorter shopping time (LT and ST, respectively). Purity helped us to determine whether the variables show in Table 2 can be used to differentiate consumers into two groups. The 12 variables were selected from our previous research, in which we used them to examine users' online shopping behavior on webpages with the best results in terms of compactness and purity metrics (Yu & Wu 2021). In the study presented here, we sought to evaluate the performance of the two sets of variables to identify types of consumers. The results show that regardless of whether the data is grouped into two, three, four, or five clusters, clustering using nine variables yields slightly better performance than twelve variables in terms of average purity and compactness, as shown in Table 3. Furthermore, the results show that clustering the consumers into five clusters (groups) using nine variables yields the best purity (i.e., 0.794) and compactness (i.e., 0.015). Thus, we further investigated the five groups of consumers via nine variables.

Table 3. Cluster compactness and purity given various numbers of indicators

# of indicators	# of clusters			
Nine indicators	#of C = 2	# of C = 3	# of C = 4	# of C = 5
Compactness	0.020	0.018	0.016	**0.015**
Purity	0.674	0.753	0.779	**0.794**
Twelve indicators				
Compactness	0.026	0.024	0.021	**0.020**
Purity	0.529	0.544	**0.574**	0.559

3.2 Consumer Segmentation

Our results indicate that five types of consumers—transactional-oriented (TO), price sensitive (PS), recommendation-adopting (RA), information-consuming (IC), and comparing—identified based on the characteristics of each cluster. Table 4 shows the details of each group of consumers with associated values for each variable, while Table 5 shows the corresponding proportional values of conducting searches, adopting recommendations, reviewing comments, and visiting the shopping cart.

Discussion 1: The TO group (C_1) of consumers, all of whom belonged to the ST group, spent the least amount of time to accomplish the task. Among the group's characteristics, all such consumers visited their shopping carts only once and made purchase decisions quickly. Of the 20 consumers in the group, 12 were goal-oriented consumers. Table 5 shows that consumers in the TO group were more likely to use search functions than the other groups, which suggests that they prefer to actively search for products. However, they did not tend to read comments on the products; according to our observations and interviews, the reason was that they tended to explore information on the website and only later shop on the app. The results also show that the TO group proceeded to the purchase stage without engaging in the pre-purchase stages of the consumer journey on the app. They were also liable to undertake cross-channel online shopping journeys.

Discussion 2: The IC group (C_4), all members of which belonged to the LT group, spent the longest amount of time to accomplish the task. Among the group's characteristics, they engaged in more types of actions at greater frequencies (e.g., checked more information on product pages, clicked more times on advertisements, and went to their shopping carts more frequently than the other groups). They also had a higher frequency of repeatedly browsing for the same brands (i.e., V6). According to our observations and interviews, unlike the TO group, the IC group tended to explore information and shop on the app simultaneously. Our results also show that the IC group engaged in the pre-purchase and purchases stages of consumer journey by using the app.

Discussion 3: The comparing group (C_5), all of whose members were in the LT group, also spent quite a long time to accomplish the task. Among the group's characteristics, they took more time to use functions provided by the app for sorting, filtering, and selection products, as well as to check product-related information and compare products.

They primarily engaged in the pre-purchase stage of the consumer journey on the app but were liable to not make any purchases due to seldom visiting their shopping carts.

Discussion 4: Of all five groups, the PS group (C_2) had the greatest range of time spent to accomplish the task, namely from 87 to 497 s. Accordingly, some of its members belonged to the ST group and others to the LT group. The characteristics of the group's members are that they used all types of functions and explored information related to brands, stores, and products. They especially focused on product-related information and had the highest proportional value for V7 (see Table 5). They also typically preferred to compare the prices of products on the app and were easy attracted by promotions. We can infer that members of the PS group wanted to save money. Similar to the comparing group, they primarily engaged in the pre-purchase stage of the consumer journey on the mobile app and seldom, if at all, visited their shopping carts.

Discussion 5: Last, the RA group (C_3) spent the second-least amount of time of the five groups to accomplish the task. They did not take time to repeatedly browse the same brands, stores, or products, and they were willing to accept recommendations as well as read reviews. They also preferred to click on advertisements provided by the app. Several consumers in the group reported shopping online with their phones and affirmed the convenience of doing so. Among other results, the RA group engaged in all stages of the consumer journey (i.e., pre-purchase, purchase, and post-purchase) on the app.

Table 4. Clustering results via nine variables

ID of Variables	C1 (n = 20) Transaction (TO)	C2 (n = 22) Price Sensitive (PS)	C3 (n = 10) Recommendation (RA)	C4 (n = 9) Information (IC)	C5 (n = 7) Comparing	Average
Total Time (Short/Long)	**20/0**	**8/14**	**6/4**	**0/9**	**0/7**	
V_1	1.05	2.55	2.80	**7.56***	2.86	2.84
V_2	1.10	2.73	2.80	**10.56***	**4.14**	3.44
V_3	1.05	1.82	**2.80**	**8.33***	**2.71**	2.69
V_4	0.25	2.86	1.30	**12.33***	**4.00**	3.24
V_5	0.00	0.82	0.00	**4.67***	**2.00**	1.09
V_6	0.00	**1.09**	0.00	**2.89***	**1.00**	0.84
V_7	0.30	**0.65***	0.40	**0.57**	**0.48**	0.48
V_8	0.10	0.23	**0.30**	**0.56***	0.14	0.24
V_9	35.71	43.85	37.66	**110.59**	**206.73***	66.15

(continued)

Table 4. (*continued*)

ID of Variables	C1 (n = 20) Transaction (TO)	C2 (n = 22) Price Sensitive (PS)	C3 (n = 10) Recommendation (RA)	C4 (n = 9) Information (IC)	C5 (n = 7) Comparing	Average
Target Variable: Total time	<u>72.45</u>	219.01	175.00	**661.61**	**438.39**	250.59
(Frequency) Shopping cart	**20**	**28**	14	**24**	6	18.4

Table 5. Features of the app used to characterize the clusters

ID of Variables	C1 (n = 20) TO	C2 (n = 22) PS	C3 (n = 10) RA	C4 (n = 9) IC	C5 (n = 7) Comparing
Time (Short/Long)	**20/0**	**8/14**	**6/4**	**0/9**	**0/7**
(Proportion) Search	0.33	0.20	0.32	0.24	**0.36**
(Proportion) Recommendation	0.06	0.04	**0.10**	0.05	0.05
(Proportion) Comments	0.03	0.17	**0.46**	0.15	0.11
Frequency/Average of using shopping cart (# of unique users)	20/1.00 (20)	28/1.27 (17)	14/1.40 (10)	**24/2.67** (9)	6/0.86 (6)

3.3 Surveys of Attitudes Toward Shopping Online

We used the 24 questionnaire questions developed by Kau et al. (2003) to investigate attitudes toward shopping online within the five groups of consumers identified. In their study, Kau et al. collected 3,712 valid responses to differentiate types of consumers in factor analysis according to attitudes toward and behaviors in online shopping. As a result, they identified six types of consumers: brand-comparing, impulsive, deal-oriented, information-seeking, advertisement-oriented, and offline shoppers. Of the 24 questions, five (i.e., Q1–Q5) passed statistical muster in our study (see Table 6) and were used to examine each group of consumers regarding their attitudes toward online shopping.

Discussion 1: Both the PS (C_2) and comparing (C_5) groups achieved higher scores for Q1 and Q2. According to the results of our evaluation, they belonged to the groups that engaged in the pre-purchase stage but did not ultimately shop on the app. Those results are consistent with the results of Kau et al. (2003), who showed that such consumers preferred to collect a large amount of information and compare product features, prices, and brands. In addition, Table 6 shows that the comparing group achieved the highest scores for Q1 and Q2 but scored lower for Q4. Such findings indicate that the comparing group was more deliberate than the PS group and thus spent more time on evaluating information on the app.

Discussion 2: The IC group (C_4) achieved the lowest scores for Q1 and Q4. The results reveal that they were deliberate consumers who were not distracted by promotions and were liable to have certain preferences for brands or products, as revealed during interviews. Thus, they scored higher on Q2 and Q5. Providing specific brands and/or product information of interest to them is quite important.

Discussion 3: The TO group (C_1) did not achieve any higher or lower scores for the five questions compared with the other groups. After checking the answers to the 24 questions, we found that their rating for each question was middling compared to the ratings of the other groups. They indicated no preference for specific brands or products but had relatively clear goals while shopping online.

Discussing 4: The RA group (C_3) achieved the lowest score for Q2, which indicates that price is not an important factor attracting them to shop online. They also received the second-highest score for Q4. Given the results of the survey, interviews, and clustering, we can infer that the RA group has the potential to accept recommendations and shop using apps.

Table 6. Questionnaires results for attitudes toward shopping online

	C1 TO	C2 PS	C3 RA	C4 IC	C5 Comparing
Q1': I do an overall comparison of different brands before I decide my most preferred brand	3.2	3.6	2.4	1.4	3.9
Q2': I do more brand price comparisons than in a traditional retail environment	3.5	4.4	3.3	4.3	4.4
Q3': I prefer to buy from an online store, compared to a firm which has both physical and online store	**2.2**	**1.3**	**1.7**	**1.0**	**3.7**
Q4' I react more sales promotions compared to that in a traditional retail environment	1.5	2.9	1.9	-0.6	0.4
Q5': I use the same search engine on a regular basis	2.7	3.4	3.9	4.3	4.1

4 Conclusions

Our results reveal that the time factor is a primary means to differentiate consumers' mobile shopping behaviors instead of need-states. There are five types of consumers can be identified based on their usage behaviors of using mobile Taobao app. Our results also reveal that each group employs its own particular features of the apps to facilitate the shopping process, and we can identify consumer types based on shopping behavior and amount of time spent on making purchase decisions while using the app. Those findings suggest that e-retailers can refine the features of the app and deploy marketing strategies tailored to different types of consumers. Regarding different types of devices for shopping online, we have found that consumers with clearly exploration and focused searches stages for online shopping while browsing webpages can be checked by their need-states. However, the condition is not clear for all types of consumers who shop on mobile apps. Our results also shed initial light on how stages of consumers' journeys and spending of time manifest in each type of consumer. Our preliminary results revels that consumers have a more complete shopping journey on webpages than on mobile apps. It seems that some of the groups prefer to purchase products efficiently by the apps without exploring and browse information carefully by using the apps. In our future work, we will compare the similarity and differences of factors that influence consumers' online shopping on mobile apps and websites by collecting their behavioral data via two channels. The issue of cross-channel online shopping journeys will be investigated in the future.

Acknowledgement. This research was supported by the Ministry of Science and Technology, Taiwan under Grant No. 108-2410-H-003-132-MY2.

References

eMarketer: Mcommerce Forecast 2021:What driving growth and what it means for retail (2021). https://www.insiderintelligence.com/content/mcommerce-forecast-2021

Pantano, E., Priporas, C.V.: The effect of mobile retailing on consumers' purchasing experiences: a dynamic perspective. Comput. Hum. Behav. **61**, 548–555 (2016)

Almarashdeh, I., et al.: Search convenience and access convenience: The difference between website shopping and mobile shopping. In: Proceedings of International Conference on Soft Computing and Pattern Recognition, pp. 33–42. Springer, Cham (2018). https://doi.org/10.1007/978-3-030-17065-3_4

Lemon, K.N., Verhoef, P.C.: Understanding customer experience throughout the customer journey. J. Mark.: AMA/MSI Spec. **80**(6), 69–96 (2016)

Rennie, A., Protheroe, J., Charron, C., Breatnach, G.: Decoding decision: Making sense of the messy middle. Think with Google (2020)

Dasgupta, S., Grover, P.: Impact of digital strategies on consumer decision journey: Special. Acad. Mark. Stud. J. **23**(1), 1–14 (2019)

Rudkowski, J., Heney, C., Yu, H., Sedlezky, S., Gunn, F.: Here today, gone tomorrow? Mapping and modelling the pop-up retail customer journey. J. Retail. Consum. Serv. **54**, 101698 (2019)

Tupikovskaja-Omovie, Z., Tyler, D.J.: Experienced versus inexperienced mobile users: eye tracking fashion consumers' shopping behaviour on smartphones. Int. J. Fashion Des. Technol. Educ. **15**(2), 178–186 (2022)

Kuehnl, C., Jozic, D., Homburg, C.: Effective customer journey design: Consumers' conception, measurement and consequences. J. Acad. Mark. Sci. **47**(3), 551–568 (2019)

Wu, I.C., Yu, H.K.: Sequential analysis and clustering to investigate users' online shopping behaviors based on need-states. Inf. Process. Manage. **57**(6), 102323 (2020)

Zhang, W., Wang, J., Xu, S.: The probing of e-commerce user need states by page cluster analysis- an empirical study on women's clothes from Taobao.com. New Technol. Library Inf. Serv. **31**(3), 67–74 (2015)

Kau, A.K., Tang, Y.E., Ghose, S.: Typology of online shoppers. J. Consum. Mark. **20**(2), 139–156 (2003)

Liu, Y., Li, H., Peng, G., Lv, B., Zhang, C.: Online purchaser segmentation and promotion strategy selection: evidence from Chinese E-commerce market. Ann. Oper. Res. **233**(1), 263–279 (2013). https://doi.org/10.1007/s10479-013-1443-z

Moe, W.W.: Buying, searching, or browsing: differentiating between online shoppers using in-store navigational clickstream. J. Consum. Psychol. **13**(1–2), 29–39 (2003)

Laran, J., Janiszewski, C., Salerno, A.: Exploring the differences between conscious and unconscious goal pursuit. J. Mark. Res. **53**(3), 442–458 (2016)

Borlund, P.: Experimental components for the evaluation of interactive information retrieval systems. J. Documentation **56**(1), 71–90 (2000)

Manning, C.D., Raghavan, P., Schütze, H. Introduction to Information Retrieval. Cambridge University Press (2008)

Wu, I.C., Yu, H.K.: Mining types of online consumers by search patterns: the influence of need states and product familiarity. J. Library Inf. Sci. Res. **16**(1), 77–123 (2021)

The Study of Different Types of Menus Layout Design on the E-Commerce Platform via Eye-Tracking

Ya-Chun Yang and Tseng-Ping Chiu[✉]

Department of Industrial Design, National Cheng Kung University, Tainan, Taiwan
{p38101516,mattchiu}@gs.ncku.edu.tw

Abstract. The role of thumb of good website design is menu layout design, it represents excellent user interface and user experience. In this study, we aim to investigate the various menu layout and layer designs on the E-commerce website by constructing two iterative prototyping e-commerce websites. There are two analyses of the experiments. First, we introduced visualization of eye-tracking data to explore the visual perception of different menu layouts and layer designs. Second, we investigate the physiological and psychological data to understand the satisfaction and usability related to eye movement. The result showed that the reach rate of the eye-tracking data on the vertical layout is significantly higher than on the horizontal layout design in the second layer of the menu on the e-commerce website. In addition, we found gaze times back and forth will influence usability but not affect satisfaction. Overall, the vertical layout focus is more effective than the horizontal layout on the menu.

Keywords: Menu layout · Visual perception · Eye-tracking · User experience · E-commerce website

1 Introduction

E-commerce combines marketing and technology perfectly. Under the epidemic in the past, the integration of virtuality and reality has become complex. E-commerce has become an indispensable factor in commercial economic growth worldwide, and it also became the spotlight during the epidemic. The digital economy under the epidemic has not decreased but increased. It seems that people's economic activities have been reduced, but in fact, it has prompted more consumption behaviors in e-commerce. In 2021, the report of the United Nations Conference on Trade and Development showed the growth trend of e-commerce in the global retail industry, and it rose sharply during the epidemic from 2019 to 2020 [1, 2]. In Taiwan, the report of the Department of Statistics also shows that the annual growth rate from 2020 to 2021 increased by 28.5% in Taiwan, which further indicates the development of the retail e-commerce platform in the future [3].

In the face of the rapid digital transformation and the maturity of technology, e-commerce platforms pay more and more attention to the application of user experience

© Springer Nature Switzerland AG 2023
F. Fui-Hoon Nah and K. Siau (Eds.): HCII 2023, LNCS 14038, pp. 314–328, 2023.
https://doi.org/10.1007/978-3-031-35969-9_22

and even attach great importance to "people." In "Marketing 5.0", it is mentioned that digital is better than marketing and suggests that the strategy should lead to customer experience and enhance user experience through information, interaction, and immersion [4]. In 2006, Hassenzahl believed that user experience is technology. However, it is not only a tool to meet business needs but also a tool for subjective, emotional, and complex interactive reactions. User experience is the inner state of the user and the result of the interaction between the state and the design system [5].

Therefore, this research takes "people" as the starting point to conduct an in-depth understanding of the menu on the e-commerce platform, and then takes "people" as the center and applies it to the "machine" to find out the effect of the menu on the e-commerce platform in human-computer interaction on "people". And how it can be applied to "machines" to achieve better interactions. The previous research is understanding the e-commerce platform by collecting data and trying to collect the user's feelings and experts' ideas to establish a questionnaire to evaluate the user's thinking and ideas. This study is expected to propose those better elements of the menu layout and hopes to optimize the user experience as a reference for future development.

1.1 Research Background

In Human-Computer Interaction (HCI), usability in user experience is generally regarded as the quality evaluation of website development, but it ignores the user's feelings, psychology, experience and so on [6]. Past research has shown that people tend to pay more attention to hedonic zones during the operation of the website than utilitarian zones [7]. Therefore, scholars in the past have proposed that by developing innovative and competitive websites, users should be more comprehensively considered and understood. Then it is a reference for future design.

In the past, eye trackers were regarded as the best tool to understand the user experience. It allowed us to know where the user "sees." With objective physiological data, we can better understand the user's eye viewing position. Some scholars have shown that users may even forget what they have watched [6], so the eye-tracking data can help users recall their behavior during the viewing process. Nielsen Norman Group (N/N Group) has also pointed out that eye-tracking is very suitable tool for understanding user experience and proposed guidelines for user experience research in 2022 [8]. It mentioned that eye-tracking has always been able to understand the physiological data of users naturally. However, In the user experience research in 2018, the actual use of eye-tracking is the lowest (Fig. 1). Although the industry and academia have many affirmations for eye-tracking, it is speculated that it is due to the burden of funding [9–12]. This has led to a decrease in the usage rate. Even though the fair and objective standard is an unshakable existence.

This research will use some methods that were more commonly used in user research in the past to gain an in-depth understanding. Through expert interviews, user questionnaires, focus groups, and card sorting to conduct a preliminary investigation of this research. Finally, through making the visual prototyping of the e-commerce platform to simulate the reality e-commerce platform. Using eye-tracking to deeply understand the iterative version, then the semi-structured interview after the eye-tracking experiment.

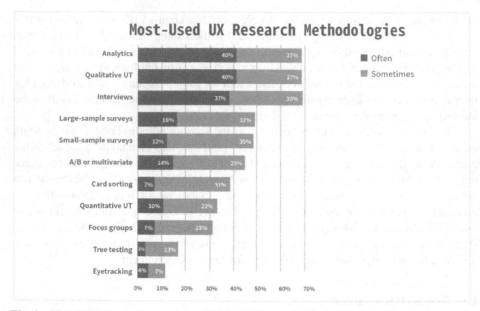

Fig. 1. 2018 N/N Group Report. Most-Used UX Research Methodologies. (Repaint from this research, Resource: N/N Group)

It is used to understand the feeling of using and seeking. The combination of physiological and psychological data was by filling in the usability questionnaire, interview, and eye-tracking data.

Through triangulation, we seek to gain a more comprehensive and richer understanding of menu layout as used in website design. It is expected to use triangulation to understand the core "people." First, through expert interviews and previous user online questionnaires to establish the questionnaire and design the iterative prototype. Second, the SUS usability scale and satisfaction survey were used to understand the user's thoughts and preferences after finishing the operation the iteration. Third, the preliminary visual understanding of the eye-tracking data and the further physiological preferences of the user were carried out through the physiological data. Finally, this research proposes that the future suggestion for development, design and even business better understand the interaction between e-commerce platforms and consumers.

1.2 Research Contribution

This research finds through past literature that information interaction is divided into user, content and system, and in the overall website [10, 11, 13, 14], and the whole system on the website is divided into organization system, navigation system, label system, and search system [11, 14]. Information architecture on the website has a significant impact on usability and user experience during navigation, and it exists as a communication bridge between designers and managers [15]. This research, through the understanding of the literatures in the past ten years, found that there is less for in-depth research on the menus on the website. The past literature pointed out that the existence of the menu is like the map of the website [16]. When the user gets lost, it can be used to find user's direction. It is tiny existence but plays an important role in the navigation system in the information architecture. The scholars proposed to optimize the user experience on the website, the functions of the search and navigation system should be enhanced [17]. In the past, the research on search systems in the literature was relatively complete, but there were few menus on the navigation system. Therefore, this study will explore the menu of the navigation system and more understand the user behavior. As the Fig. 2 and Fig. 3, the menu layers were defined in this study initially.

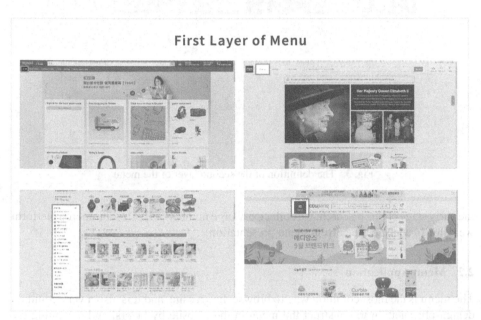

Fig. 2. The definition of the first layer of the menu

2 Literature Review

2.1 E-commerce

Through the definition of e-commerce by scholars, this research is compiled as: "The sales and purchase way of information, products and services are carried out through the Internet." [18, 19]. The business models of e-commerce platforms can be divided into B2B (Business to Business), B2C (Business to Customer), C2C (Customer to Customer), C2B (Customer to Business), B2B2C (Business to Business to Consumer) and O2O (Online to Offline), and even more business models may be derived in response to a more diverse and faster network in the future. This research will focus on the B2C business platform, the B2C e-commerce platform is the most familiar business model for the public, which means that consumers purchase goods and even services through platform companies. Then Amazon, Rakuten, and Argos are the best B2C platform displays; In

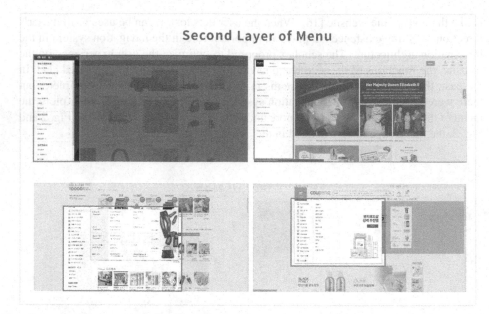

Fig. 3. The definition of the second layer of the menu

the early stage of this research, we collected a large number of B2C e-commerce platform websites to analyze the current menu presentation.

2.2 Menu Application

The menu is the most common way to browse the website, and it is also a very common design presentation to construct the menu in the website by layers, and designing the menu will balance the depth with breadth [20]. The drop-down menu is the most common and familiar, and it is mostly in the vertical form of the menu. In recent years, scholars have mentioned that in the academic websites, the drop-down menus have decreased significantly, but the menus presented horizontally have increased [21]. The vertical menus present a more in-depth hierarchical representation, while the horizontal menus present a broader hierarchical relationship. The core of the menus is the whole information on the website. How to make the menu quickly find the location of the user who is lost is also one of the main cores of the menu. Over the decades, scholars have conducted research on the details of the presentation on the menu. For example, the horizontal menus are preferable to vertical menus and their content is displayed at one time and the number of mouse clicks on the user's operation is reduced [20, 22]. In brief, the importance of the menu is undeniable, and how to achieve a balance between the horizontal and vertical presentation of the menu, and at the same time have breadth and depth. It will be an issue to be further explored in the future. This study will redesign the horizontal menu by investigating the psychological and psychology data to understand the differences.

2.3 User Experience

Why is understanding user experience important? Understanding user experience helps us gain insights into a good or service. In the field of human-computer interaction, the System Usability Scale (SUS) has been regarded as one of the best definitions of usability. However, It is quite a narrow definition of quality that neglects additional human needs and related phenomena, such as emotion, thinking and user experience [6]. In 2016, the Nielsen Norman Group (N/N Group) defined user experience as the product that ultimately makes people feel happy. In a good user experience, the first criterion is to meet the exact needs of users, and a good user experience should separate usability and discuss cross-domain integration services such as design, programming, engineering, etc. [23]. The most important thing in the user experience is the "insight" of the user, but the insight cannot be understood through the usability scale. Therefore, the user experience should be supplemented by qualitative and quantitative, and then become more user-friendly experience.

2.4 Eye-Tracking

"What you see is what you get" should be changed to "what you see is what you choose." In the past, scholars have found that visual perception is closely related to consumer behavior, and even affects the final choice [24]. The eye-tracking provides a good analysis way to understand the "insight" required in user experience [6, 25]. The data presentation of the eye-tracking allows us to understand the insights of the user more intuitively and then infer consumer behavior through visual perception behavior. In the past, some research indicated the longer the visual gaze represents the more positive the emotional response, which also provides a good reference indicator in research [7]. In addition to providing dynamic indicators for viewing, eye-tracking also provides auxiliary research instructions. In the past, it was pointed out that 47% of the subjects would forget some elements that they valued during the experiment after conducting the experiment [25]. Therefore, eye-tracking is a certain necessity in the user's experience and provides playback of the eye-tracking to allow the subjects to recall and describe their feeling of seeking during the interview. The eye-tracking is very useful for understanding the details of the user visual perception behavior [6, 26]. Through the dynamic playback eye-tracking, the researcher can observe the whole experimental process and context of the subject through the "human" perspective. It is good tool for qualitative and quantitative research.

This study will use Heat Map to understand the visual perception presented by the fixation of multiple participants, and this study will also discuss with eye-tracking expert and calculate the Focus Rate, Effective Focus, Reach Rate, Order of First Fixation and Revisit Rate by fixation duration and eye movement. Finally, exploring the correlation between usability and satisfied and Total Fixation duration (TF), Average Fixation duration (AF) and Fixation Count (FC).

3 Research Process

3.1 Interviews

This study invited a total of four experts to conduct online and physical interviews to consider the factors. Because the COVID-19 epidemic, some experts were invited online interviews and discuss through screen sharing. A total of front-end and back-end engineers, visual designers and UI/UX designers were invited. Every expert conducts an interview for 1 to 2 h with in-depth expert interviews. The interview structure starts from understanding the past experience and background of the experts and discussing the existing e-commerce platform trends, and then introduces the results of the questionnaire conducted in the previous step to discuss with the experts. Understanding the ideas and suggestions on the user and evaluating the feasibility and application on the development at the same time.

Through the questions collected from the previous 100 valid questionnaires and the expert interviews, the questions are converged and explored, and the important design elements on the B2C e-commerce platform were collected. The evaluation of the weight the importance of the design elements considered by the experts from the perspective of the user, and through this part, the ideas and opinions of the experts and the user were synthesized.

3.2 Eye-Tracking Experiment

Through the previous data collection about e-commerce platform trends and ideas on the development and user ends to gain the importance of the menu arrangement and categories was obtained. Therefore, the prototype iteration was carried out, twelve participants were invited to operate eye-tracking in the laboratory and filled in the satisfaction and SUS scales after operating the two version. Finally, they were conducted a semi-structured interview after completing the experiment. On the eye tracking experimental sample, the first layer and the second layer menu in the two tasks were drawn by Area Of Interest (AOI) respectively. AOI drawing is performed on the area that the subject may be interested in (as Fig. 4 show). As shown below. This study proposed two hypotheses on the eye-tracking experimental stage:

H1: In the first layer of the menu, the product category classification will affect the seeking behavior in the second layer of the menu.

H2: In the second-level menu, vertical layout is easier to read and more efficient than horizontal layout.

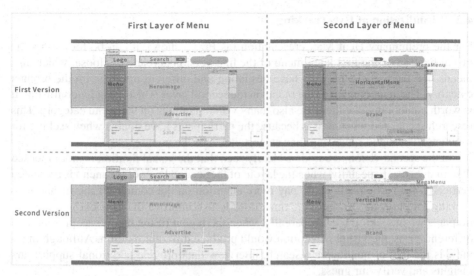

Fig. 4. The Area of Interest (AOI) on Eye-tracking

4 Result

4.1 Usability and Satisfied

While the eye-tracking is in progress, after each prototype operation, the researchers will ask the participants to take the SUS scale [26] and fill out the satisfaction scale, so that the researchers could understand the feelings of the participants when we analyze the data afterwards. In terms of satisfaction and usability tests, the second prototypes vision based on feedback from experts and users significantly showed growth in satisfaction and usability after operation, which also showed that users were satisfied during operation. The usability increased by nearly 30 score in the questionnaire feedback on ease of use (as Fig. 5 show). Looking at the two data at this stage, it also reflects that the revised version (second vision) has an impact on user experience.

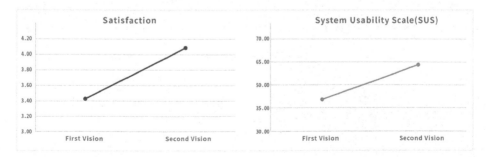

Fig. 5. The Satisfaction and SUS result of two vision

4.2   Visualization of Eye-Tracking

For the preliminary Heat Map presentation (as Fig. 6 show), it can be found that the visual fixation on the first layer menu of the first version is relatively loose, which also means that the participants may have hesitation and different opinions on the product category. The inner content of the menu on the second layer is also visually confusing. It is worth mentioning that there are also many visual fixations on the menu category. This research speculated whether it is because the participants have doubts when seeking for products that lead to reviewing the category items of the menu.

In the second vision, it could be clearly seen that the visual fixation is more focused than first vision. Especially, when the layout of the second layer of the menu is presented vertically, the participants could pay more attention on the content of the menu when seeking for products. At the same time, does it also improve the efficiency of the participants when seeking for? Apparently, when the layout of the menu was presented in different forms, the visual perception would produce different fixation. Although interview is the tool of qualitative research, interview data provide additional support and insights and verify our guess.

In this study, further data analysis was carried out the insights obtained by visualization. In the overall, AOI was preliminarily calculated the fixation duration of the participants in each area. There was a significant reduction in the fixation of the first layer of the menu. It means that the efficiency after changing the menu category classification is better. In terms of the fixation duration of the second layer as a heat map (as Fig. 6 show), the revisit rate of the menu category classification on the first vision is more than second vision. It means category classification on the first vision cause more confusion than second vision.

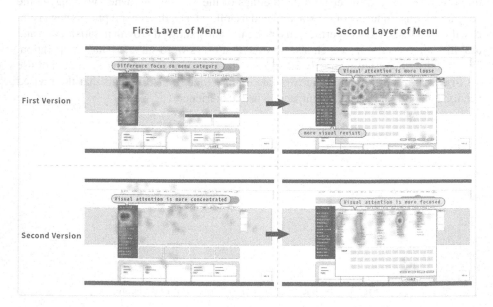

Fig. 6. Heat Map from 12 participants.

In the same area, the fixation time of the second layer of the menu of second vision is less than the first vision, which also proves that the vertical layout of the second layer of menu on second vision can improve the viewing efficiency more than the horizontal layout of the second layer of menu first vision (as Fig. 7 show). Through participants interview, it was also learned that the horizontal presentation of the first vision would lead to uncertainty in the seeking for participants, so they revisited to the menu category to reconfirm whether they selected the correct product category item. Although the interview could not support accurately quantify data, it could provide additional support and insights for this study.

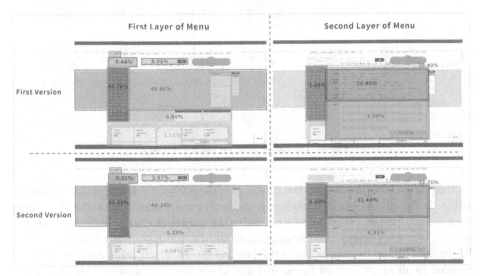

Fig. 7. Focus Rate from 12 participants

This research uses eye-tracking data to calculate the overall content reach rate in the second layer of the menu in the two versions (as shown Fig. 8), and the reach rate represents the exposure of the content. The menu of the second vision has nearly 30% more exposure than the first vision, and the two versions both achieved 100% exposure for the content of menu (MegaMenu, Horizontal/ Vertical Menu, Total Menu). In terms of the Effective Focus rate, it means the level of interest in attention. Through data analysis, it can be found that although the fixation duration of the second vision is significantly less than that of the first vision. But the overall effective focus rate on the second vision is more superior than second vision.

The effective focus rate on the overall menu and the vertical layout of menu presentation on the second vision are better than the first vision. The vertical layout of menu of the second vision is better than the horizontal layout of the first vision.

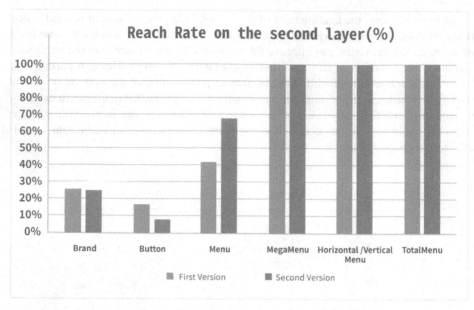

Fig. 8. The Reach Rate on the second layer

It confirmed the presentation and depiction of the previous heat map at the beginning. Although the vertical menu is different from the eye movement direction of the human eye muscles, the vertical layout of menu helps the viewer to read visually. It affects the viewing efficiency and overall experience, as shown in Fig. 9. The effective focus on the second layer on first vision is less than second vision. The fixation duration and visit duration of all menu item on the second was spent less time than first vision, but the effective focus on the second vision is well than first vision. Efficiency may also be one of the factors affecting user experience on fast networks.

4.3 Result of the Eye-Tracking

Through preliminary visualization of eye-tracking to understand how the users search for the something on the e-commerce website. This study be found that the menu of the second vision can make the participants pay more attention on the content than the menu of the first version. On the contrary, as mentioned above, the horizontal layout menu of the first vision would lead to confusion in viewing, so that the participants have uncertainty and look back to the category classification of menu items. Because of this, this study speculates that this is one of the reasons for why it declined in the efficiency focus of the first vision. Based on the data analysis and description of the eye-tracking, we can preliminarily confirm the hypothesis mentioned above. The category classification of the menu may affect the seeking behavior. To sum up, in the layout of the menu on the second layer, the vertical is more efficient in visual perception and easier to read than the horizontal.

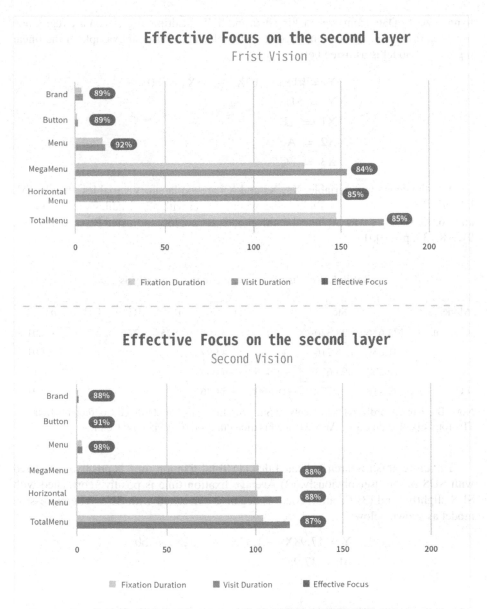

Fig. 9. The Effective Focus on the second layer of two visions

Then, this study through the eye-tracking data about Total Fixation duration (TF), Average Fixation duration (AF)and Fixation Count (FC) and System Usability Scale (SUS) score of the participants to explore by regression analysis. TF means the total time that user focus on somewhere, AF is about an average of the time is focused on and FC means the gaze somewhere back and forth. Using a repeated measures ANOVA to analyze the Data from eye-tracking data and SUS. Multiple regression is a regression in which there is more than one predictor variable. The common example of the linear regression model is as listed below:

$$Y = b1X_1 + b2X_2 + b3X_3 + b0$$
$$Y = SUS$$
$$X1 = TF$$
$$X2 = AF$$
$$X3 = FC$$

Y is the dependent variable. X1, X2 and X3 are explanatory variables, then b0, b1, b2 and b3 are constants. As Table 1, the overall model with all variables was significant and 76.6% variance was explained in this model, $R^2 = .766$, (adjusted $R^2 = .679$); $F (1, 9) = 8.739$, $p < .001$.

Table 1. Results of the multiple regression analysis.

Model	B	Std	β	t	R	R^2	F	p
Contant	174.636	28.320		6.167	.875	.766	8.739	<.001
TF	31.059	6.216	17.981	4.997				<.001
AF	−379.203	93.653	−1.366	−4.049				<.01
FC	−8.918	1.771	−18.200	−5.036				<.001

Note: B – Unstandardized Coefficients B; Std – Std Error; β- Standardized Coefficients Beta
TF: Total Fixation duration; AF: Average Fixation duration; FC: Fixation Count

Therefore, this research indicated that (1) Total fixation time is positive correlated with SUS of the user obviously. (2) Average fixation time is positive correlated with SUS slightly. And (3) Fixation count is negative correlated with SUS. The regression model as shown below:

$$Y = 17.98X_1 − 1.37X_2 − 18.2X_3 + b0$$
$$b1 = 17.98$$
$$b2 = −1.37$$
$$b3 = −18.2$$

Especially this last analysis of the result show more fixation duration may promote the usability on the e-commerce, but too gaze back and forth to reduce the usability.

At the present stage, previous user questionnaires and expert interviews, we have sorted out the elements on the e-commerce platform, compiled them and integrated them into the menu layout of the second vision. Through the description of the data, we could initially understand the overall satisfaction, usability, thought, eye movement of the user, compiling subjective and objective data of the psychology and physiological data to explain them. In the last repeated measures ANOVA analysis, we found the gaze times may greatly influence the usability of the user. At last, although the usability was affected by the eye-tracking data of the layout of the menu, user satisfied was not influenced on the eye-tracking data of the layout of the menu. According to this result, this study conjectures that perhaps usability is microscopic process, however satisfied is macroscopic process on the e-commerce website.

5 Conclusion and Discussion

"What do you think", "What do you seek for", "What do you choose", and "How do you feel". The "You" is not only the expert but user in this research.

Past studies have pointed out that participants tend to forget which parts they have seen while seeking for [6], but this study uses eye-tracking to assist in exploring visual perception in the research. The eye-tracking supply the data to explore the difference of the two visions and recall the operation to help arouse the memory of the participants at the same time, so that this study could explain the results in more detail in the analysis of the data. Although the interview cannot be presented quantitatively, it provides a good direction and recommendations for the future. The data of questionnaires, satisfaction and usability scales, and physiological data analysis of eye-tracking were described through qualitative interview assistance. Through triangulation, we seek to gain a more careful and comprehensive understanding of menu layout as used in website design and combine the thought of the consumer and the development.

This study validated the two previous hypotheses through the final data, at the same time, through repeated measures ANOVA analysis to understand the correlation between the psychology and physiological data. Through the data of eye-tracking to gain a deep understanding of the influence on visual perception. The presentation of product categories and the layout of menus not only affect the ease of use and satisfaction, but also quietly affect the feeling in which consumers see, and even perhaps become an influence on the commercial, which affect to the application of development in the future. Although this research screened out and weakened the color, advertising, and brand on the e-commerce website, it is still an indispensable existence on the e-commerce website platform and will be the focus of continued in-depth and broad discussion after this research. It is also expected that an increase in variable factors may lead to more visual possibilities.

As mentioned at the beginning, the rise of e-commerce platforms and the catalysis of the epidemic have forced the commercial end and the development end to further understand user behavior. In the face of rapid technological changes, e-commerce needs to have depth in addition to the breadth of use. Only by deeply understanding users can we return to the main core of this research, "people". User experience is often an important but easily overlooked existence. Just like the role of the menu, it plays an extremely important role on the website and may also be one of the important elements and topics to be discussed for long-term operation on the e-commerce website.

References

1. UNCTAD, Global E-Commerce Jumps to $26.7 Trillion, Covid-19 Boosts Online Retail Sales. UNCTAD: United States (2021)

2. McMullen, R.: United States. In: Hutchings, R., Suri, J. (eds.) Modern Diplomacy in Practice, pp. 189–224. Springer, Cham (2020). https://doi.org/10.1007/978-3-030-26933-3_10

3. Ministry of Economic Affairs, R.O.C., Online sales continue to grow, helping retail industry mitigate the impact of the epidemic. Ministry of Economic Affairs, R.O.C (2021)

4. Kotler, P., Kartajaya, H., Setiawan, I.: Marketing 5.0: Technology for Humanity. Wiley (2021)

5. Hassenzahl, M., Tractinsky, N.: User experience - a research agenda. Behav. Inf. Technol. **25**(2), 91–97 (2006)

6. Cyr, D., et al.: Exploring human images in website design: a multi-method approach. MIS quarterly, 539–566 (2009)

7. Cyr, D., Head, M.: The impact of task framing and viewing timing on user website perceptions and viewing behavior. Int. J. Hum Comput Stud. **71**(12), 1089–1102 (2013)

8. Gordon, K., Rohrer, C.: A Guide to Using User-Experience Research Methods, N.N. Group, Editor (2022)

9. Moran, K.: Quantitative UX Research in Practice, N.N. Group, Editor. Nielsen Norman Group (2018)

10. Isa, W.A.R.W.M., Noor, N.L.M., Mehad, S.: Towards a theoretical framework for understanding website information architecture development. **20**, . 44 (2006)

11. Toms, E.G.: Information interaction: Providing a framework for information architecture. J. Am. Soc. Inform. Sci. Technol. **53**(10), 855–862 (2002)

12. Wallas, G.: ART OF THOUGHT. Harcourt, Brace and Company New York (1926)

13. Morville, P., Rosenfeld, L.: Information Architecture for the World Wide Web: Designing Large-Scale Web Sites. O'Reilly Media, Inc., United States (2006)

14. Rosenfeld, L., Morville, P.: Information architecture for the world wide web. O'Reilly Media, Inc. (2002)

15. Ruzza, M., et al.: Designing the information architecture of a complex website: a strategy based on news content and faceted classification. Int. J. Inf. Manage. **37**(3), 166–176 (2017)

16. Webster and Ahuja: Enhancing the design of web navigation systems: the influence of user disorientation on engagement and performance. MIS Quarterly **30**(3), 661 (2006)

17. Tang, G.-M., et al.: A Cross-cultural Study on Information Architecture: Culture Differences on Attention Allocation to Web Components, pp. 391–408. Springer (2020)

18. Gregory, G., Karavdic, M., Zou, S.: The effects of e-commerce drivers on export marketing strategy. J. Int. Mark. **15**(2), 30–57 (2007)

19. Kalakota, R., Whinston, A.B.: Electronic commerce: a manager's guide. Addison-Wesley Professional (1997)

20. Naylor, S.: Breadth and depth: A comparison of search performance in hierarchical and mega menus, p. 1–64 (2016)

21. Comeaux, D.J.: Web design trends in academic libraries—a longitudinal study. J. Web Librariansh. **11**(1), 1–15 (2017)

22. Ouyang, X., Zhou, J.: Smart TV for Older Adults: A Comparative Study of the Mega Menu and Tiled Menu, pp. 362–376. Springer, Cham (2018). https://doi.org/10.1007/978-3-319-92037-5_27

23. Norman, D., Nielsen,J.: The Definition of User Experience (UX), N.N. Group, Editor. Nielsen Norman Group: U.S. (2016)

24. Clement, J.: Visual influence on in-store buying decisions: an eye-track experiment on the visual influence of packaging design. J. Mark. Manag. **23**(9–10), 917–928 (2007)

25. Schall, A.: Eye Tracking Insights into Effective Navigation Design, pp. 363–370. Springer, Cham (2014). https://doi.org/10.1007/978-3-319-07668-3_35

26. Nielsen, J.: Eartracking: A New UX-Research Method. Nielsen Norman Group: Nielsen Norman Group (2019)

Use of Disruptive Technologies to Enhance Customer Experience

Gamification in Organizational Contexts: A Systematic Literature Review

Luciana S. Assis[1] and Sergio A. A. Freitas[2]([ORCID])

[1] Brazilian Agricultural Research Corporation - Embrapa, Brasilia, Brazil
luciana.assis@embrapa.br
[2] Centro de Estudos, Desenvolvimento e Inovação em Software - CEDIS, University of Brasilia - UnB, Brasilia, Brazil
sergiofreitas@unb.br

Abstract. Getting engaged organizational environments is a competitive advantage for companies. In these contexts, the use of game design elements seems promising. Gamification is a form of motivational design which, fundamentally, is a means of encouraging individuals to behave in a certain way. This study presents a systematic literature review to investigate the existing gamified solutions applied in organizational contexts to identify best practices to get effective gamification. Before defining the research protocol, an exploratory analysis was fulfilled to provide fundamentals to conduct the review itself. A total of 844 papers were initially found, 26 of which met the eligibility criteria and other 10 studies have been added by snowballing process. The results showed in which organizational contexts gamification can be applied, the main frameworks and processes, the leading game techniques and effects observed in gamified solutions. This systematic review can serve as a basis for developing new gamification projects in organizational environments.

Keywords: gamification · review · workplace · organization · company

1 Introduction

Deterding et al. [60] investigated the historical origins of the term "gamification" in relation to precursors and similar concepts. The term originated in the digital media industry and did not see widespread adoption before the second half of 2010. The authors proposed a definition of "gamification" as the use of game design elements in non-game contexts.

Werbach and Hunter [61] described gamification as a form of motivational design which, fundamentally, is a means of encouraging individuals to behave in a certain way. Huotari and Hamari [65] presented the definition as a process of enhancing a service with affordances for gameful experiences to support user's overall value creation. For Chou [64], gamification is the craft of deriving fun and engaging elements found typically in games and thoughtfully applying them to real-world or productive activities.

© Springer Nature Switzerland AG 2023
F. Fui-Hoon Nah and K. Siau (Eds.): HCII 2023, LNCS 14038, pp. 331–352, 2023.
https://doi.org/10.1007/978-3-031-35969-9_23

According to Chou [64], there are four application fields of gamification: Product gamification is about making a product more engaging and fun through game design; Workplace gamification is the craft of creating environments and systems that inspire and motivate employees towards their work; Marketing gamification is the art of creating holistic campaigns that engage users in fun and unique experience designed for a product, service, platform, or brand; and, Lifestyle gamification involves applying game design into daily habits and activities.

Gamification can be a powerful toolkit to make business more successful. The purpose of this study is to conduct a systematic review to investigate the existing gamified solutions applied in organizational contexts to identify best practices to get effective gamification.

The remainder of this paper is structured as follows: Sect. 2 describes the systematic review methodology and the related works; Sect. 3 reports the results obtained to answer the research questions; Sect. 4 discusses the main findings to identify best practices for gamification designs in organizational environments; and, finally, Sect. 5 presents the conclusions of this study.

2   Methodology

In order to conduct this systematic literature review, we followed a procedure defined by Kitchenham and Charters [1]. The review was carried out in the following phases: a) exploratory analysis; b) planning; c) conducting; and d) reporting.

2.1   Exploratory Analysis

This phase was fulfilled in order to provide fundamentals for defining the research protocol with the following specific objectives:

- Get secondary studies examples published in the gamification area;
- Select secondary studies related to the research topic to serve as a basis for defining the SLR protocol;
- Identify relevant papers to be analyzed during the SRL conduction stage;
- Discover gaps in existing systematic review papers.

In this step, a comprehensive search was performed on ACM, Science Direct, IEEE and Scopus databases to locate secondary studies in gamification. The search string has been applied to the title description field and the string terms are shown in Table 1.

Table 1. Search keywords for Exploratory Analysis.

Main term	Alternative terms
gamification	gamified, gamify, gamifying
systematic review	literature review, literature survey, mapping study, research review, systematic mapping, systematic overview

The search result returned 251 articles. Using the Parsifal tool [2], 73 duplicate studies were excluded. The resulting 178 articles were accepted for analysis of the title, abstract content, and identification of the domain area. Figure 1 shows the analysis result.

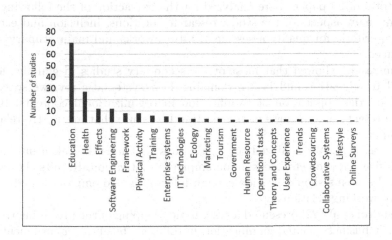

Fig. 1. Secondary studies in gamification by domain.

The studies classified in the following domain areas were selected for further analysis: effects, frameworks, enterprise systems, human resources, theoretical, user experience, trends and collaborative systems. Table 2 presents the list of secondary studies related to gamification in organizational contexts, as well as systematic reviews that address more general topics.

Table 2. Systematic reviews related to gamification in organizational contexts.

Domain	Secondary studies	Total
Effects	[3–12]	10
Frameworks	[13–20]	8
Enterprise systems	[21–24]	4
User experience	[27,28]	2
Trends	[29,30]	2
Human resources	[25]	1
Theoretical	[26]	1
Collaborative systems	[31]	1

A total of 29 papers were analyzed for the extraction of the following information: main aspects of the study, research questions, inclusion and exclusion criteria, criteria for quality assessment, future works and main primary studies referenced.

The results showed that most of the secondary studies in gamification are related to education and health domains. Few systematic reviews specifically addressed the application of gamification in organizational contexts. Related works present some reviews of the existing literature that are partly related to this paper.

Hinton et al. [21] analyzed the legal issues that must be taken into account in gamification projects, focusing on employee motivation. Results showed that gamified systems should provide enough information to employees and support positive working relationships.

Szendroi et al. [22] presented a systematic mapping of the prevailing trends in for-profit business-related gamification in different business environments. The results confirmed that there is an increase in the number of empirical studies of gamification in organizational environments. The findings suggest that gamification can have a positive impact on business processes in diverse ways.

Miranda and Vergaray [23] analyzed the use of gamified mobile applications and their impact in work productivity. The authors investigated the main gamification techniques and effects related to the use of gamified mobile applications in companies.

Encarnação et al. [24] presented a bibliometric review to investigate the main topics addressed on gamification in the workplace. The results showed that gamification process in the business context were gaining more relevance since 2019. This evolution reveals the recent and growing interest to increase motivation and performance of individuals.

Thomas et al. [25] investigated the current state of research on gamification in human resource development and the future research directions. The authors identified the contexts where gamification has been shown to be effective: employee learning, task performance, and employee wellness.

The accomplishment of the exploratory analysis phase before starting the research protocol planning provided a better direction for this study. The analysis of secondary studies related to the application of gamification in organizational contexts allowed us to identify gaps in the literature.

2.2   Planning

The planning phase consists of developing the review protocol, a plan describing the research proposal. The review protocol components are as follows: a) review's objectives; b) research questions; c) search string; d) sources; e) exclusion criteria; f) inclusion criteria; g) quality assessment criteria; and h) data extraction form. The description of each component will be presented in the following subsections.

Review's Objectives: Identify and analyze relevant results of gamification solutions applied in organizational environments.

Research Questions: The research questions attempt to provide specific insights into the relevant aspects of the existing proposals in gamification applied in organizational contexts. The research questions of this systematic literature review are:

RQ1. In which organizational contexts are gamification been applied?
RQ2. What are the main gamification frameworks or processes applied in organizational systems?
RQ3. What are the main gamification techniques applied in organizational environments?
RQ4. What are the main effects observed in gamification solutions applied in organizational environments?

Search String: The keywords "gamification" and "enterprise" were used along with the respective synonyms and alternative terms. Table 3 shows the list of search string terms.

Table 3. Search string keywords.

Main term	Alternative terms
gamification	gamified, gamify, gamifying
enterprise	business, company, organization, workplace

Sources: The following digital databases were selected:

- ACM Digital Library: http://portal.acm.org/
- IEEE Digital Library: http://ieeexplore.ieee.org/
- Science@Direct: http://www.sciencedirect.com/
- Scopus: http://www.scopus.com/

Exclusion Criteria: For the purpose to develop the study selection process, the exclusion criteria were defined based on practical issues, that is, reasons that hinder a prior analysis of each paper's content. We defined the following criteria:

EC1. Papers imported without title or abstract;
EC2. Papers with less than four pages;
EC3. Papers that were not written in English;
EC4. Papers out of the scope;
EC5. Papers without full text accessibility;
EC6. Duplicated Papers.

Inclusion Criteria: After reading the paper abstract, the following inclusion criteria must be observed:

IC1. Gamification studies applied in organizational contexts;
IC2. General studies related to gamification definitions, processes and frameworks that can be applied in organizational contexts.

Quality Assessment Criteria: In order to select the most relevant studies, we defined the following quality assessment criteria:

QC1. How relevant is the abstract content?
QC2. How relevant is the introduction content?
QC3. How relevant is the conclusion content?

The questions must be answered according to the scale shown in Table 4. The sum of the three questions results in a final score ranging from 0 to 12.

Table 4. Quality assessment scale.

Very low	Low	Average	High	Very high
0	1	2	3	4

Data Extraction Form: The form contains fields that seek to answer the research questions. Table 5 shows the data extraction form.

Table 5. Data extraction form.

Description	Values
Research method	case study, experiment, focal group, interview, etc.
Gamification Frameworks	6D, MDA, Octalysis, PCD, etc.
Gamification techniques	achievements, avatar, levels, points, badges, etc.
Organizational context	human resource, training, knowledge management, etc.
Observed effects	autonomy, collaboration, efficiency, engagement, etc.
Gamification definition	descriptive field
Purpose of the study	descriptive field
Research questions	descriptive field
Methodology	descriptive field
Results	descriptive field
Discussions	descriptive field
Future works	descriptive field

2.3   Conduction

The conduction phase consists of executing the research protocol for the purpose of identifying the main studies that answer the research questions. The activities performed in this phase are as follows: a) study selection process; b) quality assessment; and c) snowballing;

Study Selection Process: The search string was performed on the four digital databases and the search result returned 844 articles. Using Parsifal [2], a tool to perform Systematic Literature Reviews, 132 duplicate papers have been excluded. The resulting 712 papers were analyzed by reading the content of the title and abstract. Exclusion and inclusion criteria have been applied resulting in 529 rejected papers and 183 accepted papers. Figure 2 shows the study selection process. Table 6 presents a quantitative summary of studies retrieved from digital databases.

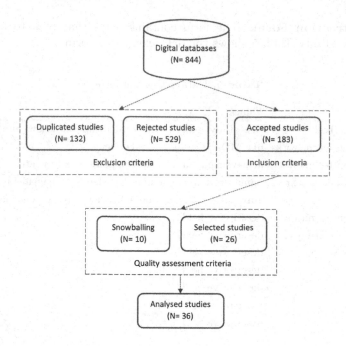

Fig. 2. Study selection process.

Table 6. Quantity of imported studies.

Database	Duplicated	Rejected	Accepted	Total	Analysed
ACM	28	40	8	76	1
IEEE	51	189	20	260	2
Science@Direct	98	3	39	186	5
Scopus	202	4	116	322	18
Snowballing	0	0	0	0	10
Total	132	529	183	844	36

Quality Assessment: Table 7 presents the list of selected studies after evaluating the quality assessment criteria. Papers with the score higher than 7.0 were selected for full reading of the content as they were considered relevant to answer at least one of the research questions.

Table 7. Selected studies according to quality assessment criteria.

Score	Studies	Total
12.0	[32,33]	2
11.0	[34]	1
10.0	[35–37]	3
9.0	[38–44]	7
8.0	[45–51], [52–57]	13
Total		**26**

Snowballing: The snowballing process aims to prevent relevant studies from being omitted. When executing the exploratory analysis, some studies were selected to be analyzed in the SLR in case they were not retrieved by the search string. In addition, during the conduction phase, other relevant studies have been identified and added to the selected studies list, shown in Table 8.

Table 8. Selected studies by the snowballing process.

Score	Studies	Total
10.0	[14, 15]	2
9.0	[58]	1
8.0	[22, 59–64]	7
Total		**10**

3   Results

This section provides the outcomes of reviewing the selected studies to answers to the research questions defined in the research protocol.

3.1   RQ1 - In Which Organizational Contexts Are Gamification Been Applied?

Among the selected works that answer the first research question, the secondary study conducted by Augustin et al. [40] presented ten gamified solutions applied in enterprise systems. The following contexts were presented: employee training, knowledge management, employee performance and social network.

The systematic literature review carried out by Obaid et al. [57] investigated the application of gamified solutions in employee recruitment and training processes. Most of the analyzed studies adopted the points, badges and leaderboards techniques. Overall, the results showed that gamification is an effective way to improve people' motivation.

Business Process Outsourcing - BPO: Neeli [45] proposed a method to apply gamification in BPO industry to increase the employee engagement. Points, levels, achievements, progress bars, bonuses, challenges and missions were some of the game mechanics used to enable the motivating factors. From a career perspective, the employees need to see their work as "challenging and meaningful". For increased clarity on work, the goals for individual or team were updated and the current state were clearly visible to employees. The proposed method has not been tested in practice and the author suggested that future work would be needed to ascertain the validity and efficacy of the solution.

Knowledge Management Systems - KMS: Friedrich et al. [33] investigated the types of motivations that support knowledge management activities and how gamification can be applied in this context. The results showed that altruism, collaboration, self-efficacy, reciprocity, fellowship and reputation were the main intrinsic motivations supported in KMS. Gamification mechanics like challenges, competition, feedback, performance graphs, rewards and status create incentives that address intrinsic as well as extrinsic motivation. Gamification elements like points, badges, and leaderboards can be used to address motivational aspects like reciprocity, reputation, and visibility of achievements.

Employee Recruitment: The study presented by Buil et al. [54] provides an understanding of how gamified recruitment tools foster positive attitudes among applicants. The recruitment process was based on a business simulation game competition. The applicants had to manage the productive plant, deal with outsourcing, purchase raw materials and quality control, among other activities. They were also expected to make decisions on marketing and to manage the financials. Results indicated that satisfaction of the participants' needs for competence and autonomy is significantly associated with their autonomous motivation to take part in the gamified recruitment process.

Employee Training: The secondary study presented by Armstrong and Landers [53] described the scientific understanding of gamification as it can be used to improve web-based employee training. The study suggested that the use of points, badges, leaderboards, challenges, narratives and immersions are the most used techniques to produce good learning outcomes. The authors also concluded that gamification is more effective when it is used in conjunction with instructional design principles.

Enterprise Social Network: Thom et al. [38] analyzed the effects of removing a points-based incentive system in an enterprise social network. The gamified solution had the main objective of encouraging the content contribution and employees received points for each photo, list and comments on a profile page. The experiment showed that removing the points system decreased the use of website activities, suggesting that there was a negative effect.

Corporate Intranet: Morschheuser et al. [51] conducted an experimental study to analyze whether gamification can improve the usage of corporate intranet. Basically, the solution included the use of points, leaderboards and quests to increase the motivation to read and learn content for banking consultants. The results showed a quantitative and qualitative increase in corporate intranet usage and provided a more efficient acquisition of knowledge.

Heldesk: Robson et al. [41] discussed how gamification can aid customer and employee engagement. The authors cited the "Freshdesk", a gamified heldesk software program for customer support. The solution involves transforming customer inquiries into virtual tickets that are then randomly assigned to employees. In this way, "Freshdesk" inspired a real-time, competitive environment via which players improve their performance.

Ideation Process: Zimmerling et al. [37] conceptualized a gamified system for companies to be used in ideation processes. The proposed model contains gamification elements to serve competition and collaboration among participants with the use of scores and virtual prizes as a function of presenting ideas and comments. Managers can put ideation challenges on the platform related to new innovation projects. Results showed that game elements in ideation processes have the potential to improve research and development.

Crowdsourcing: Afentoulidis et al. [43] conducted a study involving 101 employees from two multinational companies, adopting a user-centric approach to apply and experiment with gamification for enterprise crowdsourcing purposes. Competition and collaboration techniques were implemented based on the quantity and quality of contributions. Results showed that competition can better foster engagement than collaboration among employees.

3.2   RQ2 - What Are the Main Gamification Frameworks or Processess Applied in Organizational Systems?

Mora et al. [14] conducted a literature review on gamification design frameworks to identify the existing ones and their key features. The authors analyzed that most of the frameworks are based on a Human-Focused Design principles, taking into account the person as a main goal of the design and some of them are based on each other. Psychological related aspects are very common items of great importance in most the frameworks proposed.

In the systematic review of design frameworks conducted by Mora et al. [15], the authors identified four categories of frameworks according to the main areas of application: learning, business, health and generic. The largest number of reviewed works focused on a business environment and user-centered design principles, while generic ones can be applied to a wide range of environments.

Mechanics, Dynamics and Aesthetic - MDA: Hunicke et al. [59] proposed the MDA framework, a formal and iterative approach to build and refine gamification projects, composed by the components: mechanics, dynamics and aesthetics. Mechanics describe game components at the level of data and algorithm representation. Dynamics describe the behavior player actions and outcomes over time. Aesthetics describe the player's emotional responses when interacting with the game. From the designer's perspective, the mechanics give rise to dynamic system behavior, which in turn leads to particular aesthetic experiences. From the player's perspective, aesthetics set the tone, which is born out in observable dynamics and eventually, operable mechanics.

Kaleidoscope of Effective Gamification: Kappen and Nacke [48] proposed a design-centric model and analysis tool for effective gamification based on layers. The layers converge to a central core that represents the focal point of player experience. It represents the objectives of a game design. The outermost layer represents the players' perception of fun. In the intermediate layers, game mechanics and dynamics are integrated into the design. The model establishes an initial checklist for game designers in the form of guidelines for effective gamification.

Six Steps to Gamification - 6D: Werbach and Hunter [61] proposed a design framework customized for developing gamified systems. The 6D is a process to be performed in six steps: define business objectives, delineate target behaviors, describe the players, devise activity cycles, don't forget the fun and deploy the appropriate tools. Although the framework is proposed in stages, the gamification project must be iterative and learned by experience. For a successful project, it must be tested, monitored, evaluated and players must be interviewed to give their opinions.

Player-Centered Design - PCD: Kumar [46] introduced the PCD, a process that considers the player at the center of the design and development. The first step in the model approach is to understand the player and their context. In the next step, the current business scenario and target outcomes must be understood in order to define an appropriate mission for the gamification project. Next, the designer needs to understand human motivations for the purpose to apply game mechanics and rules to create a positive flow for the gamification project. Finally, the mission needs to be managed, the motivation needs to be monitored and the game mechanics need to be measured continuously.

Role-Motivation-Interaction - RMI: Gears and Braun [62] introduced a gamification design model aimed at improving project staffing in business. The RMI facilitate the architecting of gameful interactions. The design considers the intrinsic desires of employees, along with the human needs for autonomy, competence, and relatedness aided in the selection and customization of game design patterns. Extrinsic motivations drive corporate dynamics to offer employees a positive and engaging experience.

Sustainable Gamification Design - SGD: Raftopoulos [63] proposed the SGD framework, a conceptual model that yield a minimum viable design for gamified enterprise applications. The model is composed of four phases: discover, reframe, envision and create. The first one has the purpose to understand the context and actors of the system to be gamified. The reframe phase aims to analyze the opportunities and a range of potential solutions. The envision phase has the purpose to explore, identify and scope a preferred solution. Finally, the last one aims to design and launch a gamified solution.

Octalysis: The framework presented by Chou [64] derives from an octagonal shape, based on the following eight core drives that motivate people towards a variety of decisions and activities: epic meaning, development, empowerment of creativity, ownership, social influence, scarcity, unpredictability and avoidance. The author describes that everything we do is based on one or more of the eight core drives. If none of the core drives are present within a system, there is no motivation, and users will drop out.

3.3   RQ3 - What Are the Main Gamification Techniques Applied in Organizational Environments?

To answer this research question, a count of the gamification techniques presented in the observed studies was carried out. Figure 3 shows the 14 most used gamification techniques.

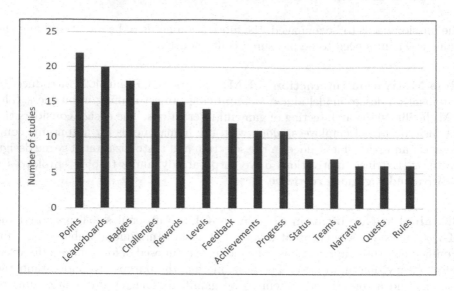

Fig. 3. Main gamification techniques.

Points, Badges and Leaderboards (PBL) were found in most of the observed studies. According to Werbach and Hunter [61], these techniques are so common within gamification and if they are used properly, they can be powerful, practical and relevant. On the other hand, they have important limitations. The PBL triad forms a useful starting point for gamification efforts.

Challenges and Quests are powerful game techniques to motivate people to action, especially if they believe they are working to achieve something great, awe-inspiring and bigger than themselves [46]. In the study presented by Zimmerling et al. [37], a challenge-based approach was used in ideation processes, resulting in a function that managers can put challenges on the platform related to innovation projects. Employees could post, comment and vote on ideas suggested by co-workers.

Rewards are mechanics that benefit players for a certain action or achievement. Robson et al. [41] presented the Freshdesk, a helpdesk software program for customer support centers. The solution involves transforming customer inquiries into virtual tickets that are randomly assigned to players. As tickets are resolved, players collect points, badges, trophies and quests and the employees are rewarded according to their efforts.

Levels, Feedback, Progress and Status are game techniques often used to show players' progress. Augustin et al. [40] cited the solution developed by Accenture to improve the internal knowledge management process. Employees performed activities and received scores, which in turn represented knowledge levels. When accomplishing a goal, players receive direct feedback, thus reinforcing important behaviors and ensuring a feeling of accomplishment.

Teams are mechanics that allow employees to see beyond personal achievements and seek to collaborate towards the organization's goals [45]. According to Chou [64], game elements that encourage competitive behavior, such as leaderboards, can be applied to combine team efforts.

Narratives are game elements that provide players some context about the game. Armstrong and Landers [53] presented a secondary study about gamified solutions in employee training. The authors mentioned the difference of gamifying the content or the method. Game elements that provide feedback such as points, badges, and leaderboards are often used to motivate people for completing a given module. When gamifying content, gains are potentially more transformative.

Rules are prescribed guides for conduct or actions allowed for a given player and how the execution of such actions map to given rewards [36]. They are important to create a core engagement loop and make the work enjoyable.

3.4 RQ4 - What Are the Main Effects Observed in Gamification Solutions Applied in Organizational Environments?

To answer this research question, a count of the effects presented in the observed studies was carried out. Figure 4 shows the 19 most reported effects.

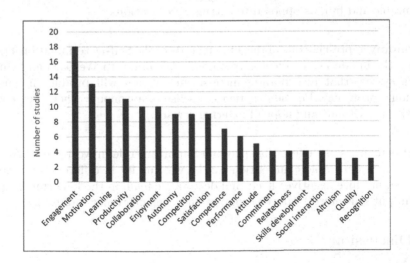

Fig. 4. Main observed effects.

Engagement and Motivation are the most observed effects. Werbach and Hunter [61] described that the same human needs that drive engagement with games are the same ones present in the workplace. Gamification can be understood as a means of designing systems to motivate people to perform certain activities to increase engagement with organizational goals.

Learning and Skill Development are effects that can be observed in solutions focused on knowledge acquisition, training and information sharing. Armstrong and Landers [53] described how gamification can improve employee training programs, making them more engaging and effective.

Productivity and Performance are effects related to the concept of flow, the mental state of operating in which a person in an activity is fully immersed in the feeling of energized focus, full involvement and the success in the process of the activity [46]. Business software can benefit from incorporating game techniques to enable their users to achieve this sense of flow.

Collaboration is a desirable effect in organizational environments. The gamified ideation system proposed by Zimmerling et al. [37] encourages the contribution of ideas for new projects. The resulting platform design is a challenge-based approach with business driven quests.

Enjoyment is an important element of gamification strategies. Dale [35] takes a critical look at the potential of gamification as a business change agent that can deliver a more motivated and engaged workforce making activities more pleasurable and fun, as opposed to extrinsic motivations.

Autonomy represents one of the human needs of the Self-Determination Theory (SDT), the innate need to feel in command. According to Werbach and Hunter [61], a system that incorporates intrinsic motivation will produce a sense of autonomy or agency. People experience a sense of autonomy when they engage in activities without any hope of external reward.

Competition is one effect that plays an important role in workplaces. According to Chou [64], while competition can be very useful in different scenarios, it can often produce negative effects and demoralize team in the long run. Employees can put self-interest above corporate and even customer interests.

4   Discussion

This section provides a discussion of the results obtained from answering the research questions, in order to identify best practices in building gamification designs in organizational environments.

4.1   Gamification in Organizational Contexts

This work provided contributions for understanding gamification as a valuable tool for improving activities in organizational environments. According to Oprescu et al. [39], gamified workplaces could be a positive and innovative solution to addressing contemporary problems in organizations. Such problems include high levels of stress, reduced sense of community, reduced loyalty and rapid changes in the workforce.

Werbach and Hunter [61] described in the first step of the design process to implement an effective gamification, a well-developed understanding of the business objectives, that means the specific performance goals for the gamified system, such as improving employee productivity. Gamification, even when effective, can produce results that don't necessarily help. To avoid this risk, it's important to identify the potential objectives and justify how they would benefit the organization.

4.2   Gamification Frameworks

This study allowed us to observe a common feature among gamification frameworks, most of them are human-centered approaches. According to Chou [64], most systems are inherently function-focused, that means the users will do their jobs because they have to complete the tasks. Human-centered design optimizes people's feelings and motivations as the basis for designing the system as well as its functions.

According to Oprescu et al. [39], the focus of a gamified system must be on the user's psychosocial experience and wellbeing. Regardless of the framework to be adopted in the gamification project design, understanding what can motivate and demotivate players is an important step to have an effective gamification.

4.3   Gamification Techniques

The results showed that the triad Points, Badges and Leaderboards (PBL) are present in most solutions. When properly used, these elements are practical and relevant to gamified designs. According to Chou [64], adding these elements in a scalable manner can result in generic game mechanics and poorly designed applications. Studies have shown that stimulating people's intrinsic motivations makes activities much more engaging.

In a gamification solution, the design must be built based on people's motivational profile. Game elements must be carefully selected for a truly engaging user experience. The Octalysis framework, introduced by Chou [64], indicates that different types of game techniques push us forward differently and adding them according to players' motivational core drives can result in an effective gamification.

5   Conclusion

This study investigated the use of gamification in organizational environments. Before starting the systematic review itself, the exploratory analysis stage provided us a better direction for this research. Analysing the secondary studies related to gamification applied in organizational contexts allowed us to identify gaps in the literature and to guide in conducting the remaining of the work.

Answering the research questions, we could observe the importance of adopting a framework or a process to construct the gamification design. The application of game techniques without clear goals can bring negative results. The third research question showed that the triad PBL are present in most applications. It is common for game designers to start implementing points, badges and leaderboards in their solutions, But gamification is so much more than PBLs. Studies that referenced the use of frameworks or processes for building gamified solutions had a common feature, most of them mentioned the importance of understanding what can motivate and demotivate players to have an effective gamification.

The results of this systematic review can serve as a basis for developing new gamification designs in organizational environments. The first research question showed us several possibilities for using gamification to improve engagement in business processes. In fact, gamification can provide more employee engagement, job satisfaction, worker performance, skill development, and organization changes.

References

1. Kitchenham, B., Charters, S.: Guidelines for performing systematic literature reviews in software engineering (Vol. 5). Technical report, ver. 2.3 ebse technical report. ebse (2007)
2. Parsifal. https://parsif.al. Accessed 27 Dez 2022
3. Hamari, J., Koivisto, J., Sarsa, H.: Does gamification work?-a literature review of empirical studies on gamification. In: 2014 47th Hawaii International Conference on System Sciences, pp. 3025–3034. IEEE, January 2014
4. Darejeh, A., Salim, S.S.: Gamification solutions to enhance software user engagement-a systematic review. Int. J. Hum.-Comput. Interact. **32**(8), 613–642 (2016)
5. Hervás, R., Ruiz-Carrasco, D., Mondéjar, T., Bravo, J.: May. Gamification mechanics for behavioral change: a systematic review and proposed taxonomy. In: Proceedings of the 11th EAI International Conference on Pervasive Computing Technologies for Healthcare, pp. 395–404 (2017)
6. Looyestyn, J., Kernot, J., Boshoff, K., Ryan, J., Edney, S., Maher, C.: Does gamification increase engagement with online programs? a systematic review. PloS One **12**(3), e0173403 (2017)
7. Hassan, L., Hamari, J.: Gamification of e-participation: a literature review. In: Proceedings of the 52nd Hawaii International Conference on System Sciences (2019 January)

8. Lopes, S., Pereira, A., Magalhães, P., Oliveira, A., Rosário, P.: Gamification: Focus on the strategies being implemented in interventions: a systematic review protocol. BMC. Res. Notes **12**(1), 1–5 (2019)

9. De Croon, R., Geuens, J., Verbert, K., Vanden Abeele, V.: A systematic review of the effect of gamification on adherence across disciplines. In: Fang, X. (ed.) HCII 2021. LNCS, vol. 12789, pp. 168–184. Springer, Cham (2021). https://doi.org/10.1007/978-3-030-77277-2_14

10. Jayawardena, N.S., Ross, M., Quach, S., Behl, A., Gupta, M.: Effective online engagement strategies through gamification: a systematic literature review and a future research agenda. J. Global Inf. Manage. (JGIM) **30**(5), 1–25 (2021)

11. Rachad, T., Idri, A. and Zellou, A.: Gamified Mobile Applications for Improving Driving Behavior: A Systematic Mapping Study. Mobile Information Systems (2021)

12. Xu, J., Lio, A., Dhaliwal, H., Andrei, S., Balakrishnan, S., Nagani, U., Samadder, S.: Psychological interventions of virtual gamification within academic intrinsic motivation: a systematic review. J. Affect. Disord. **293**, 444–465 (2021)

13. Schlagenhaufer, C., Amberg, M.: A descriptive literature review and classification framework for gamification in information systems (2015)

14. Mora, A., Riera, D., Gonzalez, C., Arnedo-Moreno, J.: A literature review of gamification design frameworks. In: 2015 7th International Conference on Games and Virtual Worlds for Serious Applications (VS-Games), pp. 1–8. IEEE (September 2015)

15. Mora, A., Riera, D., González, C., Arnedo-Moreno, J.: Gamification: a systematic review of design frameworks. J. Comput. High. Educ. **29**(3), 516–548 (2017). https://doi.org/10.1007/s12528-017-9150-4

16. Azouz, O., Lefdaoui, Y.: Gamification design frameworks: a systematic mapping study. In: 2018 6th International Conference on Multimedia Computing and Systems (ICMCS), pp. 1–9. IEEE (May 2018)

17. Bouzidi, R., De Nicola, A., Nader, F. and Chalal, R., 2019. A systematic literature review of gamification design. In Proceedings of the 20th Annual Simulation and AI in Games Conference (GAME-ON 2019)(Breda, The Netherlands). EUROSIS, Ostend, Belgium

18. Rozi, F., Rosmansyah, Y., Dabarsyah, B.: A systematic literature review on adaptive gamification: components, methods, and frameworks. In: 2019 International Conference on Electrical Engineering and Informatics (ICEEI), pp. 187–190. IEEE (July 2019)

19. Rodrigues, L., Toda, A.M., Palomino, P.T., Oliveira, W., Isotani, S.: Personalized gamification: a literature review of outcomes, experiments, and approaches. In: Eighth International Conference on Technological Ecosystems for Enhancing Multiculturality, pp. 699–706 (October 2020)

20. Krath, J., von Korflesch, H.: Designing gamification and persuasive systems: a systematic literature review. GamiFIN, pp. 100–109 2021

21. Hinton, S., Wood, L.C., Singh, H., Reiners, T.: Enterprise gamification systems and employment legislation: a systematic literature review. Ajis: Australasian J. Inf. Syst. **23**, 1–24 (2019)

22. Szendrői, L., Dhir, K.S. and Czakó, K.: Gamification in for-profit organisations: a mapping study. Business: Theory Practice **21**(2), pp. 598–612 (2020)

23. Miranda, M.A.C., Vergaray, A.D.: Mobile Gamification Applied to Employee Productivity in Companies: A Systematic Review (2021)

24. Encarnação, R., Reuter, J., Ferreira Dias, M., Amorim, M.: Gamification as a driver of motivation in the organizations: a Bibliometric Literature Review. In: Ninth International Conference on Technological Ecosystems for Enhancing Multiculturality (TEEM'21), pp. 167–172 (October 2021)

25. Thomas, N.J., Baral, R., Crocco, O.S.: Gamification for HRD: Systematic Review and Future Research Directions. Human Resource Development Review, p. 15344843221074859 (2022)

26. Rodrigues, L.F., Oliveira, A., Rodrigues, H.: Main gamification concepts: a systematic mapping study. Heliyon 5(7), e01993 (2019)

27. Azouz, O., Karioh, N., Lefdaoui, Y.: A systematic mapping study: how can UX design be adapted to improve the design of meaningful gamified solutions? Int. J. Innov. Technol. Manag. 18(06), 2130006 (2021)

28. Oliveira, W., Pastushenko, O., Rodrigues, L., Toda, A.M., Palomino, P.T., Hamari, J., Isotani, S.: Does gamification affect flow experience? a systematic literature review (2021). arXiv preprint arXiv:2106.09942

29. Kasurinen, J., Knutas, A.: Publication trends in gamification: a systematic mapping study. Comput. Sci. Rev. 27, 33–44 (2018)

30. Bozkurt, A., Durak, G.: A systematic review of gamification research: In pursuit of homo ludens. Int. J. Game-Based Learn. (IJGBL) 8(3), 15–33 (2018)

31. Ayastuy, M.D., Torres, D., Fernández, A.: Adaptive gamification in Collaborative systems, a systematic mapping study. Comput. Sci. Rev. 39, 100333 (2021)

32. Cardador, M.T., Northcraft, G.B., Whicker, J.: A theory of work gamification: something old, something new, something borrowed, something cool? Hum. Resour. Manag. Rev. 27(2), 353–365 (2017)

33. Friedrich, J., Becker, M., Kramer, F., Wirth, M., Schneider, M.: Incentive design and gamification for knowledge management. J. Bus. Res. 106, 341–352 (2020)

34. Fathian, M., Sharifi, H., Nasirzadeh, E., Dyer, R., Elsayed, O.: Towards a comprehensive methodology for applying enterprise gamification. Decision Sci. Lett. 10(3), 277–290 (2021)

35. Dale, S.: Gamification: making work fun, or making fun of work? Bus. Inf. Rev. 31(2), 82–90 (2014)

36. Stanculescu, L.C., Bozzon, A., Sips, R.J., Houben, G.J.: Work and play: an experiment in enterprise gamification. In Proceedings of the 19th ACM Conference on Computer-Supported Cooperative Work and Social Computing, pp. 346–358 (February 2016)

37. Zimmerling, E., Höflinger, P.J., Sandner, P. and Welpe, I.M.: Increasing the creative output at the fuzzy front end of innovation-A concept for a gamified internal enterprise ideation platform. In: 2016 49th Hawaii International Conference on System Sciences (HICSS), pp. 837–846. IEEE (January 2016)

38. Thom, J., Millen, D., DiMicco, J.: Removing gamification from an enterprise SNS. In: Proceedings of the ACM 2012 Conference on Computer Supported Cooperative Work, pp. 1067–1070 (February 2012)

39. Oprescu, F., Jones, C., Katsikitis, M.: I PLAY AT WORK-ten principles for transforming work processes through gamification. Front. Psychol. 5, 14 (2014)

40. Augustin, K., Thiebes, S., Lins, S., Linden, R. and Basten, D., 2016. Are we playing yet? A review of gamified enterprise systems

41. Robson, K., Plangger, K., Kietzmann, J.H., McCarthy, I., Pitt, L.: Game on: Engaging customers and employees through gamification. Bus. Horiz. 59(1), 29–36 (2016)

42. Swacha, J.: Gamification in enterprise information systems: what, why and how. In: 2016 Federated Conference on Computer Science and Information Systems (FedCSIS), pp. 1229–1233. IEEE (September 2016)
43. Afentoulidis, G., Szlávik, Z., Yang, J., Bozzon, A.: Social gamification in enterprise crowdsourcing. In: Proceedings of the 10th ACM Conference on Web Science, pp. 135–144 (May 2018)
44. Hammedi, W., Leclercq, T., Poncin, I., Alkire, L.: Uncovering the dark side of gamification at work: impacts on engagement and well-being. J. Bus. Res. **122**, 256–269 (2021)
45. Neeli, B.K.: A method to engage employees using gamification in BPO industry. In: 2012 Third International Conference on Services in Emerging Markets, pp. 142–146. IEEE, December 2012
46. Kumar, J.: Gamification at work: designing engaging business software. In: Marcus, A. (ed.) DUXU 2013. LNCS, vol. 8013, pp. 528–537. Springer, Heidelberg (2013). https://doi.org/10.1007/978-3-642-39241-2_58
47. Webb, E.N.: Gamification: when it works, when it doesn't. In: Marcus, A. (ed.) DUXU 2013. LNCS, vol. 8013, pp. 608–614. Springer, Heidelberg (2013). https://doi.org/10.1007/978-3-642-39241-2_67
48. Kappen, D.L. and Nacke, L.E.: The kaleidoscope of effective gamification: deconstructing gamification in business applications. In: Proceedings of the First International Conference on Gameful Design, Research, and Applications, pp. 119–122 (October 2013)
49. Herranz, E., Colomo-Palacios, R., Amescua-Seco, A.: Towards a new approach to supporting top managers in SPI organizational change management. Procedia Technol. **9**, 129–138 (2013)
50. Robson, K., Plangger, K., Kietzmann, J.H., McCarthy, I., Pitt, L.: Is it all a game? understanding the principles of gamification. Bus. Horiz. **58**(4), 411–420 (2015)
51. Morschheuser, B., Henzi, C., Alt, R.: Increasing intranet usage through gamification-insights from an experiment in the banking industry. In: 2015 48th Hawaii International Conference on System Sciences, pp. 635–642. IEEE (January 2015)
52. Perryer, C., Celestine, N.A., Scott-Ladd, B., Leighton, C.: Enhancing workplace motivation through gamification: transferrable lessons from pedagogy. Int. J. Manage. Educ. **14**(3), 327–335 (2016)
53. Armstrong, M.B., Landers, R.N.: Gamification of employee training and development. Int. J. Train. Dev. **22**(2), 162–169 (2018)
54. Buil, I., Catalán, S., Martínez, E.: Understanding applicants' reactions to gamified recruitment. J. Bus. Res. **110**, 41–50 (2020)
55. Ulmer, J., Braun, S., Cheng, C.T., Dowey, S., Wollert, J.: Human-centered gamification framework for manufacturing systems. Procedia CIRP **93**, 670–675 (2020)
56. Wolf, T., Weiger, W.H., Hammerschmidt, M.: Experiences that matter? the motivational experiences and business outcomes of gamified services. J. Bus. Res. **106**, 353–364 (2020)
57. Obaid, I., Farooq, M.S., Abid, A.: Gamification for recruitment and job training: model, taxonomy, and challenges. IEEE Access **8**, 65164–65178 (2020)
58. Deterding, S., Sicart, M., Nacke, L., O'Hara, K., Dixon, D.: Gamification. using game-design elements in non-gaming contexts. In: CHI'11 Extended Abstracts on Human Factors in Computing Systems, pp. 2425–2428 (2011)
59. Hunicke, R., LeBlanc, M., Zubek, R.: MDA: a formal approach to game design and game research. In: Proceedings of the AAAI Workshop on Challenges in Game AI, vol. 4, No. 1, p. 1722 (July 2004)

60. Deterding, S., Dixon, D., Khaled, R., Nacke, L.: From game design elements to gamefulness: defining "gamification". In: Proceedings of the 15th International Academic MindTrek Conference: Envisioning Future Media Environments, pp. 9–15 (September 2011)
61. Werbach, K., Hunter, D.: For the win: How game thinking can revolutionize your business (2012)
62. Gears, D., Braun, K.: Gamification in business: designing motivating solutions to problem situations. In: Proceedings of the CHI 2013 Gamification Workshop (April 2013)
63. Raftopoulos, M.: Towards gamification transparency: a conceptual framework for the development of responsible gamified enterprise systems. J. Gaming Virtual Worlds 6(2), 159–178 (2014)
64. Chou, Y.K.: Actionable Gamification-Beyond Points, Badges, and Leaderboards. Octalysis Media (2015)
65. Huotari, K., Hamari, J.: Defining gamification: a service marketing perspective. In: Proceeding of the 16th International Academic MindTrek Conference, pp. 17–22 (October 2012)

The Impact of Gender and Visual Presentation of Advertising on User Experience in Mobile Shopping Apps

Yan Cao[1] , Weimin Zhai[2](✉), and Weiren Zhai[3]

[1] College of Economics and Management, Northeast Forestry University, Harbin 150040, China
[2] Department of Design, National Taiwan University of Science and Technology, Taipei 106, Taiwan
zwm0908480013@gmail.com
[3] Computer Science and Technology, Jiangsu University of Science and Technology, Zhenjiang 212100, China

Abstract. The visual presentation of mobile shopping ads as a visual presentation provides a good user experience for the user when performing online shopping operations. The usability of the shopping app user interface is a significant design issue in the user experience. This study aimed to explore the usability of gender and mobile shopping ad visuals in a shopping website's operation and to suggest future design improvements. A 2 x 2 between-subjects experiment (Between-subjects design) was planned to help explore whether gender (i.e., male and female) and different visual presentation types of mobile shopping ads (i.e., static and dynamic ads) affect users' task performance and their subjective evaluations. This experiment used a purposive sampling method, and a total of 32 participants were recruited to participate in the experiment. Data collection for the experiment included participants' task performance and subjective ratings on a 7-point Likert scale and semi-structured interviews. The generated results revealed that: (1) A significant interaction between gender and ad presentation was found in operational performance, with males significantly faster between task operational performance during static ads than between task operational performance during dynamic ads. However, the opposite was true for females at operation time; (2) Gender influenced users' preference and trustworthiness level in shopping apps, and females generally rated higher in preference and trustworthiness level than males; (3) In the subjective measure of reasonable, the reasonable level of static ads was generally higher than that of dynamic ads; (4) There was a significant interaction between gender and ad presentation in the subjective measure of satisfaction. Males rated significantly higher in static ads than in dynamic ads. (5) There was a significant interaction between gender and ad presentation in the subjective measure of satisfaction, with males significantly more satisfied with static ads than dynamic ads. However, the opposite result is obtained for females. The findings generated from the research can improve the understanding of gender differences in ad usability in shopping application interfaces and provide a reference for the visual design of ads in future shopping APPs, especially for gender-specific shopping APPs.

© Springer Nature Switzerland AG 2023
F. Fui-Hoon Nah and K. Siau (Eds.): HCII 2023, LNCS 14038, pp. 353–364, 2023.
https://doi.org/10.1007/978-3-031-35969-9_24

Keywords: Shopping app · Gender · Advertising · Visual attention · User experience

1   Introduction

More and more users in China are using smartphones. At the same time, e-commerce is gradually shifting to mobile commerce (Lin et al. 2021). Shopping apps are becoming increasingly popular in China, relying on smartphones as a user shopping vehicle. Mobile shopping is becoming the mainstream of online shopping. Unlike website shopping, mobile shopping is independent of time and environment and allows for faster payments on smartphones (Sukhani, Qomariyah, & Purwita, 2022). Especially in the epidemic era, online shopping has gradually become a part of people's daily life. Users can perform shopping behaviors such as product information search, information comparison, order, and payment on the shopping app's interface (Patel and Pandit 2021). The study by Patel et al. (2020) highlighted the shopping APP interface and pointed out that the interface quality of shopping APPs influences users' purchase intention to some extent. Furthermore, Fu et al. (2018) insisted that visual appearance in shopping APPs is also an important factor influencing users' shopping attitude, and therefore, how to cue the usefulness and ease of use of the shopping APP interface to improve user experience becomes the key for shopping APP merchants to compete (Hu et al. 2020).

As more and more consumers begin to use mobile applications for shopping, the interface design of mobile shopping APPs is constantly pushing the boundaries, among which advertising is an important channel for merchants to market their products. A good advertisement background can bring a perfect visual experience to users and stimulate consumers' desire to shop (Zhang, Luo, Wu, & Deng, 2021). The design elements of mobile ads include visuals, location, interaction, content, and presentation time (Jung and Yoo 2020), and app ads are the primary source of revenue for many mobile apps. Moreover, with the continuous development of shopping apps, mobile advertising has recently experienced tremendous growth (Gao et al. 2021). One study found that an app with ad insertion is 30% more expensive to develop than an app without ads (Gui et al. 2015). Therefore, the main focus of this study is to explore the impact of the visual presentation of ads on user experience in mobile shopping APPs.

It was found that dynamic banner ads evoke higher emotional engagement, and users invest more attention and attraction (Cassioli 2019). White et al. (2021) found in their study that the color of banner ads affects users' purchase intention. Palcu et al. (2017) stated in their study that dynamic ads are effective in attracting users' Chiu et al. (2017) found that a more extended period of gaze and a higher number of gaze increases the user's attitude towards the advertised brand. Dynamic advertising as a form of visual salience presentation enhances the user's ability to perceive the memory aspect of the advertisement. Users' attention is precious, and the small size of the mobile screen should avoid information overload. Dynamic ads can cause users to devote too much visual attention. Nevertheless, users have limited cognitive resources to process interface information, which can also cause inattention and interfere with users' task completion, among other things (Resnick and Albert 2016). so this study explores the visual effect of shopping ads (i.e., static versus dynamic) on user experience in a mobile shopping app.

With the rapid development of online shopping, users are exposed to a large number of advertising messages in different online and offline channels (Baek and Morimoto 2012), and it has been the case that consumers mostly use an avoidance approach to deal with advertising messages (Cho and as- 2004). Therefore many retailers use personalization in their advertisements to increase customer awareness and interest in their products (Schreiner et al. 2019). In addition, some studies have found that advertising has a more significant impact on women than men(Apeagyei 2011) and that women are more likely to actively seek out information of interest before making a purchase. (Jackson et al. 2011). Past research has found significant differences in shopping behavior and motivation between males and females. For example, males prefer smaller recommendation sets than females (Schreiner et al. 2019). However, few studies have shown whether there is an effect of gender on the visual effect of mobile shopping ads, so this study explores the effect of gender (i.e., male and female) and the visual effect of shopping ads (i.e., static versus dynamic) on user experience in a mobile shopping app.

2 Method

This study aimed to explore the effects of gender and ad-motion type on users' operational experience in a shopping app. The experiment was a two-way between-groups experimental design with two experimental prototypes. Each subject operated only one experimental prototype and completed three operational tasks. The experiment was simulated on a restaurant shopping APP with a banner advertisement at the top of the interface, the content of the advertisement reflected the thematic elements of the restaurant shopping APP, and the overall interface was color coordinated and unified. In order to avoid too fast dynamic presentation interfering with the user and unable to complete the task, we set the animation presentation speed within a reasonable range.

2.1 Participants

We invited 32 participants aged 18–30 years to experience different shopping APPs using purposive sampling; their education level was at least a bachelor's degree, they had experience in using shopping websites, and 28 of them (88%) spent on average within 2 h per day on online shopping activities, 4 of them (12%) spent an average of 2 to 6 h per day on online shopping activities, all participants had normal or corrected normal vision, and all were right-handed. All these participants gave full informed consent and fully understood the experimental tasks and the questionnaire. There were no barriers to using the application, and each participant could complete the experiment independently. The duration of the experiment was approximately 20 min, and participants received a participation fee of approximately 50 RMB.

2.2 Materials and Apparatus

The experimental prototype was done with Proto.io. Illustrator was used for graphic design and drawing in this experimental design. The experimental prototype was designed to simulate the Burger King shopping app and was done with Proto.io. The experiment is equipped with an IOS 5.5-inch screen (i.e., iPhone 7 Plus) with 1920 × 1080 pixels and 401 PPI. The experimental site is a laboratory free from noise and external interference.

2.3 Experimental Design and Procedure

This experiment used a two-factor intergroup experimental design of 2 (gender) × 2 (advertising kinetic type). The two levels of gender factor are male and female, and the two levels of ad kinetic type are static and dynamic. The prototype of this experiment is shown in Fig. 1. Before the experiment, participants were informed of the purpose. Then participants were asked to complete a questionnaire and consent form with basic information about their individuals. Screen recording software recorded each participant's task completion time for further analysis. Participants were asked to subjectively evaluate the questionnaire when the experiment was completed. Participants were asked to complete questionnaires regarding their subjective evaluations of the overall task operation. These questions were created on a 7-point Likert scale, with the lowest score for each question being 1 (strongly disagree), the highest score being 7 (strongly agree), and the medium score being 4.

Fig. 1. The prototype of this experiment.

In addition, three tasks of this experiment were determined (including visual search and information comparison tasks). A simulation of one of the almost most frequently used shopping software in China was used to help participants participate in the experiment. The subjects of this study were all students from Hunan University in China. In addition, the controlled variables were the same environmental settings with stable WiFi speed.

Table 1. Experimental task designs of this study.

Task number	Descriptions
Task 1	Find the price of the Double Original Crispy Chicken Casserole Medium Set inside the Daily Value section
Task 2	Find the product with a price of $19 inside the snack dessert section
Task 3	Compare which is the highest priced one inside the children's package section

3 Results

In this study, task performance, and subjective evaluations were statistically counted, and a two-way ANOVA was conducted using SPSS software on the main effects of subjective evaluations collected regarding participants' task completion time (seconds), as well as interactions.

3.1 Analysis of Task Completion Time

Table 2. Two-way ANOVA analysis of variance for time to completion of each task.

	Source	SS	df	MS	F	p	η^2	LSD test
Task 1	Gender	22.05	1	22.05	2.52	0.124	0.08	
	Ad presentation	1.46	1	1.46	0.17	0.686	0.01	
	Gender × Ad presentation	61.16	1	61.16	6.98	0.013*	0.20	
Task 2	Gender	1.00	1	1.00	0.05	0.822	0.00	
	Ad presentation	4.24	1	4.24	2.20	0.150	0.07	
	Gender × Ad presentation	1.21	1	1.21	0.62	0.436	0.02	
Task 3	Gender	3.56	1	3.556	1.26	0.271	0.04	
	Ad presentation	7.02	1	7.02	2.49	0.126	0.08	
	Gender × Ad presentation	3.97	1	3.97	1.41	0.245	0.05	

* Significantly different at the $\alpha = 0.05$ level (* $p < 0.05$);
** Significantly different at the $\alpha = 0.01$ level (* $p < 0.01$)

The first task: "Find the price of the Double Original Crispy Chicken Casserole Medium Set inside the Daily Value section" was an identifiable task. Table 2 shows the results of a two-way analysis of variance (ANOVA) for task 1 completion time, showing no significant main effect for gender (F = 2.52, p = 0.124 > 0.05; $\eta^2 = 0.08$). Likewise, there was no significant main effect for ad presentation (F = 0.17, p = 0.686 > 0.05; $\eta^2 = 0.01$). In addition, there was a significant interaction effect between gender and ad presentation (F = 6.98, p = 0.013 < 0.05; $\eta^2 = 0.20$). Figure 2, shows that the inter-task operational performance of males during static ads (M = 8.53, SD = 2.47) was significantly faster than the task operation performance interval during dynamic ads (M

= 10.87, SD = 2.59). However, females were significantly slower in task 1 performance during static ads (M = 12.96, SD = 4.36) than during dynamic ads (M = 9.77, SD = 1.81).

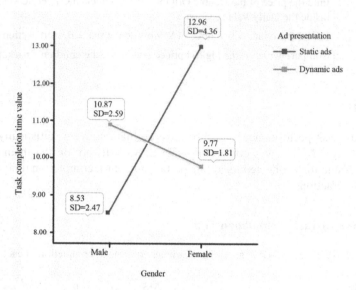

Fig. 2. Graph of the interaction between gender and ad presentation on task 1operational performance.

The second task: "Find the product with a price of $19 inside the snack dessert section" was an identifiable task. Table 2 shows the results of the two-way analysis of variance (ANOVA) for Task 2 completion time, showing no significant main effect for gender (F = 0.05, p = 0.822 > 0.05; η^2 = 0.00). Likewise, there was no significant main effect for ad presentation (F = 0.17, p = 0.686 > 0.05; η^2 = 0.01). In addition, there was no significant interaction effect between gender and ad presentation (F = 2.20, p = 0.150 > 0.05; η^2 = 0.07).

The third task: "Compare which is the highest priced one inside the children's package section" was an identifiable task. Table 2 shows the results of a two-way analysis of variance (ANOVA) for Task 3 completion time, showing no significant main effect for gender (F = 1.26, p = 0.271 > 0.05; η^2 = 0.04). Likewise, there was no significant main effect of ad presentation (F = 2.49, p = 0.126 > 0.05; η^2 = 0.08). In addition, there was no significant interaction effect between gender and ad presentation (F = 1.41, p = 0.245 > 0.05; η^2 = 0.05).

3.2 Analysis of Subjective Evaluations

According to a 7-point Likert scale, the results of participants' subjective evaluation after completing the operational tasks (i.e., 1: least agree, 7: most agree) are presented as follows. The results of the two-way analysis of variance (ANOVA) regarding participants' subjective evaluations are provided as follows.

Table 3. Two-way ANOVA analysis of variance on participants' subjective evaluations.

	Source	SS	df	MS	F	p	η^2	LSD test
The degree of effortlessness	Gender	8.00	1	8.00	5.70	0.024*	0.17	Female > Male
	Ad presentation	1.13	1	1.13	0.80	0.378	0.03	
	Gender × Ad presentation	3.13	1	3.13	2.23	0.147	0.07	
The degree of subjective preference	Gender	6.13	1	6.13	4.43	0.045*	0.14	Female > Male
	Ad presentation	4.50	1	4.50	3.25	0.082	0.10	
	Gender × Ad presentation	0.13	1	0.13	0.09	0.766	0.00	
The degree of physical demand	Gender	2.53	1	2.53	2.61	0.117	0.09	
	Ad presentation	7.03	1	7.03	7.26	0.012*	0.21	Static > Dynamic
	Gender × Ad presentation	2.53	1	2.53	2.61	0.117	0.09	
	Gender	1.13	1	1.13	1.36	0.254	0.05	
	Ad presentation	3.13	1	3.13	3.76	0.063	0.12	
	Gender × Ad presentation	4.50	1	4.50	5.42	0.027*	0.16	

* Significantly different at the $\alpha = 0.05$ level (* p < 0.05);
** Significantly different at the $\alpha = 0.01$ level (* p < 0.01)

The degree of preference data was analyzed to find out which shopping app was the most satisfying. The questionnaire was designed based on a 7-point scale with the two endpoints labeled strongly dissatisfied and strongly satisfied, respectively. Table 3 shows the results of the two-way analysis of variance (ANOVA) for the degree of preference, revealing a significant main effect of gender (F = 5.70, p = 0.024 < 0.05; η^2 = 0.17). More specifically, although both mean scores were higher than the medium score of 4, the degree of preference of female (M = 6.13, SD = 0.96) was significantly higher than that of male (M = 5.13, SD = 1.41). However, there was no significant main effect of ad presentation (F = 0.80, p = 0.378 > 0.05; η^2 = 0.03). In addition, there was no significant interaction between gender and ad presentation (F = 2.23, p = 0.147 > 0.05; η^2 = 0.07).

Data on the degree of trustworthiness was analyzed to find out which shopping site was the most satisfying. The questionnaire was designed based on a 7-point scale with the two endpoints labeled strongly dissatisfied and strongly satisfied. Table 3 shows the results of the two-way analysis of variance (ANOVA) for the degree of trustworthinesst, revealing a significant main effect of gender ($F = 4.43$, $p = 0.045 < 0.05$; $\eta^2 = 0.14$). More specifically, despite the fact that both mean scores were higher than the medium score of 4, the degree of trustworthinesst was significantly higher for female ($M = 6.06$, $SD = 0.68$) than for male ($M = 5.18$, $SD = 1.56$). However, there was no significant main effect of ad presentation ($F = 3.25$, $p = 0.082 > 0.05$; $\eta^2 = 0.10$). In addition, there was no significant interaction between gender and ad presentation ($F = 0.09$, $p = 0.766 > 0.05$; $\eta^2 = 0.00$).

Data on the degree of reasonableness were analyzed to find out which shopping site was the most satisfying. Table 3 shows the results of the two-way analysis of variance (ANOVA) for the degree of reasonableness, revealing no significant main effect of gender ($F - 2.61$, $p = 0.117 > 0.05$; $\eta^2 = 0.09$). However, there was a significant main effect of ad presentation ($F = 7.26$, $p = 0.012 < 0.05$; $\eta^2 = 0.21$). More specifically, although both mean scores were above the medium score of 4, the degree of plausibility was significantly higher for static ($M = 5.18$, $SD = 0.25$) than for dynamic ($M = 6.13$, $SD = 0.25$). In addition, there was no significant interaction between gender and ad presentation ($F = 0.21$, $p = 0.117 > 0.05$; $\eta^2 = 0.09$).

Table 3 shows the results of the two-way analysis of variance (ANOVA) for satisfaction, revealing no significant main effect of gender ($F = 1.36$, $p = 0.254 > 0.05$; $\eta^2 = 0.05$). In addition, there was no significant main effect of ad presentation ($F = 3.76$, $p = 0.063 > 0.05$; $\eta2 = 0.12$). However, there was a significant interaction between gender and ad presentation ($F = 5.42$, $p = 0.027 < 0.05$; $\eta^2 = 0.16$). Figure 3, shows that males were significantly more satisfied during static ads ($M = 6.00$, $SD = 0.53$) than during dynamic ads ($M = 4.63$, $SD = 1.51$). However, females were significantly less satisfied

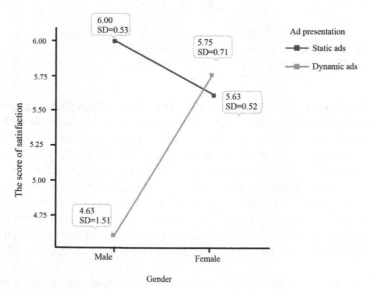

Fig. 3. Graph of the interaction between gender and ad presentation in terms of the degree of satisfaction.

during static ads (M = 5.63, SD = 0.52) than during dynamic ads (M = 5.75, SD = 0.71).

4 Discussions

We found that the main effects of gender and ad presentation were insignificant, probably because the task was relatively easy and a simple visual search. The users opened the experimental prototype when The first thing users looked for was the location of task one. The overall operation is speedy visually, and there is no significant main effect difference for either gender or ad presentation. We then found a significant interaction between gender and ad presentation variables. It shows that males operate significantly faster between task performance during static ads than during dynamic ads. However, the opposite is true for females during operations. The possible reason is based on the fact that there is a significant visual attention allocation difference between males and females on the shopping interface, with females paying more visual attention to shopping information than males Hwang and Lee (2018), based on the selectivity model (N. Li et al. 2001), females are better at processing comprehensive information, while males are biased towards being good at selective processing of information (Meyers-Levy 1989). Thus, females are easily attracted to dynamic advertisements and do not feel distracted. In contrast, males have more robust attentional selection mechanisms when performing visual searches. Males are better at ignoring irrelevant visual cues and have more executive control than females (Deaner et al. 2007). However, dynamic advertisements are always in the peripheral vision of the eye.

Regarding the subjective scale studies of preference and trustworthiness levels, we found a significant main effect of gender, i.e., females generally rated them higher than males, possibly because images in ads are more visually appealing to females (Oyibo et al. 2016), while males will not have disproportionate visual attention to images popping in. In semi-structured interviews, we generally analyzed that males are less attracted to the presentation of dynamic ads. Very few males feel that constant dynamic ads affect the mood of the shopping experience or even the willingness to shop; however, females generally accept the presence of ads. Regarding the subjective scale of reasonableness, we found a significant main effect of ad presentation, i.e., the reasonableness of static ads is generally higher than dynamic ads, mainly because static ads play a role in regulating the interface atmosphere. Although continuous dynamic ads are attractive to a certain extent, as users become more familiar with the interface in the process of completing tasks, although the presentation speed of dynamic ads does not affect users' task completion to a greater extent, to a particular extent user may also become visually bored and tired of continuous dynamic ads. This is because users prefer static ads to a reasonable degree.

Regarding the subjective measure of satisfaction, we found a significant interaction between gender and ad presentation, i.e., males were significantly more satisfied with static ads than with dynamic ads. However, females were significantly less satisfied with static ads than with dynamic ads. This result is because studies based on gender differences in information processing have found inherent differences between males and females (Simon 2000), with females being more emotionally affected by the shopping environment than males (disastrous, 2000). When dynamic advertisements are presented, females are drawn to the information as they navigate through it, influencing female's positive emotions toward the interface. Females are more concerned with the interface's visual design than males, especially in terms of images, fonts, thematic elements, and colors within the mobile medium. However, males devote more attention to information usability (Oyibo et al. 2016). Males, in general, focus more on the content of the interface and less on the visual appearance (Richard et al. 2010). Moreover, constant dynamic ad presentation affects male users' sense of visual search fluency. Thus, we have evidence of a significant interaction between gender and the visual presentation of mobile shopping ads in viewership.

5 Conclusions

This study focused on the differences caused by gender and different types of ad presentation in users' task performance and their subjective evaluations. The main conclusions are as follows: (1) A significant interaction between gender and ad presentation was found in operational performance, with males significantly faster between task operational performance during static ads than between task operational performance during dynamic ads. However, the opposite was true for females at operation time; (2) Gender influenced users' preference and trustworthiness level, and females generally rated higher in preference and trustworthiness level than males; (3) The reasonable level of static ads was generally higher than that of dynamic ads; (4) There was a significant interaction between gender and ad presentation in the subjective measure of satisfaction. Males rated significantly higher in static ads than in dynamic ads. (5) There is a significant interaction between gender and ad presentation in the subjective measure of satisfaction, with males significantly more satisfied with static ads than dynamic ads. However, the opposite result is obtained for females.

Studying the effects of different types of ad presentation and gender on user experience can help us to reveal more insights about visual design and user attention allocation in-app interfaces. It is hoped that the results generated from this study can provide a reference for the visual design of ads in future shopping APPs, especially for gender-specific shopping APPs. In addition to the visual presentation of advertisements in mobile shopping APPs, factors such as presentation position and interaction mode can affect the user shopping experience. In addition to the differences between genders, different age groups may also have different user experiences; for example, shopping APPs designed for senior citizens should consider more factors, which we will continue to explore in depth in future studies.

References

Apeagyei, P.R.: The impact of image on emerging consumers of fashion. Int. J. Manage. Cases **13**(4), 242–251 (2011)

Baek, T.H., Morimoto, M.: Stay away from me. J. Advert. **41**(1), 59–76 (2012)

Cassioli, F.: Contents, animation or interactivity: neurophysiological correlates in App advertising (2019)

Chiu, Y.P., Lo, S.K., Hsieh, A.Y.: How colour similarity can make banner advertising effective: Insights from Gestalt theory. Behav. Inf. Technol. **36**(6), 606–619 (2017)

Cho, C. H., as-, U. O. T. A. A. I. A.: Why do people avoid advertising on the internet? J. Advertising **33**(4), 89–97 (2004)

Deaner, R.O., Shepherd, S.V., Platt, M.L.: Familiarity accentuates gaze cuing in women but not men. Biol. Let. **3**(1), 65–68 (2007)

Fu, Y., Zhang, D., Jiang, H.: Comparison of Users' and Designers' Differences in Mobile Shopping App Interface Preferences and Designs. In: Companion Proceedings of the The Web Conference 2018, pp. 31–32 (April 2018)

Gao, C., Zeng, J., Sarro, F., Lo, D., King, I., Lyu, M.R.: Do users care about ad's performance costs? Exploring the effects of the performance costs of in-app ads on user experience. Inf. Softw. Technol. **132**, 106471 (2021)

Gui, J., Mcilroy, S., Nagappan, M., Halfond, W G.: Truth in advertising: the hidden cost of mobile ads for software developers. In: 2015 IEEE/ACM 37th IEEE International Conference on Software Engineering, vol. 1, pp. 100–110. IEEE (May 2015)

Hu, T.W., Feng, C.S., Wen, Y.H.: Process design and evaluation of app interface for delivery platform-take foodpanda as an example. In: 2020 the 6th International Conference on Communication and Information Processing, pp. 1–6 (November 2020)

Hwang, Y.M., Lee, K.C.: Using an eye-tracking approach to explore gender differences in visual attention and shopping attitudes in an online shopping environment. Int. J. Hum.-Comput. Inter. **34**(1), 15–24 (2018)

Jackson, V., Stoel, L., Brantley, A.: Mall attributes and shopping value: differences by gender and generational cohort. J. Retail. Consum. Serv. **18**(1), 1–9 (2011)

Jung, H.S., Yoo, S.H.: Mobile Advertising UX Design Framework based on Users' Cognitive (2020)

Li, N., Kirkup, G., Hodgson, B.: Cross-cultural comparison of women students' attitudes toward the Internet and usage: China and the United Kingdom. Cyberpsychol. Behav. **4**(3), 415–426 (2001)

Lin, X., Qiu, Y., Chaveesuk, S., Chaiyasoonthorn, W.: The acceptance model of mobile shopping apps in China. In: 2021 3rd International Conference on Information Technology and Computer Communications, pp. 69–72 (2021, June)

Meyers-Levy, J.: Gender differences in information processing: A selectivity interpretation in cognitive and affective responses to advertising. W: P. Cafferata i A. Tybout (red.) Cognitive and affective responses to advertising (s. 219–260). Lexington. Exploring Differences in Males' and Females' Processing Strategy," Journal of Consumer Research **18**, 63–70 (1989)

Oyibo, K., Ali, Y.S., Vassileva, J.: Gender difference in the credibility perception of mobile websites: a mixed method approach. In: Proceedings of the 2016 Conference on User Modeling Adaptation and Personalization, pp. 75–84 (2016, July)

Palcu, J., Sudkamp, J., Florack, A.: Judgments at gaze value: gaze cuing in banner advertisements, its effect on attention allocation and product judgments. Front. Psychol. **8**, 881 (2017)

Patel, V., Pandit, R.: Impact of Quality of Unfamiliar Shopping App on Initial Trust Formation: A Moderated Mediation of Risk Attitude. Vision (2021). 0972262920984542

Patel, V., Das, K., Chatterjee, R., Shukla, Y.: Does the interface quality of mobile shopping apps affect purchase intention? an empirical study. Australasian Market. J. (AMJ) **28**(4), 300–309 (2020)

Resnick, M.L., Albert, W.: The influences of design esthetic, site relevancy and task relevancy on attention to banner advertising. Interact. Comput. **28**(5), 680–694 (2016)

Richard, M.O., Chebat, J.C., Yang, Z., Putrevu, S.: A proposed model of online consumer behavior: assessing the role of gender. J. Bus. Res. **63**(9–10), 926–934 (2010)

Schreiner, T., Rese, A., Baier, D.: Multichannel personalization: Identifying consumer preferences for product recommendations in advertisements across different media channels. J. Retail. Consum. Serv. **48**, 87–99 (2019)

Simon, S.J.: The impact of culture and gender on web sites: an empirical study. ACM SIGMIS Database: The Database for Advances in Information Systems **32**(1), 18–37 (2000)

What Do User Experience Professionals Discuss Online? Topic Modeling of a User Experience Q&A Community

Langtao Chen(✉) (iD)

Department of Business and Information Technology,
Missouri University of Science and Technology, Rolla, MO 65409, USA
`chenla@mst.edu`

Abstract. Questioning and answering (Q&A) communities have been widely used by user experience (UX) professionals to exchange knowledge online. As online content has been increasingly generated by professional users in the community, it becomes infeasible for human experts to manually analyze those discussions. Thus, automatic and intelligent analysis of the online content generated by UX professionals is needed to understand how UX knowledge is created and maintained by professionals in online communities. This research offers a comprehensive understanding of user-generated content in a UX Q&A community through topic modeling, an intelligent text-mining approach for discovering hidden topics from textual documents. Specifically, this research identifies 40 important latent discussion topics from a dataset containing 31,314 questions and 80,579 answers posted by UX professionals. Those topics are classified into 8 major categories, followed by popularity and content cohesion analysis of those categories and their mapping to a well-established design thinking process model. Overall, this research contributes a systematic exploration of user-generated content by UX professionals in a Q&A community and highlights opportunities for UX researchers and practitioners.

Keywords: User Experience (UX) · Q&A Communities · User-Generated Content · Topic Modeling · Latent Dirichlet Allocation

1 Introduction

Online questioning and answering (Q&A) communities such as Yahoo! Answers, Zhihu, and sites in the Stack Exchange network have been popularly used by people to share and exchange knowledge in particular domains. Like other professionals, user experience (UX) designers have adopted Q&A communities to discuss UX-related topics and solve problems encountered in UX design [1]. A user can submit a question regarding a specific UX topic. Other users with relevant expertise or knowledge can provide answers to the question. Such professional Q&A communities have become an innovative form of community of practice (CoP) for producing and disseminating professional knowledge [2]. Online Q&A communities are especially important for shaping and transforming

© Springer Nature Switzerland AG 2023
F. Fui-Hoon Nah and K. Siau (Eds.): HCII 2023, LNCS 14038, pp. 365–380, 2023.
https://doi.org/10.1007/978-3-031-35969-9_25

the volatile practice of UX design that lacks a coherent body of disciplinary knowledge and a clear path for beginners to become experienced UX professionals [2].

As an academic field, human-computer interaction (HCI) has seen numeric studies of UX problems and topics such as UX design and development, UX evaluation, user training, user cognitive belief and behavior, and user attitude [3]. Although HCI is a well-established academic field addressing all sorts of important UX design issues, we know little about UX practitioners and their problems or challenges in UX design. The availability of a large volume and variety of user-generated content related to UX practice in social media sites provides an excellent opportunity for both UX practitioners and researchers to understand how practical UX knowledge is produced and disseminated on online platforms. Specifically, in the setting of Q&A communities dedicated to practical UX knowledge exchange, it is crucial to know: (1) what do UX practitioners discuss online? (2) what are the primary problems or issues they have encountered in UX design practice? and (3) what is the nature of those online discussions? Answering those questions has important implications for both UX practice and research.

The increasing use of Q&A communities has led to tremendous growth in the volume and variety of user-generated content on such social media platforms. Therefore, traditional manual content analysis approaches become intractable given the large scale and variety of user-generated content in online communities [4]. Consequently, automatic content analysis based on natural language processing and text mining methods is needed. This research applies latent Dirichlet allocation (LDA) method [5] to model topics embedded in the user-generated content posted to a UX Q&A community. A set of comprehensive analyses based on the identified topics are conducted to comprehensively understand user online discussions related to UX.

Analyses conducted in this research contribute a comprehensive understanding of how the body of UX knowledge is generated and maintained by UX professionals in online communities. In doing so this research addresses the call by the HCI community for increasing relevance to practice [6]. Findings of this research offer insights for both UX practitioners and researchers.

This paper is organized as follows. The next section explains the LDA-based topic modeling method used to identify latent topics from a textual dataset collected from a Q&A community. Then, subsequent analyses of the latent topics extracted from LDA are presented in Sect. 3. Section 4 concludes the current work and discusses future directions.

2   LDA-Based Topic Modeling Method

Latent Dirichlet allocation (LDA) was used to extract latent topics from user-generated content in a community-based Q&A website for UX designers and professionals. The following subsections explain the details of the topic modeling method.

2.1   Research Setting and Data Collection

Data were collected from User Experience Stack Exchange, a US-based Q&A website hosted in the Stack Exchange network for UX designers and human-computer interaction professionals. Users can register in the web-based online community for free. A typical process of questioning and answering starts with a user submitting a UX question to the community. Then other users can provide answers to the question. Users involved in the problem-solving process can discuss issues and collaborate in the problem-solving process by submitting comments to both the original question and its answers. Questions and answers can also be voted down or voted up by other peers in the community. Those questions with the highest up votes are displayed on top of the question list. If a provided answer solves the question, the original question poster can accept the answer. Figure 1 shows a sample question posted to the Q&A community.

Fig. 1. A sample question asked in the community.

The raw dataset collected from the Q&A community contains 31,314 questions and 80,579 answers posted from September 22nd, 2008 to November 30th, 2022. Figure 2 presents monthly activities in the community during the data collection period. The community was officially launched in September 2008 and then has maintained a relatively high-level user participation since August 2010. However, a general decreasing trend of content posting activities can be seen from August 2015 to November 2022.

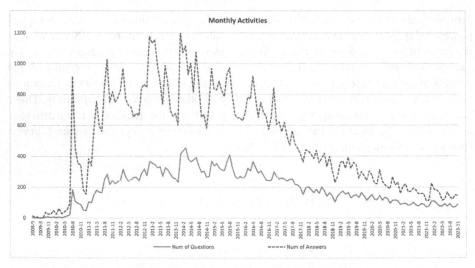

Fig. 2. Monthly activities in the community.

2.2 Data Preprocessing

Both questions and answers posted to the community were analyzed to find thematic topics of user discussion. Since a discussion thread containing a question and its answers focuses on a particular topic, topic modeling was conducted at the thread level. Thus, a discussion thread was treated as an individual document containing a sequence of words or terms used in the question and its answers. As a general process of text mining, data need to be preprocessed before fed into the topic modeling algorithm. Specifically, data preprocessing used in this study includes the following steps:

- Remove code tags and HTML tags that are only used to format the textual content;
- Filter out all punctuation characters contained in documents;
- Filter out all numeric digits as well as possible leading symbols (i.e., "+" or "−");
- Filter out all terms with less than 3 characters;
- Filter out all stop words such as "a", "an", "the", and "in" that only have little meaning in English;
- Convert all terms to lower case;
- Reduce all terms to their stem by using the Snowball stemming algorithm. For example, terms "using" and "used" were converted to their common stem "use."

2.3 Latent Dirichlet Allocation (LDA)

The preprocessed data were then fed into the latent Dirichlet allocation (LDA) algorithm to extract thematic topics discussed in the textual content. LDA was initially developed by Blei et al. [5] as a generative probabilistic model for modeling text corpora. LDA has been widely used by prior studies to discover latent topics in a variety of settings such as consumer online reviews [7] and consumer complaints [8]. Figure 3 illustrates the LDA-based topic modeling method.

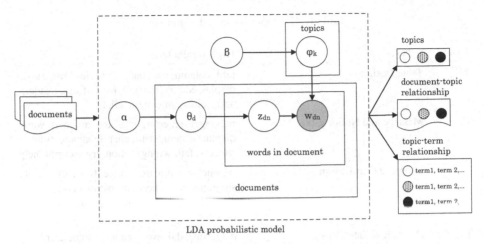

LDA probabilistic model

Fig. 3. Topic modeling using latent Dirichlet allocation (LDA).

As shown in Fig. 3, w_{dn} ($d = 1, 2, ..., D, n = 1, 2, ..., N$) is the observed word (the n-th word in the d-th document), which is generated from other components including word distribution for topic k (φ_k), per-word topic assignment z_{dn}, and topic distribution per document θ_d. α and β are Dirichlet hyperparameters used to generate θ_d and φ_k respectively. A set of documents is the input of the LDA model, which produces three outputs including: (1) a list of topics; (2) a document-topic relationship table showing the probability of each document belonging to each topic; and (3) a topic-term relationship table showing the terms and their weight per topic.

The LDA analysis requires the specification of k, the number of topics. A too small k usually results in themes that are too general and thus lack capability needed to reveal specific topics, whereas a too large k often leads to themes that do not have a sufficient level of abstraction. Given the exploratory nature of this analysis, the number of topics was determined by a trial-and-error procedure, which found that a 40-topic solution ($k = 40$) led to a satisfactory result that includes a comprehensive set of important and meaningful UX topics and at the same time maintains a sufficient level of abstraction that can summarize broader UX discussion themes.

3 Data Analysis and Results

3.1 Latent Topics Extracted

Table 1 shows the 40 topics identified from LDA and their high-loading terms. Each topic was labeled based on both its high-weight terms (in format of stems) and high-probability documents associated with the topic. For example, topic T1 was named as *"Table Design"* because its high weight terms (refer to the first row in Table 1) and documents with a high probability of belonging to this topic both described UX issues related to data table design.

All extracted thematic topics are about UX related issues in computer systems, except for topic T34 that is about UX issues related with cars. To focus on UX issues in computer systems, topic T34 was removed from further analysis.

Table 1. UX Topics extracted from LDA.

Topic	Label	High-weight terms
T1	Table Design	tabl, column, row, data, sort, view, user, grid, displai, edit, cell, detail, list, inform, header, add, solut, name, item, click
T2	Message and Notification	messag, user, error, notif, inform, alert, warn, displai, system, statu, chat, feedback, time, success, fail, wrong, action, try, exampl, help
T3	Shopping Cart Design	product, custom, user, price, bui, cart, servic, shop, purchas, checkout, payment, page, offer, detail, item, inform, option, pai, store, process
T4	Datetime Related Design	date, time, dai, event, month, hour, user, calendar, week, start, displai, minut, rang, picker, format, schedul, period, exampl, timelin, current
T5	Usability Survey	question, answer, rate, vote, survei, system, review, score, student, star, scale, cours, respons, user, feedback, teacher, class, school, posit, rank
T6	Multimedia System Design	video, game, plai, player, sound, control, audio, music, watch, start, volum, youtub, song, paus, movi, time, screen, camera, voic, record
T7	Page Loading Design	user, time, load, progress, anim, process, bar, wait, indic, step, complet, updat, task, chang, refresh, delai, displai, screen, start, page
T8	UX Design Resources	content, site, websit, articl, http, page, inform, web, read, help, look, exampl, post, link, user, blog, new, topic, book, design
T9	Report and Dashboard Design	data, inform, report, user, record, dashboard, displai, databas, detail, applic, system, etc., employe, time, collect, info, store, import, export, look
T10	UX Prototyping	design, code, develop, prototyp, wirefram, look, tool, materi, http, creat, interact, compon, style, element, us, visual, web, html, pattern, build

(*continued*)

Table 1. (*continued*)

Topic	Label	High-weight terms
T11	Names and Tags	name, term, tag, word, call, exampl, refer, label, mean, us, person, titl, describ, type, context, peopl, domain, specif, phrase, suggest
T12	UX Research	design, user, experi, interfac, peopl, usabl, research, interact, system, exampl, look, question, mean, human, learn, effect, understand, time, answer, studi
T13	User Account and Authentication	user, password, account, login, log, sign, secur, email, regist, usernam, registr, site, creat, requir, access, facebook, enter, option, rememb, authent
T14	User Input	keyboard, kei, touch, shortcut, mous, press, screen, devic, hand, us, type, move, finger, control, input, gestur, time, button, interfac, enter
T15	Color Design	color, colour, background, red, contrast, white, green, text, light, blue, black, dark, chang, visual, us, look, grei, design, yellow, theme
T16	Social Media Site Design	user, peopl, site, comment, time, person, feel, experi, share, reason, post, facebook, answer, social, try, actual, featur, question, us, ad
T17	Mobile Screen Design	mobil, size, screen, devic, design, desktop, respons, width, resolut, site, layout, displai, websit, pixel, tablet, browser, web, version, phone, us
T18	User Selection Design	select, option, user, button, checkbox, dropdown, radio, choic, control, list, default, check, box, click, toggl, choos, item, set, switch, multipl
T19	Email Related Design	email, card, address, user, contact, send, phone, inform, confirm, person, account, credit, mail, e-mail, servic, provid, receiv, client, compani, code
T20	Menu and Navigation Design	menu, navig, bar, item, top, option, main, user, nav, hamburg, action, bottom, level, sidebar, design, toolbar, dropdown, tab, content, left

(*continued*)

Table 1. (*continued*)

Topic	Label	High-weight terms
T21	Text Design	text, font, read, charact, word, line, letter, space, titl, readabl, sentenc, look, style, size, print, write, us, type, format, paragraph
T22	Form Design	field, form, user, input, enter, fill, text, step, requir, valid, type, submit, label, data, inform, edit, box, complet, exampl, option
T23	Layout Design	left, top, space, look, design, layout, bottom, align, element, line, posit, visual, read, vertic, label, screen, text, horizont, content, place
T24	Page Link Design	page, link, user, site, websit, click, navig, url, home, browser, logo, web, content, land, redirect, exampl, visit, breadcrumb, look, main
T25	Mobile App Design	app, mobil, screen, user, android, io, applic, devic, design, tap, phone, guidelin, appl, swipe, view, iphon, us, googl, platform, web
T26	Slider Design	valu, slider, rang, user, calcul, amount, displai, exampl, unit, set, chang, control, us, percentag, scale, measur, total, numer, input, bar
T27	File and Document	file, window, applic, document, browser, user, folder, upload, web, version, softwar, command, copi, system, pdf, program, download, instal, mac, support
T28	Icon Design	icon, indic, text, arrow, label, symbol, mean, repres, us, look, visual, exampl, mark, circl, button, imag, suggest, action, help, standard
T29	User Change and Editing	user, chang, set, edit, applic, featur, system, allow, creat, option, task, function, add, help, current, view, interfac, manag, approach, app
T30	Tab and Scrolling Page Design	tab, scroll, page, user, content, section, view, screen, panel, top, navig, bottom, accordion, horizont, pagin, expand, bar, scrollbar, displai, vertic

(*continued*)

3.2 Latent Topic Categorization

The extracted topics except T34 were further grouped into eight major categories, as summarized in Table 2. The number of online threads with document-topic probability $>= 0.2532$ was also calculated for each topic. The threshold value (0.2532) was chosen

Table 1. (*continued*)

Topic	Label	High-weight terms
T31	User Search	search, filter, result, user, list, type, box, option, queri, googl, look, match, categori, suggest, criteria, exampl, displai, select, provid, sort
T32	Hierarchical Display	categori, level, chart, tree, data, graph, visual, hierarchi, exampl, parent, node, inform, structur, displai, repres, bar, child, type, line, map
T33	UX Profession and Career	design, product, team, project, develop, compani, user, process, client, persona, experi, busi, custom, help, job, creat, manag, requir, start, understand
T34	User Experience Related with Cars	car, door, peopl, driver, switch, drive, light, machin, time, elev, devic, floor, power, system, control, reason, vehicl, seat, road, requir
T35	Mouse Click/Motion	click, user, hover, drag, mous, interact, element, map, tooltip, button, cursor, object, move, drop, clickabl, action, zoom, text, touch, indic
T36	Language and Region	languag, countri, locat, english, translat, local, code, user, citi, flag, region, map, cultur, list, currenc, name, address, exampl, format, intern
T37	Button Design	button, action, user, save, click, delet, modal, close, dialog, chang, cancel, option, confirm, edit, popup, press, window, undo, add, screen
T38	Item and List Design	item, list, user, download, creat, sourc, mockup, add, bmml, wirefram, balsamiq, select, ad, option, displai, solut, suggest, view, top, sort
T39	Image Design	imag, access, text, photo, user, screen, reader, pictur, us, content, element, upload, disabl, visual, wcag, thumbnail, read, requir, provid, galleri
T40	Usability Test	test, user, usabl, task, research, time, particip, question, feedback, studi, result, peopl, measur, us, data, product, perform, try, method, complet

to retain $1/k$ of the document-topic probabilities for a k-factor solution such that each thread would just load on one topic on average [9, 10].

Table 2. Categorization of topics.

Category	Topic	Label	Thread count
1. UX Design Guide and Research	T8	UX Design Resources	549
	T12	UX Research	1,343
2. UX Prototyping and Evaluation	T10	UX Prototyping	850
	T5	Usability Survey	347
	T40	Usability Test	1,144
3. UI Page and Layout Design	T22	Form Design	1,571
	T23	Layout Design	965
	T30	Tab and Scrolling Page Design	1,023
	T7	Page Loading Design	690
	T24	Page Link Design	932
	T1	Table Design	796
	T32	Hierarchical Display	647
4. UI Element Design	T20	Menu and Navigation Design	766
	T21	Text Design	577
	T28	Icon Design	631
	T39	Image Design	395
	T37	Button Design	1,424
	T38	Item and List Design	575
	T21	Slider Design	527
	T18	User Selection Design	1,339
	T11	Names and Tags	382
	T15	Color Design	779
	T4	Datetime Related Design	627
	T36	Language and Region	451
5. User Input	T14	User Input	431
	T29	User Change and Editing	1,172
	T35	Mouse Click/Motion	736
6. UX Design for Mobile	T25	Mobile App Design	836
	T17	Mobile Screen Design	729

(continued)

Table 2. (*continued*)

Category	Topic	Label	Thread count
7. UX Design for Specific Functions	T13	User Account and Authentication	1,111
	T2	Message and Notification	757
	T19	Email Related Design	557
	T3	Shopping Cart Design	733
	T6	Multimedia System Design	196
	T16	Social Media Sit Design	1,439
	T9	Report and Dashboard Design	152
	T27	File and Document	574
	T31	User Search	908
8. UX Profession and Career	T33	UX Profession and Career	1,315

As shown in Table 2, the top six most popularly discussed UX topics include:

- *Form Design* (1,571 online threads)
- *Social Media Site Design* (1,439 online threads)
- *Button Design* (1,424 online threads)
- *UX Research* (1,343 online threads)
- *User Selection Design* (1,339 online threads)
- *UX Profession and Career* (1,315 online threads)

Not surprisingly, those common technical issues in UX design such as *Form Design*, *Button Design*, and *User Selection Design* are hot topics in the Q&A community. Many UX designers and professionals encounter those problems very frequently in their routine work and thus often need help regarding those design issues. The popular adoption of social media sites such as Facebook, Instagram, Twitter, and various online forums also leads to many discussions on *Social Media Site Design* in the Q&A community.

Interestingly, users in the community also oftentimes discuss UX Research. The academic field of human-computer interaction (HCI) has a tradition to investigate practical UX design issues [11]. Thus, it is natural that UX professionals in the online community often discuss UX research to seek guidelines for UX design.

In addition to those technical UX design issues, users also share knowledge regarding UX profession and career. Discussions such as "As a programmer, how do I move into User Experience Design?" have triggered insightful conversations on UX as a professional career. Those conversations can provide useful job and education/training advice for UX professionals, especially those beginners.

Figure 4 shows the dynamic change of the percentage of discussion threads in each topic category. In the early stage of the Q&A community in 2009, discussions more focused on general topics including *UX Design Guide and Research*, *UX Prototyping and Evaluation*, and *UX Profession and Career*. However, since 2010, the primary topics discussed in the community have been more technical issues including *UI Element Design*, *UI Page and Layout Design*, and *UX Design in Specific Functions*.

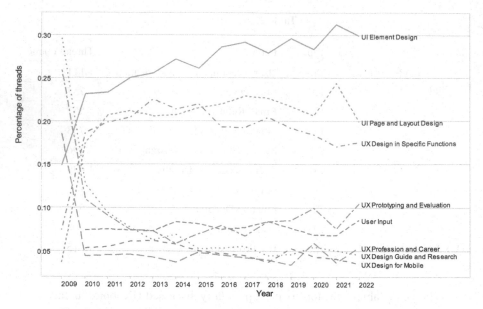

Fig. 4. Percentage of discussion threads in each topic category across years.

3.3 Popularity and Content Cohesion of Topic Categories

This study considers two important measures at topic category level including population and content cohesion. Popularity of a topic category is defined as the number of online discussion threads that have a high probability of belonging to this topic category. Following the similar method used by Chen et al. [10], content cohesion is defined as the extent to which semantics of latent topics are common across online discussion threads. Specifically, content cohesion of a topic category is measured as the average probabilities of discussion threads belonging to the topic category. Figure 5 compares the popularity and content cohesion of eight topic categories. We notice that topic categories with a higher level of content cohesion tend to have a lower level of popularity.

Based on the comparison depicted in Fig. 5, all eight topic categories were classified into the following three major groups.

- **Group 1: High popularity but low content cohesion**. The most popular topics discussed in the Q&A community are technical issues including *UI Element Design* (total 8,069 discussion threads), *UI Page and Layout Design* (total 6,373 threads), and *UX Design for Specific Functions* (total 6,061 threads). However, those most popular topics have a relatively low level of content cohesion (0.382, 0.372, and 0.379 for *UI Element Design*, *UI Page and Layout Design*, and *UX Design for Specific Functions* respectively).
- **Group 2: Low popularity and low content cohesion**. This group is comprised of three topic categories, namely *User Input*, *UX Design Guide and Research*, and *UX Design for Mobile* (total number of threads is between 1,545 and 2,306, content cohesion ranges from 0.365 to 0.383). We notice that the number discussions on *UX Design Guide and Research* is relatively small, suggesting that UX design guidelines

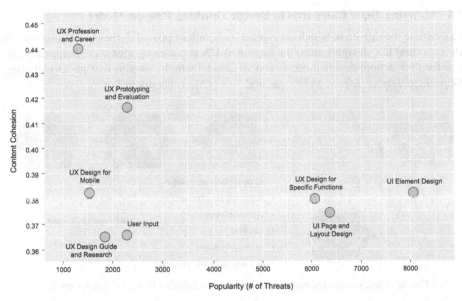

Fig. 5. Popularity and content cohesion of topic categories.

and research related issues might be more academic and thus might be less relevant to UX design practices. Given the wide use of mobile devices, it is counter intuitive that the online community does not have many discussions related to UX design for mobile devices.

- **Group 3: Low popularity but high content cohesion**. Within this group, *UX Profession and Career* has the highest level of content cohesion (0.44) but the lowest level of popularity (total 1,315 discussion threads). This suggests that user discussions related to UX as a profession or career have intensively addressed the core theme but such discussions are relatively rare. Although user discissions on *UX Prototyping and Evaluation* have a moderate level of content cohesion (0.416), there are also not many such discussions (total 2,295 threads) in the community.

What does the above analysis suggest for the UX Q&A community? It is evident that this Q&A community has become an excellent platform for producing and disseminating those more technical UX design knowledge such as *UI Element design, UI Page and Layout Design*, and *UI Design for Specific Functions*. However, the community could encourage users to discuss more issues beyond those traditional technical UI design. Topics such as UX design guideline and research, UX prototyping and evaluation methodologies, and UX job and career are great issues that can potentially enrich discussions in the online community. This might be helpful to increase the diversity of online discussions in the community and thus hopefully help maintain the continuity and sustainability of the community.

3.4   Mapping Topic Categories to Design Thinking Process Model

To evaluate the degree to which user-generated content posted to the community covers all important UX design issues, the identified UX topic categories were mapped to the Stanford's d. school design thinking process model which contains five stages including *Empathize*, *Define*, *Ideate*, *Prototype*, and *Test* [12], as shown in Fig. 6.

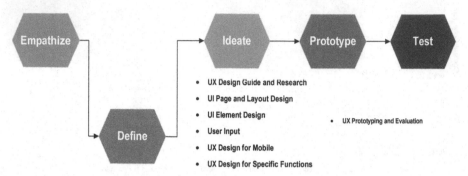

Fig. 6. Topic categories mapped to the five-stage design thinking process model.

As Fig. 6 presents, current topics discussed in the Q&A community are not evenly distributed across the five design stages. Most issues discussed in the community occur in the *Ideate* stage of the UX design thinking process. Technical design issues such as *UX Design Guide and Research, UI Page and Layout Design, UI Element Design, User Input, UX Design for Mobile*, and *UX Design for Specific Functions* are all about coming up with creative solutions to solve relevant problems. There are also discussions on *UX Prototyping and Evaluation* (such as *Usability Survey* and *Usability Test*) mainly encountered in the *Prototype* and *Test* stages.

However, discussions in the Q&A community have not comprehensively addressed issues in the *Empathize* and *Define* stages, probably because those issues are very specific to a particular system setting and thus might be difficult to be meaningfully discussed online with people who usually do not know each other offline. Such issues might be more suitable to be thoroughly discussed with design audience or team members. Hence, there are opportunities for reshaping discussions in the Q&A community and expanding the scope of online discussions to cover important issues in the *Emphasize* and *Define* stages of the design thinking process model.

4   Discussion

Online Q&A communities have greatly enhanced the efficiency and effectiveness of professional problem-solving and knowledge sharing by connecting people with specialized expertise free from temporal and spatial constraints. An automated and intelligent approach to analyzing user-generated content posted in Q&A communities is needed, especially when a large volume and variety of user-generated content has been accumulated through a long time range. The LDA-based topic modeling method applied in this study has been found useful in discovering and tracing emerging topics discussed in online communities.

This research contributes a systematic investigation of user-generated content in a UX Q&A community. Specifically, this research has identified 40 important online discussion topics in the Q&A community and classified them into 8 major categories. Then this study compares popularity and content cohesion of those topic categories. Furthermore, topic categories are mapped to a five-stage design thinking process model to assess whether users have discussed all important issues encountered in UX design processes.

Findings of this research offers practical and theoretical implications for understanding important and emerging UX topics discussed in large scale online communities. The latent topics and topic categories identified in this study can be used to build or refine an ontology or taxonomy for effectively organizing or tagging user-generated content in online UX knowledge repositories. HCI researchers can pay more attention to research that can bridge the gap between popular practical topics discussed by UX professionals and research topics addressed traditionally by the academic community. Focusing on those topics that have been frequently discussed by UX practitioners not only helps increase the relevance of UX research but also brings innovative ideas from practice to enhance the value and contribution of UX research. The HCI community shall spend effort in aligning research with UX practice to further consolidate and expand the UX knowledge base.

Future work can extend the current research by incorporating more UX discussion content from other social media platforms. This can reveal an even more comprehensive and complete view of what UX professionals discuss on online platforms. This research can also be extended to conduct higher level analyses such as identifying the evolution of important discussion topics, detecting leaders of specific discussion topics, and recommending experts for specific discussions. This can yield even richer insights into designing better user experiences that help address today's grand challenges of effective user experience and human-computer interactions. In addition, future work can apply the similar topic modeling approach to other social media platforms such as online health communities [4, 13] to model large-scale user-generated content in other domains.

References

1. Chen, L., Baird, A., Straub, D.: Why do participants continue to contribute? Evaluation of usefulness voting and commenting motivational affordances within an online knowledge community. Decis. Support Syst. **118**, 21–32 (2019)
2. Kou, Y., Gray, C.M., Toombs, A.L., Adams, R.S.: Understanding social roles in an online community of volatile practice: a study of user experience practitioners on reddit. ACM Trans. Soc. Comput. **1**, 1–22 (2018)
3. Zhang, P., Li, N., Scialdone, M., Carey, J.: The intellectual advancement of human-computer interaction research: a critical assessment of the MIS literature (1990–2008). AIS Trans. Hum.-Comput. Interact. **1**, 55–107 (2009)
4. Chen, L., Baird, A., Straub, D.: Fostering participant health knowledge and attitudes: an econometric study of a chronic disease-focused online health community. J. Manag. Inf. Syst. **36**, 194–229 (2019)
5. Blei, D.M., Ng, A.Y., Jordan, M.I.: Latent Dirichlet allocation. J. Mach. Learn. Res. **3**, 993–1022 (2003)

6. Benbasat, I.: HCI research: future challenges and directions. AIS Trans. Hum.-Comput. Interact. **2**, 1 (2010)
7. Kwon, H.-J., Ban, H.-J., Jun, J.-K., Kim, H.-S.: Topic modeling and sentiment analysis of online review for airlines. Information **12**, 78 (2021)
8. Bastani, K., Namavari, H., Shaffer, J.: Latent Dirichlet allocation (LDA) for topic modeling of the CFPB consumer complaints. Expert Syst. Appl. **127**, 256–271 (2019)
9. Sidorova, A., Evangelopoulos, N., Valacich, J.S., Ramakrishnan, T.: Uncovering the intellectual core of the information systems discipline. MIS Q. **32**, 467-A420 (2008)
10. Chen, L., Baird, A., Straub, D.: An analysis of the evolving intellectual structure of health information systems research in the information systems discipline. J. Assoc. Inf. Syst. **20**, 1023–1074 (2019)
11. Hevner, A., Zhang, P.: Introduction to the AIS THCI special issue on design research in human-computer interaction. AIS Trans. Hum.-Comput. Interact. **3**, 56–61 (2011)
12. Hasso Platner Institute of Design at Standford: An introduction to design thinking process guide (2010)
13. Chen, L., Baird, A., Straub, D.: A linguistic signaling model of social support exchange in online health communities. Decis. Support Syst. **130**, 113233 (2020)

The Study of User Experience Within Advertising in Virtual Reality

Sara Dieter⬛, Ben Mark(✉)⬛, Matt Childress⬛, Anthony Anderson⬛,
Andrea Mower⬛, and Max Harberg⬛

PricewaterhouseCoopers, New York, NY 10017, USA
US-ASR-MetaXRInnovationHub@PwC.onmicrosoft.com, ben.mark@pwc.com

Abstract. The growing adoption of virtual reality (VR) introduces marketing campaigns that are unlike traditional advertising techniques. The immersion of VR technology allows advertisers to create simulated interactions for products with a more significant impact than two-dimensional (2D) alternatives. Participants in this study underwent a series of games, embedded with advertisements, to better understand user sentiment between static and immersive advertisements. User surveys spatial data and time-based data show that users prefer and engage more with immersive advertisements rather than 2D banners which are more commonly represented in Web 2.0 environments. Gamifying brand interactions and rewarding user engagement yielded positive user feedback. In addition, this paper highlights key concerns around data security and privacy users may have with advertising marketing agencies should consider. This study outlines potential key mechanisms in driving experiential marketing tactics for virtual environments and increasing advertisement conversions.

Keywords: Immersive advertising · Customer engagement · Brand recognition · Product placement · Gamification · Virtual reality · Incentivized ads · Customer behavior research

1 Introduction

Virtual reality (VR) has been around since 1956 when Morton Heilig created the first VR machine called Sensorama, which rendered 360 views from a computer [1]. In recent years, advancement in the form factor and user experience, combined with increased availability and a dip in the price point, has accelerated consumer adoption. Today nearly 14.8 million Quest headsets have been sold, primarily for gaming [2]. Meta, the leading device manufacturer, positions its device alongside the concept of the "metaverse," the next digital platform for customer engagement. While enterprise-level applications for the metaverse are emerging, most VR device usage today is for gaming. It is important

© Springer Nature Switzerland AG 2023
F. Fui-Hoon Nah and K. Siau (Eds.): HCII 2023, LNCS 14038, pp. 381–400, 2023.
https://doi.org/10.1007/978-3-031-35969-9_26

to note that virtual reality (VR) and the metaverse often go hand in hand. Although this study focused mainly on VR, many concepts can be applied to the metaverse as well.

Commercial companies focusing on digital are signifying their preparation for leveraging the metaverse. Publishers from Unsupervised AI stated that 60% of their respondents admitted they do not know the metaverse very well, or at all, but still plan to launch metaverse marketing strategies this year [3]. These companies are exploring new enterprise applications and revenue opportunities in the metaverse and are seeking to identify how the metaverse may disrupt their industries.

Today several groups are experimenting with using VR for marketing and sales. Oculus (Meta) has previously tested using in-headset ads [4]. The game engine Unity has advertised new tools for easy implementation of in-game ads [5]. In more formal contexts, academic researchers have been studying advertising in VR. Anna Borawska and her fellow researchers look at the effectiveness of using virtual reality games for marketing social good. Furthermore, researchers led by Mariano Alcañiz present a framework for presenting research in virtual marketing in a more rigorous way[6]. Joshua Lupinek and fellow researchers funded through the Feliciano School of Business have proposed a set of factors to leave users with a favorable attitude toward in-game advertisements [7]. PwC's team of Extended Reality (XR) researchers and metaverse specialists wanted to contribute to this body of research.

In this paper, the research team utilizes the framework and considerations presented by Alcaniz and Lupinek to describe their research on in-game advertising in VR. The paper starts by classifying types of commercial advertising in VR gaming and provides several examples from the market research conducted by the researchers. Then the Alcaniz framework is used to describe the VR experience made for this study. Next, the paper explores how the advertisements align with the recommendations for congruity proposed by Lupinek. The research team shares observations of how human subjects engaged with our in-game advertisements and discusses results that point to validation of user preference results.

2   Industry Analysis

Before interpreting the study data, it is essential to understand innovation in the advertising industry from a larger scale. Brands have come a long way from McDonald's Happy Goggles released in 2016 [8]. The team identified three main familiar approaches brands take when engaging with consumers in virtual reality: subliminal advertising, product utilization, and brand expansion. By putting a metaverse lens on these advertising strategies, brands can explore the convergence of commerce, content, and entertainment.

As seen in video games, movies, TV, experiential, radio, and other forms of advertising, subliminal advertising is when the brand engages with consumers subconsciously. With the required technology and necessary digital production, this marketing strategy can be replicated in a virtual reality environment with a hyper-state of immersion providing the opportunity for an even stronger impression on the user. An example of this would be "Beat Saber," one of the top-selling Oculus games of all time [9]. Like Guitar Hero's effect on the music industry, Beat Saber [re]introduces players to new and old songs, giving artists an additional outlet to reach new and existing audiences. This technique triggers a subconscious psychological recall when looking for new music and products. There is no limit to the continual flow of new free and premium music packages for players.

Product placement and utilization is another strategy with significant opportunities in virtual reality. A 3D, 360-degree immersive environment gives players a more realistic sense of interacting with the brand. Notable racing games, such as "DiRT Rally" and "Live For Speed," are prime examples of players seamlessly using products in-game. Players select from a lineup of cars ranging from common to luxury brands in order to race against others in time trials. With a heightened VR immersive experience, players have a greater sense of driving the car, thus leading to subconscious brand recognition and the desire to explore the brand when purchasing vehicles in real life.

Expanding a product outside its primary medium is not a new concept for brands. Brands develop VR gaming experiences to enable continued interaction and build a deeper connection with the brand. Vader Immortal, another highly rated Oculus game, is an example of this strategy. VR technology allows players to transport directly into the Star Wars Universe. Experiences like these can increase loyalty: keeping fans excited for new movies, content, and more.

Brands use these strategies together for a more effective and immersive experience. Taking these strategies into consideration, the research team conducted a user study to uncover how users prefer to interact with brands using VR.

3 Methodology

3.1 Virtual Reality Experience

Using the Virtual Experience in Marketing (VEM) characterization introduced by Mariano Alcaniz, this research was conducted using solely virtual reality technology given its high relative level of adoption and technical maturity as compared to augmented reality and mixed virtuality. The research team created a virtual reality archery game and deployed it to a commonly used VR headset for testing. This game included four levels of play, with each level increasing in difficulty. For each level, the player's objective is to hit a certain number of targets using a limited number of arrows. New challenges are introduced throughout the game (i.e., timed pop-up targets, flying drone targets). Players can earn coins by hitting targets which can then be used to buy in-game items such as arrows and bows. In addition to the game, the development team created a fake soft drink brand, Splat. Splat branding was used for all embedded advertising experiences within the VR app (Fig. 1).

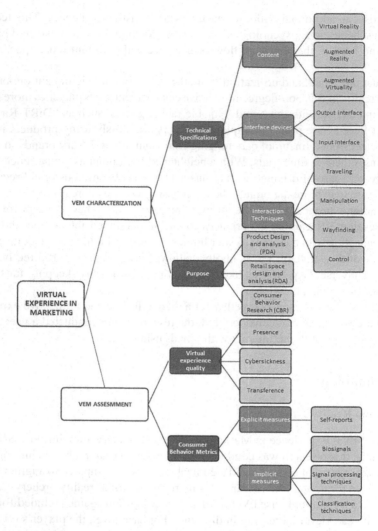

Fig. 1. An image of the VEM framework referenced from Alcaniz's work. "Virtual Experience in Marketing" bifurcates into "VEM Characterization" and "VEM Assessment". Branches under "VEM Characterization" include "Technical Specification" and "Purpose". Branches of "VEM Assessment" include "Experience Quality" and "Consumer Behavior Metrics".

Two versions of the VR application were developed; one was gamified and the other non-gamified. The gamified version allowed participants to earn coins by enabling ads. After every level, users are presented with the option to enable 2D billboard ads for Splat in the archery arena, explore an immersive Splat experience, or decline ads altogether. If users chose to enable the 2D billboard ads, they were awarded in-game coins, and the 2D billboards were shown in the bow and arrow game arena as users played the next level. If users chose the immersive ad, they were transferred to a speakeasy Splat-branded bar, where they could find hidden interactions and earn in-game coins for each interaction. The non-gamified version of this app forced users into an ad after each level, and they earned no coins. For the non-gamified version, the type of ad they would enable after each level (2D billboard or fully immersive) was randomized, but the same type of ad was never repeated. The team decided to take a randomized approach to the ad type so there was a higher chance of getting representative data from both ad types. For example, if users were more likely only to pass one level, and that level was only showing 2D ads, then there would have been little supporting data for immersive ads.

As noted above, the research team provided two different types of advertising: 2D billboard ads and the fully immersive ad. The 2D billboard ads consisted of three different sets of three billboards that would appear in the bow and arrow arena as static images that allow users to observe the ads while playing. The three sets of 2D billboards represented different art styles, one with real-world humans with the Splat soda, one with game art familiar to the game, and finally, one with just Splat branding (Figs. 2 and 4).

Fig. 2. Three images of 2D Billboard Splat ad consisting of real-world humans. One is a woman in shorts sitting on a large splat can, a man on a rearing horse next to a Splat logo, and two girls clinking drinks in front of a Splat logo.

Fig. 3. Three images of 2D billboard Splat ad consisting of just the Splat Soda. Additional text in these ads includes: "It also removes rust", "Best enjoyed cold", "It tastes better than water."

Fig. 4. Three side by side images of 2D billboard Splat ads consisting of robots. The left side image is of a bar scene with robotic arms holding out drinks in the foreground. In the central image, a lounging humanoid robot sits in a night beach scene with a Splat logo illuminating the night sky. In the right-side image, an android with a large Splat logo painted on the slide rolls over sand dunes.

The user experience for the immersive ad involved users being transported to a speakeasy Splat bar with several interactable objects. These interactions included a branded Splat bow, a vending machine, a soda dispenser, a jukebox, and a piano. The entire bar and all interactable objects were fully branded as Splat objects. The interactions with these branded objects are an example of product utilization. Users are given a more realistic sense of interacting with the brand. According to Alcaniz, transference measures are standard in VR studies and compare user behavior in a real environment to behavior when interacting with a virtual replica of the environment [6]. Although the research team's main goal did not involve measuring user behavior in the real environment to the virtual, interestingly, many users did simulate real-world actions like bringing soda bottles to their mouths to "drink." (Fig. 5).

Fig. 5. An image of the fully immersive speakeasy Splat advertisement room. All the furniture faces the middle of a shabby looking room. The foreground is occupied with a piano on the left and purple vending machine on the right. The middle space includes a bar with stools and soft drink dispenser on the left opposite a couch on the right. Splat art appears on the walls. In the background is the rear wall of the room and a green door marked "exit".

Each interactable object within the immersive ad was designed to resemble a real-world interaction. The piano allowed users to press the keys, and the sound of the piano key would play while that key lit up. A music sheet was shown on the piano, and if users played the piano keys to match the music note, they were awarded in-game coins in the gamified version of the app. The jukebox allowed users to press one of three buttons, and the room's music would change. The vending machine allowed users to press a coin and then choose a soda button which would cause that soda to fall from the vending machine which in turn users could pick up the soda and open the lid. For the soda dispenser, users could bring the cup from the bar to any of the dispensers, and they would see the cup fill up. If users completed any of these interactions in the gamified version of the app, they would earn in-game coins (Fig. 6).

Fig. 6. User earning coins with piano interaction within the Splat Immersive Ad. This image shows a close-up of the piano keys and a notification of " + 50 coins!".

3.2 Participant User Flow

Participants of the user study, who were all PwC employees, had either prior experience in virtual reality or no experience. The research team's requirements for the study included a testing population of at least 15 females and males for representation and diversity within the results. After giving written consent to participate, users attended in-person sessions in Chicago to volunteer for the study. Users were asked to provide their written consent and complete a pre-survey that captured their general sentiment towards advertisements and virtual reality. This data served as the control results for the analyses. After completing the survey, users were instructed by facilitators who helped them fit the headset and were instructed not to interfere with the remainder of the experience. The beginning of the game provided onboarding exercises to familiarize participants with the controls to play the game.

The game experience component consisted of two separate applications, the gamified-advertising app, and the non-gamified ad app. All users started with the gamified- ad app. After either losing in that game or completing all four levels, users were asked to complete the non-gamified version. After losing or completing the non-gamified version of the game, users were asked to take the headset off and complete the post-game survey where post-sentiment analysis was collected along with some additional questions.

3.3   Data Collection and Surveys

The goals of this research fell into the category of Customer Behavior Research (CBR). The research team used two data collection methods to obtain the necessary evidence to disprove or prove the hypotheses—the first method leveraged back-end data collection, which was embedded into the virtual reality application. The data collected includes game statistics and data about the advertisements. The second method of data collection was the pre-and post-game surveys. It was disclosed what data would be collected, and user consent was requested before the study began.

The research team used a third-party, back-end data collection platform, allowing the research team to collect different spatial data types. Most of this data falls into the category of implicit measures following the VEM framework [6]. Using human behavior tracking, the team collected time intervals to test how long users were in the game and in the immersive ad, and how long they engaged with particular objects [6]. The data platform allowed the team to further understand product placement and interaction within VR games. In addition, the platform allowed data tracking of what objects were being picked up or what buttons were pressed. The platform also enabled implicit measures of eye tracking [6] by generating heat maps of what users were looking at and recorded spatial sessions allowing the researchers to go back and study user behavior (Fig. 7).

Signals	What is measured?	How is it measured?	Which metrics can be derived	Related psychological constructs
ET (eye tracking)	Corneal reflection and pupil dilation	Infrared cameras point toward eyes	Eye movements (gaze, fixation, saccades), blinks, pupil dilation	Visual attention, engagement, drowsiness and fatigue, emotional arousal
GSR (galvanic skin response)	Changes in skin conductance	Electrodes attached to fingers, palms or soles	Skin conductance response (SCR)	Emotional arousal, engagement, congruency of self-reports
FEA (facial expression analysis)	The activity of facial muscles	Camera points toward the face	Position and orientation of the head. Activation of action units (aus). Emotion channels	Emotional valence, engagement, congruency of self-reports
HRV (heart rate variability)	Variability in heart contraction intervals	Electrodes attached to chest or limbs or optical sensor attached to finger, toe or earlobe	Heart rate (hr). Interbeat interval (IBI). Heart rate variability (HRV)	Emotional arousal, stress, physiological activity
EEG (electroencephalogram)	Changes in electrical activity of the brain	Electrodes placed on the scalp	Frequency band power, frontal lateralization, event-related potentials, wavelets	Attention, emotional arousal, motivation, cognitive states, mental workload, drowsiness and fatigue
fNIRS (functional near-infrared spectroscopy)	Relative changes in hemoglobin concentration	Electrodes placed on the scalp	Frequency band power, frontal lateralization, event-related potentials, wavelets on prefrontal cortex	Attention, emotional arousal, motivation, cognitive states, mental workload, drowsiness, and fatigue
fMRI (functional magnetic resonance imaging)	Relative changes of cerebral blood flow	Magnetic resonance imaging	Blood-oxygen-level-dependent (BOLD) contrast	Several cognitive and emotional responses
HBT (human behavior tracking)	Body movements (head, hands, rest of the body) and product movements	Cameras placed in front of the subject	Cinematics and dynamics of biomechanical joint movements	Visual attention, engagement, cognitive states, mental workload

Fig. 7. Image of implicit measures of consumer behavior referenced from Alcaniz's Work. Measure categories includes: Signals, What is measured, How it is measured, Which metrics can be derived, and Related psychological constructs. Example details are below each category.

Using the VEM framework the research team used explicit measurements with the pre-game survey as a baseline for key questions and compared them to the post-game survey to see how users' opinion or sentiment changed after the experience. A sample question from the survey included, "Rate your feelings towards advertisements." The scoring methodology allowed users to rate on a scale of 1–5, 1 being negative and 5 being positive. This yielded a quantifiable output with a degree of positive or negative sentiment users had towards advertisements. In addition, the level of user trust was assessed and gathered from this question in both surveys; "Rate your trust towards advertising." The scoring methodology allowed users to rate on a scale of 1–5, 1 being not trustworthy and 5 being trustworthy. This numeric output shows the degree of trust users may have gained from the experience (Fig. 9).

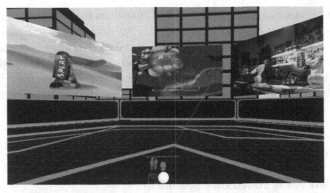

Fig. 8. Image of the heat map of user gaze on 2D advertisements containing robots. This picture shows the VR game setting, an outdoor black and neon archery arena with views to skyscrapers and sky, partially blocked by three large billboard ads. The centrally located billboard ad is the only image with indications of user gaze. This billboard is of a lounging humanoid robot in a night beach scene with a Splat logo illuminating the night sky, half a dozen birds appear to fly in a line above the horizon. A large pink heat map blob with gradient rainbow border highlights the birds and the horizon. The "L-A-T" letters of the Splat logo are covered in a yellow-green heat map blob.

Rate your feeling towards advertisements.

	1	2	3	4	5	
Negative	○	○	○	○	○	Positive

Rate your trust towards advertisements.

	1	2	3	4	5	
Do not Trust	○	○	○	○	○	Trust

What components of advertisements could break your trust? (you can choose more than one)

- ☐ Data Collection
- ☐ Misinformation or misrepresentation of products.
- ☐ No disclosure if content contains ads
- ☐ No disclosure around data collection
- ☐ No permissions around data collection

Rate your feelings towards advertisements that provide incentives for choosing to watch/experience the advertisement.

	1	2	3	4	5	
Negative	○	○	○	○	○	Postive

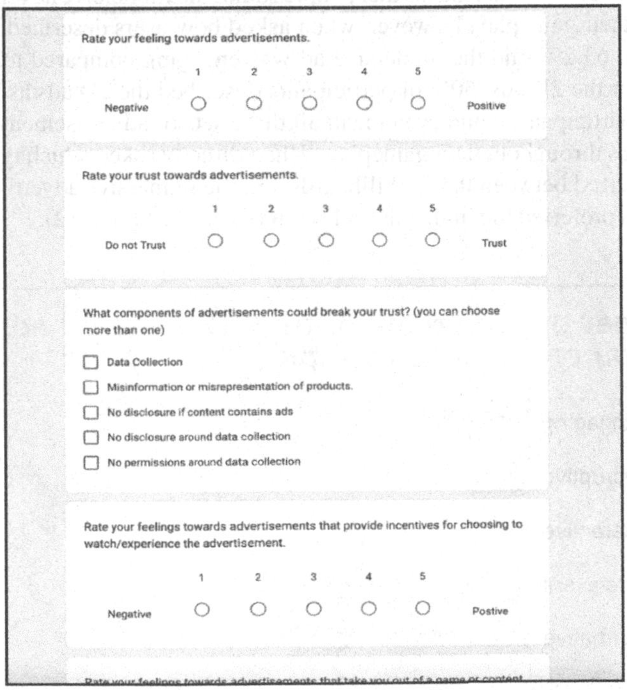

Fig. 9. An image depicting a portion of the Pre-Game Survey Questions. The blank form shows four visible questions and the response options. The top of a 5th question is also partially visible. Response types include rating on a scale of 1–5 for three of the questions, and places to indicate agreement with statements for another question.

The post-game survey included additional questions that were not in the pre-game survey. These questions helped to understand what drives users to make decisions in the game. For example, we found in the back-end data that users chose to go into the immersive ad more than the 2D Ad. In the post-game survey, players were asked, "During this Virtual Reality experience which type of advertising did you prefer?" Their choices included: being a fully immersive ad (speakeasy bar) or a 2D billboard. These questions unlocked more concrete answers about user behaviors.

4   Results

4.1   Immersive Ad vs 2D Ad User Sentiment

A primary goal of the study was to determine whether or not users would willingly engage with 2D billboard advertisements or a fully immersive 3D space embedded with branded products. The post-survey results gave insight into user sentiment towards each type of advertisement. Users were asked how disruptive they believed each advertisement to be on a scale from 1 to 5, with 5 being the most disruptive. On average, out of the 38 participants, users rated the disruptiveness of the immersive ad at 3.58/5. Whereas the 2D ad was rated at just 1.81/5. This means that people found the immersive advertisement more disruptive to their gameplay. However, when asked how users described the immersive ad experience, 63.2% said the immersive ad was engaging compared to just 19.4% of participants for the 2D ads. 50% of participants described the 2D ads as irrelevant, and only 26% of participants could even recall all three sets of advertisements displayed on giant billboards throughout their gameplay. When directly asked which type of advertising users preferred between the 2D billboards and the immersive advertisement, 63.2% of participants preferred the immersive ad room (Figs. 10, 11 and 12).

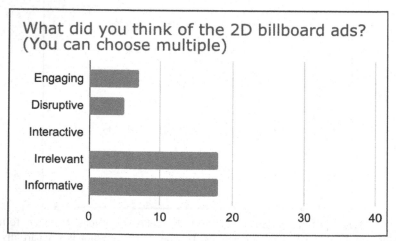

Fig. 10. Post-Survey Results for 2D Ad User Sentiment (38 respondents). What did you think of the 2D billboard ads? (You may choose multiple). A bar chart of responses shows equal selection of "Irrelevant" and "Informative" as the most popular answers. Interactive was never selected, "Disruptive" and "Engaging" were selected by a few respondents.

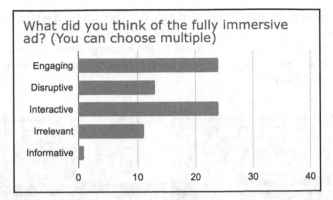

Fig. 11. Post-Survey Results for Immersive Ad User Sentiment (38 respondents). What did you think of the fully immersive ad? (You can choose multiple). A bar chart of responses shows equal selection of "Engaging" and "Interactive" as the most popular answers. "Informative" had the lowest score, with some responders marking "Disruptive" and "Irrelevant".

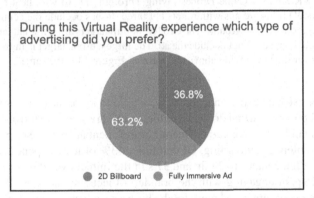

Fig. 12. Post-Survey User Preference Between Types of Advertisements. During this Virtual Reality experience, which type of advertising did you prefer? A Pie chart shows a large orange slide indicating 63.2% responded with "Fully Immersive Ad", a pink segment of half the size shows 36.8% said the "2D Billboard".

The data collected from the experience provided insights based on time spent between the game and the immersive ad and time spent gazing/interacting with dynamic objects. Dynamic objects were items tagged as branded products. When a user looked at one of the branded billboards within the game or looked at/interacted with an object in the immersive ad, that time was recorded.

The findings show that the average time spent gazing at any of the 2D billboards throughout their gameplay was just 6.16 s. It should be noted that this advertisement was placed in the center of the game experience, where there are targets that also spawn in the middle of the screen. The maximum average between all the billboards on either the left or right sides was just 1.09 s. This is to say that users spent about 5.65x the time gazing at the middle advertisement versus the advertisements on either side. From this, it could be implied that the length of gaze time is correlated with the placement of in game objects. Referencing Figs. 15 and 16, there is a clear correlation between the drone movement to the user gaze outlined by the heatmap shown in pink on the center ad. Simply, 2D advertisements were looked at for a more extended period if targets were placed in front of them.

a   b

Fig. 13. & 14. VR Archery Game Drones Flying Through 2D Ad with heat map comparison. Figure 13 shows 2D billboard ads with robots but three drones occlude the central billboard and three pop up targets are on the ground. The two left popup targets are lying flat, and the rightmost target has popped up but does not occlude the ad. The largest drone target is in front of the central ad specifically over the birds flying above the horizon. Figure 14 is the same as Fig. 8.

In the immersive ad experience, users were able to leave the room at any given time by simply grabbing the virtual doorknob which would remove them from the immersive ad and bring them back to the game. Overall, users spent an average of 6.08 min in each game after completing onboarding. Of that time, 43% of it was spent in the immersive ad experience. That equates to 2 min and 37 s in the immersive ad room of which 1 min and 18 s were spent engaging with the branded objects/interactions. Which is 12.66x longer than the maximum gaze length for the billboard advertisements. This shows users spent much more time willingly engaging with branded Splat products than gazing at any of the 2D billboards.

The gameplay was restricted to a certain number of targets which needed to be hit before advancing to the next level. Once that was satisfied, the user was given the option to choose either advertisement. The user could spend an unlimited amount of time in the immersive ad. Each time users entered the immersive ad, users were only allowed to interact with the piano, soda machine, soda fountain, and juke box up to one time per object. Each interaction gave the user 50 coins in the gamified version of the game, meaning the user was able to earn up to 200 coins in the immersive ad whereas the 2D advertisements only granted the user 50 coins in the gamified version.

However, in the non-gamified version of the game users were not granted any coins for interacting with the branded objects. Users in the gamified version spent on average 78.09 s interacting with branded objects and 77.88 s in the non-gamified version. This observation could indicate that despite not being rewarded for interacting, users still chose to interact with the objects. One caveat to this observation is that users could have interacted with objects thinking they may receive coins, as there was no formal indication that users would not receive coins in the non-gamified version. 60% of participants found the immersive advertisement to be interactive while 0% said the same for the 2D advertisements. Interaction was rated as highly important to users; when asked in the post survey to "Rate your feelings towards this statement 'Advertising should be interactive'" on a scale of 1–5, with 5 being agree, 50% of users rated a 4 or 5.

Researchers in the field of advertising count impressions on social media sites as significant after only a matter of seconds or even milliseconds [10]. Virtual reality enables users to spend uninterrupted time engaging with branded products for minutes at a time. Virtual reality users had a greater level of recognition of brands when interacting in a VR environment. Interactivity with in-game advertisements enhanced brand awareness as opposed to lower levels of brand awareness when users were simply "looking at" or "exposed" to in-game advertisements [7]. Branded interactions within a virtual reality experience are a significant alternative to traditional advertising techniques. However, by additionally incentivizing users to engage with the branded interaction, the impact and results can be amplified.

4.2 Gamification

When investigating user preference towards gamification: 90% of respondents said they preferred gamified advertisements; meaning consumers received in-game coins for interacting with or enabling the ads. Additionally, 75% chose to enable an ad (either 2D or immersive), showing the respondents preferred the gamified experience over the non-gamified version. This result stems from gamification providing an opportunity for a more personalized, organic, and non-disruptive experience. In Splat's instance, the consumer could choose how they interact with the ad, disguising it as an interaction point to choose from while playing the game. This concept relates back to subliminal advertising, which we see in the industry: the experience might not feel like an ad, per se, as the user doesn't feel a disruption to the regular content; rather, it's simply the next level or room of the game. A respondent to the post-survey stated, "[r]ather than your typical infomercials or demos, ads will be more embedded into the environment," when asked about how they see the future application of advertising within virtual reality. The respondent further stated, "[i]t didn't feel like an actual ad but just a component of your surroundings."

Matthew Pierce, Founder and CEO of Versus Systems, further validated this finding in an independent discussion he had with Forbes about the benefits and application of gamification in advertising. As a user, the ability to choose what you want to pay for as opposed to being forced to watch an ad is an important distinction, Pierce said to Forbes. "That [option] opens up different pathways in your brain, and you stop thinking about it as an ad. You see it as a reward, a prize [and] something you earned, and that's materially different." [11] Pierce further elaborated that one of the gamified advertisements which Versus Systems created for White Castle had a 36% conversion rate to purchase from the campaign–a game in which consumers could win free products after playing. He explained humans have an inherent desire to win and are significantly more engaged when challenged. Our respondents echoed a similar sentiment, stating they do see an application for immersive advertising where users can win awards by playing the seamlessly integrated branded activities.

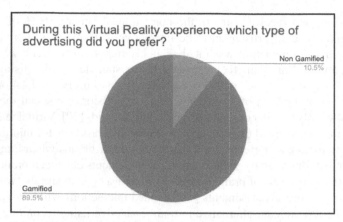

Fig. 15. Post-Survey User Opinion on Preference: Gamified Vs Non-Gamified. During this Virtual Reality experience which type of advertising did you prefer? A pie chart shows the majority selected gamified. An orange slide indicates 10.5% chose non gamified.

4.3 Game Art

Another area of exploration included variation of game art and how it can affect the feeling of immersion. The hypothesis was if the game art of the advertisement did not match the game art of the game experience (bow and arrow game), it would break the feeling of being fully immersed in the experience.

Users were asked to rate the 2D billboard advertisements to see which they felt belonged in the space. One set of ads was intended to match the game art more than the other two. One of the sets of advertisements included real-life humans. The results were inconclusive, as many participants voted similarly for each set of 2D ads. Participants were also asked if the VR advertising art should match the game art style. 57.9% of participants rated a 5/5, agreeing that it should match. Although this is a large percentage of the population, these results are inconclusive without behavioral data and more exploration (Fig. 17).

4.4 Additional Findings

In addition to the above insights around the hypothesis, the research team explored and discovered some additional findings. These areas fell into the following categories: data privacy and user trust, product placement, and in-game purchases through ads.

Data privacy and user trust were significant areas of interest for the study. The team's goal was to understand user sentiment around these areas in virtual reality and immersive advertising.

Regarding trust, a goal of the study was to explore where users found advertising in VR acceptable. How would users respond to ads for free content versus paid content? Through the user survey, it was found that 84% of users believe advertising is acceptable in free VR content, whereas 82% of users believe advertising is unacceptable in paid VR content. Users were asked whether they believe virtual reality resembles a gaming console, a cell phone, or unsure. An overwhelming number (80%) of participants voted

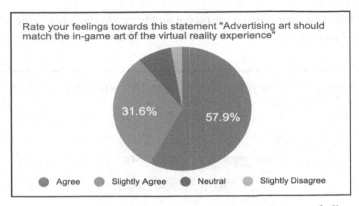

Fig. 16. Post-Survey User Opinion on Matching Ad Art Styles. Rate your feelings toward this statement "Advertising art should match the in-game art of the virtual experience. Pink, orange, red, and yellow slides show agree, slightly agree, neutral, and slightly disagree. The pink slice is the largest 57.9%. (Color figure online)

Fig. 17. Image of colorful splat ads from Fig. 3 positioned in the background of a black and neon archery arena in VR.

that VR headsets resembled gaming consoles. Although technically VR headsets are mobile devices, if users see and feel this device is a gaming console, this should be considered when curating the modality's advertising content. One participant stated, "On certain free games, it makes sense to advertise in the game. But on paid games, I wouldn't recommend this. Like the Xbox, I cannot remember ever seeing an ad on a paid game." Based on these stats, it can be concluded that users are more willing to participate in advertising when they receive something in return, in this case, free game content. If users are forced to view an advertisement after they have paid for their game content, their trust can be broken, and they could form a negative sentiment towards a brand.

The research team was also curious about how users felt about ad disclosure. Should advertisers and game creators disclose if there are advertisements in-game content? 47% of users believe there should be disclosure if the content contains ads. Finally, the research team wanted to compare users' trust in day-to-day advertising to VR advertising. 29% of users had a higher trust towards advertising in VR versus typical day-to-day advertising. Additionally, when comparing the pre-game survey to the post-game survey, it was also found that 89% kept their original trust score for the day-to-day advertising or increased their score, which asked, "Rate your trust towards advertising in VR." (Fig. 18).

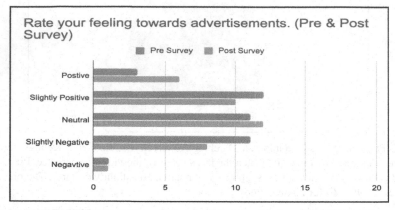

Fig. 18. Pre- and Post-Survey Results for User Sentiment (38 respondents) Rate your feelings toward advertisements. (Pre & Post Survey) Pink presurvey and Orange Post survey bars indicate aggregated ratings of "Positive" (3,6), "Slightly Positive" (12,10), "Neutral" (11,12), "Slightly Negative" (11,8), and "Negative" (1,1).

Data privacy is another important area of interest for us. "Roughly six-in-ten U.S. adults say they do not think it is possible to go through daily life *without having data collected about them* by companies or the government" [12]. The user survey showed 66% of users stated if there is no disclosure around data collection, their trust is broken. The survey also showed 50% users said their trust would be broken if data was being collected in the first place. By disclosing data collection, the survey showed there is a higher probability to gain user trust (Fig. 19).

Throughout the study researchers were curious about how they could use VR advertising to explore product placement. As mentioned before, the backend data collection system in VR allowed the team to track what products were picked up, how long they were interacting with the objects and even created heat maps of what users were looking at. Within the immersive ad researchers used the vending machine to explore product placement. The vending machine consisted of 4 buttons which correlated to 4 soda options (diet Splat, regular Splat, skinny Splat, and orange Splat). If users interacted with the Splat vending machine and received a Splat soda, results showed that 78% of users choose Regular Splat. Of the four options, this button was at eye level and the first in the vertical row of the button options. This shows how important product placement is, but also shows it is possible to study where the best product placement is on the backend of these advertisements. The team did test product placement within 2D advertisements in VR as well, and found the users spent roughly 5.65x the amount of time looking at the middle 2D ad versus the left and right. This could be due to the location of the 2D advertisement being placed where targets are coming from that direction, or the fact that users are naturally positioned to look that way at the start of the game.

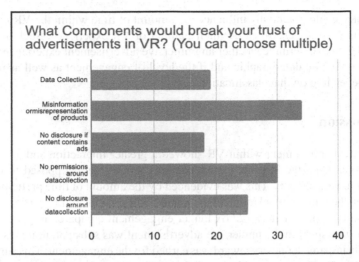

Fig. 19. Post-Survey Results for User Trust (38 respondents) What Components would break your trust of advertisements in VR? (You can choose multiple). Pink bars show most responses were "Misinformation or misrepresentation of products", followed by "No permissions around data collection" and "No disclosure around data collection".

The final area of interest in this study was in game branded purchases. In today's online world, users can make in-game purchases which include in-game coinage, skins, and other assets. "DMarket, which has a platform for trading skins, estimates the skin market is $40 billion a year." [13] Clearly there is a large market for in-app purchases. The game included a store with purchasable items to explore how users felt about the possibility to buy branded content in an advertisement and the ability to use that branded content in their game. Users had the ability to purchase a Splat bow from the immersive ad and use that Splat bow in their game experience. 68% of users who went into the immersive ad purchased the Splat Bow. It was also found in the survey that 76% of users liked the idea of being able to buy branded assets like the Splat Bow within an advertisement and bring to their in-game experience. This is a great example of the advertising approach, brand expansion. The soda brand Splat *expanded their product* outside the primary medium(soda) into a 3D asset branded bow that users could purchase in game.

5   Limitations and Variables

Best efforts were used throughout the study to limit outside influence on participants and to capture their actions within the VR app. Throughout the study, the researchers did determine potential limitations or constraints that may have influenced the current study or provide an area for future study. The participants of the study were all employees of the same company and as such, an amount of bias may have been introduced. Although the researchers did not find any evidence of such bias, a broader demographic sample could enforce this study's findings. Similarly, many of the users had not experienced VR before which could potentially influence the amount of time within the VR experience. Participants were not classified per their gaming or VR experience. Such a classification could help to determine whether an immersive advertisement experience should be catered to a specific demographic and if the level of engagement as well as motivation changes depending on that classification.

6   Discussion

Advertisement engagement within VR showed a greater interaction and a higher level of immersion when the in-game experience was immersive as opposed to traditional 2D banner advertisements. This was evidenced by the amount of time participants spent engaging in the immersive ad room as compared to the amount of time gazing at in-game 2D billboards, approximately 12.66x longer engagement compared to gaze. While the amount of time spent in the immersive advertisement was substantial, it was not unexpected as an in-game monetary reward was earned for the engagement. This suggests that a crucial element of engagement is a reward-based system returning value to participants in exchange for the time engaging with the brand, i.e., gamification.

While the gamification impact was apparent, the study also observed engagement in a non-gamified version of the VR experience where no coins were earned for the immersive experience. Participants still spent a similar amount of time within the immersive ad (on average less than one second difference). Notable here is that the participants experienced the non-gamified version of the VR app after first participating in the gamified version. The researchers provided no context to the participants prior to entering the app so they had no understanding of the difference in the two experiences. It could therefore be inferred that the participants expected game coins for interacting with the immersive advertisement similar to the gamified version of the app. This may explain why engagement was so high without coin rewards. There were no survey questions to capture expectations between the gamified and non-gamified versions of the app from participants. Data was not captured to demonstrate whether this affected the results of the study. Future work could explicitly investigate user expectations of such experiences, and if those expectations are not met, what impact on the brand that may have.

Although not directly explored, the study did examine the concept whereas users overwhelmingly (84%) found advertising acceptable in free games and likewise felt the inverse (82%), that advertising is unacceptable in paid content. The choice of paid vs. unpaid was not a part of this study and therefore potential future research into this area could expand on how and when immersive ad experiences are appropriate and how they can be properly deployed. Our results show that an immersive ad experience is much more popular than a 2D banner ad and is almost universally preferred with 89.5% of participants choosing the gamified advertisements over the non-gamified.

7   Conclusion

Through this user study the research team was able to deduce that there is potential to make advertising in VR a positive experience for users. The team concluded that users preferred interacting in the 3D immersive ad opposed to the 2D billboard ad. This study also showed many users see VR headsets as gaming consoles rather than mobile devices. Advertisers need to take this into consideration when creating content for VR to keep the experience positive. One user stated, "The idea of on-screen pop ups [in VR] is a huge deterrent." This study also showed gamification is an important aspect to user satisfaction. As mentioned above, an overwhelming number of participants preferred gamified advertisements as opposed to non-gamified advertisements.

Referring to the industry analysis mentioned above, product utilization and product expansion are a major opportunity within advertising in VR. Users responded positively to interacting with a brand's product within an immersive ad; one participant explained, "I get to be in the ad sipping coke instead of just watching it on a big screen." Product expansion can also be met with a positive reception from users, in addition to providing a large market for content creators. The ability to digitize real world branded products and embed those products into widely played metaverse games will unlock new brand interactions. This study showed that users will purchase in-game branded assets, and the market size for these types of purchases is only growing. Grand View Research Inc. Estimated that the Metaverse market size would be worth $678.8 Billion by 2030 [14].

In addition, this study showed that users are concerned about data privacy in advertising but also show a higher percentage of trust towards advertisements in VR than other day-to-day types of advertising. Users expect clear disclosures if advertising will be present in content, and to be asked for consent on any data collection in advertising material.

Data collection within advertising in VR presented a unique opportunity for product placement research. Researchers were able to determine users choose the top button on the vending machine more often than any of the other three options. The team also discovered that the 2D billboard ad in the middle of the set of three was looked at overwhelmingly longer than the left or right option. This could present a new way to use advertising, not just as a form of brand recognition, but as a new way to test product placement for real world scenarios like store setups.

As the metaverse continues to grow and more content for virtual reality is produced, the concept of advertising in VR creeps closer to fruition. The research team as well as the users in this study are excited to see what advertising in VR will look like. 66% of users from this user study said that they were excited about the possibilities of advertising in VR.

Acknowledgements. We thank the PwC Innovation Hub Meta (XR) leadership team for funding this project. Thanks to all the PwC staff who contributed to the development, testing, and deployment of VR bow game and embedded advertisements, and especially to all PwC staff who participated in the user study. Thank you to our diligent team of editors and subject matter specialists for helping take our research and paper to the next level.

References

1. Barnard, D.: History of VR - timeline of events and Tech Development (2022). https://virtua lspeech.com/blog/history-of-vr
2. Sutrich, N.: Looks like the oculus quest 2 is still selling better than the Xbox (2022). https://www.androidcentral.com/gaming/virtual-reality/quest-2-units-sold-spring-2022

3. Unsupervised: Marketers on the metaverse (2022). https://unsupervised.com/resources/blogs/marketers-on-the-metaverse/
4. Meta Quest, Oculus: Testing in-headset VR ADS (2021). https://www.oculus.com/blog/testing-in-headset-vr-ads/
5. Unity Technologies: Advertising in VR (2022). https://create.unity.com/vrads
6. Alcañiz, M., Bigné, E., Guixeres, J.: Virtual reality in marketing: a Framework, review, and research agenda. In: Frontiers (2019). https://doi.org/10.3389/fpsyg.2019.01530
7. Lupinek, J., Yoo, J., Ohu, E., Bownlee, E.: Congruity of virtual reality in-game advertising. Frontiers (2021). https://doi.org/10.3389/fspor.2021.728749
8. DaveGian, D.: McDonald's is now making happy meal boxes that turn into virtual reality headsets. In: Adweek (2016). https://www.adweek.com/creativity/mcdonalds-now-making-happy-meal-boxes-turn-virtual-reality-headsets-169907/
9. Hector, H.: Best oculus quest 2 games 2022: A guide to wireless VR gaming and oculus link titles. In: TechRadar. (2021). https://www.techradar.com/news/best-oculus-quest-games
10. Sajjacholapunt, P., Ball, L.J.: The influence of banner advertisements on attention and memory: human faces with averted gaze can enhance advertising effectiveness. In: Frontiers (2014). https://doi.org/10.3389/fpsyg.2014.00166
11. Prossack, A.: How gamification is Changing Advertising. In: Forbes (2021). https://www.forbes.com/sites/ashiraprossack1/2021/05/27/how-gamification-is-changing-advertising/?sh=4bd76871d4e3
12. Auxier, B., Rainie, L., Anderson, M., Perrin, A., Kumar, M., Turner, E.: Americans and privacy: concerned, confused and feeling lack of control over their personal information (2020). https://www.pewresearch.org/internet/2019/11/15/americans-and-privacy-concerned-confused-and-feeling-lack-of-control-over-their-personal-information/
13. Takahashi, D.: Newzoo: U.S. gamers are in love with skins and in-game cosmetics. In: VentureBeat (2020). https://venturebeat.com/games/newzoo-u-s-gamers-are-in-love-with-skins-and-in-game-cosmetics
14. Grand View Research: Metaverse market size worth $678.8 billion by 2030: Grand View Research, inc. (2022). https://www.bloomberg.com/press-releases/2022-03-09/metaverse-market-size-worth-678-8-billion-by-2030-grand-view-research-inc

Increasing Customer Interaction of an Online Magazine for Beauty and Fashion Articles Within a Media and Tech Company

Christina Miclau[1], Veronika Peuker[2], Carolin Gailer[3], Adrian Panitz[1(✉)], and Andrea Müller[1]

[1] Hochschule Offenburg – University of Applied Sciences, Badstraße 24, 77652 Offenburg, Germany
{christina.miclau,adrian.panitz,andrea.mueller}@hs-offenburg.de
[2] Burda Verlag, Arabellastraße 23, 81925 München, Germany
Veronika.peuker@gmx.de
[3] Burda Forward GmbH, St.-Martin-Straße 66, 81541 München, Germany
Carolin.Gailer@burda-forward.de

Abstract. The present paper addresses the research question: What recommendations for action and potential adjustments should an online magazine for beauty and fashion implement in order to make affiliate articles in these sections even more appealing to the target group and provide added value for them?

To be able to answer this research question, three hypotheses were defined and tested with using qualitative and quantitative research. The qualitative research consisted of user experience testings, where four affiliate articles in the fields of beauty and fashion were tested with 13 participants. The quantitative research involved collecting, analyzing and evaluating data from the four affiliate articles conducted with the company's real-life target group. Based on these results, recommendations for action were derived, which should not only improve the quality of the content in the future, but also increase the efficiency of the implementation of those articles.

Keywords: Empirical Research · Online Magazine · Target Group Oriented Design · User Experience

1 Relevance

Due to the lack of knowledge on the exact target group of the fashion and beauty columns, it is important to know how the content is perceived by the target group and what they expect in terms of content and design. Since an online magazine competes with many publishers and aims to be directly relevant to the Google algorithm as well as to the readership as a single point of truth through valuable content, the facts regarding the use of different devices must also be taken into account. About 87% of users access articles via smartphone, [1] which is why responsive design, i.e. adapting articles to the respective device, is also an important factor for the user experience. Yet so far, only A/B tests on headlines have been carried out for affiliate articles in the beauty and fashion section at the company under study.

F. Fui-Hoon Nah and K. Siau (Eds.): HCII 2023, LNCS 14038, pp. 401–420, 2023.
https://doi.org/10.1007/978-3-031-35969-9_27

This study's premise is to ensure that readers receive reliable, relevant, and solution-oriented information that provides them with support in their everyday lives. The intention is to write beauty and fashion articles that can offer added value to the readership. In addition, the design and structure of the articles should be pleasant and inspiring. Furthermore, the optimization of the articles should build loyalty so that the readership returns regularly as well as the single point of truth, i.e. that the company becomes the first choice or point of contact for the two sections when readers want to take some time off or require advice. On the way to obtaining all the necessary information, data can thus be gained on the exact target group, their needs and expectations.

Qualitative and quantitative research methods will be used to investigate the challenge, which will then be tested on the target group by means of a partial survey. Based on the findings of the research, recommendations for action are made for the online magazine as to how it can design and plan affiliate articles for the fashion and beauty sections even more efficiently and in a more target group-oriented approach with regard to the future. Therefore, the following hypotheses were defined for the study, which had to be verified:

H1: A list of products at the beginning of the article generates more click outs to the linked online store.
H2: Fewer text (a text block of about 50 words), but more than just the teaser image, generates more click outs.
H3: Call-to-action blocks generate more click outs than image and button as single elements.

2 Status Quo and Latest Developments

Fashion and cosmetics products in particular were frequently added to the virtual shopping carts of online shoppers in 2021. The purchase frequency, i.e. how many orders a customer submits in a given time period, amounted to 6.66 million purchases per month for fashion items, shoes were purchased 3.38 million times, and cosmetics 3.05 million times [2]. The global pandemic caused by Corona also had a positive impact on affiliate sales. In a survey of 1,100 marketers responding to the survey by performance marketing agency XPose360, for example, 65% said they expected sales to be higher than in 2021. Only 7% expect their revenue to decrease. In 2021, a similar number of marketers stated (68%) that they were anticipating more sales, but 15% also predicted a decline in sales at that time [3]. In terms of technology, there are several ways to track visitors and their actions back to specific partners. A few well-known methods will now be discussed below. On the one hand, the affiliate ID can be written directly into the HTML code [4] and thus it can be found out via URL how often the link was clicked on (click out) and how often the product was bought and paid for, what sales were generated with this ID and how much commission was generated. An example URL might look like as follows:

https://www.awin1.com/cread.php?awinmid= [Affiliate ID]&clickref = [Click reference with link to article]&ued = [Link to product].

It first shows the selected affiliate network "Awin", followed by an affiliate ID that identifies the affiliate, followed by the click reference with a link to the product of the online store. This method has the advantage of consistently and securely tracking how the user behaves, regardless of browser settings. The disadvantage is that affiliate commission is only paid if the user buys this specific product directly [4]. The "Amazon.de Associates Central" network also considers, with a lower commission, all other products in the shopping cart that are purchased along with the tracked product. Moreover, there is a possibility to collect information about the purchase behavior to a linked product via cookie.

The European Union's General Data Protection Regulation (DSGVO), which came into law in 2018, and the various browsers' anti-cookie policies, however, make cookie tracking considerably more difficult. Affiliates can also collect data with a session ID, which records and evaluates transactions and data per opened session. The particular tracking method is specified by the affiliate network [4].

Advantages and Key Success Factors Compared to Other Marketing Activities: From a marketing perspective, affiliate marketing has great potential for merchants compared to the quite costly banner or keyword ads. The merchant has the opportunity to increase awareness of its brand without paying anything in advance [4]. As such, affiliate marketing most closely resembles the term "performance marketing" [5]. Through affiliate marketing, one receives not only additional sales, but also greater visibility and more or less free branding, as usually rewarded only when a purchase is completed [4]. Affiliate marketing also has an impact on the visibility of paid or organic results in search engines. Even as a booster of response, affiliate marketing can achieve immediate demand effects, such as participation in lotteries [6].

2.1 Emotions Involved in Customer Interaction with Online Magazines

In order to influence customers in their purchasing decisions, emotional speech has been a central aspect of marketing and advertising for years [7, 8]. By evaluating consumer behavior, market research and analysis programs can be used to draw conclusions about customers, wishes and needs and can be analyzed and interpreted. Research into the value system of customers has also gained in importance, especially since the value system is said to influence purchasing decisions [9].

Human actions and behavior are largely controlled by emotions, so that more than 70% of all decisions are made unconsciously. The mind controls the remaining 30%, because in the course of evolution man has learned which laws he should follow to act successfully [10]. Emotions are a complex web of interactions of subjective and objective factors that are transmitted through neural/hormonal systems and can lead to affective experiences such as excitement, pleasure, or displeasure. Moreover, emotions, on the one hand, generate cognitive processes that produce emotionally relevant perceptual effects and evaluations and, on the other hand, lead to behavior that is often expressive, goal-directed, and adaptive [11].

Ekman and Friesen defined six basic emotions, disgust, sadness, anger, surprise, and fear, in 1977 to categorize emotions [12]. Figure 1 shows four of these basic emotions.

In order to make these categorized emotions measurable, they are each assigned to physical changes, e.g., facial expressions, gestures and voice [13, 14] and are measured and interpreted with the help of facial expression or voice analysis. In this context, facial expression analysis is considered a reliable tool due to its independence from cultural,

Fig. 1. Four basic emotions: joy, anger, sadness, disgust [22].

gender, and age-related factors, and it is used primarily in UX [13, 15]. It has also been demonstrated cross-culturally that individuals from different cultures can recognize Ekman's six basic emotions through facial expressions [20, 21]. Nonverbal expressions are often measured by means of video recordings, which are analyzed and assigned to similarities of prototypical expressions using a coding system. The Facial Action Coding System (FACS) by Paul Ekman, which defines 44 different facial expressions and movement units, is particularly frequently used here [18, 22].

The quality in which emotions develop can only be determined after the interpretation of the physiological stimulation. According to this, emotions are dependent on cognitive processes, because according to the theory of emotions in psychology, the reward system is primarily attributed as the driving function for human action [16].

In order to better understand customers and users and to create the best possible (shopping) experience for them, it is necessary to take a closer look at usability, joy of use and user experience or customer experience and to use measurement methods for those experiences. Numerous procedures exist with which emotions can be measured both very precisely to imprecisely and subsequently leave room for interpretation. Since the dimension of facial expressions and their evaluation is not developed in the same way in every person, the measurement usually requires manual coding by specialists. However, distortion effects can still be partially estimated and avoided [17].

2.2 Methods of the Qualitative and Quantitative Study

To examine customer experiences in the field of marketing, the multi-stage, modular and scalable Customer Experience Tracking (CXT) method is often used [14]. The focus of the method lies on the examination of emotions [23]. In recent years, many methods for UX measurement and investigation have been developed, but their different procedures and results make them difficult to compare with each other [24]. Figure 2 shows different methods of user experience analysis.

Analytical User Experience (UX) methods refer to the evaluation of experts, while empirical methods involve the target audience [24].

In addition, the CXT process also uses technical solutions, such as eye tracking, skin conductance measurement and facial expression measurement, to measure usability and

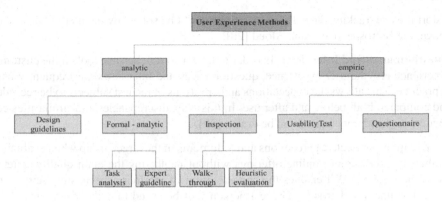

Fig. 2. User Experience Methods [24].

take into account the emotions that arise [23]. With the CXT method, it is possible to simultaneously examine the subject's facial expressions, gestures, voice, and eye movement during testing. In practice, as well as in this study, a combination of analytical and empirical methods is often used [14]. For this paper, the main modules for measuring UX are eye tracking, facial expression measurement, Think Aloud, and interviews.

Eye Tracking: During the CXT method, it is possible to make the course of gaze, irritations and user behavior measurable with the help of eye tracking. Using mobile and stationary sensors, the data can be measured and analyzed in real time. The gaze of the test person(s), also known as the fixation point, provides information about the length of time spent [25]. The connecting lines between the fixation points, the saccades, do not transmit any information. So-called "blind" areas are created. With the help of user comments and the results of eye tracking and facial expression and voice analysis, recommendations for action can be generated. [14].

Facial Expression Measurement: Facial expression measurement can provide targeted conclusions about the user experience in detail, so it is an important method to illustrate the user experience [23]. For thousands of years, facial expressions have been regarded as a mouthpiece, because emotions are reflected as psychological arousal in facial expressions. They are subjectively perceived by the counterpart and show as so-called "relevance detectors" what is essential and significant for the person [19].

Think Aloud: Another frequently used method is Think Aloud, in which the participant communicates all thoughts about the tasks of the usability test out loud. A distinction is made between two types: Concurrent Think Aloud, in which the subject thinks out loud while solving the task, and Retrospective Think Aloud, in which the subject thinks out loud after the task has been completed [26]. It may happen that some subjects have to be continuously encouraged to think aloud, which reduces the spontaneity of the answers [27]. Also the combination with eye tracking can influence spontaneity and reactivity. Through thinking aloud, the fixation point remains longer on the test website or online store, because the expression of the thought takes longer than the thought itself. This can distort the eye tracking data. These irritations could be solved by the method described above, the Retrospective Think Aloud [28].

Questionnaires and Interview: In order to take a comprehensive look at the customer experience and the user experience, questionnaires and interviews are frequently used in practice. This allows the expectations and emotions of the test subjects to be recorded and compared both before and after use. In this way, discrepancies and similarities can be identified and conclusions can be drawn [29].

To capture subjective perceptions and judgments of the target group when evaluating aesthetics and design of online offerings without neglecting the main quality criteria, the online tool VisAWI enables the consideration of the four aspects simplicity, variety, colorfulness and artistry. The test person will be asked to make statements about characteristics of the design using a scale of 1 (strongly disagree) to 7 (strongly agree) [30] In interviews, it is possible to gain deeper insights into the overall impression of the application, the didactic design, the operation and navigation, the comprehensibility and the motivation. In this way, the interviewer can discuss the participant's answers directly and gain a deeper qualitative understanding of the participant's insights [31].

Web Analytics: Web analytics allows analyzing a user's behavior while interacting with a website or mobile application [32] by collecting data. In general, a distinction is made between on-site and off-site analytics. The on-site analysis refers to the data collected within the website, such as the number of visits to a website and the average time the user spends on the site. Off-site analytics relates to information that uses data from usually external, sources, such as surveys, market reports, and comparisons [33]. A limitation of web analytics is the lack of qualitative understanding of the data collected when the user interacts with a website. In order to evaluate the data quality, the source is of high relevance. A difference is made between 1st party data, which is collected by the company itself, 2nd party data, which is collected by other providers, and 3rd party data, which is collected by companies that have also obtained it from service providers. Due to stricter privacy regulations, such as the DSGVO, 1st party data is gaining importance, but in practice it becomes apparent that this often has to be enriched with external data and therefore 2nd or 3rd party data cannot be avoided [34].

3 Findings

Chapter 3 focuses on the results of the UX testing and deals with the quantitative findings on the one hand and the qualitative findings on the other. Based on these results, suitable recommendations for action will be made in chapter 5.

3.1 Findings of the Qualitative Data in the Context of UX Testing

The qualitative data were collected in June 2022 and lasted between 31 to 83 min. The 13 test persons were divided into two groups in advance by the testing management; six test persons tested two beauty articles (see Fig. 3), and seven test persons tested two fashion articles (see Fig. 5). The odd number of participants was due to low participation and the persons were randomly divided into the respective groups.

The questions asked were the same for each of the four articles, so it is possible to be able to compare the answers. At the beginning of the testing, the general research and its purpose were explained. Then the participants signed a data protection statement on the confidentiality of the information. In the next step, the eye tracking method was introduced, the scenario was explained, and the test subjects were then given the opportunity to read the first article. They were asked to express their thoughts out loud using the Think Aloud method. The aim of this first task was to familiarize the test subjects with the article and to take away their nervousness. Through the test leader, the test subjects were repeatedly encouraged to speak their thoughts out loud. They were then asked about their first impression and given a grade from 1 (very good) to 6 (very bad), as well as which three key words could be used to describe the article.

In article 1 of the beauty section, the company's advertising block with the most-read articles after the first text block was noticed negatively by 50% of the test persons. This "disturbs the flow of reading" or has a "disturbing effect, as it distracts the reader". This advertising block was also actively addressed in the second beauty article by 50% of the test persons with the same arguments. In Beauty 1, four participants did not fully watch the embedded TikTok video because it was too long or the protagonist talked too much. For Beauty 1, only two respondents mentioned that the background information on the brand's founder, who was mentioned in the article, was interesting. The article received an average grade of 2.6, which implies as "good".

Fig. 3. The examined beauty articles 1 and 2 (from left to right)

Also, the advertisements between the first and second text block were rated negatively and the TikTok video was also aborted by three test persons, because it took too long for them. In the case of Beauty 2, four test persons were also the opinion that the article was better than the first, but two test persons found the alternating product presentation confusing. On average, the article was given a school grade of 1.9, which implies as "good" and the keywords "informative" and "confusing" were mentioned most often.

After each of the first and second articles and the questions on first impressions, the test persons received an online questionnaire from VisAWI with 20 questions on the four main categories "simplicity", "versatility", "colorfulness" and "artistry". This resulted in the following ratings on a scale of 1 (strongly disagree) to 7 (strongly agree) which can be seen in Fig. 4:

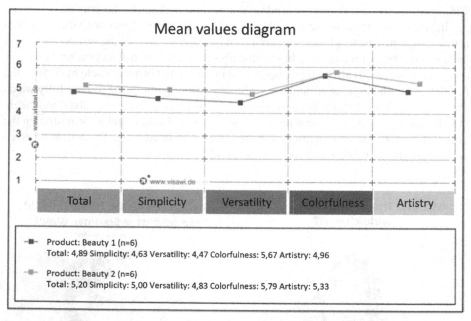

Fig. 4. Mean values of the responses from the VisAWI questionnaire for Beauty 1 and Beauty 2

When asked whether a video should be inserted instead of the cover picture, 50% of the test persons answered that they prefer a cover picture, two test persons thought that a video is better because it is unusual and one could save reading through the entire article. One participant thought that a video would only be useful if it did not play automatically when the article was opened and if it matched the content of the article.

Matching the hypothesis H1, it was asked about the usefulness of the listing in article Beauty 1 after the first text block. Only two test persons found the listing helpful, but only if there are different prices, as it is the case in the article. The remaining four respondents found the listing unhelpful because the listing is already displayed too early and the participants are still missing important information before they would skip to the store or buy the product. The Test persons especially value clear price labeling, comparable products, reviews by real customers, expert opinions, test comparisons, and information on how to use the product when buying beauty products.

In addition, all test persons agree that linking to a second online store is advantageous, as it allows them to choose the store they trust or give them the opportunity to choose a lower-priced store, where they can save on shipping costs, for example. The test persons tend to do their shopping online at Douglas or Flaconi or visit stationary stores.

It was also asked for the hypothesis H2 whether the test persons would prefer to have an additional image displayed instead of a text block. Here, the answers varied. Three respondents even thought that Beauty 1 lacked text, and two participants stated that more images were generally perceived as better than text. One person commented that Beauty 2 shows too many identical images and that more varied product images would be desirable. It was also mentioned once that the current distribution between images and text is balanced.

To test hypothesis H3, the call-to-action blocks and the combination of image and button were compared. Two participants stated that they found the CTA blocks more appealing and two participants found the combination of image and button more appealing. Three participants found the CTA block too empty and would have liked more information, such as the product's contents.

For article Beauty 2, it can be noted that it was generally rated more positively. "A clear indication that money is being made with the article" as well as "informative" and "professional" were most frequently stated as "Agree" and " Fully agree".

Fig. 5. The examined beauty articles 1 and 2 (from left to right)

At the end of the testing, the test persons were shown both articles side by side in the mobile view, since most users of the online magazine and also the test persons mainly read articles with their smartphones. Thereby, four test persons found that the block with the most-read articles takes up too much space. Three participants found the font and cover size suitable. Three respondents each stated that the Beauty 2 article had a more appealing design and a more pleasant structure. Two respondents said that the product images were too large for the mobile view. In line with hypothesis H3, two respondents found that the CTA block was more compact and therefore better than in the desktop version.

The procedure for the seven test persons who tested the two fashion articles was the same as for the Beauty test persons. In the case of article Fashion 1, four test persons were annoyed by the other articles that were displayed after the first text block. Two test persons looked intensively at the embedded Instagram posts and clicked on them. However, they were then shown the Instagram login page, which is why they then left the page again.

The affiliate notice after the teaser text was read carefully by two test persons. It was mentioned five times that the test persons had expected different products and were therefore disappointed with the selection. In this context, one person showed an angry facial expression after Ekman's FACS, visibly shown by the contracted eyebrows (see Fig. 6).

Fig. 6. Angry facial expression during an irritation. Source: Test recording, 06/21/2022

Four participants did not look at the follow-up articles at the end. Two participants each clicked on one of the follow-up articles and two did not find them appealing and would leave the page afterwards.

The first impression of the Fashion 2 article was confusing for four respondents, three of them stated that they were disappointed by the selection of clothes. Two respondents each stated that they did not like the clothes or that they had only skimmed the text. Two participants found the cover so appealing that it encouraged them to read on. The participants gave the article an average grade of 3, which implies as "satisfying". The keywords "inspiration" and "appealing" were mentioned twice each.

Similar to the beauty articles, "variety" was rated most negatively in the VisAWI questionnaire (see Fig. 7): Mean values of the answers from the VisAWI questionnaire for Fashion 1 and Fashion 2.

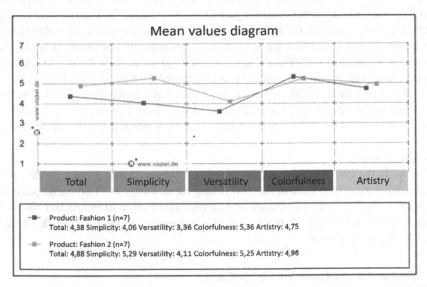

Fig. 7. Mean values of the answers from the VisAWI questionnaire for Fashion 1 and Fashion 2

In the category versatility, the article Fashion 1 has an average rating of 3.63 and Fashion 2 of 4.11, which means a rather neutral opinion. Similar to the beauty items, the second item performed better, Fashion 1 was more convincing than Fashion 2 in the colorfulness category with 5.36. This means that the test persons tended to agree that they liked the design and layout. As with the beauty articles, the artistry category was rated very positively. It is noticeable that Fashion 2 scored significantly better in simplicity, with 5.29, than Fashion 1. As with the beauty articles, both articles were then shown next to each other and the hypotheses and suggestions for improvement were addressed with questions.

It was also asked whether it would be useful to enumerate the products of the article. This question was asked to answer hypothesis 1. Here the statement was also not consistent: three test persons did not find the enumeration important, three times it was stated that it was a helpful overview for the article. One participant would only find the list useful if a product image were also shown next to the product. The participants attach particular importance to the brand and suitable images of the products in the case of fashion articles. One participant also wished that the entire outfit should be linked when tips are given. Half of the participants find it useful if a second online store were linked, as this would allow them to make their own selection. Three participants did not find this information helpful.

In order to be able to evaluate hypothesis 2, five test persons would find it useful if another image were included instead of a text block, since the text is repeated in the tested articles. Only two participants would find it confusing if another image were inserted instead of a text block.

In order to be able to test hypothesis 3 for the fashion articles, the call-to-action blocks and the images with buttons were also displayed next to each other. Two respondents found the CTA blocks rather disturbing and therefore found the combination of image and button more beautiful and modern in design, because the image is displayed larger. Three participants stated that the CTA blocks instinctively encourage them to click. Twice it was stated that all relevant information is given in the CTA blocks. A second questionnaire with the same questions was also tested with this group of subjects.

In the case of article Fashion 2, the responses tended to agree. Five participants found the article rather "inspiring", four fully agreed that there was "a clear indication that money is being made with the article". Again, four participants found the article "logical" and three participants each agreed that the article contained "appealing pictures" and was "descriptive".

Towards the end, the two articles were also displayed side by side in the mobile version. Five participants found the advertising block annoying. Two participants each said that the layout was more attractive, that the combination of image and button was more attractive than the CTA block, and that the images in the CTA blocks were too small for the mobile view.

3.2 Findings of the Quantitative Methods

The quantitative data was analyzed via Google Analytics, Google Datastudio and On.the.io. In each case, the period of 14 days from publication of the article was considered. The selected KPIs consist of demographic key figures that provide insight into the target group, as well as performance key figures.

One of the most common KPIs are **page views**. These can be divided into unique page views and total page views. Comparing the two KPIs, it can be seen that both beauty articles achieved a total of 6,425 page views and both fashion articles achieved a total of 56,158 page views. Further analysis shows that both beauty articles generated fewer **page views** than the fashion articles. This is due, among other things, to whether the article is recorded as valuable by the Google algorithm, evaluated and shown to Google users. In Google Analytics, it can also be recognized that both beauty articles were not recorded as valuable content by the Google algorithm and therefore were not shown at the top search ranks for certain keywords. Furthermore, it is important that the title and the cover image are designed to be appealing to the readership. This can be supported by certain incentives, such as particularly low prices or a reference to a celebrity. Nine of the top ten articles in terms of reach in 2022 contain at least one of these stimulating words.

It can also be seen that neither of the two beauty articles generated **sales** and **turnover**. In both beauty articles, two products from a well-known online store were linked, and the more expensive trend product was clicked more often than the cheaper alternative product. On average, the conversion rate for this online store is 1.49% and €0.03 revenue per click. [35] As a general guideline, conversion rates between 0.5% and 1% are considered average in affiliate marketing. [36] The two fashion articles have been significantly more successful. There were 74 sales made, which generated €6,770.81 in revenue.

Therefore, the conversion rate for Fashion 1 is 0.08% and for Fashion 2 is 0.17%. Both KPIs are below the industry average and the average values of the well-known online store.

Another key figure that is important in affiliate marketing are the **click outs**, which means how often the affiliate link was clicked on, so that the reader gets to the partner store. These can also be found in Table 2, which comes from the affiliate network Awin. From this, the click through rate (CTR) can be calculated, which tells how many people have clicked on affiliate links in relation to the page views. Accordingly, the CTR for Beauty 1 is 0.11, which means that 11% of readers clicked on the link. For Beauty 2, the CTR is 0.17, for Fashion 1 0.35 and for Fashion 2 0.14. The higher the CTR, the greater the chance that a purchase will be made there. [37] With the data from Google Analytics and the affiliate network, another performance indicator can be calculated: the effective Cost per mille (eCPM). It indicates how much commission can be generated per 1,000 page views. [38].

The average **time spent** on the Beauty 1 article was 51 s, while the total reading time of the article was approximately 3:10 min. On average, readers spent 1:52 min. The short time spent can be explained by the fact that many readers clicked on the links and therefore reached the store page. But even if the readership does not like the article, they can leave the page to either click on another article or leave the page completely. The **bounce rate**, which is derived from this, provides information about the percentage of visits to a single page without interaction with the page. [45] For Beauty 1, this rate is 16.12%, for Beauty 2 it is 15.57%, for Fashion 1 it is 65.26%, and for Fashion 2 it is 78.40%. High bounce rates are not necessarily bad. For content websites, bounce rates of up to 60% are normal. This can also mean that the reader found the answer to their question in the content or that the article inspired them to make a purchase. The **exit rate**, in comparison to the bounce rate, measures the exits from the specific article without knowing if this article was the only page visited. The exit rate is very high for all articles, ranging from 72.49% to 90.21%. Therefore, all bounces are automatic exits. [39].

Also, the **readability**, meaning how many percent of the readers read through the article to the end, reveals a lot about where exactly the reader jumped off. [40] On average, only 6% of readers read through to the end of articles. For article Beauty 1, the readability was 26%, for Beauty 2 16%. These increased values are similar to the low bounce rates. It is possible that the readers did not find the answer to their question or that the content did not encourage them to make a purchase. For the fashion articles, the readability averaged 6%. The **scroll depth** for Beauty 1 shows that 75% of the readers read the article up to halfway. In comparison, the scroll depth funnel for Fashion 1 is relatively sharp, with just under half of the readers reading the article halfway through. These values also match the KPIs conversion rate, bounce rate and click outs.

4   Critical Consideration

In the course of the study, internal communication with two UX experts revealed findings indicating how the approach regarding the study was not ideally suited to the UX. Accordingly, the original plan to list several products in the articles was not practicable enough, as both experts were of the same opinion that the same product should be compared at best. However, since it was not possible to publish two articles featuring the same product, the only option was linking similar products, which is why any influence on the part of the readership must be expected. In the case of different products, too much information, e.g. on the brand or which personality also uses the product, can also result in bias.

Even the choice of the online shop is important for the UX measurement. Both experts were of the opinion of Amazon being the most advantageous due to its better measurability. Not only is it possible to measure the interaction of the linked product but also the other products that were purchased. Since Amazon as an online shop does not offer the same products as fashion stores, we have not included Amazon in our analysis. As a result, less information can be used with regard to performance and possible up-selling or cross-selling effects.

In retrospect, the selected group of the UX testing should be critically evaluated as well, given its difference from the actual readership. On the one hand, the test group is on average nine years younger than the actual target group of the media company. They also differ in gender, as the test group consists only of female participants, although the actual readership is also partly male. There was also a difference in terms of their interest in general and their personal behavior toward online shopping and reading online articles. Due to the limited possibilities of finding participants, we could not take these differences into account within the time available for the study.

The data analysis using Google Analytics has also been causing difficulties for years, on the one hand from a data protection point of view [41], and on the other hand as Google requires cookies to be activated in order to access the data. In particular, this type of risk occurred during the analysis of the quantitative data for the demographic data of age and gender. Thus, it is not certain whether the data is sufficient to conclude on the entire target group. [42].

The quality of certain values depends on the type of article and the user access. In order to be able to compare valid data, there should be more than these four articles compared with each other. Data would also need to be collected and analyzed over a longer period of time. Given the limited project duration of this study, we could not provide sufficient data for this purpose.

Evaluating the device used by the target group to read articles in the online magazine is also important. 87% of users access the website via smartphone, 63% via tablet, and 52% via laptop or desktop PC. [1] The majority of beauty articles were also accessed via smartphone by 65%, whereas more than 90% of the users accessed the fashion articles via smartphone.

5   Recommendations

The following recommendations for action serve to further developments of a target group-oriented design of articles and are merely suggestions based on the UX testings.

Title: The title of the article Beauty 2, which was posed as a question, was perceived positively, as it encourages further reading. Keywords such as "figure-flattering", "garden party" or "summer glow" - words that readers associate with feelings or experiences - were also positively noted.

Affiliate Reference: The reference that money (commission) is earned with the affiliate articles was perceived as positive and accurate by most of the test persons. Only two participants of the beauty articles and one participant of the fashion articles were startled at the beginning, since they could not assign the reference and it was taken as an advertising reference. Since it is legally obligatory, no recommendations for change can be given as a result. [43].

Teaser Image: The first image of the article is the teaser image, which has to be designed in an appealing way and match the theme of the article. To avoid disappointment, it is essential to already present the product through your teaser image. By displaying the product, e.g. in a compilation, the other products in the article can also be shown, so that the reader already knows what can be expected in the article. Including a video instead of the teaser image would not be an option for the fashion test persons. For beauty articles, 50% said they would prefer a video, but only if it were followed by no more text. However, this would be disadvantageous for companies, since no products can be linked in a video and thereby no visibility can be generated through SEO amongst search engines.

Teaser Text: The text shouldn't be too complicated to read, as was noted in the case of Beauty 1. The participants were positively impressed in both article categories when the price and web shop were already mentioned in this brief section of text. The formulation of an (everyday) problem situation, e.g., in Fashion 1, and implying solutions as part of the article were also seen as positive and should be implemented.

Advertising: The first advertising element in the article, which appears after the first text block and shows the current articles of the online magazine, was found to be particularly annoying by half of both groups of participants. Since online publishers are largely financed by advertising and internal links [44], these articles are important for building readership. Hence, it is recommended to add articles from the same column to this article block, i.e., in this case, the latest beauty or fashion articles. The latest articles of all categories could then be shown at the end of the article.

Listing: The listing of products as a second text block was tested in the first article of both categories. Four of the six users tested for the beauty articles found it unhelpful, as they were still missing too much information about the product. In the case of the fashion articles, four participants equally stated that the listing was not useful at this point. A recommended action that bulleted lists only make sense if the article is about more than one product was also mentioned. In addition, the indication of the brand is important for the testees.

Text: The expert opinions in Beauty 1 were well perceived by the participants, and also the background information on the brand mentioned. When it comes to expert opinions, the users can also imagine having a sort of product review summary at the end of the article, in which, for example, check marks or "plus" icons are used to show how certain features of the product appear when used. Regarding Beauty 2, all six respondents cited that they liked the tips and tricks as well as the product reviews, and half of the respondents also found the bulleted lists on how to use the product informative. These three text modules should therefore be considered for beauty articles in the future. In the case of fashion articles, the respondents also appreciated the information on the possible combinations of the product with accessories and other items of clothing. It was also important to them e.g. that more attention is paid to the type of the product's fabric or that products for different body shapes are shown when "figure-flattering" is mentioned. In the beauty articles, a new series of articles could be published in which the editors test products for the readership.

Images: Also, for the product images within the CTA-blocks/combination of images and buttons, it was noted that switching between the two products should be avoided, as the product images are always the same. As a recommendation for action, it is therefore helpful to show pictures of the texture of the product or how the product looks on different skin colors in the case of beauty, for example, and to include pictures of the product being worn on a model in the case of fashion, for example. The large images in the combination of image and button often seemed too large to the test persons and they would like the images to be linked to the online store as well. Cropping the product images is also preferred as it makes the article look more professional.

Social Media Embeds: Linking to posts or videos from social media was well liked by the test persons. With videos, especially from TikTok, it should be ensured that they are not too long, as the attention period of the test persons was not especially long and the videos were often stopped earlier. It was requested that you can fast-forward the videos. The technical implementation should be tested here. In general, the test persons would like to see more social media embeds.

Call-to-Action Blocks/Combination of Image and Button: The opinion about the call-to-action blocks or the combination of image and button was different for both categories. For the CTA blocks, it is recommended to be able to enlarge the image, either by the article creator or by the user clicking on the image. The image should also be linked and show more information about the available colors or sizes. In addition, the price should be displayed more present and centered. Color accents, such as the button in the CD purple of the Beauty section or the CD pink of the Fashion section would make the block more appealing. In the combination of image and button, it is helpful to mention the price, the product name and a short product description.

Second Online Shop: Regarding the question of whether a second online store should be listed, all of the beauty test persons thought it made sense, as this would allow them to compare the price or buy from their favorite store. Among the fashion test persons, four stated that they saw this as positive, while three thought that only the cheapest store should be listed. Therefore, for both categories, it should be tested whether more click outs can be generated with the specification of two stores, more sales and therefore more commission can be earned.

More Articles at the End: The further articles, which are manually listed by the creator of the article at the end, were clicked only three times. Three further test persons stated that these did not sound interesting enough to them to click on these articles. The recommended action is to continue to indicate these articles, since there were no negative statements regarding the presentation or position and therefore internal traffic can be generated.

Figure 8 shows what optimized articles could look like on the basis of the recommendations for actions that have been identified.

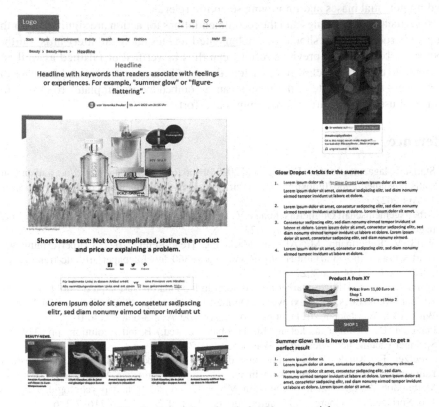

Fig. 8. Examples of optimized beauty articles

6 Final Evaluation and Outlook

This paper addresses recommendations for action to a media company for its online magazine in the field of beauty and fashion to improve its respective articles in terms of UX.

Hypothesis H1 is falsified, i.e. a list of products at the beginning of the article does not generate more click outs to the referenced online shop. To answer hypothesis H2, which states that fewer text (a text block of about 50 words) but more than one teaser image generates more click outs, requires more data in the future. The third hypothesis H3, which deals with the CTA blocks and the combination of image and buttons can be verified for both categories, fashion and beauty. Collecting more data is also useful since the hypothesis could only be verified by a narrow margin.

Future developments will require earlier involvement of UX experts in the planning process and the additional use of, for example, A/B testing. Consequently, the privacy issues associated with Google Analytics can also be prevented. It's equally important to ensure that participants in future UX testing are more closely matched to the effective target group that was identified in the quantitative part of the study. This will allow avoiding potential biases and improving scientific relevance.

Nevertheless, the findings and the recommendations for action are significant for the company's success. They should be implemented and tested again afterwards. Further hypotheses, based on the previous results, can thus be verified or falsified as well and contribute to enhance articles. In order for editors to be more efficient and increase the effectiveness of their content, the results can be embedded in templates to ensure the creation of user-friendly articles with minimal effort.

References

1. Statista: Magazine: Bunte in Deutschland 2022 Brand Report, Statista. https://de.statista.com/statistik/studie/id/98670/dokument/magazine-bunte-in-deutschlandbrand-report/. Accessed 15 Apr 2022
2. Arbeitsgemeinschaft Onlineforschung e.V.: Digital Report 2021, agof. https://www.agof.de/?page_id=18852. Accessed 29 May 2022
3. Weiß, V., Kellermann, M.: Affiliate Marketing Trend-Report 2022, Outperforming with heart & brain - xpose360 GmbH. https://www.xpose360.de/whitepaper/affiliate-trend-report-2022/. Accessed 24 Apr 2022
4. Lammenett, E.: Affiliate-marketing: Hintergründe, Funktionsprinzipien und Formen des Affiliate-Marketings. In: Praxiswissen Online-Marketing, pp. 47–79. Springer Fachmedien Wiesbaden, Wiesbaden (2021). https://doi.org/10.1007/978-3-658-32340-0_2
5. Petersen, D.: Affiliate marketing. In: Theobald, E. (ed.) Brand Evolution, pp. 329–346. Springer Fachmedien Wiesbaden, Wiesbaden (2017)
6. Kreutzer, R.: Kundendialog online und offline: Das große 1x1 der Kundenakquisition, Kundenbindung und Kundenrückgewinnung, Springer Fachmedien, Wiesbaden (2021)
7. Mattenklott, A.: Emotionale Werbung. In: Moser, K. (ed.) Wirtschaftspsychologie. S, pp. 85–106. Springer, Heidelberg (2007). https://doi.org/10.1007/978-3-540-71637-2_6
8. Müller, A., Miclau, C., Demaeght, A.: Customer experience: Die Messung und Interpretation von Emotionen im Dialogmarketing. In: Holland, H. (ed.) Digitales Dialogmarketing, pp. 603–625. Springer, Wiesbaden (2021). https://doi.org/10.1007/978-3-658-28959-1_26
9. Solomon, M.: Konsumentenverhalten. https://elibrary.pearson.de/book/99.150005/9783863267193. Accessed 17 May 2022
10. Häusel, H.: Limbic Success - So beherrschen Sie die unbewussten Regeln des Erfolgs - Die besten Strategien für Sieger. Haufe, Freiburg/Berlin/Planegg/München (2002)
11. Kleinginna, P., Kleinginna, A.: A categorized list of emotion definitions, with suggestions for a consensual definition. Motiv. Emot. 5(4), 345–379 (1981)
12. Ekman, P., Rosenberg, E.: What the Face Reveals Basic and Applied Studies of Spontaneous Expression Using the Facial Action Coding System (FACS). Oxford University Press, Oxford (2005)
13. Schmidt-Atzert, L., Stemmler, G., Peper, M.: Lehrbuch der Emotionspsychologie, Kohlhammer, Stuttgart, Germany (1996)
14. Müller, A., Gast, O.: Smart Big Data Management Customer Experience Tracking - Online-Kunden conversion-wirksame Erlebnisse bieten durch gezieltes Emotions management, pp. 313–343 (2014)

15. Miclau, C., Gast, O., Hertel, J., et al.: Nutzerprobleme beim ECommerce sehen und hören: Wie KI die Analyse der User Experience unterstützt. In: Verband, E.V. (ed.) Deutscher Dialogmarketing: Dialogmarketing Perspektiven 2019/2020, p. 60. Springer Fachmedien Wiesbaden, Wiesbaden (2020)

16. Trommsdorff, V., Teichert, T.: Konsumentenverhalten, 8th edn. Kohlhammer, Stuttgart (2011)

17. Gast, O., Müller, A.: Entscheidungsfindung: Die Rolle der Kundenemotionen – Was Mimik über Emotion und Entscheidung verrät. In: Keuper, F., Schomann, M., Sikora, L. (eds.) Homo Connectus, pp. 141–159. Springer Fachmedien Wiesbaden, Wiesbaden (2018)

18. Wulf, C.: Emotion. In: Wulf, C., Zirfas, J. (eds.) Handbuch Pädagogische Anthropologie, pp. 113–123. Springer, Wiesbaden (2014). https://doi.org/10.1007/978-3-531-18970-3_9

19. Scherer, K.: Neuroscience projections to current debates in emotion psychology. Cogn. Emot. **7**(1), 1–41 (1993)

20. Ekman, P.: Facial expression and emotion. Am. Psychol. **48**(4), 384–392 (1993)

21. Pease, A.: Body language: how to read others' thoughts by their gestures (2014)

22. Ekman, P., Matsumoto, D., et al.: (1997)

23. Thüring, M., Mahlke, S.: Usability, aesthetics and emotions in human–technology interaction. Int. J. Psychol. **42**(4), 253–264 (2007)

24. Sarodnick, F., Brau, H.: Methoden der Usability Evaluation: wissenschaftliche Grundlagen und praktische Anwendung, 2nd edn. Verlag Hans Huber, Bern (2011)

25. Miclau, C., Gast, O., Hertel, J., et al.: Nutzerprobleme beim ECommerce sehen und hören: Wie KI die Analyse der User Experience unterstützt. In: Verband, E.V. (ed.) Deutscher Dialogmarketing: Dialogmarketing Perspektiven 2019/2020, p. 61. Springer Fachmedien Wiesbaden, Wiesbaden (2020)

26. Nisbett, R., Wilson, T.: Telling more than we can know: verbal reports on mental processes. In: Psychological Review, pp. 231–259 (1977)

27. Ericsson, K., Simon, H.: Protocol Analysis: Verbal Reports as Data, p. 426. The MIT Press, Cambridge (1984)

28. Jo, Y., Stautmeister, A.: Don't make me think aloud! – Lautes Denken mit Eye Tracking auf dem Prüfstand (2011)

29. Demaeght, A., Müller, A., Miclau, C.: Empirische Forschung als Herausforderung im laufenden Lehrbetrieb – Lösungsansätze aus der Praxis. In: Breyer-Mayländer, T., Zerres, C., Müller, A., et al. (eds.) Die Corona-Transformation, pp. 419–430. Springer Fachmedien Wiesbaden, Wiesbaden (2022)

30. Thielsch, M., Moshagen, M.: Manual zum VisAWI (Visual Aesthetics of Websites Inventory) und der Kurzversion VisAWI-S (Short Visual Aesthetics of Websites Inventory), (2014)

31. Mueller, A., et al.: Hidden champions: a study on recruiting top-level staff in rural areas. In: Nah, F.-H., Siau, K. (eds.) HCII 2019. LNCS, vol. 11589, pp. 393–407. Springer, Cham (2019). https://doi.org/10.1007/978-3-030-22338-0_32

32. Beasley, M.: Chapter 3 - How web analytics works. In: Beasley, M. (ed.) Practical Web Analytics for User Experience, pp. 25–48. Morgan Kaufmann, Boston (2013)

33. Kaushik, A.: Web Analytics 2.0: The Art of Online Accountability & Science of Customer Centricity. Wiley Publishing, Inc. Indianapolis (2010)

34. Halfmann, M., Schüller, K. (eds.): Marketing Analytics. Springer, Wiesbaden (2022). https://doi.org/10.1007/978-3-658-33809-1

35. Internal evaluation of the Awin data

36. Sherer, S.: How to use affiliate marketing to drive low-cost conversions for your brand, Awin. https://www.awin.com/ca/how-to-use-awin/how-to-use-affiliate-marketing-to-drive-low-cost-conversions-for-your-brand. Accessed 07 July 2022

37. Wandiger, P.: CTR steigern – Mit höherer Klickrate mehr verdienen!, Affiliate Marketing Tipps. https://www.affiliate-marketing-tipps.de/tipps/ctr-steigern-mit-hoeherer-klickrate-mehr-verdienen/1005048/. Accessed 07 July 2022

38. Hanseranking GmbH: Kennzahlen im Online Marketing ⇒ Die wichtigsten Begriffe! https://www.hanseranking.de/kennzahlen-im-online-marketing/. Accessed 07 July 2022

39. Großmann, P.: Absprungrate in Google Analytics verstehen & verbessern [2022 Update], keyperformance.de. \https://keyperformance.de/absprungrate. Accessed 07 July 2022

40. Ostapchuk, B.: Understand the scrolling map, IO Technologies. https://help.iotechnologies.com/en/articles/489629-understand-the-scrolling-map. Accessed 07 July 2022

41. Siebert, S.: Ist Google Analytics legal oder illegal: Was Webseitenbetreiber jetzt tun müssen. http://www.e-recht24.de/artikel/datenschutz/6843-google-analytics-datenschutz-rechtskonform-nutzen.html. Accessed 07 Sept 2022

42. Doll, M.: Demografische Merkmale in Google Analytics I lunapark, Online Marketing Agentur - lunapark. https://www.luna-park.de/blog/9012-google-analytics-demografische-merkmale/. Accessed 07 Sept 2022

43. § 5a UWG

44. Röper, H.: Zeitungsfinanzierung I bpb.de. https://www.bpb.de/themen/medien-journalismus/lokaljournalismus/151250/zeitungsfinanzierung/. Accessed 08 Nov 2022

45. Internal evaluation of the Google Analytics data

Achieve Your Goal Without Dying in the Attempt: Developing an Area-Based Support for Nomadic Work

Guillermo Monroy-Rodríguez[1], Sonia Mendoza[1]([✉])[iD],
Luis Martín Sánchez-Adame[2][iD], Ivan Giovanni Valdespin-Garcia[1],
and Dominique Decouchant[3]

[1] Computer Science Department, CINVESTAV-IPN, Mexico City 07360, Mexico
{guillermo.monroy,sonia.mendoza,ivan.valdespin}@cinvestav.mx
[2] Computer Engineering Department, UAEM Valle de México,
State of Mexico 54500, Mexico
lmsancheza@uaemex.mx
[3] Information Technologies Department, UAM-Cuajimalpa,
Mexico City 05348, Mexico
decouchant@cua.uam.mx

Abstract. Organizations are constantly navigating a complex landscape of users, information, and security policies. Nomadic workers and visitors move throughout the organization, accessing resources and services to accomplish their goals. However, existing access control solutions often follow a general-purpose centralized approach, which can lead to limitations such as system bottlenecks and failures, and misunderstanding and confusion in people. In some cases, access control is performed in a manual way by the own workers. In this paper, we propose an innovative solution: an Area Management System (AMS) that addresses the unique needs of nomadic users. AMS takes into account the goals and requirements of nomadic users, the security policies of organizations, and the permissions and restrictions of specific areas. By distributing access control management, our solution can provide a more efficient and comprehensive approach to administrating resources, services, and information within an area.

Keywords: Resource access control · Area-based services · Intelligent buildings · Nomadic users

1 Introduction

People move between different areas within any kind of organization (e.g., universities, hospitals, companies, or government departments). By interacting with other people and computer systems, they can access services and perform specific actions in these areas to accomplish their partial and global goals [7]. Each area provides a part of the final solution by supplying the correct information to

© Springer Nature Switzerland AG 2023
F. Fui-Hoon Nah and K. Siau (Eds.): HCII 2023, LNCS 14038, pp. 421–438, 2023.
https://doi.org/10.1007/978-3-031-35969-9_28

the right persons, according to that zone's administration and security policies. Thus, nomadic users are people that gather pieces of a solution through the areas they visit to fulfill their assigned tasks [31]. However, in some cases, these tasks are not carried out efficiently, as there are no clear guidelines for resource and service management.

Under this perspective, every organization can be structured and modeled as an area hierarchy [24], i.e., a tree structure where each area is represented by a node, either encapsulating (ascendant) or encapsulated (descendant) [8]. This model represents the organizational relationships between areas rather than physical features, such as size or shape. Areas at higher levels help nomadic users to carry out general tasks, whereas those at lower levels, help them to perform more specific or specialized tasks [9].

As an example of the previous modeling approach, let us suppose a university campus. At the root node, it can be found the General Administration, which establishes both general organization security policies and access permissions for the included departments such as: Computer Science, Automatic Control, Biology, and Chemistry. These departments, being lower nodes, inherit the general rules of the General Administration, however, they set their policies, as they may also have specific areas (even lower nodes) with stricter rules, e.g., the warehouse of the Chemistry Department has restricted access as it houses hazardous substances. All of these policies can have a negative impact on making the most of the university's resources and services as, although they are intended to maintain security, their application is cumbersome and can be confusing, especially if they are to be applied on a temporary basis, e.g., to a delivery person or a visiting student.

Through our AMS proposal, we aim to provide support to develop applications for nomadic users within organizations logically subdivided into areas. This proposal presents a decentralized solution in which each area manages, in an autonomous way, the access to its resources and workflows, which establish the activity plan that nomadic users will carry out within an area. Instead of defining a centralized global workflow for each user role, our solution defines a distributed area-based workflow system, where each area has the responsibility to define the roles played by the different nomadic users it wants to provide work support. Besides, the workflow security rules of each area are inherited from upper areas following the social and institutional structure. The general workflow is not statically predetermined, and each sub-workflow area can be dynamically redefined and adapted to specific (possibly temporary) cases. The different areas collaborate by sharing information to ensure the accomplishment of the nomadic users' goals, while observing the security policies of the organization. The first prototype of our proposal has been successfully developed, validated and deployed within a real institution.

This paper is organized as follows. In Sect. 2, we provide an overview of related work pertaining to the main topics of our area management proposal, including control access systems and workflows. In Sect. 3, we present the model and key contributions of this paper. Section 4 includes a case study to demonstrate the application of the concepts outlined in the previous sections. Section 5 details the testing of our prototype with end-users. Finally, in Sect. 6, we discuss conclusions and future work.

2   Related Work

Most research works approach access control and workflows in a separately way, i.e., these topics are not usually combined into a single solution proposal as our work does it. As far as access control systems are concerned, the Role Base Access Control (RBAC) [12,22] model has been recognized as a fundamental model for building access control systems both in academic circles and the software engineering community. Although there are multiple works focused on the RBAC model (such as those proposed by Kim and Park [19], van der Laan [20], Cao et al. [5], Pasquale et al. [26], and Ben Fadhel et al. [1]) this model does not tackle the autonomous area management schema. On the other hand, in mobile and nomadic computing applications, the RBAC model has been extended to deal with user requirements inside smart spaces [28].

In recent years, various extensions have been proposed to the RBAC model to address specific constraints and requirements. For instance, T-RBAC [2,33] extends the RBAC model to support temporal constraints for enabling and disabling roles. This model is further extended in Generalized T-RBAC [18] to support temporal conditions on *user-role* and *role-permission* assignments. Another extension, S-RBAC [16,34], supports spatial constraints on enabling and disabling roles but does not address area hierarchy structure. GEO-RBAC [10] extends the RBAC model to incorporate spatial and location-based information, introducing the concept of *geographically bounded organizational function*. LoT-RBAC [4,8] adds time and space dimensions to the RBAC model and introduces the concept of *location hierarchy*. The Or-BAC (Organizational Based Access Control) model [9,14] addresses the management of virtual communities securely, adapting the RBAC model to deal with sub-organizations. RBAC Administration in Distributed Systems (RBAC-ADS) [11] and Distributed Role-Based Access Control for Dynamic Coalition Environments (dRBAC) [13] propose distributed access control mechanisms. Although these models have added useful dimensions to RBAC, they do not adequately address the specific requirements of our proposed area management schema.

Some works have been proposed regarding workflow systems and the RBAC model. Bertino et al. [3] develop a method for assigning roles and users to tasks in a workflow. Task-Role Based Access Control [25] focuses on permissions that are not directly assigned to roles but to related tasks first and then to roles. The W-RBAC model [32] presents two models incorporating RBAC into workflow systems: W0-RBAC, which includes separation of concerns, and W1-RABC, which extends the basic model by incorporating exception handling capabilities. Sun et al. [29] propose the design and architecture of a flexible workflow system that incorporates RBAC. AW-RBAC [21] is an extended RBAC model for adaptive workflow systems that includes change operations and various objects subject to change within workflow systems. Despite their contribution, these works do not address the specific needs of auto-manageable areas where autonomous administration is carried out.

Finally, several studies have been conducted on the intersection of workflow systems, and mobile and ubiquitous computing. Maurino and Modafferi [23] propose a model to divide a single workflow into multiple autonomous workflows, supporting disconnected features for wireless networks in mobile scenarios. uFlow [15] is a workflow service framework that integrates, manages, and executes services in ubiquitous environments. However, these works need to address the specific requirements of an area structure approach, where the management of workflows assigned to nomadic users is accomplished autonomously.

3   Distributed Management of Areas

In this section, we describe AMS, a system to support area-based nomadic work that follows an autonomous management approach to allow each area to perform the subsequent three actions: 1) managing its services and resources; 2) exchanging information with other areas; and 3) making context-based decisions to provide nomadic users with suited functions that allow them to accomplish their goals. This autonomous management approach deals with the well-known disadvantages of most related works that propose a centralized solution [5,6,30], such as slow response, network overhead, restrictions to access and manage shared resources and services, belated resource updating, among others. This autonomy constitutes a keystone to efficiently support nomadic user work within organizational environments.

AMS is based on a distributed model that considers a self-management of services and resources provided by each area of an organization. Each area administrates the objects under its domain (e.g., area context information, roles, workflows, and data exchange among areas) while observing the general security policies of the organization. Consequently, the following issues are considered in our solution:

- Actions and tasks to be performed by nomadic users;
- Roles that can be assigned to them;
- Resources that nomadic users can manipulate;
- Permissions that allow executing operations on resources;
- Constraints which specify the access types such as obligations, prohibitions, and permissions;
- Area context information storage and recovery;
- Inter-relations and interactions between areas to distribute and share information, services and resources.

To our knowledge, there is no research in the fields of ubiquitous computing and human-computer interaction that fulfills such requirements in a comprehensive way.

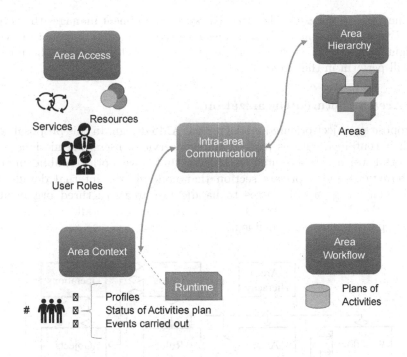

Fig. 1. Components of the AMS Architecture

3.1   The AMS Architecture

An area is a logical entity that supplies specialized services and resources. In addition, an area manages its own information, in an independent way, to help nomadic users to accomplish their partial or global objectives. Figure 1 shows the AMS architecture with several components such as: *Area Access*, *Area Hierarchy*, *Area Workflow*, *Area Context Information*, and *Intra Area Communications*.

The *Area Hierarchy* component establishes the relationship among the current area and the area hierarchy structure and provides operations that modify this structure, such as creating sub-areas (descendants), joining a sub-area into another one, deleting a sub-area, copying a sub-area, and moving a sub-area. The *Area Context* component is responsible for managing area context information, which includes current nomadic users within a given area, their profile, the status of their activity plan, and the list of events being carried out within the area. The *Intra Area Communications* component sends and receives area context information to and from the others areas in the area hierarchy.

On the other hand, the *Area Access* component is responsible for allowing nomadic users to access the resources and services managed by an area. The access control system is based on the RBAC model. The main tasks carried out by this component are: assigning and modifying the nomadic user's role; activating and deactivating the nomadic user's role; assigning a set of permissions to each role; assigning a set of operations to each resource; and establishing obligations from the users to the area. The *Area Workflow* component manages the activity plans. Every time a nomadic user comes into the area, this component retrieves the right workflow for such a user, i.e., the list of activities that the nomadic user will perform in this area.

3.2   Area-Structured Organization

We propose an object-oriented model for the AMS design. In this proposal, areas and their contained entities (e,g., resources, services, users, permissions, roles, and workflows) are represented as objects that have both a public interface for interactions and a private section that conceals the internal details. Our model defines a group of classes to handle the area-structured organization, including the classes User, Area Hierarchy, Access and Workflow. Additionally, some supporting classes are defined.

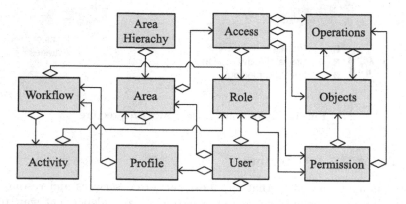

Fig. 2. Classes of the AMS Architecture

"User" Class

The User class models the nomadic user within the area-structured organization by managing information such as user identification, current role, user profile, and smartphone identification. It is responsible for assigning both external (at the organization level) and internal (at the area level) workflows, saving and retrieving roles, determining the current area, identifying the current activity performed by the user, and obtaining the current activity plan assigned to the user.

The user profile is an instance of the Profile class, which is based on the BDI (Beliefs, Desires, and Intentions) theoretical model [27]. This class provides information on the user's goals within the organization (Desires), their activity plans to achieve these goals (Intentions), and their behavior and interests (Beliefs).

Taking the user's interests into account, AMS can make recommendations, such as inviting a visiting student to an academic event. If the user accepts the invitation, AMS will guide them to the event venue. The user's behavior is also an important aspect considered by the Profile class. For example, at the end of a visit to the organization, the user's behavior can be rated as "excellent", "regular", or "bad", depending on whether they comply with the organization's security policies. This will impact future access to the organization.

"Area Hierarchy" Class

The Area Hierarchy class models the organizational area hierarchy structure. It encompasses all the area hierarchies within the organization. This class defines a node (area or subarea) as the root of the structure in the current area. It supports creating, deleting, copying, moving a node, and displaying the area hierarchy. The area hierarchy structure consists of instances of the Area class.

The Area class links to its parent node and descendant nodes and stores information about activity plans (via the Workflow class), roles, permissions, operations, and resources (through the Access class) within its domain. It is responsible for assigning both workflows and roles to nomadic users (using the User class).

"Access" Class

The Access class models the management of access within an area, drawing on the RBAC model, which includes Users, Roles, Permissions, Operations, Objects, Separation of Duties, and Role Hierarchy. This class uses multiple data structures to represent the RBAC model elements. For example, it uses the Objects class to specify the resources and services within an area, such as a wireless network, a projector, or a printer. The Operation class defines operations that can be performed on these objects, such as connecting to a wireless network or printing a file. The Permission class holds a data structure that defines the permissions available within an area, linking operations and objects. The Role class defines the roles available within the area and establishes a list of permissions that each role allows. It also considers the static and dynamic separation of duties within the organization's security policies. Additionally, the Access class keeps track of a user's role within the area and their translated role from a neighboring area.

This class offers operations for managing access, including creating permissions, assigning permissions to roles, assigning users to roles, translating external roles to internal area roles, and checking for separations of duties. It also provides operations for retrieving specific permissions from a role and specific roles from a user.

"Workflow" Class

The Workflow class manages the activity plan for a nomadic user within an area. The Activity class defines a single task and provides the tools to create and complete it. Multiple activities can be grouped using the SetOfActivities class (not depicted in Fig. 2), which outlines the execution order of the activities, and whether they should be performed in parallel, sequentially, or a combination.

The Activity class holds information about the list of tasks, the authorized roles for executing the workflow, and the maximum time allotted for the user to complete their activity plan. This class offers functions for assigning a workflow to a role, starting and stopping it, removing it from a role, and determining the user's current activity.

4 Application: Delivery Within a Research Institution

To illustrate how AMS provides users with an area-based support for nomadic work, let us consider the case study of a deliveryman that has to deliver a parcel to a researcher in the Computer Science Department. This department is located within a large university domain that includes the following areas: *Administration, Reception, Computer Science Department, CSCW Laboratory* of the Computer Science Department, *Accountancy & Cashier*, and of course the rest of the university campus (i.e., other departments, laboratories, libraries, auditoriums, coffee shops, corridors, parking lots, and outdoors places). These areas are represented and logically structured in a hierarchy that is directly related to the socio-labor organization of the institution.

Arrival of the Deliveryman at the Reception

At the beginning of the process, the deliveryman arrives to the Reception of the research institution. In this initial area, he first has to provide an official identification to be scanned by a member of the security staff. Once identified, the Reception AMS subsystem assigns him the *Visitor* role and activates the associated permissions, as well as the obligations required for that role. Consequently, the deliveryman has to fill out a registration form on a touchscreen, where he has to select the purpose of his visit and, depending on it, a series of options will be displayed. In this case, these options are the department to which the deliveryman will go and the person to whom he will deliver the parcel. Thus, the Reception AMS subsystem identifies his purpose as a *Delivery of Parcel* to the researcher *Tom* in the *Computer Science Department*.

Then, the Reception AMS subsystem sends a query to the involved areas (Computer Science Department and Accountancy & Cashier) to find out the presence and availability of persons and resources required for the purpose of the deliveryman. If everything is fine, the Reception AMS subsystem displays the global security policies imposed by the *Administration* of the research institution, so that he can be aware of them. Examples of security policies are: «Salesmen are not allowed to come into the institution domain» or «Visitors are allowed to come in only if they have a programmed visit». To make sure he has read the security policies, the Reception AMS subsystem will not let him move on to the next step before the average time it takes a person to read them. Once the time has passed, the deliveryman must accept them using his fingerprint as an acknowledgment signature. The deliveryman's identification and fingerprint as well as the name of the person requested to deliver the parcel are sent to the AMS subsystems of the Computer Science Department and Accountancy & Cashier, so that these areas can identify the deliveryman and give him the necessary support to achieve his partial goals.

To determine whether the requested person, i.e., Tom, is present in the Computer Science Department, each member goes in and out of the building using their fingerprint to register their stay in the department. If the presence of Tom in the department's facilities is confirmed, it is also necessary to know whether Tom is in their office, i.e., in the CSCW Laboratory. By using one of the presence detectors located on the light switches in all rooms, the Computer Science Department AMS subsystem not only obtains presence information, but also the lights will turn on or off automatically in the presence or absence of a person. To ensure that the requested person is the one in the laboratory, it is necessary to verify the fingerprint registration and the presence of people inside the CSCW laboratory to send a positive answer to the Reception AMS subsystem.

Once the presence of the requested person is confirmed, the deliveryman is provided with a mobile device that has all the necessary accessories, such as a camera with QR code reading and Internet access, so that he can carry out his activities within the institution (step #0 of Fig. 3). In case the deliveryman enters with a vehicle, a camera located in the entrance area of vehicles identifies the license plate number.

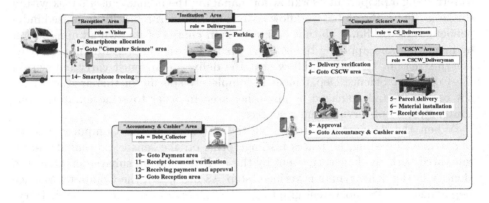

Fig. 3. Delivery of a Parcel within a Hierarchically Organized Institution

Once outside the Reception area, the Institution AMS subsystem takes control to guide the deliveryman to his first stop: the Computer Science Department. This AMS subsystem establishes his profile by assigning the *Deliveryman* role and the global workflow that defines the general activity plan. In this case, this workflow downloaded to the deliveryman's mobile device is reduced to three sequential activities: the journey of the deliveryman to the Computer Science Department (step #1 of Fig. 3), the journey of the deliveryman to the Accountancy and Cashier (step #9 of Fig. 3), and the return of the deliveryman to the Reception (step #12 of Fig. 3). These areas will help the deliveryman to achieve his global goal. Areas not included in this workflow (i.e., other faculties of the campus) will not grant access to the deliveryman. In this way, these areas participate to the enforcement of the security policies by denying access to this temporary nomadic user.

Journey from Reception to Computer Science Department

As previously mentioned, the Institution assigned to the deliveryman a global workflow whose first part consists in reaching the Computer Science Department. The journey between these two areas includes all the steps (possibly passing through other areas) that the deliveryman can take. To do this route, the deliveryman uses an interactive map of the institution enriched with relevant contextual information to guide him.

The Institution AMS subsystem checks whether the space designated for the parking of parcel vehicles is available by sending a query to the Computer Science Department AMS subsystem (step #2 of Fig. 3). This information is obtained by monitoring the parking spaces, which are equipped with proximity sensors strategically placed around each parking space in addition to an identifier.

At any level, the task of each area consists in: a) activating the associated permissions, b) establishing the obligations from the nomadic user to the organization, c) informing to other institution areas about the user's profile, and d) retrieving a proper sub-workflow for managing the nomadic user's task within this area. Thus, the global workflow is divided into hierarchically organized independent sub-workflows, so that each area is in charge of its own administration by inheriting and applying the rules, rights and duties imposed by upper areas. For this reason, the sub-workflow that the deliveryman must carry out within the Computer Science Department is completely specific of this area and cannot be known and managed by any other area in order to achieve autonomous administration of services and resources.

When the deliveryman arrives at the entrance of the Computer Science Department building, he places his fingerprint on the sensor, so that it can be compared with the fingerprint sent by the AMS Reception subsystem. Once the identity of the deliveryman is verified (step #3 of Fig. 3), the Computer Science Department AMS subsystem grants the *CS_Deliveryman* role to the deliveryman and creates the corresponding sub-workflow, which includes the following activities: obtain authorization of the requested person to go to the CSCW Laboratory (step #4 of Fig. 3), receive an approval once successfully delivered the parcel (step #8 of Fig. 3) and exit the building to go to the Accountancy & Cashier area (step #9 of Fig. 3).

The Computer Science Department AMS subsystem also sends the deliveryman's fingerprint to the CSCW AMS subsystem, which notifies the person to whom the parcel should be delivered, i.e., the researcher Tom, about the arrival of the deliveryman. This notification is made by sending a text message to Tom's mobile device and by displaying the deliveryman's identification photo along with the *authorization* and *denial* options on a touchscreen placed in the CSCW laboratory to show important information.

Once Tom has confirmed the *authorization* option, the Computer Science Department AMS subsystem allows the deliveryman to access the facility and displays the building map on his mobile device to guide him inside the Computer

Science Department. When the deliveryman arrives at the CSCW Laboratory, he must use his fingerprint to enter. At this moment, the CSCW AMS subsystem grants him the *CSCW_Deliveryman* role and creates the corresponding sub-workflow, which contains the following activities: effectively deliver the parcel (step #5 of Fig. 3), install and test the parcel content (step #6 of Fig. 3) and obtain a receipt document (step #7 of Fig. 3).

After checking the deliveryman's fingerprint, the CSCW AMS subsystem authorizes him to enter and displays, on the touchscreen of the CSCW Laboratory, a positive and a negative icons to indicate if the parcel is in good or bad condition, respectively. In case Tom selects the positive icon, two more options are displayed to ask Tom to select whether the parcel content should be installed or not. In case it does not require installation, a QR code containing a receipt document from the CSCW Laboratory will be sent to the deliveryman's mobile device. This code will be later scanned in the Accountancy & Cashier area. In the event that an installation is required, two options will be displayed on the touchscreen to indicate whether it was successful or not. If so, the QR code will be sent to the deliveryman's mobile device. If there exists a problem with the parcel content or installation, the role and corresponding sub-workflow will be updated to send the deliveryman to the corresponding area for clarification, and the permissions obtained in the CSCW laboratory and in the Computer Science Department will be canceled.

Once the QR code has been received on the deliveryman's mobile device, he can exit the CSCW Laboratory and his permissions will be removed, so he can start wrapping up his activity in the Computer Science Department. Then, the CSCW AMS subsystem sends the Computer Science Department AMS subsystem an approval that will allow the deliveryman to leave the building (step #8 of Fig. 3). In case the deliveryman does not have such an approval, he cannot leave the building, and the Computer Science Department AMS subsystem will send the Reception AMS subsystem a warning message, so that members of the security staff can handle the case. As soon as the deliveryman receives the approval notification, he can exit the building using his fingerprint to go to his next stop: the Accountancy & Cashier are (step #9 of Fig. 3).

It is important to notice that the different nested areas define their own work-flow specifying what they require and what they produce. The AMS subsystem of each area assigns a specific role to the nomadic user, allowing him to obtain permission to access some local resources (e.g., the CSCW Laboratory where requested person works). Moreover, the security is enforced denying the access to some well identified parts of these areas.

Journey from Computer Science Department to Accountancy & Cashier

Being outside the Computer Science Department, the deliveryman has to reach the Accountancy & Cashier area to perform the required paperwork to initialize and obtain the payment. In this case, the Institution AMS subsystem again helps him reach this service and guides his movements within the institution's domain.

Given that we have supposed that the deliveryman arrived in a vehicle, when he leaves the parking space of the Computer Science Department, the proximity sensors system detects the liberation of that space and notifies it to the Computer Science Department AMS subsystem to change the status of the parking space to free.

When the deliveryman approaches the Accountancy & Cashier building, the Institution AMS system requests the Accountancy & Cashier AMS subsystem to assign a space to the deliveryman's vehicle. Like in the case of the parking lot in the Computer Science Department, each parking space has the same proximity sensors system and an identifier. Thus, the mobile application provides the deliveryman with the parking space number that he must use. In case there is no free parking space, it is processed as an exception as in any workflow system.

Already being at the entrance of the Accountancy & Cashier building, just like the deliveryman did in the Computer Science Department, he places his fingerprint on the sensor to gain access. After having verified the deliveryman's identity, the Accountancy & Cashier AMS subsystem will allow him to know the security policies of the building through his mobile device. Being a building where economic resources are managed, security inside is greater, so wearing items like hats or dark glasses is prohibited.

Once the deliveryman has accepted these security policies, a camera placed at the entrance will verify compliance with security policies to enter the building. This camera is part of a system that uses an artificial intelligence algorithm to recognize objects that are not allowed. Once this is confirmed, the Accountancy & Cashier AMS subsystem will give him the *Debt_ Collector* role, which allows him to access the building, and creates the corresponding sub-workflow, which includes the following activities: go to payment area (step #10 of Fig. 3), put the QR code on the scanner and his fingerprint to verify that he is the correct person to pay for the work performed (step #11 of Fig. 3), sign the received payment to get an exit approval (step #12 of Fig. 3), and finally leave the building to return to the Reception area (step #13 of Fig. 3).

On the mobile device, the deliveryman will also visualize the map of the building that contains only the route to the Payment area, since he will not have access to any other part of the building, except the toilets. Once in this area, the deliveryman must show the QR code (obtained in the Computer Science Department) to be scanned and place his fingerprint on the sensor in order to receive his payment, which can be in cash, by check or bank deposit. Then the deliveryman must sign an acknowledgment on a touchscreen, and the payer must approve his exit from the building, just like in the Computer Science Department. This sub-workflow finishes when the deliveryman leaves the Accountancy & Cashier building using his fingerprint.

Deliveryman's Exit of the Research Institution

Once outside the Accountancy & Cashier building, the Institution AMS subsystem again updates his role and sub-workflow, so that the deliveryman can go to the Reception and complete his general objective. This subsystem presents new services in the deliveryman's mobile device, mainly an interactive map to guide him back to recover his vehicle and return to the Reception area.

Since the purpose of his visit has been fulfilled when the deliveryman arrives at the Reception area, he will return the mobile device (step #14 of Fig. 3) that was provided to him at the entrance and will put his fingerprint back on the sensor to record the end of his activities (i.e., global workflow). It is important to mention that the usage of the fingerprint also allows the AMS system to determine whether a visitor tried to enter some other facility that was not specified in the different workflows.

If all is correct, he recovers his official identification, while the Reception AMS subsystem deactivates the *Visitor* role and workflows associated to this temporary visitor and stores the information of this visit to the persistent archiving system. Finally, the deliveryman can leave the institution.

5 Tests

In order to evaluate the effectiveness of our area-based support for nomadic work, we conducted a user study using the *Task Load Index (NASA-TXL)* [17] tool with five participants. This tool rates the workload perceived by end users to evaluate the efficacy of a task, a system, a group or other performance aspects. The NASA-TXL questionnaire is composed of six scales:

1. *Mental demand*: How much mental activity was necessary? (e.g., think, decide, calculate, recall, search for, and investigate) Is it about an easy or difficult, simple or complex, heavy or light task?
2. *Physical demand*: How much physical activity was necessary? (e.g., push, pull, spin, press, and activate) Is it about an easy or difficult, slow or fast, relaxed or tired task?
3. *Temporal demand*: How much time pressure did the user feel due to the pace in which the tasks or their elements occurred? Was the rate slow and deliberate, or fast and frenetic?
4. *Effort*: To what extent has the user had to work (physically or mentally) to achieve the level of results?
5. *Performance*: What is the satisfaction degree with the execution level?
6. *Frustration*: During the task, to what extent did the user feel insecurity, discouragement, irritability, stress, worry or, otherwise, security, liking, relaxation, and satisfaction.

To carry out our tests, we had briefly explained to our participants the objective of the system, as well as the intention of the questionnaire and the meaning of each scale. The participants were asked to perform a series of tasks related to their specific roles while using our prototype system and then complete the NASA-TLX questionnaire by assigning a score between 0 and 100 to each scale. The roles taken by the participants were:

1. A delivery man who was going to drop off a parcel at the Toxicology Department.

2. An applicant participating in the entry process for the Master in Computer Science.
3. A professor that was invited as a jury to a thesis defense in the Mathematics Department.
4. A participant in the Electrical Engineering congress.
5. An air conditioning technician carrying parts to repair a unit in school services.

The tasks were designed to simulate real-world scenarios that nomadic users may encounter, such as finding a specific location, accessing important documents, and taking advantage of local resources.

Table 1 shows the results of the questionnaire. In particular, it shows the weighting per user as well as the average ($\bar{x}$) and standard deviation (σ) for each scale. As seen from Table 1, the mean score for mental, physical, and temporal demand are low, indicating that the system was able to significantly reduce the mental, physical, and temporal demands experienced by users. Additionally, the participants reported high levels of performance, low levels of effort, and minimal frustration while using the system.

Table 1. Results of the questionnaire for the area-based support for nomadic work

User	Mental demand	Physical demand	Temporal demand	Performance	Effort	Frustration
1	5	5	5	100	5	0
2	2.5	0	5	90	2.5	2.5
3	5	2.5	5	100	5	0
4	10	5	10	80	10	5
5	5	0	2.5	95	5	0
Mean	5.5	2.5	5.5	93	5.5	1.5
SD	2.73	2.5	2.73	8.36	2.73	2.23

The results of our study suggest that our area-based support for nomadic work effectively reduces the mental, physical, and temporal demands experienced by users, regardless of their specific roles. This is likely due to the system's ability to provide personalized support based on the user's location and work needs. The low mean scores for mental demand, physical demand, and temporal demand indicate that the system was able to alleviate the cognitive and physiological demands of the tasks, allowing the participants to focus on their work. Additionally, the high levels of performance and low levels of effort reported by the participants indicate that the system is user-friendly and easy to use. The low mean score for frustration also indicates that the system is intuitive and straightforward for the users. However, it should be noted that the study was conducted with a small sample size of five participants, and further validation with a larger sample size is needed. Additionally, it would be interesting to see how the system performs in different scenarios and contexts to see if the results are generalizable to other roles and environments.

6   Conclusion and Future Work

We proposed AMS to aid in developing nomadic applications within organizations that are logically divided into areas. Our decentralized approach allows each area to independently manage access to its resources and services and create workflows that establish the activities that nomadic users will carry out within the area.

The case study presented in this proposal emphasizes the importance of our solution for organizing and administering a global workflow system. Rather than defining a unique workflow for each user role, our solution is based on a distributed, area-based workflow system where each area is responsible for defining the roles and activities associated with the different types of users it wishes to support.

The workflow security policies of each area can be inherited from upper areas, following the organization's social and institutional structure. As a result, workflows are not predetermined and can be dynamically redefined and adapted to specific, temporary cases. Different areas within the institution collaborate by sharing information to ensure the accomplishment of the nomadic users' goals while adhering to the organization's security policies.

We have successfully designed and validated a prototype of AMS with end-users, who found the system usable, user-friendly, easy to use, intuitive, and straightforward. Additionally, we are currently in the process of deploying the prototype within a larger real-world institution.

Implementing AMS in a realistic nomadic working environment has the potential to support multiple applications. For example, we plan to add groupware facilities such as a system for managing the availability of distributed resources at the user level and a face recognition system for detecting and locating users, even if they are temporarily separated from their smartphones. These facilities will create an advanced working environment to support well-organized, nomadic, and cooperative work. Finally, we also plan to replace nomadic users' mobile devices with smart badges, using e-ink screens or other low-power technologies.

References

1. Ben Fadhel, A., Bianculli, D., Briand, L., Hourte, B.: A model-driven approach to representing and checking RBAC contextual policies. In: Proceedings of the Sixth ACM Conference on Data and Application Security and Privacy, CODASPY 2016, pp. 243–253. Association for Computing Machinery, New York (2016). https://doi.org/10.1145/2857705.2857709
2. Bertino, E., Bonatti, P.A., Ferrari, E.: TRBAC: a temporal role-based access control model. ACM Trans. Inf. Syst. Secur. 4(3), 191–233 (2001). https://doi.org/10.1145/501978.501979

3. Bertino, E., Ferrari, E., Atluri, V.: The specification and enforcement of authorization constraints in workflow management systems. ACM Trans. Inf. Syst. Secur. **2**(1), 65–104 (1999). https://doi.org/10.1145/300830.300837

4. Bertolissi, C., Fernandez, M.: Time and location based services with access control. In: 2008 New Technologies, Mobility and Security, pp. 1–6 (2008). https://doi.org/10.1109/NTMS.2008.ECP.98

5. Cao, Y., Huang, Z., Yu, Y., Ke, C., Wang, Z.: A topology and risk-aware access control framework for cyber-physical space. Frontiers Comput. Sci. **14**(4), 1–16 (2020). https://doi.org/10.1007/s11704-019-8454-0

6. Cao, Y., Ping, Y., Tao, S., Chen, Y., Zhu, Y.: Specification and adaptive verification of access control policy for cyber-physical-social spaces. Comput. Secur. **114**, 102579 (2022). https://doi.org/10.1016/j.cose.2021.102579. https://www.sciencedirect.com/science/article/pii/S016740482100403X

7. Carruthers, A.: Role-based access control (RBAC), pp. 123–149. Apress, Berkeley, CA (2022). https://doi.org/10.1007/978-1-4842-8593-0_5

8. Chandran, S.M., Joshi, J.B.D.: *LoT-RBAC*: a location and time-based RBAC model. In: Ngu, A.H.H., Kitsuregawa, M., Neuhold, E.J., Chung, J.-Y., Sheng, Q.Z. (eds.) WISE 2005. LNCS, vol. 3806, pp. 361–375. Springer, Heidelberg (2005). https://doi.org/10.1007/11581062_27

9. Cuppens, F., Miege, A.: Modelling contexts in the Or-BAC model. In: Proceedings of the 19th Annual Computer Security Applications Conference, pp. 416–425 (2003). https://doi.org/10.1109/CSAC.2003.1254346

10. Damiani, M.L., Bertino, E., Catania, B., Perlasca, P.: GEO-RBAC: a spatially aware RBAC. ACM Trans. Inf. Syst. Secur. **10**(1), 2-es (2007). https://doi.org/10.1145/1210263.1210265

11. Dekker, M., Crampton, J., Etalle, S.: RBAC administration in distributed systems. In: Proceedings of the 13th ACM Symposium on Access Control Models and Technologies, SACMAT 2008, pp. 93–102. Association for Computing Machinery, Estes Park, CO, USA (2008). https://doi.org/10.1145/1377836.1377852

12. Ferraiolo, D.F., Sandhu, R., Gavrila, S., Kuhn, D.R., Chandramouli, R.: Proposed NIST standard for role-based access control. ACM Trans. Inf. Syst. Secur. **4**(3), 224–274 (2001). https://doi.org/10.1145/501978.501980

13. Freudenthal, E., Pesin, T., Port, L., Keenan, E., Karamcheti, V.: dRBAC: distributed role-based access control for dynamic coalition environments. In: Proceedings 22nd International Conference on Distributed Computing Systems, pp. 411–420 (2002). https://doi.org/10.1109/ICDCS.2002.1022279

14. Guesmia, K., Boustia, N.: OrBAC from access control model to access usage model. Appl. Intell. **48**(8), 1996–2016 (2017). https://doi.org/10.1007/s10489-017-1064-3

15. Han, J., Cho, Y., Kim, E., Choi, J.: A ubiquitous workflow service framework. In: Gavrilova, M.L., et al. (eds.) ICCSA 2006. LNCS, vol. 3983, pp. 30–39. Springer, Heidelberg (2006). https://doi.org/10.1007/11751632_4

16. Hansen, F., Oleshchuk, V.: SRBAC: a spatial role-based access control model for mobile systems. In: Proceedings of the 7th Nordic Workshop on Secure IT Systems (NORDSEC 2003), pp. 129–141 (2003)

17. Hart, S.G., Staveland, L.E.: Development of NASA-TLX (task load index): results of empirical and theoretical research. In: Hancock, P.A., Meshkati, N. (eds.) Human Mental Workload, Advances in Psychology, vol. 52, pp. 139–183. North-Holland (1988). https://doi.org/10.1016/S0166-4115(08)62386-9. https://www.sciencedirect.com/science/article/pii/S0166411508623869

18. Joshi, J., Bertino, E., Latif, U., Ghafoor, A.: A generalized temporal role-based access control model. IEEE Trans. Knowl. Data Eng. **17**(1), 4–23 (2005). https://doi.org/10.1109/TKDE.2005.1
19. Kim, J., Park, N.: Role-based access control video surveillance mechanism modeling in smart contract environment. Trans. Emerging Telecommun. Technol. **33**(4), e4227 (2022). https://doi.org/10.1002/ett.4227. https://onlinelibrary.wiley.com/doi/abs/10.1002/ett.4227. e4227 ETT-20-0572.R2
20. van der Laan, J.: Incremental verification of physical access control systems, January 2021. http://essay.utwente.nl/85634/
21. Leitner, M., Rinderle-Ma, S., Mangler, J.: AW-RBAC: access control in adaptive workflow systems. In: 2011 Sixth International Conference on Availability, Reliability and Security, pp. 27–34 (2011). https://doi.org/10.1109/ARES.2011.15
22. Masoumzadeh, A., van der Laan, H., Dercksen, A.: BlueSky: physical access control: characteristics, challenges, and research opportunities. In: Proceedings of the 27th ACM on Symposium on Access Control Models and Technologies, SACMAT 2022, pp. 163–172. Association for Computing Machinery, New York (2022). https://doi.org/10.1145/3532105.3535019
23. Maurino, A., Modafferi, S.: Workflow management in mobile environments. In: Baresi, L., Dustdar, S., Gall, H.C., Matera, M. (eds.) UMICS 2004. LNCS, vol. 3272, pp. 83–95. Springer, Heidelberg (2004). https://doi.org/10.1007/978-3-540-30188-2_7
24. Mohamed, A.K.Y.S., Auer, D., Hofer, D., Küng, J.: A systematic literature review for authorization and access control: definitions, strategies and models. Int. J. Web Inf. Syst. (2022). https://doi.org/10.1108/IJWIS-04-2022-0077
25. Oh, S., Park, S.: Task-role based access control (T-RBAC): an improved access control model for enterprise environment. In: Ibrahim, M., Küng, J., Revell, N. (eds.) DEXA 2000. LNCS, vol. 1873, pp. 264–273. Springer, Heidelberg (2000). https://doi.org/10.1007/3-540-44469-6_25
26. Pasquale, L., et al.: Topology-aware access control of smart spaces. Computer **50**(7), 54–63 (2017). https://doi.org/10.1109/MC.2017.189
27. Rao, A.S., Georgeff, M.P.: Modeling rational agents within a BDI-architecture. In: Proceedings of the Second International Conference on Principles of Knowledge Representation and Reasoning, KR 1991, pp. 473–484. Morgan Kaufmann Publishers Inc., San Francisco, CA, USA (1991)
28. Sánchez, Y.K.R., Demurjian, S.A., Conover, J.C., Agresta, T., Shao, X., Diamond, M.: Role-based access control for mobile computing and applications. In: Information Diffusion Management and Knowledge Sharing (2020)
29. Sun, Y., Meng, X., Liu, S., Pan, P.: Flexible workflow incorporated with RBAC. In: Shen, W., Chao, K.-M., Lin, Z., Barthès, J.-P.A., James, A. (eds.) CSCWD 2005. LNCS, vol. 3865, pp. 525–534. Springer, Heidelberg (2006). https://doi.org/10.1007/11686699_53
30. Tsigkanos, C., Pasquale, L., Ghezzi, C., Nuseibeh, B.: Ariadne: topology aware adaptive security for cyber-physical systems. In: 2015 IEEE/ACM 37th IEEE International Conference on Software Engineering, vol. 2, pp. 729–732 (2015). https://doi.org/10.1109/ICSE.2015.234
31. Vijayalakshmi, K., Jayalakshmi, V.: A study on current research and challenges in attribute-based access control model. In: Hemanth, D.J., Pelusi, D., Vuppalapati, C. (eds.) Intelligent Data Communication Technologies and Internet of Things. LNDECT, vol. 101, pp. 17–31. Springer, Singapore (2022). https://doi.org/10.1007/978-981-16-7610-9_2

32. Wainer, J., Barthelmess, P., Kumar, A.: W-RBAC - a workflow security model incorporating controlled overriding of constraints. Int. J. Coop. Inf. Syst. **12**(04), 455–485 (2003). https://doi.org/10.1142/S0218843003000814
33. Wang, Y., Yang, Y., Wang, B., Ran, Q., Ju, X.: Research on improved access control model based on T-RBAC. J. Phys. Conf. Ser. **1453**(1), 012011 (2020). https://doi.org/10.1088/1742-6596/1453/1/012011
34. Zou, Z., Chen, C., Ju, S., Chen, J.: The research for spatial role-based access control model. In: Taniar, D., Gervasi, O., Murgante, B., Pardede, E., Apduhan, B.O. (eds.) ICCSA 2010. LNCS, vol. 6019, pp. 296–308. Springer, Heidelberg (2010). https://doi.org/10.1007/978-3-642-12189-0_26

Digital Showroom in 3DWeb, the Scene Effect on Object Placement

Giorgio Olivas Martinez[1]([✉]) [iD], Valeria Orso[1], and Luciano Gamberini[1,2]

[1] Department of General Psychology, University of Padova, Padua, Italy
giorgio.olivasmartinez@studenti.unipd.it
[2] Human Inspired Technologies Research Centre, University of Padova, Padova, Italy
http://www.springer.com/gp/computer-science/lncs

Abstract. The COVID-19 pandemic has had a profound effect on the world, particularly the retail industry, which had to adapt to the new reality of exclusive remote shopping, determined by the lockdown condition of its customers. The impossibility of accessing physical stores has strengthened the need to improve the online journey. Over the past few years, the majority of online shops have been organized as plain grids in which products were displayed as static images; they lacked the ability to deliver the brick-and-mortar shop experience. 3DWeb, Augmented Reality, and Virtual Reality are the new technologies that can assist retailers in dealing with these new challenges. Reality-enhancing technologies can innovate the consumer journey, transforming the purchase path into a complex, omnichannel purchase experience. Moreover, they can also make research more ecologically valid. Through immersion and noninvasive data collection, which records information in the background, a richer understanding of purchase journeys can be gained.

Keywords: Digital Showroom · 3DWeb · Object placement

1 Introduction

Retailing has evolved significantly over the past two decades, largely due to the introduction of new technologies and the increasing digitalization of the industry [24].

E-commerce, among other digital solutions, enabled consumers to make purchases online in a secure and convenient way, without any space or time limitations [7]. Digital catalogs, for example, offered a quick way to access product information, with attractive designs and a wide selection [3]. However, they lacked product interaction, which has been replaced with lengthy descriptions, creating an obstacle to user flow and engagement [16]. In recent years, retailers have also used virtual showrooms in an attempt to simulate the customer shopping experience online. More specifically, virtual showrooms can be defined as

Supported by organization x.

"the use of virtual environment technology on the World Wide Web for visualization and 3-dimensional (3D) interaction with the products" [16]. Their goal is to provide a realistic, interactive experience that allows customers to browse products, compare prices, and make purchases without ever leaving their homes. However, typically virtual showrooms offer slow navigation and product searching.

In rare cases, retailers have also adopted more advanced technologies to improve the customer experience, such as virtual dressing rooms. The advanced technological solutions described above are much more affordable and manageable by larger retailers [8]. While the majority of online retailers settle for 2D graphical user interfaces for their online stores [13].

This phenomenon is linked to the issue of new technology adoption which is a common problem in the retail industry, especially for small businesses [2,17].

This has generated frustration and disillusionment with the implementation of new technologies [5,22]. This is also due to many failures of e-tail businesses because of a lack of technical know-how [25]. This is a problem not only because of lost business opportunities but can also become a survival issue, like in the case of pandemic events during which online can become the only viable solution for reaching new and habitual customers. There are a number of risks involved for retailers when investing in new technologies, which have been identified [2] as follows:

- Digital connectivity
- Unavailability of IT Skills
- Routine automation and upgrade cost
- Risk of failure
- Concentration on Operational Improvement
- Cybersecurity risks
- Data privacy

We, as other researchers [25], believe that small retailers can significantly benefit from a specialized e-tailing solution that requires minimal technical effort and takes care of data privacy and security issues. This likely reduces the need for advanced IT skills and the risk of failure.

This research intends to help the development of such a specialized platform that considers not only the consumer's perspective but also the retailer's. More specifically, the overall goal of the platform will be to enable small retailers to autonomously manage their e-tail, thereby bypassing the slow procedures of in-house development [9].

In the next section we will report our exploratory study on the impact of store layout and product display on consumer behavior. Moreover, we will look into the role of 3D objects in customer experience.

2   Background

In the following sections we will outline the research background on which the present study was based.

2.1   Influence of Store Layout on Consumer Behavior

Store layout has been identified as a powerful factor, capable to influence consumer behavior both in physical and digital environments [4,15]. In fact, store layout has been shown to affect search efficiency in traditional shops [23], while the type of layout used in 2D online stores has been found to influence customers' attitudes [26].

More specifically, [14] identified three layout types for traditional retail stores, each suiting specific shopping experiences:

– "grid" for planned shopping (e.g., grocery stores);
– "free-form" for unplanned purchases and impulse buying;
– "racetrack-boutique" for a unique atmosphere and quality products.

In the attempt to build online stores, the layouts already employed in physical shops were transposed to online environments [14]. Notably, also in virtual store the layout affects consumer behavior. Specifically, it was found that hierarchical structures increase ease of navigation, free-form layouts increase ease of use, perceptions, and entertainment, while mixed grid/free-form layouts increase consumer experience. Additionally, both racetrack and free-form layouts increase the time spent in online stores [14].

2.2   Product Display in Retail

"Bricks-and-mortar retailers are subject to the law of gravity so that despite variations in shape, height, width, material, and color shelves are bound to be horizontal in order to prevent products from falling." [18]. For online retailers, shelves are virtual, allowing for more creative displays, impossible to obtain in physical stores. This flexibility has enabled retailers to overcome the typical constraints of physical shelves, such as height, location, and product accessibility [18].

In their analysis, [18] highlight two streams of literature regarding how to organize products on physical shelves in traditional stores: one focusing on the effect of the visual display without considering assortment size, and the other focusing on the effect of assortment organization without considering the physical location of products on the shelf.

Research on the impact of assortment size has yielded contradictory results. Some studies suggest a direct relationship with consumer behavior (e.g., Skallerud et al., 2009), while others point to an inverse relation (e.g., Diehl and Poynor, 2010).

[18] have shown how online things can be even more complicated, discovering that the same assortment can be perceived as small or large depending on organization criteria used to show products on a web page.

2.3   The Role of 3D in the Customer Experiences

In order to increase the engagement of the user in recent years has been highlighted the importance of image-based interaction [1,28]. In this sense, there have been different experiments, such as virtual changing rooms that did not find a favorable response from consumers [20,21]. On the other hand, the importance of realistic 3D experiences has increased, because they can deliver a clear and vivid representation of the product as a way to reduce the perception of risk in online purchases [10,11] and to increase purchase intentions [12]. In 3DWeb research has been found that many variables can influence the consumer response to the product, for example [27] has shown that the context can have a role in increasing the level of immersion of the user. They also found the also size, and variety of categories can be factors influencing user involvement.

3   Study

The overall research question we will try to answer is: how can reality-enhancing technologies improve the online experience in retail? 3DWeb technologies customers can explore 3D product models and better understand product information and characteristics than with traditional e-commerce.

Previous studies have investigated the factors affecting customers' perception of the product arrangement in physical stores reporting mixed conclusions [6,19].

The present research targets retailers and investigates how they would naturally arrange digital showrooms in a 3D Web environment.

Literature often considers the consumer part of the problem, discussing the limited spread of innovative solutions such as 3D objects in e-tailing. However, it is common knowledge that retailers must overcome a series of barriers before they can adopt new technologies [14]. This can be achieved by the use of simple technical solutions that enable a fast and cost-effective way to implement an innovation, which is a big vector of adoption and diffusion [25].

We decided to focus on the retail side of the problem and their needs when creating a virtual showroom. We did not mean to push the technical capabilities of these technologies to their limits, but rather provide the simplest possible solution to enable object positioning in virtual environments. We did then observe if potential retailers were initially influenced by simple factors such as the type of background presented (or the category type) and if any spontaneous demand arises from the use of a minimal system in a free-choice condition.

More specifically, we studied the factors related to the context (e-store background and type of product) that affect the arrangement of the products.

The goal of the study is to understand what are the context-related aspects that affect the arrangement of products in a 3D Web store.

The experiment follows a 2×2 mixed design, with the background being the between-participant variable and the product's category being the within-participant one. Each independent variable has two levels, being the Background (high vs. low number of shelves), and the Category of products (food vs. fashion).

The order of presentation of the different backgrounds and product categories was counterbalanced across participants.

3.1   Method

Questionnaire. The questionnaire was custom-made and comprised three sections.

One section for screening purposes contains demographic information counting five items (age, gender, previous experience with product placement, product category of the product placement experience in physical places and digital environments).

One section related to the experiment included nine items assessed using a 7-point Likert scale. The dimensions included: likeness of the configuration, satisfaction with the configuration, easiness/difficulty of placing products in the environment, capacity to place the product in the desired place, and the influence of product features on placement.

The last section contains four open-ended questions about the participants' evaluation of their own stores, and a general evaluation of opportunities and threats for this kind of innovative solution.

Software. The participant used a custom software developed for the experiment with Game Maker Studio 2. The software enabled them to place products on a flat background. The software interface allowed them to move products from a static collection of 3 products at the bottom of the screen and drag them into the background area freely (without grid snapping or visual guides). They could also remove a placed item from the showroom by dragging it to the bottom section.

The software interface could be presented in four versions: two for the category type (food and fashion) and two for the background type (high and low number of shelves). With a certain combination of the dependent variables (e.g., food - low), the interface automatically displayed the other levels of those variables (e.g., fashion - high) during the second session.

Fig. 1. A screenshot of the interface the participants had to use to complete the experiment.

The software automatically tracked the position, the type, and the number of products placed by each participant, recording it in a data log which will be explained in the next section.

Measures. The software records as log data the dependent variables being:

- the number of products positioned (measured as the main factor for a statistical analysis investigating the effect of the dependent variables of this metric).
- time spent (measured to verify if the number of products positioned correlates with the time spent in the interaction and also the level of satisfaction).
- coordinates of the positioned products (measured to verify if there are some regions of the space where the participants tend to position the product and also to check if these regions are affected by the independent variables).
- type of positioned product (measured to verify if there are preferred products more frequently positioned than others, and, in combination with their coordinates, to check if some products tend to be placed by the participant in similar regions of the space).

Procedure. The participant filled out a consent form and explicitly agreed to the use of the data collected within the environment and the questionnaire. They were asked to design a digital showroom to be aesthetically appealing to attract new users and increase sales.

In the first session, the participants were presented with an empty store and were free to place up to 30 products (the total number available). They could freely choose where and which products to insert. There was no specific timeline for this phase. It ended when the participant was content with the showroom they created. To finish, they clicked the "Done" button, which directed them to the questionnaire page.

Next, the participants were asked to repeat the same task in the second session. They were presented with another empty store with the other product category and then were asked to fill in the second questionnaire. On average the experimental session lasted 22.57 min.

3.2   Analysis

Descriptive Statistics. The research included 24 Italian-speaking participants, with an average age of 27.91 years. Of them, 79.16% were male, and 16.67% had no prior experience with product placement.

Participants reported being fairly satisfied with the configurations they had produced ($M = 5.1$, $SD = 1.96$). They also reported that they were able to place the items exactly where they wanted ($M = 6.4$, $SD = 1.10$). However, the task seemed not that easy for them ($M = 4.9$, $SD = 1.81$).

Finally, they reported that the characteristics of the products somewhat influenced their choices ($M = 4.7$, $SD = 2.01$).

Fig. 2. In the Figure is showed the plotted position of all objects in the experiments with their respective backgrounds

Results. A preliminary analysis was conducted to ensure that there was no difference in the number of objects placed between the food and the fashion condition.

Fashion vs Food. The dependent variable was distributed normally in both groups. More specifically, for "Fashion" the Shapiro test revealed that the dependent variable was normally distributed (W = 0.93403, p-value = 0.12); and also for "Food" the Shapiro test revealed that the dependent variable was normally distributed (W = 0.9456, p-value = 0.2172).

Therefore, a t-test has been carried out to test any difference between conditions.

The test revealed there was no difference between the food (M = 17.50) and fashion (M = 16.29) conditions (t = 0.63213, df = 45.813, p-value = 0.5304).

Table 1. Results of t-test analysis on the product category effect on the number of products placed in the showroom.

Fashion (mean)	Food (mean)
17.50	16.29

High vs Low Number of Shelves. We conducted an analysis to test if the number of objects placed in the showroom was influenced by the number of shelves present in the scene.

The dependent variable was distributed normally in both groups. More specifically, for "Low n. of shelves" the Shapiro test revealed that the dependent variable was normally distributed (W = 0.95258, p-value = 0.3308); and also for "High n. of shelves" the Shapiro test revealed that the dependent variable was normally distributed (W = 0.9503, p-value = 0.2547).

Therefore, a t-test has been carried out. The test revealed there was no difference between the Low (M = 18.12) and High (M = 15.56) conditions (t = 1.3637, df = 45.888, p-value = 0.1793).

Table 2. Results of t-test analysis on the number of shelves effect on the number of products placed in the showroom.

Low N. of shelves (mean)	High N. of shelves (mean)
18.12	15.56

High vs Low Number of Shelves in Fashion. The dependent variable was distributed normally in both groups for Fashion. More specifically, for "Low n. of shelves" the Shapiro test revealed that the dependent variable was normally distributed (W = 0.93182, p-value = 0.4296); and also for "High n. of shelves" the Shapiro test revealed that the dependent variable was normally distributed (W = 0.93611, p-value = 0.4086).

Therefore, a t-test has been carried out to verify the presence of significant differences between conditions. The test revealed there was no difference between the Low (M = 17.92) and High (M = 17.00) conditions (t = 0.34743, df = 21.821, p-value = 0.7316).

High vs Low Number of Shelves in Food. The dependent variable was distributed normally in both groups for Food. More specifically, for "Low n. of shelves" the Shapiro test revealed that the dependent variable was normally distributed (W = 0.92873, p-value = 0.3669); and also for "High n. of shelves" the Shapiro test revealed that the dependent variable was normally distributed (W = 0.93249, p-value = 0.4073).

Therefore, a t-test has been carried out. The test revealed there was no difference between the Low (M = 17.92) and High (M = 17.00) conditions (t = 1.5041, df = 20.889, p-value = 0.1475).

Table 3. Results of t-test analysis on the number of shelves effect on the number of products placed in the showroom split for the product category.

Product Category	Low N. of shelves (mean)	High N. of shelves (mean)
Fashion	17.00	17.92
Food	14.25	18.33

Open Comments. Analysis of the open-ended questions revealed that two-thirds of participants liked the possibility to order and organize products. Furthermore, 10% expressed appreciation for the freedom enabled by the platform. Others liked the configuration colors (3.3%), creativity (3.3%), perspective (3.3%), and 3D object representation (3.3%). The remaining 10% disliked their configuration entirely. The analysis of the open-ended questions also revealed that 50% of the participants did not appreciate the inability to change the size and shape of objects. Additionally, 3% found their configurations too chaotic and lacking organization. Others desired the ability to rotate objects (6.7%) and improved usability (3.3%), while the remaining 10% were satisfied with their configuration.

3.3 Discussion

Nudging Retailers Through Complexity. Retailers have used virtual showrooms to simulate shopping behavior in physical stores for several years [16]. Many of their attempts failed due to several barriers hampering the adoption of innovative technologies [25]. Several of these barriers involve the very nature of the platform used by retailers to implement these technologies [9]. For this reason, it is considered that a specialized platform solution that doesn't requires high technical abilities would, increase the adoption of innovative solutions [25].

We adopted the retailer perspective for the present study. We chose a general approach, by asking participants to take the retailer's point of view and to design their showroom in a free environment with minimal features. We wanted to explore whether it was possible to influence and nudge the number of positioned objects by manipulating the number of perceived shelves in the background. Our results suggest a different story. Participants mostly neglected the number of evident shelves present in the showrooms, and simply tried to place as many items as possible, thereby trying to maximize the space in the virtual shopping window.

Nevertheless, participants were also quite satisfied with their showroom, showing good results in like (4.9/7) and in satisfaction (5.1/7)items. They had some problems positioning the object in the environment (3/7 in easiness), but they managed to position the products where they wanted (4.9/7). Background nudge did not have a significant effect also in split conditions, considering fashion and food by themselves. We concluded that background nudge had no significant effect on the decisions of the participants, regardless of the product category. This was true for both food and fashion. Taken together our findings suggest that the tendency to maximize product visibility, thereby calling for a virtual extension of the space available to the retailer. A multi-showroom configuration where the seller can freely place products in different virtual showrooms would possibly be more suitable.

We have to acknowledge that we proposed two pre-defined backgrounds to participants, thereby strongly driving their experience.

We also believe in the importance of adopting the retailer perspective and that this should be replicated with a larger sample of online retailers.

References

1. Ashley, C., Tuten, T.: Creative strategies in social media marketing: an exploratory study of branded social content and consumer engagement. Psychol. Market. **32**(1), 15–27 (2014). https://doi.org/10.1002/mar.20761
2. Bashar, A., Singh, S., Pathak, V.: Technology Adoption by Retailers in Response to COVID-19: Online Impulse Buying Perspective. SSRN Electronic Journal (2021). https://doi.org/10.2139/ssrn.3927090
3. Chan, G., Cheung, C.M.K., Kwong, T., Limayem, M., Zhu, L.: Online consumer behavior: a review and agenda for future research. Bled eConference, pp. 43 (2003). https://aisel.aisnet.org/bled2003/43
4. Diehl, K., van Herpen, E., Lamberton, C.: Organizing products with complements versus substitutes: effects on store preferences as a function of effort and assortment perceptions. J. Retailing **91**(1), 1–18 (2015). https://doi.org/10.1016/j.jretai.2014.10.003
5. Doherty, N.F., Ellis-Chadwick, F.E.: New perspectives in internet retailing: a review and strategic critique of the field. Int. J. Retail Distrib. Manage. **34**(4/5), 411–428 (2006). https://doi.org/10.1108/09590550610660305
6. Fazri, A., Afiff, A.Z., Balqiah, T.E.: The influence of display organization and product quantity on consumer purchase: the role of aversion of disorderly and scarcity effect. Adv. Sci. Lett. **23**(8), 7266–7268 (2017). https://doi.org/10.1166/asl.2017.9346
7. Haryanti, T., Subriadi, A.P.: Factors and theories for e-commerce adoption: a literature review. Int. J. Electr. Comm. Stud. **11**(2), 1910 (2020). https://doi.org/10.7903/ijecs.1910
8. Heiman, A., Reardon, T., Zilberman, D.: The effects of COVID-19 on the adoption of "on-the-shelf technologies": virtual dressing room software and the expected rise of third-party reverse-logistics. Serv. Sci. **14**(2), 179–194 (2022). https://doi.org/10.1287/serv.2022.0300
9. Hong, W., Zhu, K.: Migrating to internet-based e-commerce: factors affecting e-commerce adoption and migration at the firm level. Inf. Manag. **43**(2), 204–221 (2006). https://doi.org/10.1016/j.im.2005.06.003
10. In Shim, S., Lee, Y.: Consumer's perceived risk reduction by 3D virtual model. Int. J. Retail Distrib. Manag. **39**(12), 945–959 (2011). https://doi.org/10.1108/09590551111183326
11. Kim, J., Forsythe, S.: Factors affecting adoption of product virtualization technology for online consumer electronics shopping. Int. J. Retail Distrib. Manag. **38**(3), 190–204 (2010). https://doi.org/10.1108/09590551011027122
12. Kim, S., Baek, T.H., Yoon, S.: The effect of 360-degree rotatable product images on purchase intention. J. Retail. Consumer Serv. **55**, 102062 (2020). https://doi.org/10.1016/j.jretconser.2020.102062
13. Krasonikolakis, I., Vrechopoulos, A., Pouloudi, A., Dimitriadis, S.: Store layout effects on consumer behavior in 3D online stores. Eur. J. Market. **52**(5/6), 1223–1256 (2018). https://doi.org/10.1108/ejm-03-2015-0183
14. Levy, M., Weitz, B., Beitelspacher, L.: Retailing Management. McGraw-Hill Education, New York, United States (2012)
15. Mallapragada, G., Chandukala, S.R., Liu, Q.: exploring the effects of "what" (product) and "where" (website) characteristics on online shopping behavior. J. Market. **80**(2), 21–38 (2016). https://doi.org/10.1509/jm.15.0138

16. Omar, H.M., Hooi, Y.K., Sulaiman, A.: Design, implementation and evaluation of a virtual showroom. In: 2008 International Symposium on Information Technology (2008). https://doi.org/10.1109/itsim.2008.4631609
17. Orso, V., Portello, G., Pierobon, L.R.P., Bettelli, A., Monaro, M., Gamberini, L.: Rethinking the shopping experience: a qualitative explorative study on 3d web and AR for displaying and selling goods. In: Proceedings of the 16th International Conference on Interfaces and Human Computer Interaction 2022 and 15th International Conference on Game and Entertainment Technologies 2022 (2022)
18. Pizzi, G., Scarpi, D.: The effect of shelf layout on satisfaction and perceived assortment size: An empirical assessment. J. Retail. Consumer Serv. **28**, 67–77 (2016). https://doi.org/10.1016/j.jretconser.2015.08.012
19. Reynolds-McIlnay, R., Morrin, M., Nordfält, J.: How product-environment brightness contrast and product disarray impact consumer choice in retail environments. J. Retail. **93**(3), 266–282 (2017). https://doi.org/10.1016/j.jretai.2017.03.003
20. Shoolapani, B., Jinka, P.: Virtual simulation and augmented interfaces for business models with focus on banking and retail. In: 2011 Fourth IEEE International Conference on Utility and Cloud Computing (2011). https://doi.org/10.1109/ucc.2011.77
21. Sullivan, P., Heitmeyer, J.: Looking at Gen Y shopping preferences and intentions: exploring the role of experience and apparel involvement. Int. J. Consumer Stud. **32**(3), 285–295 (2008). https://doi.org/10.1111/j.1470-6431.2008.00680.x
22. Teo, T.S.H., Pian, Y.: A contingency perspective on Internet adoption and competitive advantage. Eur. J. Inf. Syst. **12**(2), 78–92 (2003). https://doi.org/10.1057/palgrave.ejis.3000448
23. Titus, P.A., Everett, P.B.: The consumer retail search process: a conceptual model and research agenda. J. Acad. Market. Sci. **23**(2), 106–119 (1995). https://doi.org/10.1177/0092070395232003
24. Verhoef, P.C., Kannan, P., Inman, J.J.: From multi-channel retailing to omni-channel retailing. J. Retail. **91**(2), 174–181 (2015). https://doi.org/10.1016/j.jretai.2015.02.005
25. Vize, R., Kennedy, A., Coughlan, J.P., Ellis-Chadwick, F.E.: Investigating the factors impacting retailers evaluations of web solution providers (2009). https://oro.open.ac.uk/27789/
26. Vrechopoulos, A.P., O'Keefe, R.M., Doukidis, G.I., Siomkos, G.J.: Virtual store layout: an experimental comparison in the context of grocery retail. J. Retail. **80**(1), 13–22 (2004). https://doi.org/10.1016/j.jretai.2004.01.006
27. Wodehouse, A., Abba, M.: 3D visualisation for online retail: factors in consumer behaviour. Int. J. Market Res. **58**(3), 451–472 (2016). https://doi.org/10.2501/ijmr-2016-027
28. Wooldridge, D., Schneider, M.: The business of iPhone and iPad app development. Apress (2011)

Exploiting 3D Web to Enhance Online Shopping: Toward an Update of Usability Heuristics

Valeria Orso[1]([⊠]) [iD], Maria Luisa Campanini[1], Leonardo Pierobon[1],
Giovanni Portello[2], Merylin Monaro[1,2] [iD], Alice Bettelli[1] [iD],
and Luciano Gamberini[1,2] [iD]

[1] Department of General Psychology, University of Padua, Padua, Italy
{valeria.orso,marialuisa.campanini,leonardo.pierobon,
merylin.monaro,alice.bettelli,luciano.gamberini}@unipd.it
[2] Human Inspired Technologies Research Center, University of Padua, Padua, Italy
giovanni.portello@unipd.it

Abstract. Online shopping has become a widespread habit for millions of users all over the world. E-stores are very convenient for both customers and retailers, because they are accessible 24/7 from anywhere in the world. However, they also convey relevant drawbacks for both parties: customers can only explore endless grids of flat items, and retailers are likely to struggle to manage and update their e-commerce autonomously. Here we present an e-commerce platform levering on 3DWeb technology that is meant to enhance the sopping experience for customers on the one side, and to retailers to independently operate on their e-store. The experience of user with Hybrid Consumer Interface and the Hybrid Seller Interface were evaluated in two interdependent sessions with representative end users (N = 57 in total). Even of both interfaces are still at a prototypical stage, results from the experimental sessions show promising outcomes.

Keywords: 3D Web · retail · usability · user experience

1 Introduction

The shopping experience has substantially been the same for decades: a seller and a buyer met in person and exchanged goods and money. The advent of Internet has radically changed this paradigm with the introduction of e-commerce, which turned out to be beneficial for both parties.

Indeed, the shopping experience is not limited to the physical store within fixed opening hours anymore, rather it can be done at anytime from anywhere. Thanks to these features it took very little to online shopping to become very popular all over the world. The outbreak of the Covid-19 pandemic and the consequent strict lockdowns in many Western countries pushed even further consumers to rely on e-commerce. It is reported that many online retailers significantly increased their business. For instance, Amazon's annual revenue almost doubled between 2017 and 2021 [1, 2]. Likewise,

F. Fui-Hoon Nah and K. Siau (Eds.): HCII 2023, LNCS 14038, pp. 450–460, 2023.
https://doi.org/10.1007/978-3-031-35969-9_30

the restaurant delivery sector increased 67% globally [3]. However, small and medium retailers with a less pervasive online presence, did not follow the same trend [4].

Online shopping has downsides also for customers. Despite very popular and convenient, the experience of online shopping cannot fully resemble the experience one can have in brick-and-mortar shops. Indeed, the costumer cannot manipulate and touch the product, thereby failing to appreciate the tiny details of it. However, novel technologies such as Virtual Reality (VR) and Augmented Reality (AR) can be exploited to compensate this lack of sensorial stimulation [5].

In the present research project, we designed and evaluated an e-commerce platform that has the twofold aim of enhancing the costumer experience by using 3D technology (Costumer interface) and facilitating the launch and maintenance of online stores for small retailers (Seller interface).

2 Background

Even if the large majority of the e-commerce are conceived and organized as grids of flat images, some brands have started to offer on their websites 3D models of their products to explore, in the attempt to make the online shopping experience more than just convenient [6]. For example, for the 2018 Spring/Summer collection Gucci launched the #Hallucination 3DWeb experience[1]. Also, Nike has implemented a 3D section within the website to allow users customize their sneakers and visualize the tridimensional output[2].

Still the research on this topic is still limited. Early work by [7] explored usability-related aspects of displaying 3D models of kitchens and clothes, finding promising feedback by users. [6] tested an interactive 3D product interface displaying fashion items, i.e., shoes. The interface was well-received, but participants noted that they still missed the experience of trying on the items or at least seeing them on a model. The store layout is a further aspect that was found to impact on the UX in 3D web. More specifically [8] found that store layouts with minimal architecture elements and attractive materials are more likely to foster a positive customer experience, because they put the 3D products on the forefront. A recent study [9] compared three fruit and vegetables stores, being a physical store, a non-immersive and an immersive virtual store. Not only they found that participants' product perception was similar in the three environments, but also that respondents tended to buy more in the virtual stores (either immersive or not). Consistently, [10] found that 3D models of food packages allowing to see the content increase the purchase intention as compared to opaque packed food package models. [11] experimented multimodal interaction modalities (i.e., speech and gestures) to manipulate the products displayed in an e-commerce. In a preliminary study, they found that participants had positive attitudes toward 3D product representations as compared to traditional 2D view. [12] implemented a 3D virtual showroom of kitchen appliances with embedded multimedia content to further illustrate the products' features. However, no evaluation with end users was reported.

---

[1] http://springsummer.gucci.com/.
[2] https://www.nike.com/.

3   The Hybrid Shopping Environment

The Hybrid shopping environment consists of two complementary interfaces, one addressed to sellers, namely the Hybrid Seller Interface (HSI) and one addressed to customers, that is the Hybrid Consumer Interface (HCI). The HSI allows users to create and customize their e-shop and autonomously manage the process of uploading the products and arrange the exhibition of the products. Notably, the seller can also add product-related information (e.g., technical characteristics, description, multimedia contents) and arrange it in special info-box, namely templates. One of the most innovative features of the system is that the set up and arrangement flow are managed as a preview. In other user words, the seller can constantly preview how her/his e-shop will look like for potential customers, thereby making the management immediate and easy. Notably, within each e-shop the seller and create and manage as many showrooms as s/he likes.

On the other side, the HCI is built in the attempt to recreate the experience of exploring products in brick-and-mortar shops. To this end, the products are not displayed in grids, rather they are shown as tridimensional models in a 3D shop. Here the user can select and explore one product, rotate it, zoom it in and out, and also access to product-related information. The user can navigate through the various showrooms available and finally ass the product to the cart. The HCI is responsive and can be explored both by computer and smartphone (Fig. 1).

Fig. 1. The schematic representation of the Hybrid Platform enabling sellers to directly manage their store online by uploading 3D objects. On the other side potential consumers can access and interact with 3DWeb store using their desktop and mobile devices

4   User Evaluation of the Hybrid Seller Interface

In the following, a detailed description of the materials and methods deployed for the evaluation of the HSI is provided. The results of the evaluation follow.

4.1   Task Lists

A task scenario was devised to help participants better contextualize the tasks. More specifically, they were invited to imagine themselves as a shop owner willing to open an e-shop using the HSI and to perform a series of actions to renovate an existing e-shop or to create a completely new one. They were asked to interact with two experimental stores within the HSI, that is the Hybrid Food Store and the Hybrid Fashion Store. A list of tasks was also prepared, to make participant try out all the available functions. The list included 14 tasks in total[3]. More specifically, in each store they were asked to.

- change the name of a showroom,
- upload new products,
- set up the showroom,
- replace a product,
- change the place of a product,
- edit product-related information.

4.2 Materials

A total of two questionnaires were devised to investigate participants' experience.

One questionnaire, comprising 8 items, was meant to collect personal information (i.e., age, education) and data related to respondent's professional experience (current job, commercial sector, years of experience, experience with e-commerce).

A second questionnaire was devised *ad hoc* to investigate the usability of the system, namely the Usability Questionnaire. It consists of a total of 25 items to be answered on a 5-point Likert scale (1 = totally disagree; 5 = totally agree). The questionnaire explored the following dimensions:

- usefulness (2 items), refers to the extent to which user thinks the interface could be helpful for her/his job;
- ease of use (6 items), measures to what extent the user found the interface effortless to use;
- learnability (2 items), explores how effortful was to learn to operate the interface;
- satisfaction (1 item), refers to how much the user was satisfied about her/his interaction with the interface;
- responsiveness (3 items), assesses the extent to which the system promptly reacted to the commands;
- intention of use (2 items), explores the extent to which the user is willing to use the interface again;
- transparency (4 items), refers to the extent to which the interface clearly shows the status and the progress;
- consistency (1 item), measures how consistent the interface is in the different parts;
- readability (1 item), evaluates the extent to which textual contents are easy to read;
- navigability (1 item), refers to the ease of exploring the interface;
- clarity of information (2 items), measures the clarity of information content.

4.3 Equipment and Procedure

Only individuals with current professional experience in the commercial sector were eligible to take part in the study. On the day of the evaluation, participants were welcomed

---

[3] Some tasks were asked twice to make the task scenario appear consistent and meaningful.

at the University premises and debriefed about the aim of the experiment. After they had read and signed the informed consent form, they were asked to fill in a demographic questionnaire. Next, they were invited to sit at a desk in front of a laptop computer (15″) and they were trained the use of the HSI. Next, the session started with the experimenter reading aloud the tasks that the user had to perform. During the execution of the tasks, the experimenter encouraged users to freely comment and give voice to all his/her thoughts (thinking aloud protocol). After they had completed the list of tasks, they were asked to fill in the post-experience questionnaire.

To accommodate participants' time constraints, the experimental session could be run both in presence and from remote. In that case, the experimental procedure unfolded the same way, with the user and the experimenter communicating in video-call using the zoom.us platform. Overall, the experiment lasted about one hour.

4.4 Participants

A total of 10 participants (F = 5) aged between 25 and 64 (M = 32,1, SD = 11,57) participated in the experiment. With respect to their area of employment, 20% of participants (n = 2) work in the health care field, 20% (n = 2) in the food and catering industry, 20% work as traders respectively in the fashion and electronic field, and 40% (n = 4) reported experience in shops. Three participants worked in a company which also has an online store.

4.5 Results

Notably, all participants were able to complete the assigned tasks successfully and autonomously. Only in one case the user could not recall how to move one item between two different showrooms and needed the experimenter's help. Moreover, the interaction modality to move an item from the carousel to the showroom (i.e., first click on the item, then click on the landing location, release) was learnt by all participants, but they commented that it would have been more natural to simply drag and drop the item.

The scores assigned to the dimensions assessed to the usability questionnaire are overall positive. Indeed, participants indicated that the HSI was useful (M = 4.2 DS = .14), easy to use (M = 3.86 DS = .63) and to learn (M = 3.6 DS = .84). Overall, they were satisfied by their interaction with it (M = 4 DS = .81), even if they reported the interface not to respond promptly to their inputs (M = 3.1DS = .14). Yet, they were positive regarding the possibility to use it in the future (M = 3.75 DS = .2). The way in which the HSI worked was considered transparent (M = 4.37 DS = .47), consistent (M = 4.1DS = .87) and easy to navigate (M = 4.6 DS = .51). The textual information was easily readable (M = 4.5 DS = .7) and clear (M = 3.8 DS = .78) (Fig. 2).

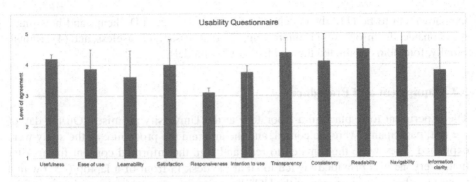

Fig. 2. Mean values and standard deviations of the scored assigned by participants to the items assessing the usability of the Hybrid Seller Interface

5 Consumer Interface

In the following, a detailed description of the materials and procedure used to conduct the evaluation of the Hybrid Consumer Interface both for the desktop and mobile versions.

5.1 Task Lists

Also for the HCI, a task list was devised. It was organized to resemble a realistic inter-action with an e-commerce platform and to make the user try all the functions available. In particular, s/he was asked to (i) access one shop, (ii) explore a product, (iii) read the info-boxes, (iv) go through the related multimedia contents, (v) browse the showrooms, (vi) add a product to the cart, (vii) change the quantity of products in the shopping cart, and (viii) proceed to the checkout. The same task list was presented to participants using both the desktop and the mobile version of the interface.

5.2 Questionnaires

A total of five questionnaires were devised.

The first one investigated participants' demographics, including age, level of edu-cation and professional role. It further assessed their purchasing habits, comprising the frequency of online purchases, the goods that they usually buy online, and their attitudes towards brick-and-mortar stores.

A second questionnaire was meant to assess the usability of the interface, and it was the same of the one used to assess the usability of the HSI[4].

A further questionnaire was meant to explore the perceived quality of both the prod-ucts and the e-shops. It consisted of 12 items devised ad hoc. The first three items asked the user to rate on a 10-point scale (1 = Not at all; 10 = Very much) the extent to which they believed that the 3D models were (1) of high graphical quality, (2) realistic, (3) inaccurate, and (4) cheap. Moreover, participants were asked how much they perceive

---

[4] The phrasing of the items was adjusted to make it consistent with the experience under investigation.

the showroom to be (1) tidy, (2) chaotic, (3) sophisticated, (4) cheap; and how much they consider the info boxes (1) interesting, (2) unnecessary, (3) useless, and (4) original. Finally, four items evaluated the vividness of 3D models.

5.3 Equipment and Procedure

The experiment took place in a laboratory at the University premises. On the day of the test, participants were welcomed, and the aim and the procedure of the study were explained. They were then invited to read and sign the informed consent form. After that, everyone was provided invited to sit at the desk in front of a laptop (13″) with a mouse for the desktop version of the HCI. Participants involved in the evaluation of the mobile version of the HCI were provided a smartphone (6.1″).

Before starting the experimental procedure, participants were trained to use the system; they were also given time to become familiar with the interface. As soon as they felt ready, the experimenter read the tasks and invited the participant to carry them out. During the session, the experimenter invited participants to verbalize their thoughts aloud (thinking aloud protocol). Finally, participants were asked to fill in the post-experience questionnaire. The experiment lasted about 30 min.

5.4 Participants

A total of 47 individuals accepted to participate in the study. Of them, 26 (F = 13) used the desktop version of the interface (hereafter Desktop Group), while the remaining 21 (F = 13) interacted with the mobile Hybrid Consumer Interface (hereafter Mobile Group). Group assignment was conducted randomly. The average age of participants in the Desktop Group ranged between 21 and 60 (M = 29.5, SD = 12.74). As for online shopping experience, they all participants reported previous experience with e-commerce purchase. The majority (42%, n = 11) of participants claimed to often shop online, while a smaller number (15%, n = 4) to always do it. Otherwise, the 27% (n = 7) do it sometimes, and the 15% (n = 4) just rarely. Moreover, the most frequently purchased product categories on online platforms (n = 21) are related to travels (i.e., train tickets) and culture (i.e., concert tickets), followed by household appliances (n = 16). The least chosen categories were food and musical products (n = 6). The Mobile Group was 24.85 years old on average (age range 21–31). They all reported to have previous experience with online shopping, and more specifically, 4 of them reported to buy online rarely, 9 of them sometimes, 7 of them often and only 1 regularly. The most reported category for online purchases are travel (n = 20), followed by books (n = 18), clothes (n = 17) and items in the cultural sector (n = 16). They also report to buy music items and beauty products (n = 8), food (n = 7) and appliances (n = 6).

5.5 Results

All participants in both groups were able to complete all the tasks assigned without the help of the experimenter and they were all able to learn how the interface worked in a little time and operate it proficiently.

Data from the usability questionnaire reflect performance data for both groups (Fig. 3). Indeed, both the desktop (M = 4.37, SD = .48) and the mobile (M = 4.41, SD = .35) interface were considered ease to use and to learn (desktop: M = 4.34, SD = .58; mobile: M = 4.61, SD = .31). The way in which information was organized and displayed was rated as clear (desktop: M = 4.37, SD = .53; mobile: M = 4.15, SD = .61). As for speed with which the interface responded to inputs, namely responsiveness, participants were more neutral about it, especially for the desktop version (M = 3.07SD = 1.68; mobile: M = 3.52, SD = 1.41). The same trend emerged for the intention to use the app in the future (desktop: M = 3.53, SD = 1.13; mobile: M = 3.33, SD = 1.23). Nevertheless, both versions were rated as consistent (desktop: M = 4.3, SD = .67; mobile: M = 4.14, SD = .72), textual contents were easily readable (desktop: M = 4.48, SD = .29; mobile: M = 4.45, SD = .30), and the various sections were considered casy to navigate (desktop: M = 4.48, SD = .24; mobile: M = 4.23, SD = .4). Finally, they considered both versions useful (desktop: M = 4.03, SD = .82; mobile: M = 4.09, SD = .62).

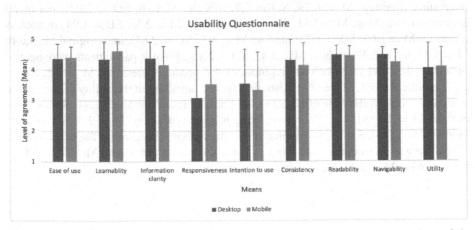

Fig. 3. Mean values of the scores attributed by participants to the various dimensions of the questionnaire assessing usability

Participants were also asked to rate on a 10-point scale the perceived quality of the 3D models, the virtual showrooms and the info-boxes (Fig. 4 and Fig. 5). They were not particularly well impressed by 3D models, which were considered of average quality (desktop: M = 5.5, SD = 2.23; mobile: M = 5.23, SD = 2.11) and realistic (desktop: M = 6.15, SD = 2.27; mobile: M = 5.23, SD = 2.11). They were neutral about the level of accuracy (desktop: M = 4.6, SD = 3.05; mobile: M = 4.19, SD = 2.58), but they disagreed with the model being cheap (desktop: M = 2.57, SD = 2.6; mobile: M = 2.9, SD = 2.8). On the other hand they were well-impressed by the virtual showrooms that were considered tidy (desktop: M = 6.38, SD = 2.15; mobile: M = 5.8, SD = .2.58) and sophisticated (desktop: M = 4.37, SD = 2.23; mobile: M = 3.09, SD = 1.86). They disagreed on the showrooms being chaotic (desktop: M = 2.88, SD = 2.77; mobile: M

= 2.57, SD = 2.56) and cheap (desktop: M = 2.15, SD = 22.66; mobile: M = 3.19, SD = 2.44).

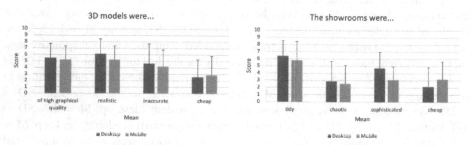

Fig. 4. The mean values and standard deviations of the items assessing the perceived quality of the 3D models and the virtual showrooms

Notably the informative boxes were particularly appreciated. They were considered interesting (desktop: M = 6.79, SD = 1.7; mobile: M = 6, SD = 2.02), not at all unnecessary (desktop: M = 2.11, SD = 1.65; mobile: M = 2.52, SD = 1.94) or useless (desktop: M = 1.65, SD = 1.74; mobile: M = 1.8, SD = 2.65), and original (desktop: M = 4.88, SD = 2.32; mobile: M = 4.42, SD = 2.18). Finally, participants were neutral regarding the level of perceived vividness of the products (desktop: M = 2.8, SD = 0.74; mobile: M = 3, SD = .77). Also, they were neutral about the ability of the virtual showroom to help them imagine using the product (desktop: M = 2.84, SD = 1.08; mobile: M = 2.9, SD = .74) or visualizing it in use (desktop: M = 3.3, SD = 1.01; mobile: M = 3.04, SD = .9). Similarly, they were neutral regarding the ability of the virtual showroom to make them think to touch the product (desktop: M = 2.88, SD = 1.03; mobile: M = 2.71, SD = 1.05).

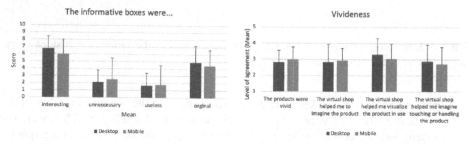

Fig. 5. The means and standard deviations of the items assessing the perceived quality of the informative boxes and the perceived vividness of the products

6 Discussion and Conclusions

In the present work we have described the overall architecture of an innovative e-commerce platform addressing both retail professionals and potential customers leveraging 3DWeb technology. The choice of revolving on 3DWeb is motivated by the expected benefits for both parties involved. Sellers, especially those running small businesses, have the opportunity to make their e-shop unique and more attractive for the general audience, thereby increasing their level of competitiveness. On the other hand, customers are offered an enhanced shopping experience and, more importantly, a tool for building a more comprehensive understanding of the item that they are about to purchase, in the attempt to overcome the lack of sensory stimulation that is typical of online shopping [5].

We have conducted a user experience evaluation with participants employed in the retail sector to investigate the extent to which the Hybrid Seller Interface could actually meet their needs. Overall, the HIS was well received and all participants were able to smoothly operate it. However, it is worth noting that they all had the natural tendency to move the items by dragging them from the carousel to the virtual showroom, instead of clicking with the mouse on one item and then release it onto the target location in the virtual showroom. Even if they adapted quickly to the correct strategy, this suggests that more spontaneous interactions should be favored, especially for the HSI that entails numerous and complex functionalities. Indeed, the very aim is to build a fluid and natural action-flow, so as to relieve the user from the mental overload of remembering how to operate the interface. Moreover, participants noted that the HIS lacked in responsiveness. This is likely due to the weight of the 3D environment and objects that required some loading time. However, this aspect needs to be optimized, to improve the user experience with the Hybrid Seller Interface.

For assessing the experience of use with the Hybrid Consumer Interface we have conducted two independent tests: one for the desktop version and one for the mobile one. In both cases, the outcomes were very positive. Also for the HCI, an issue related to a delayed responsiveness emerged. Again, this is likely due to the high quality, and hence size, of the 3D models chosen for the study. Notably, despite the high graphical quality, and the consequent lagged loading, participants were not really impressed by them. However, this did not hampered the overall experience. Participants were particularly impressed by the tidy and sophisticated appearance of the virtual showrooms. Interestingly, they appreciated the informative boxes, thereby suggesting their potential to promote specific contents relevant for the seller (e.g., fairness of the production process). We have to acknowledge that the evaluation run with the HCI was limited to two experimental stores (namely, food and apparel), with a limited number of items. It is likely that participants were not particularly motivated to interact with them, thereby limiting the generalizability of our results. Future studies should make an effort in matching the participants' actual shopping interests and the products available in the e-stores, thereby investigating a more genuine experience.

Acknowledgement. The work was partially funded by the POR-FESR Project Hybrid Sustainable Worlds (ID. 10290827).

References

1. Worldwide retail e-commerce sales of Amazon from 2017 to 2021. Statista (2020)

2. Guthrie, C., Fosso-Wamba, S., Arnaud, J.B.: Online consumer resilience during a pandemic: an exploratory study of e-commerce behavior before, during and after a COVID-19 lockdown. J. Retail. Consum. Serv. **61**, 102570 (2021). https://doi.org/10.1016/j.jretconser.2021.102570

3. Lock, S.: Online restaurant delivery growth worldwide 2019–2020, by country. Statista (2021). https://www.statista.com/statistics/1238955/digital-restaurant-food-delivery-growth-in-selected-countries-worldwide/. Accessed 10 May 2022

4. Orso, V., Portello, G., Pierobon, L., Bettelli, A., Monaro, M., Gamberini, L.: Rething the shopping experience: a qualitative explorative study on 3D Web and AR for displaying and selling goods. In: Proceedings of the International Conference on Interfaces and Human Computer Interaction, pp. 142–148 (2022)

5. Hilken, T., Chylinski, M., Keeling, D.I., Heller, J., de Ruyter, K., Mahr, D.: How to strategically choose or combine augmented and virtual reality for improved online experiential retailing. Psychol. Mark. **39**(3), 495–507 (2022)

6. Zhou, H.: Fashion E-Commerce in the 3D Digital Era: A 3D Interactive Web User Interface for Online Products (2018)

7. Moritz, F.: Potentials of 3D-web-applications in E-commerce-study about the impact of 3D-product-presentations. In: 2010 IEEE/ACIS 9th International Conference on Computer and Information Science, pp. 307–314. IEEE (2010)

8. Krasonikolakis, I., Vrechopoulos, A., Pouloudi, A., Dimitriadis, S.: Store layout effects on consumer behavior in 3D online stores. Eur. J. Mark. **52**, 1223–1256 (2018)

9. Lombart, C., Millan, E., Normand, J.M., Verhulst, A., Labbé-Pinlon, B., Moreau, G.: Effects of physical, non-immersive virtual, and immersive virtual store environments on consumers' perceptions and purchase behavior. Comput. Hum. Behav. **110**, 106374 (2020)

10. Petit, O., Javornik, A., Velasco, C.: We eat first with our (digital) eyes: enhancing mental simulation of eating experiences via visual-enabling technologies. J. Retail. **98**(2), 277–293 (2022)

11. Hewawalpita, S., Perera, I.: Multimodal user interaction framework for e-commerce. In: 2019 International Research Conference on Smart Computing and Systems Engineering, SCSE, pp. 9–16. IEEE (2019)

12. Sobociński, P., Strugała, D., Walczak, K., Maik, M., Jenek, T.: Large-scale 3d web environment for visualization and marketing of household appliances. In: Paolis, L.T.D., Arpaia, P., Bourdot, P. (eds.) AVR 2021. LNCS, vol. 12980, pp. 25–43. Springer, Cham (2021). https://doi.org/10.1007/978-3-030-87595-4_3

Booking Shore Excursions for Cruises. The Role of Virtual 360-Degree Presentations

Jenny Wagner, Christopher Zerres^(✉), and Kai Israel

Offenburg University, Badstraße 24, 77652 Offenburg, Germany
christopher.zerres@hs-offenburg.de

Abstract. Complex tourism products with intangible service components are difficult to explain to potential customers. This research elaborates the use of virtual reality (VR) in the field of shore excursions. A theoretical research model based on the technology acceptance model was developed, and hypotheses were proposed. Cruise passengers were invited to test 360° excursion images on a landing page. Data was collected using an online questionnaire. Finally, data was analyzed using the PLS-SEM method. The results provide theoretical implications on technology acceptance model (TAM) research in the field of cruise tourism. Furthermore, the results and implications indicate the potential of virtual 360° shore excursion presentations for the cruise industry.

Keywords: Virtual 360° Presentations · Cruise Tourism · Booking Intention · Shore Excursions

1 Introduction

In the evolution from a niche product to mainstream vacation, cruise tourism is one of the fastest growing sectors of the tourism industry [29]. Even though the COVID-19 pandemic had a huge impact on the cruise industry [31], it is now recovering. According to the Cruise Lines International Association [5], growth in passenger volume is projected to surpass 2019 levels in 2023.

Like most tourist products, cruises are difficult to explain to potential customers before a purchase. A complex network of individual services and experiences make up the overall product of a cruise [37]. This highlights the importance of explaining the product as precisely and informatively as possible before a booking is made. According to Peručić and Greblički [31] in recent years the success of the cruise industry was based on effective marketing strategies.

Due to the difficulties mentioned above, the cruise industry is looking for innovative and modern approaches in all aspects of the customer journey to promote their products and services. Recent research related to cruise tourism highlighted the importance of integrating smart technologies into the customer journey [4]. One of those technologies is VR, which is described to be an optimal tool to provide destination information to potential tourists [10]. Especially in the pre-purchase experience of a cruise, it is

© Springer Nature Switzerland AG 2023
F. Fui-Hoon Nah and K. Siau (Eds.): HCII 2023, LNCS 14038, pp. 461–475, 2023.
https://doi.org/10.1007/978-3-031-35969-9_31

crucial to provide the potential customer with detailed information. Research on the pre-purchase experience provided interesting findings on the improvement of the customer experience as well as the facilitation of the consumers' information gathering through VR technology [27, 37]. However, this research only considered VR in regard to the overall concept of a cruise. The studies investigated VR as an experience in the customer journey as a marketing tool and compared VR to other media consumed in a pre-purchase information gathering [27, 37]. Nevertheless, they did not focus on user acceptance factors or individual components of a cruise, such as shore excursions.

This study aims to explore the user's acceptance of non-immersive virtual shore excursion presentations utilizing 360° images. According to Beck et al. [3, p. 12] non-immersive VR can be defined *"as synthetic or 360° real-life captured content on a conventional (computer) screen, enabling virtual touristic experiences [...] either prior to, during, and/or after travel"*. Users can interact with the content by using input devices like a mouse or a touchscreen; however, they are not immersed into the virtual world. Nevertheless, telepresence - a feeling of being there, was found to be more positively influenced by virtual web-mediated information compared to traditional offline travel information [18]. This study aims to contribute to VR research in the cruise industry. Previous research in the industry has already emphasized the importance of smart technology in cruising [4] and the application of VR in the cruise industry [1, 27, 37]. However, Sharples [34] and Simoni et al. [37] emphasize the lack of research regarding the critical touchpoints in the pre-purchase stage of a cruise. Therefore, a better understanding of the drivers of booking intention in this context could provide interesting findings for the industry. Even though Arlati et al. [1] tested a VR application for booking shore excursions, there is still a big question mark regarding the acceptance of virtual shore excursion presentations.

2 Research Model and Hypotheses Development

2.1 Overview Model

To investigate the acceptance of virtual 360° presentations in the context of shore excursions a research model was built by the authors based on the TAM by Davis [6]. In addition to perceived ease of use, perceived usefulness, and behavioral intention, perceived enjoyment and curiosity were added to the model (Fig. 1). Studies in the context of VR and tourism could prove that the perceived ease of use and the perceived usefulness are important factors influencing the behavioral intention (i.e., booking intention) [20, 21]. Perceived enjoyment and curiosity have been used as important factors in previous research to investigate the acceptance of technology in the field of tourism [19, 21]. With virtual 360° presentations, interesting and exciting worlds can be created that offer users an entertaining new experience [21, 32]. This presentation format is perceived as enjoyable by the users [38]. On the other hand, novel and exciting stimuli of the virtual environment, such as 360° panoramic images, can arouse users' curiosity [15, 30]. Moon and Kim [28] describe curiosity as a combination of inquisitiveness and technical competence while engaged in an activity. Users can escape into a virtual world and freely explore it on their own. Research has shown that curiosity and perceived enjoyment influence the perceived usefulness and the intention to use a technology [19, 21].

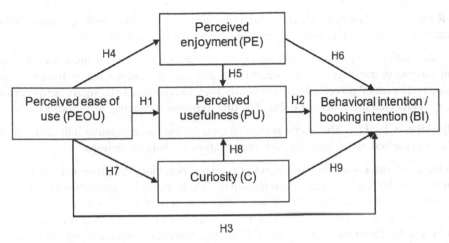

Fig. 1. Research model.

2.2 Traditional TAM Factors

Perceived Ease of Use. Perceived ease of use is defined by Davis [6, p. 320] *"as the degree to which a person believes that using a particular system would be free of effort"*. Ease of use reflects the degree of effort perceived by the user in using the information system [6, 7]. In terms of VR applications, this means that the interaction with the application on the hardware and software side must be easy for the consumer to learn, understand, and implement. Accordingly, both the design and the usability of the application must be attractively designed and intuitive to operate so that the users can orient themselves immediately in the application.

In their extension of TAM for a world-wide-web context, Moon and Kim [28] found perceived ease of use to positively influence the users' attitudes. Tourism related acceptance research for augmented reality (AR) proposed a research model incorporating perceived ease of use and perceived usefulness [23]. Furthermore, Huang et al. [17] highlighted the importance of perceived ease of use in the context of virtual destination promotion. However, the relevance of perceived ease of use is controversial in research since studies also found it had a less consistent effect on the intention to use than perceived usefulness [14, 40].

Perceived Usefulness. Davis [6, p. 320] defined perceived usefulness *"as the degree to which a person believes that using a particular system would enhance his or her job performance"*. The perceived usefulness is seen as a crucial determinant influencing users' intention to use an information system [7, 40]. In the context of shore excursion bookings VR is especially important in the *pre-booking* and *pre-visit phase* of the customer journey. In this phase, the customer is building an understanding through searching and planning, increasing the interest and reducing risk [36]. 360° images as a form of presentation could enhance the customer experience. The cruise line MSC implemented a

VR-based virtual catalog, which enabled customers to customize their upcoming cruise experience, and this was found to improve the customer experience [37].

According to Venkatesh and Davis [40], perceived usefulness is influenced by perceived ease of use, because the ease of using a system determines the usefulness of the system. The users' perceived ease of use was previously found to be significantly related to the perceived usefulness [14, 28]. The first hypothesis can thus be proposed:

Hypothesis 1 (H1): The user's perceived ease of use has a positive influence on the user's perceived usefulness of a 360° presentation of shore excursions.

Behavioral Intention. In the context of tourism, behavioral intention can include the intention to book a hotel [21], the intention to visit a virtual 3D destination [16, 17], the intention to visit a heritage site [2], or in the context of this research, the intention to book a shore excursion. Perceived ease of use and perceived usefulness are described to be the fundamental determinants influencing behavioral intention [6]. Also, in the research of VR and AR in tourism, perceived ease of use and perceived usefulness were often found to influence the technology adoption intention and also the intention to visit the destination [41].

Perceived usefulness has been found to be the most relevant determinant of behavioral intention [7]. It was found to influence behavioral intention significantly and positively in various virtual technology-related tourism studies. If a 3D tourism site was found useful as a trip planning tool, it predicted behavioral intention [16]. Haugstvedt and Krogstie [14] found that perceived usefulness had a strong positive effect on the behavioral intention towards an AR application. Furthermore, the perceived usefulness of a VR application was also found to have a positive influence on the booking intention in a hotel context [21]. It is likely that these findings can also be applied to the booking intention of shore excursions. Consequently, the hypothesis states the following:

Hypothesis 2 (H2): The user's perceived usefulness of a 360° presentation of shore excursions has a positive influence on the user's excursion booking intention.

The influence of perceived ease of use on the behavioral intention was originally established to be the secondary determinant of behavioral intention by Davis et al. [7]. However, research has obtained different results when analyzing the direct effect of perceived ease of use on behavioral intention [40]. In some tourism-related research, perceived ease of use was not found to have a significant influence on behavioral intention [17] whereas in other research it was found to have a direct impact [14]. Since for VR applications ease of use can be considered as a crucial aspect the following hypothesis is proposed:

Hypothesis 3 (H3): The user's perceived ease of use of a 360° presentation of shore excursions has a positive influence on the user's excursion booking intention.

2.3 Additional Factors

Perceived Enjoyment. Venkatesh [39, p. 351] defines perceived enjoyment *"as the extent to which the activity of using a specific system is perceived to be enjoyable in its* *own right, aside from any performance consequences resulting from system use*". In our research, perceived enjoyment is the degree to which a person enjoys using 360° images for shore excursion bookings. As most tourism products and experiences are referred to as fun and enjoyable, it is crucial to include enjoyment in this research. As previously stated, the pre-booking phase in a tourist's customer journey is extremely relevant, as it is usually the phase where the customer gathers information [35]. If the excursion research and booking in this phase were found to be enjoyable, it could enhance the customer experience.

Perceived enjoyment was incorporated in various tourism studies. The perceived ease of use was found to positively influence the users' perceived enjoyment [14, 17]. However, researching travel information can be frustrating. If the information is presented clearly, the user might enjoy the process of gathering information. This results in the following hypothesis:

Hypothesis 4 (H4): The user's perceived ease of use of a 360° excursion presentation has a positive influence on the user's perceived enjoyment of such an application.

The connection between perceived enjoyment and perceived usefulness can be seen from two different angles. On one hand, usefulness can have an impact on enjoyment. Huang et al. [16] found that useful information regarding virtual tourism sites was found to enhance the enjoyment. On the other hand, enjoyment can be seen as an influencing factor on perceived usefulness. Lee et al. [22] found in their study on the adoption of VR devices that enjoyment has a positive influence on perceived usefulness. If enjoyment can be seen as a key determinant in a leisure tourism context, one might argue that an enjoyable information and booking experience can be seen as useful regarding the whole process. Consequently, hypothesis 5 states the following:

Hypothesis 5 (H5): The user's perceived enjoyment of a 360° excursion presentation has a positive influence on the user's perceived usefulness of such an application.

Perceived enjoyment was found to positively influence the booking intention [21]. Haugstvedt and Krogstie [14] and Lee et al. [22] emphasize the importance of enjoyment as a core factor influencing the behavioral intention. In the context of this study, shore excursion sales are of high relevance to revenue generation. If enjoyment is found to positively influence booking intention, the factor is of high relevance when considering a relaunch of shore excursion presentations. Hence, the following hypothesis is proposed:

Hypothesis 6 (H6): The user's perceived enjoyment of a 360° excursion presentation has a positive influence on the user's excursion booking intention.

Curiosity. Litman and Jimerson [24] describe curiosity in two different ways. On the one hand, there is interest curiosity, which evokes positive feelings by acquiring information. On the other hand, there is deprivation curiosity, which refers to unpleasant feelings associated with uncertainty. Israel et al. [21] stated that in the VR and hotel booking context a lack of information awakens the curiosity of an individual to remedy this deficiency by acquiring new knowledge. Knowledge and information are especially relevant in the pre-booking phase of traveling, as it is the phase of searching and planning in the customer journey [35]. One could argue that a cruise passenger, booking shore excursions for an upcoming cruise, is curious in nature, as he or she must overcome a lack of information to decide about shore excursions.

According to Moon and Kim [28], curiosity can be seen as an intrinsic motive that is closely related to the user's perceived playfulness. Playfulness was found to be positively influenced by the user's perceived ease of use of the world wide web [28]. In the context of this research, the following hypothesis is proposed with regard to curiosity:

Hypothesis 7 (H7): The user's perceived ease of use of a 360° excursion presentation has a positive influence on the user's curiosity regarding such an application.

Curiosity can also have a positive influence on the perceived usefulness as it is possible that through curiosity the unpleasant feeling of uncertainty can be reduced. In a study in the VR and hotel booking context, Israel et al. [21] found that curiosity positively influenced the perceived usefulness. It can be assumed that the influence is similar in the case of shore excursions. The following hypothesis is proposed:

Hypothesis 8 (H8): The user's perceived curiosity regarding a 360° excursion presentation has a positive influence on the user's perceived usefulness of such an application.

Especially when booking shore excursions, the customer decides on an experience that is supposed to represent a whole destination experience in a cruise context. Cruise ships usually stay in port for only one day, allowing passengers to experience the destination in a limited amount of time. The feeling of curiosity can support the acquisition of information and reduce uncertainty. Therefore, it can have a positive influence on the booking intention. The related hypothesis is as follow:

Hypothesis 9 (H9): The user's perceived curiosity regarding a 360° excursion presentation has a positive influence on the user's excursion booking intention.

3 Empirical Research

3.1 Research Design and Data Collection

This research aimed to investigate the acceptance of 360° images for the presentation of shore excursions. In our study we focus on experienced German-speaking cruisers. Germany is Europe's leading market in passenger volume and therefore highly relevant in the cruise industry [5]. Participants were reached on cruise-related social media sites through the support of several German-speaking cruise bloggers and influencers. They were asked to follow instructions on a landing page, specifically developed for the study. The landing page consisted of a short introductory text followed by the description of three steps explaining the process of participation. The design, layout, and descriptive texts of the website were pre-tested, using different devices.

The survey included different steps. After reading the introduction on the landing page, participants were asked to select a destination that they had not previously visited by cruise ship. The destinations to choose from were popular cruise destinations in Europe: Iceland, Portugal, and Scotland [5]. After selecting the destination, the participants were forwarded to a presentation of a shore excursion. The presented excursion for each region was selected upon popularity of the destination. The excursions selected and presented were the Golden Circle Tour (Reykjavik, Iceland), Sintra including Palácio Nacional and Cabo da Roca (Lisbon, Portugal) and Loch Ness and Inverness (Invergordon, Scotland). These excursions all have a medium to full day length and require a certain amount of surefootedness, either due to path conditions, walking distances or stairs on site. The descriptions of the excursions (i.e., key points, duration and price) were based on a review of various descriptions from four German-speaking cruise lines (AIDA, TUI Cruises, Hapag Lloyd Cruises and Phoenix Reisen). Subsequently, three to four essential stops of the excursion were shown in a 360° image. In addition to the excursion description, the landing page included a description for starting and navigating the 360° images. Participants were asked to look around, navigate and zoom in and out. To enable a responsive view of the images on different devices, the software VirtualTours by FusionWorks was used to integrate the 360° images in the landing page (Fig. 2). The landing page was designed to be viewed on different devices and screen sizes.

Fig. 2. 360° image of Urquhart Castle at Loch Ness, Scotland [9].

Afterwards participants were asked to fill out a questionnaire. All items in the questionnaire were derived from previous research [1, 14, 21, 35] and were adapted to the context of this study. A seven-point Likert scale ranging from (1) strongly disagree to (7) strongly agree was used for all items. In addition, respondents were asked: how many cruises they had participated in so far, if they had to consider any physical limitations when deciding on a shore excursion, demographic questions, and three landing page related questions.

The research was conducted from September 30th to October 7th, 2022. A total of 332 participants were recorded.

3.2 Results

Twenty questionnaires of the 332 total showed missing values and had to be excluded from further analysis. Moreover, nine respondents did not have any previous cruise experience and therefore had to be excluded. In addition, 20 outliers in the data set (z-score limit $\pm$ 2.58) were identified and removed [8]. This results in a total of 283 valid responses. The sample is illustrated in Table 1. The sample is well distributed across multiple generations and ranges from cruise newcomers to more experienced cruisers. Furthermore, around 90% of participants had never visited the presented destination before.

Table 1. Characteristics of respondents sample (n = 283).

Characteristics	Frequency (*n*)	Percentage (%)
Gender		
Male	56	19.8
Female	226	79.9
not specified	1	0.4
Generation		
Post-war generation (1946–1955)	15	5.3
Baby boomer generation (1956–1965)	46	16.3
Generation X (1966–1980)	131	46.3
Generation Y/Millennials (1981–1995)	80	28.3
Generation Z (1996–2009)	11	3.9
Previous cruise experience		
1 - 4 cruises	93	32.9
5 - 9 cruises	102	36.0
10 or more cruises	88	31.1
Handicap to consider		
Yes	56	19.8
No	227	80.2
Device used		
Smartphone	214	75.6
Tablet	36	12.7
Desktop PC/Laptop	32	11.3
Others	1	0.4
Selected destination		
Iceland	137	48.4
Portugal	77	27.2
Scotland	69	24.4
Been there already?		
Yes	26	9.2
No	257	90.8

To evaluate the proposed research model, partial least squares structural equation modeling (PLS-SEM) was used with the software SmartPLS4. The data meets the sample size requirements for the application of PLS-SEM [13]. PLS-SEM is suitable for testing a theoretical framework and exploring theoretical extensions of established theories in exploratory research [13]. PLS-SEM has been applied in different tourism and VR related studies [21, 35].

Measurement Model. The measurement model is evaluated by analyzing the measures for internal consistency, convergence validity and discriminant validity [13]. Internal consistency is measured using Cronbach's alpha and composite reliability [13]. The results of the measurement model are shown in Table 2.

Table 2. Validity and reliability of the constructs.

Constructs & items	α (>0.6)	ρ_A (>0.7)	ρ_C (>0.7)	AVE (>0.5)	Loading (>0.7)
Perceived ease of use	0.622	0.703	0.791	0.567	
PEOU1: Looking at the 360° shore excursion images was easy for me					0.854
PEOU2: It was clear to me how to look around while viewing the 360° images					0.832
PEOU3: I don't require any specific knowledge to use 360° images					0.529
Perceived usefulness	0.858	0.859	0.914	0.780	
PU1: The 360° images are useful to get a quick impression of the excursion					0.892
PU2: The 360° images help me assess whether the excursion is suitable for me					0.895
PU3: The 360° images enhance the quality of shore excursion presentations					0.862
Perceived enjoyment	0.909	0.909	0.943	0.845	
PE1: I enjoyed viewing the shore excursion using 360° images					0.928
PE2: Viewing 360° shore excursion images was entertaining for me					0.924
PE3: It was pleasant for me to view shore excursions using 360° images					0.906
Curiosity	0.925	0.925	0.964	0.931	
C1: The 360° images make me curious to experience the excursion in person					0.965
C2: Viewing shore excursions using 360° images increases my interest in this place					0.965
Booking intention	0.884	0.891	0.945	0.896	
BI1: If I could view the excursions for my next cruise using 360° images, it would encourage me to book more than normal images					0.952
BI2: If I could view the excursions for my next cruise using 360° images, it would increase my willingness to book excursions in advance					0.941

All criteria for internal consistency were met in this research [11–13]. Convergence validity considers the loadings on the construct, as well as the average variance extracted

(AVE). Loadings should be above 0.708 as this results in a communality of 50%, indicating that more than half of the items' variance defines the construct [13]. However, if a loading is below 0.7, the item does not necessarily have to be removed as its content might be valid for the overall content [13]. Therefore, PEOU3 was not removed from the data. Another measure for convergence validity is the AVE. The AVE should be higher than 0.5 as this indicates that on average, the construct explains at least half of the variance of its indicators [13]. All criteria for convergence validity were met.

Another crucial factor to consider in the evaluation of the model is the discriminant validity [13]. Discriminant validity can be determined using the Fornell-Larcker criterion and the heterotrait–monotrait ratio (HTMT). In Table 3, HTMT values are shown in italics. A conservative threshold value for HTMT is 0.85. The discriminant validity can be confirmed by the HTMT, as well as the Fornell-Larcker criterion.

Table 3. Discriminant validity (Fornell-Larcker and HTMT in italic).

	PEOU	PU	PE	C	BI
PEOU	**0.753**	*0.457*	*0.390*	*0.329*	*0.276*
PU	0.345	**0.883**	*0.760*	*0.761*	*0.799*
PE	0.315	0.673	**0.919**	*0.845*	*0.737*
C	0.261	0.679	0.775	**0.965**	*0.781*
BI	0.227	0.699	0.662	0.707	**0.947**

Structural Model. The first step when evaluating the structural model is to test the collinearity. In this research, all variance inflation factor values are below 5 and therefore do not indicate any collinearity issues [13]. The next step is to check the path coefficients in the structural model. A bootstrapping with 5,000 subsamples [13] was performed in SmartPLS to determine the significance of the path coefficients [13]. Path coefficients and their significance can be viewed in Fig. 3. Detailed results of the hypothesis testing and the bootstrapping procedure are presented in Table 4.

To assess the model's predictive power, the coefficient of determination (R^2) is examined. The R^2 value usually lies between 0 and 1. The higher the value, the better the latent variable is explained [13]. The R^2 values for perceived enjoyment and curiosity indicate a rather low predictive power, whereas the values for perceived usefulness and booking intention indicate a moderate predictive power [13].

R^2 refers to the model's in-sample-power; however, it is not suitable for predicting the out-of-sample power, referring to the prediction of new observations [13]. Therefore, Hair et al. [13] recommend using the $PLS_{predict}$ algorithm. The algorithm replaces the previous blindfolding procedure but still indicates $Q^2_{predict}$ values that should be above zero [13]. All $Q^2_{predict}$ values were above zero, indicating a smaller prediction error than the benchmark. Moreover, Hair et al. [12] suggest testing the root mean square error (RMSE) to assess the model's predictive power. The value of RMSE is then compared to a linear regression model (LM) benchmark. The model indicates high predictive

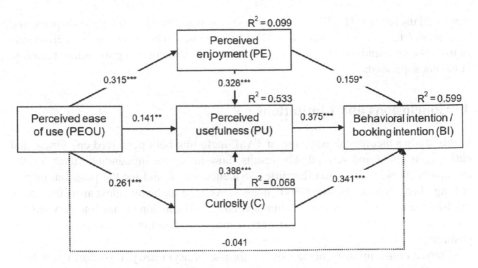

***significant at p<0.001 **significant at p<0.01 *significant at p<0.05

Fig. 3. Results for the structural model.

power, if the PLS-SEM values have a lower RSME than the LM benchmark [12]. In this research, the majority of RMSE values (8/10) are lower that the LM benchmark and indicate a medium predictive power of the model.

Table 4. Results of hypothesis testing.

	Hypothesis	Path coefficient	Standard deviation	T statistics	p-values	Study results
H1	PEOU → PU	0.141**	0.046	3.089	0.002	Supported
H2	PU → BI	0.375***	0.068	5.544	0.000	Supported
H3	PEOU → BI	−0.041	0.034	1.192	0.233	Not supported
H4	PEOU → PE	0.315***	0.066	4.784	0.000	Supported
H5	PE → PU	0.328***	0.083	3.944	0.000	Supported
H6	PE → BI	0.159*	0.078	2.036	0.042	Supported
H7	PEOU → C	0.261***	0.064	4.090	0.000	Supported
H8	C → PU	0.388***	0.086	4.506	0.000	Supported
H9	C → BI	0.341***	0.075	4.547	0.000	Supported

***significant at $p < 0.001$, **significant at $p < 0.01$, *significant at $p < 0.05$

Overall, statistical significance was found for eight out of nine path coefficients. Perceived ease of use shows a positive and significant relationship with the perceived usefulness, perceived enjoyment, and curiosity. Therefore, H1, H4 and H7 can be supported. Moreover, perceived enjoyment and perceived curiosity both positively influence perceived usefulness (H5, H8). Booking intention is positively influenced by perceived usefulness (H2), perceived enjoyment (H6) and curiosity (H9). However, perceived ease of use was not found to have a significant influence on the booking intention, therefore H3 is not supported.

4 Conclusions and Limitations

In this study a theoretical extension of TAM, including both perceived enjoyment and curiosity, is tested and verified. The results show important influencing factors on the customers' booking intention. Empirical evidence was found for the potential of presenting shore excursions using 360° images. As well as being relevant to the cruise tourism, the results have interesting implications for destination management, where the promotion of excursions and experiences within a destination is a crucial part of the business.

From a cruise industry perspective, information asymmetry, misinformation from crew, and accessibility were previously detected as problems and challenges in marketing shore excursions [25, 33]. Displaying the main spots of an excursion using 360° images allows cruise passengers to assess whether a certain excursion is suitable for their needs. Moreover, a clear and helpful presentation could benefit the excursion crew when briefing themselves about the excursions so they can correctly inform passengers. An insightful presentation using VR technology can also solve the problem of accessibility [10]. Presenting shore excursions using 360° images can benefit passengers and crew in different ways and can therefore potentially result in a better destination experience in general.

A further implication is to consider the multiple ways in which the customer experience management can be enhanced [34] by providing an enjoyable excursion booking process. During the pre-travel phase, the customer can prepare for the upcoming cruise in a pleasant way and enjoy the process of information gathering. Furthermore, the aroused curiosity and interest in the destination clearly has a positive influence on the cruise passenger's booking intention.

In line with the findings of Pleyers and Poncin [32], using virtual 360° presentations on classical standard itineraries can lead to a competitive advantage and an increased positive perception of the cruise line. Moreover, it should be considered that it is local tour agencies that provide the shore excursions for cruise lines and they often struggle with local competition [26]. Offering high quality 360° images of the local excursions to cruise lines could be a competitive advantage.

In general, it might be worth innovating the presentation of shore excursions, changing from a text-based booklet to a virtual presentation, especially when considering the results of this study. This research has implications for the cruise industry on how passengers perceive 360° images when booking shore excursions. Furthermore, it provides insights on which aspects to consider when implementing such an excursion booking tool.

However, there are also some limitations and areas of further research. This study was limited to a non-immersive setting. Further research could test a fully-immersive virtual excursion presentation. In such a setting, the theoretical framework of this research could be complemented by incorporating further theoretical constructs of VR research (e.g. telepresence) [18, 21].

Moreover, this research investigated the acceptance of virtual shore excursion presentations in a pre-travel phase. However, speaking of VR as an onboard experience, it would be interesting to investigate a fully-immersive excursion presentation as an informative and entertaining tool onboard a cruise ship. From practical experience, movement of the ship and seasickness would have to be considered. However, such an onboard excursion presentation could also enhance passengers' excursion booking intentions. Furthermore, it could be interesting to consider a fully-immersive excursion presentation as a means of post-travel memory storage, as well. This could give the opportunity to relive the experience.

References

1. Arlati, S., Spoladore, D., Baldassini, D., Sacco, M., Greci, L.: VirtualCruiseTour: an AR/VR application to promote shore excursions on cruise ships. In: De Paolis, L.T., Bourdot, P. (eds.) AVR 2018. LNCS, vol. 10850, pp. 133–147. Springer, Cham (2018). https://doi.org/10.1007/978-3-319-95270-3_9

2. Atzeni, M., Del Chiappa, G., Pung, J.: Enhancing visit intention in heritage tourism: the role of object-based and existential authenticity in non-immersive virtual reality heritage experiences. Int. J. Tour. Res. 24(2), 240–255 (2021)

3. Beck, J., Rainoldi, M., Egger, R.: Virtual reality in tourism: a state-of-the-art review. Tour. Rev. 74(3), 586–612 (2019)

4. Buhalis, D., Papathanassis, A., Vafeidou, M.: Smart cruising: smart technology applications and their diffusion in cruise tourism. J. Hosp. Tour. Technol. 13(4), 626–649 (2022)

5. CLIA: Europe Passenger Report 2021. https://cruising.org/en-gb/news-and-research/research/2022/july/clia-europe-passenger-report-2021. Accessed 18 Oct 2022

6. Davis, F.: Perceived usefulness, perceived ease of use, and user acceptance of information technology. MIS Q. 13(3), 319–339 (1989)

7. Davis, F., Bagozzi, R., Warshaw, P.: User acceptance of computer technology: a comparison of two theoretical models. Manag. Sci. 35(8), 982–1003 (1989)

8. Field, A.: Discovering Statistics Using IBM SPSS Statistics. SAGE Publications (2018)

9. Google Street View: 360-degree image of Urquhart Castle at Loch Ness, Scotland, Sven W. (2015)

10. Guttentag, D.A.: Virtual reality: applications and implications for tourism. Tour. Manag. 31(5), 637–651 (2010)

11. Hair, J.F., Black, W.C., Babin, B.J., Anderson, R.E.: Multivariate Data Analysis: Pearson New International Edition PDF eBook, Pearson Education (2013)

12. Hair, J.F., Hult, G., Ringle, C., Sarstedt, M.: A Primer on Partial Least Squares Structural Equation Modeling (PLS-SEM). SAGE Publications, Thousands oaks (2016)

13. Hair, J.F., Hult, G.T.M., Ringle, C.M., Sarstedt, M., Danks, N.P., Ray, S.: Partial Least Squares Structural Equation Modeling (PLS-SEM) Using R: A Workbook. Springer, Cham (2021)

14. Haugstvedt, A.-C., Krogstie, J.: Mobile augmented reality for cultural heritage: a technology acceptance study. In: ISMAR 2012 - 11th IEEE International Symposium on Mixed and Augmented Reality 2012, Science and Technology Papers (2012)

15. Huang, M.-H.: Designing website attributes to induce experiential encounters. Comput. Hum. Behav. 19(4), 425–442 (2003)

16. Huang, Y., Backman, K., Backman, S., Chang, L.-L.: Exploring the implications of virtual reality technology in tourism marketing: an integrated research framework: the implications of virtual reality technology in tourism marketing. Int. J. Tour. Res. **18**(2), 116–128 (2015)
17. Huang, Y., Backman, S., Backman, K., Moore, D.: Exploring user acceptance of 3D virtual worlds in travel and tourism marketing. Tour. Manag. **36**, 490–501 (2013)
18. Hyun, M.Y., O'Keefe, R.M.: Virtual destination image: testing a telepresence model. J. Bus. Res. **65**(1), 29–35 (2012)
19. Israel, K., Buchweitz, L., Tscheulin, D.K., Zerres, C., Korn, O.: Captivating product experiences: how virtual reality creates flow and thereby optimize product presentations. In: Nah, F.-H., Siau, K. (eds.) HCII 2020. LNCS, vol. 12204, pp. 354–368. Springer, Cham (2020). https://doi.org/10.1007/978-3-030-50341-3_28
20. Israel, K., Tscheulin, D.K., Zerres, C.: Virtual reality in the hotel industry: assessing the acceptance of immersive hotel presentation. Eur. J. Tour. Res. **21**, 5–22 (2018)
21. Israel, K., Zerres, C., Tscheulin, D.K.: Presenting hotels in virtual reality: does it influence the booking intention? J. Hosp. Tour. Technol. **10**(3), 443–463 (2019)
22. Lee, J., Kim, J., Choi, J.Y.: The adoption of virtual reality devices: the technology acceptance model integrating enjoyment, social interaction, and strength of the social ties. Telematics Inform. **39**, 37–48 (2019)
23. Leue, M. C., Jung, T., tom Dieck, D.: A theoretical model of augmented reality acceptance. e-Rev. Tour. Res. **5** (2014)
24. Litman, J.A., Jimerson, T.L.: The measurement of curiosity as a feeling of deprivation. J. Pers. Assess. **82**(2), 147–157 (2004)
25. London, W.R.: Shore-side activities. In: Vogel, M. P., Papathanassis, A. and Wolber, B. (eds) The Business and Management of Ocean Cruises, Wallingford, CABI (2012)
26. Lopes, M.J., Dredge, D.: Cruise tourism shore excursions: value for destinations? Tour. Plan. Dev. **15**(6), 633–652 (2017)
27. Martínez-Molés, V., Jung, T., Pérez-Cabañero, C., Cervera-Taulet, A.: Gathering pre-purchase information for a cruise vacation with virtual reality: the effects of media technology and gender. Int. J. Contemp. Hosp. Manag. **34**(1), 407–429 (2022)
28. Moon, J.-W., Kim, Y.-G.: Extending the TAM for a World-Wide-Web context. Inf. Manag. **38**(4), 217–230 (2001)
29. Papathanassis, A.: Cruise tourism management: state of the art. Tour. Rev. **72**(1), 104–119 (2017)
30. Pelet, J.-É., Ettis, S., Cowart, K.: Optimal experience of flow enhanced by telepresence: Evidence from social media use. Inf. Manag. **54**(1), 115–128 (2017)
31. Peručić, D., Greblički, M.: Key factors driving the demand for cruising and challenges facing the cruise industry in the future. Tourism **70**, 87–100 (2021)
32. Pleyers, G., Poncin, I.: Non-immersive virtual reality technologies in real estate: how customer experience drives attitudes toward properties and the service provider. J. Retail. Cons. Serv. **57**, 102175 (2020)
33. Samarathunga, W.: Challenges in cruise tourism in relation to shore excursions: the case of Sri Lanka. Colombo Bus. J. Int. J. Theory Pract. **7**(2), 4–21 (2016)
34. Sharples, L.: Research note: customer experience management in cruise pre-consumption. Int. J. Cult. Tour. Hosp. Res. **13**(2), 235–243 (2019)
35. Shen, S., Sotiriadis, M., Zhang, Y.: The influence of smart technologies on customer journey in tourist attractions within the smart tourism management framework. Sustainability **12**(10), 4157 (2010)
36. Shen, S., Xu, K., Sotiriadis, M., Wang, Y.: Exploring the factors influencing the adoption and usage of augmented reality and virtual reality applications in tourism education within the context of covid-19 pandemic. J. Hosp. Leisure Sport Tour. Educ. **30**(4), 100373 (2022)

37. Simoni, M., Sorrentino, A., Leone, D., Caporuscio, A.: Boosting the pre-purchase experience through virtual reality: insights from the cruise industry. J. Hosp. Tour. Technol. **13**(1), 140–156 (2022)
38. Song, K., Fiore, A.M., Park, J.: Telepresence and fantasy in online apparel shopping experience. J. Fash. Mark. Manag. **11**(4), 553–570 (2007)
39. Venkatesh, V.: Determinants of perceived ease of use: integrating control, intrinsic motivation, and emotion into the technology acceptance model. Inf. Syst. Res. **11**(4), 342–365 (2000)
40. Venkatesh, V., Davis, F.: A Theoretical extension of the technology acceptance model: four longitudinal field studies. Manag. Sci. **46**(2), 186–204 (2000)
41. Wei, W.: Research progress on virtual reality (VR) and augmented reality (AR) in tourism and hospitality. J. Hosp. Tour. Technol. **10**(4), 539–570 (2019)

Analyzing Customer Experience and Willingness to Use Towards Virtual Human Products: Real Person Generated vs. Computer Program Generated

Mingling Wu[1], Michael Xu[2], and Jiao Ge[1]([✉])

[1] Harbin Institute of Technology Shenzhen, Shenzhen, China
jiaoge@hit.edu.cn
[2] Basis International School Park Lane Harbour, Huizhou, China

Abstract. Virtual human is the research focus in the field of human-computer interaction under the trend of metaverse. The current research on virtual human focuses on the influence of virtual human driven by computer program on the user's attitude and behavior, and the virtual human generated by real person through wearable sensing devices presents us a new form of virtual human. However, there are no studies to examine the different effects of these two virtual human concepts on customer experience and willingness to use preference. We study the differences on customer experience and willingness to use for these two kinds of virtual human products through an online experiment. Our results show that customers have higher willingness to use real person generated virtual human products than computer program generated virtual human products. The intrinsic reason can be explained by flow theory. Compared with computer program generated virtual human, real person generated virtual human has a higher "Explicit Feedback" experience and a higher "Ability vs. Challenge Balance" experience, which affects their willingness to use the products.

Keywords: Virtual Human · Real Person Generated · Computer Program Generated · Customer Experience · Willingness to use

1 Introduction

In the research of artificial intelligence (AI) marketing, there are three kinds of AI generally studied, that is mechanical AI for automating repetitive marketing functions and activities, thinking AI for processing data to make decisions, and feeling AI for analyzing interactions and human emotions to achieve user resonance [1]. Plangger et al. [2] divides the research on artificial intelligence marketing into three themes: decentralized marketing, marketing mechanization and metamodern customer experiences, among which metamodern customer experience refers to the use of technology to create malleable realities to connect and interact with customers and satisfy customer preferences and delight them. In the metaverse, artificial intelligence as a virtual creator plays an increasingly important role in the creation of digital space and digital products [3].

© Springer Nature Switzerland AG 2023
F. Fui-Hoon Nah and K. Siau (Eds.): HCII 2023, LNCS 14038, pp. 476–486, 2023.
https://doi.org/10.1007/978-3-031-35969-9_32

Virtual human is completely generated by computer program with the help of computer vision technique and natural language processing, but the appearance and voice have no differences from real human. One of the main research areas in AI marketing focus on the effect of virtual human on customers' attitudes and behaviors. For instance, virtual human with chattering function can change customers' emotions and non-verbal behaviors [4]. The features of virtual human will affect customers' willingness to learn with virtual humans [5].

Recently a very popular and the world's first virtual reality game show in China has shown us a different concept of virtual human, which is a virtual human generated in real time by real person through wearing sensing devices. This virtual human responds to a real person's real action and voice in real time. The main difference for this new virtual human concept is that it requires real person to participate in real time. However, no research has examined the different effects of these two virtual human concepts on either customer experience or willingness to use the associated products. Which one better meets customer preferences and enhances customer experience? Why? Therefore, this study mainly discusses two questions: (1) Whether there are differences in customer experience and willingness to use brought by different virtual human products? (2) What is the influencing mechanism by which different virtual human products affecting customers' willingness to use them?

2 Literature Review

2.1 Virtual Human

Current research on virtual human mainly focuses on the type of computer program generated virtual human. Virtual human generated by computer programs can affect the productivity of human subjects [6], synthetic emotions expressed by virtual humans can cause positive or negative emotions of human conversation partners and affect the satisfaction of the conversation [7], and the interaction between individuals and virtual humans can affect the anxiety level of individuals [8]. Many researchers have planned to use computer program generated virtual human as a therapeutic tool for social anxiety disorder [9]. The effect of virtual human on customer attitude and behavior is one of the key marketing issues. Studies have shown that virtual people with chattering function can change users' emotions and non-verbal behaviors [4], and the features of virtual human (such as voice) can affect users' willingness to learn with virtual human [5]. And what about customer experience and willingness to use the real person generated virtual human which is generated under the background of the metaverse? It is worth further study.

2.2 Flow Theory

Flow is used to describe a state of mind in which people are deeply engaged in an activity. When people are in a state of flow, time seems to stand still and nothing else seems to matter. Flow is often described as a pleasurable experience [10]. In an online environment, flow leads users to be fully engaged in online tasks and interested in continuing those activities [11]. Flow affects customers' experience and cognition, perception and behavioral response [12, 13]. In the virtual environment, Rauschnabel et al. [14] pointed out that flow has a positive impact on users' attitude towards playing AR games, and flow experience or virtual experience will affect customers' attitude and behavior [15]. Users' participation in mobile augmented reality science games creates a flow experience, which leads to a positive attitude towards AR technology [16]. The best experience a person can obtain in the state of flow can enhance the attitude towards the brand and the willingness to use it in the virtual environment [17]. In other words, AR technology creates a flow experience that promotes a positive attitude towards and higher trust in the brand [18]. With the development of virtual reality technology and the gradual clarity of the concept of metaverse, virtual humans have become the focus of virtual reality research. How do different types of virtual human products generate different consumer flow experience? is investigated in this paper.

Although many scholars have studied flow in computer-mediated environment, the factors and methods used to measure flow are different. Hoffman and Novak [19] proposed that multi-dimensional measurement methods should be used to measure flow by combing and analyzing existing literatures. Academic studies on the antecedents of flow are mostly based on the original nine dimensions: clear goals, loss of self-consciousness, focus, lost sense of time, explicit feedback, sense of control over activity, ability vs. challenge balance, intrinsic pleasure of action and automaticity [17]. Due to significant differences between computer program generated and real person generated virtual humans in terms of real-time response and human participation, different flow experiences may be generated. In order to compare the experiencing differences between two different virtual human products, in this paper, three dimensions of "explicit feedback", "sense of control", and "ability vs challenge balance" are selected to measure flow experience.

3   Hypothesis Development

The conceptual framework of this research is presented in Fig. 1.

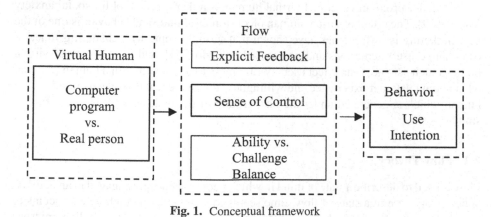

Fig. 1. Conceptual framework

3.1   Different Virtual Human Products and Willingness-to-Use

Companies are interested in understanding how to use different technological channels to shape customer attitude [20]. Hooker et al. [17] verified that flow experience can have a positive impact on brand attitude and purchase intention in three-dimensional virtual environment (3-DVEs). Therefore, we believe that in the virtual human environment, due to the differences in operation and presentation of different virtual humans, compared to computer program generated virtual human products, real person generated virtual human products could bring customers better flow experience, so customers have higher willingness to use them. Thus, we propose that

H1. Compared to computer program generated virtual human products, consumer willingness to use for real person generated virtual human products is higher.

3.2   Different Virtual Human Products and Consumer Flow Experience

Individuals must have quick and clear feedback when participating in activities so as to make appropriate responses and adjust performance [21]. When the feedback is clear, individuals do not need to exert additional cognitive efforts due to confusion, thus increasing the acceptance of the activity [22, 23]. Thus, the perception of clear feedback during an activity is expected to affect an individual's flow level because it indicates to participants that their actions are consistent with the game or environment without distracting them from the activity [10]. Although computer program generated virtual human programs can provide different appearance, actions and voices, the formation of these commands must be based on fixed virtual human database, while real person generated virtual human products through wearable sensing devices can jointly generate virtual human images based on real-time movements, voices and virtual human databases, which enable more explicit and timely feedback. Thus, we hypothesize that:

H2a. Compared to computer program generated virtual human products, real person generated virtual human products provide consumers higher "explicit feedback" flow experience.

Perception of control describes the feeling that individuals have when they are in control of their actions, without feeling as if they are consciously trying to exert control. Without a sense of control over one's own behavior, an individual may not be able to achieve flow experiences [24]. When a person feels "in control", they can focus more on the activity rather than maintaining control, which should lead to a flow experience in a three-dimensional reality environment where participants do not exert excessive mental effort to control their avatar and guide their environment [17]. Although computer program generated virtual human can be customized according to the instructions of customers, due to the particularity of customers' personal characteristics. They may not be able to meet the personalized needs of customers, for instance, the real-time simulation of human images according to the appearance of customers, such as body shape and hair style. Customers need to spend time on debugging of virtual human images. The real person generated virtual person can customize personalized image of customers through vision and motion capture technology, and realize the "sense of control" experience. Thus, we propose that.

H2b. Compared to computer program generated virtual human products, real person generated virtual human products provide consumer higher "sense of control" flow experience.

Ability and challenge balance is the situation where a person's skills and abilities are properly aligned with the demands of the challenge or situation at hand. Achieving this balance and maximizing the use of skills may be necessary for participants to achieve a flow experience. If the challenges individuals faced and their abilities are not properly balanced, the activity must be re-constructed, perhaps even during the activity, which will interrupt or prevent the flow from occurring [21]. If the activity is too easy, participants will get bored, but if the task is too difficult, they may have to stop and read the instructions or simply drop out of the activity altogether. This prevents the flow experience from being achieved because it distracts the participant from the activity at hand [17]. Computer program generated virtual human is characterized by automation and convenience. Users only need to select and set attributes on the operation interface to generate virtual human images easily. Users' personalized ability is difficult to embody. Users with basic dance ability are more able to produce pleasing movement effects and have a better flow experience. Thus, we suppose that:

H2c. Compared to computer program generated virtual human products, real person generated virtual human products provide consumers higher "ability vs challenge balance" flow experience.

3.3   The Mediating Effect of Flow Experience

In the study of human-computer interaction in virtual environment, existing studies have shown that flow experience of augmented reality (AR) devices can generate a good attitude towards AR applications and more trust in AR applications, thus triggering customers' stickiness towards AR applications and improving brand attitudes and brand usage intentions [10]. Barhorst et al. [25] show that AR produces a flow state, which indirectly affects customers' satisfaction with AR experience. Studies have shown that, in a highly realistic virtual learning environment, the immersive inquiry environment created and presented by the metaverse can greatly stimulate learners' various senses. Just like role immersion in role-playing games, it provides learners with a strong sense of engagement in learning, which is incomparable to previous learning environments or Spaces [26]. Virtual human is a new stage in the development of human-computer interaction in virtual environment. The optimal psychological state achieved by a person in the process of flow experience may provide a level of deepening, enhancing and subconscious contact with the brand [17], just as advertising creativity influences advertising attitude and then brand attitude [27]. If customers have a positive attitude towards a brand, customers will be more inclined to buy products sold through the brand [28]. Therefore, the flow experience generated by using virtual human products becomes an important selling point of virtual human products and promotes the willingness to use. Thus, we suppose that:

H3a. The effect of different virtual human products on consumer willingness to use is mediated by "explicit feedback" flow experience.

H3b. The effect of different virtual human products on consumer willingness to use is mediated by "sense of control" flow experience.

H3c. The effect of different virtual human products on consumer willingness to use is mediated by "ability vs challenge balance" flow experience.

4   Method and Data

Most of the measures of the constructs were developed based on prior literature. The measurement of flow experience of the three dimensions is based on the scale of Hooker et al. [17] and adjusted for the specific situations of virtual human. The measurement of willingness to use is based on the scale of Liu et al. [29] and adjusted appropriately.

This experiment was designed as an independent sample t-test experiment, with the independent variable being "virtual human product (computer program generated/real person generated)" and the dependent variable being "willingness to use". According to the calculation of G*Power [30], at least 128 subjects are needed for the experiment to achieve the statistical efficacy of $1 - \beta = 0.8$. There are 259 participants from China Wenjuanxing platform completed the experiment online.

Participants were randomly assigned to one of two groups (computer program generated or real person generated) to watch videos of virtual human online. When the participant watches the video, it is assumed that the virtual human image in the video is generated by the participant. The participant in scenario 1 can generates the virtual human by computer program. The face of the virtual human is generated based on the neural network rendering technology, the voice is generated based on the super natural voice technology, and the song is generated based on the artificial intelligence song synthesis technology, which is as shown left in Fig. 2; In scenario 2, the participant could generate virtual human images through herself wearing sensing devices and conduct real-time voice and action interactions. The actions and expressions of such virtual human were generated in real time through sensing devices and visual capture technology, as shown right in Fig. 2.

Fig. 2. Computer program generated (left) vs. Real person generated (right)

The viewing page was set for a lockout time to ensure that participants watched the virtual human video in its entirety. At the end of the viewing process, to ensure data quality, "Is the virtual human in the video computer program generated or real person generated?" question was used as an attention screening question, and those who failed to answer correctly were automatically rejected by the system and re-recruited. The participants were asked to evaluate their willingness to use the virtual human products displayed in the video. The experiment adopted a 7-point Likert scale consisting of three items (1 = "strongly disagree", 7 = "strongly agree") to evaluate their willingness to use [29].

For the mediating variables, the participants were first asked to recall the virtual human video and think about the reasons and psychological processes for deciding whether to use virtual human products. Then, the participants answered questions related to flow experience, including three dimensions of flow experience [17]. Finally, the subjects answered questions such as "Have you ever used a similar type of virtual human product", "Have you heard or know about virtual human products?", "What do you think of your physical abilities" (1 = "very bad", 7 = "very good"), and completed measures of demographic variables (age, gender, monthly consumption).

5   Results and Discussion

5.1   Summary Statistics

A total of 259 samples were collected in the experiment, and 43 samples that answered "no" in "Have you heard or known about virtual human products" and "yes" in "Have you used this type of virtual human products" were eliminated. Finally, 216 samples were obtained in this experiment, among which 103 were computer program generated virtual human samples and 113 were real person generated virtual human samples. The 216 respondents are (49.5%) male and (50.5%) female, showing that both genders are practically balanced in this research. About (8.8%) are less than18 years old, (30.1%) are 18–25 years old, (19.4%) are 26–30 years old, (27.3%) are 31–35 years old, and (14.4%) above 36 years of age. Approximately (56.1%) of the respondents have a monthly consumption of less than 2000 RMB. Regarding education levels, (21.3%) have not completed high school and (20.8%) have completed and about (57.9%) have a college-educated degree. The mean value of physical ability is 4.63 and about (59.7%) of the respondents are above mean value.

5.2   Hypothesis Testing

The reliability and validity analysis of the scale data involved in this paper is carried out. Questions of the four variables "Explicit Feedback, Sense of Control, Ability Vs Challenge Balance, Willingness to use" all have good reliability (Cronbach-α > 0.8). The KMO coefficient of the scale was 0.833, the data validity met the requirements.

According to the independent sample t-test in Table 1, The mean value of Willingness to use (UI) towards real person generated virtual products (RP), Explicit Feedback (EF), Ability versus Challenge Balance (ACB) was higher than that of computer program generated virtual human products (CP). Results in Table 1 show that there are significant differences between real person generated virtual human and the computer program generated virtual human in UI, EF, and ACB ($p < 0.05$), that is, compared with computer program generated virtual human products, real person generated virtual human products has higher consumer willingness to use, a higher experience of "explicit feedback" and a higher experience of "ability vs challenge balance", which verifies hypothesis H1, H2a and H2c.

Table 1. Independent sample t-test

	VH	Sample	Mean	S.D.	F.sig	t	p
UI	CP	103	4.495	1.841	0.022	−3.873	0.000
	RP	113	5.386	1.506			
EF	CP	103	4.735	1.587	0.004	−2.918	0.004
	RP	113	5.305	1.244			
SC	CP	103	4.830	1.290	0.497	−1.205	0.230
	RP	113	5.058	1.467			
ACB	CP	103	4.745	1.295	0.981	−2.281	0.024
	RP	113	5.159	1.366			

Note: VH = Virtual Human, EF = Explicit Feedback, SC = Sense of Control, ACB = Ability vs. Challenge Balance, UI = Willingness to use, CP = Computer Program Generated, RP = Real Person Generated.

There was no significant difference in Sense of Control (SC) ($p > 0.05$) in Table 1, the hypothesis in this paper that H2b "compared with computer program generated virtual human, real person generated virtual human has a higher 'sense of control' flow experience" is not supported. The possible explanation is that the production of computer program generated virtual human products also requires the selection of functional attributes by users. Users can customize by controlling related properties, which is similar to the sense of control generated by controlling virtual human through their own real-time movements.

The mediation hypothesis of Explicit Feedback (EF), Sense of Control (SC) and Ability Vs Challenge Balance (ACB) were verified by bootstrapping intermediate test. VH was taken as the independent variable, EF, SC, ACB as the intermediary variable, and UI as the dependent variable. The bootstrap sample number was set to 5000 and the confidence interval was set to 95%. The results are shown in Table 2. The indirect effect of EF and ACB mediating paths is significant, indicating that the influence of different virtual human products on consumer willingness to use is mediated by the experience of "Explicit Feedback" and the experience of "Ability vs. Challenge Balance", which supports hypothesis H3a and hypothesis H3c, while the path mediated by SC is not supported and confirms that H3b, "'he effect of different virtual humans on willingness to use is mediated by "sense of control" flow experience' is not valid, which is consistent with the above invalid hypothesis that H2b, Compared to computer program generated virtual human products, real person generated virtual human products have higher "sense of control" flow experience.

The results of customer experience and willingness to use of virtual human products can provide a reference for enterprises to make decisions in developing virtual human products. In the process of product design and marketing, "explicit feedback" and "ability vs. challenge balance" flow experience of users can be mainly considered for both computer program and real person generated virtual human products. For instance, the interface of the virtual human game timely reflects the operation instructions of the players, and at the same time, the players have clear tips and rewards at different stages when carrying out tasks. In addition, the virtual human players get corresponding points according to their abilities, and the virtual human game ranking is updated in real time. The unique abilities of the players are displayed, which encourages the players to immerse themselves in the competition and continue to use the virtual human products. And the use of real person generated virtual human for livestreaming retailing, livestreaming chat, and news broadcast are also encouraged.

Table 2. Bootstrapping intermediate test

	Effect	BootSE	BootLLCI	BootULCI
VH > EF > UI	0.1344	0.0751	0.0179	0.3086
VH > SC > UI	0.0617	0.0590	−0.0399	0.1938
VH > ACB > UI	0.1063	0.0553	0.0091	0.2240

Note: VH = Virtual Human, EF = Explicit Feedback, SC = Sense of Control, ACB = Ability vs. Challenge Balance, UI = Willingness to use

6 Conclusion

This paper focuses on the investigation of consumer willingness to use of different virtual human products and the associated influencing mechanism. The study confirms that customers have a higher willingness to use for real person generated virtual human products than computer program generated virtual human products. The reason can be explained by flow theory. Compared to computer program generated virtual human products, customers of real person generated virtual human products have higher "explicit feedback" and "ability vs. challenge balance" flow experience, which affects their willingness to use virtual human products.

This research has two main contributions. We first define a new virtual human type, real person generated virtual human compared to computer generated virtual human, and further examine the differences of consumer willingness to use towards these two virtual human products, which fills the research gap of the existing research field of virtual human products. In addition, the consumer flow experiences are investigated as mediators in the difference of consumer willingness to use towards two virtual human products. In addition, the research conclusions of this paper provide a reference for enterprises to make better use of virtual human products for social media e-commerce marketing communications.

There are still some limitations in this study, which also provide possible directions for future research. First, the experimental design of this paper is to let the participants watch the virtual human video that has been made, and imagine that the virtual human video is made by the participants themselves, which lacks the sense of personal experience and may cause certain deviation in the use experience. The follow-up research can try to make virtual human products by the participants themselves. Besides, this paper focuses only on the intention to use virtual human products at customer level. However, different virtual human products also play an important role in marketing communication. At the enterprise level, the influence of the use of different virtual human products on brand awareness and brand sales is worth further investigation.

Funding. The research was financially supported by the national natural science foundation of china under grant [number 71831005], and Natural Science Foundation of Shenzhen under grant [number JCYJ20220531095216037].

References

1. Plangger, K., Grewal, D., de Ruyter, K., et al.: The future of digital technologies in marketing: a conceptual framework and an overview. J. Acad. Mark. Sci. **50**(6), 1125–1134 (2022)
2. Huang, M.-H., Rust, R.T.: A strategic framework for artificial intelligence in marketing. J. Acad. Mark. Sci. **49**(1), 30–50 (2020). https://doi.org/10.1007/s11747-020-00749-9
3. Mei, X.Y., Cao, J.F.: From the information to the value interconnection: the transformation of the knowledge economy in the metaverse and the reconstruction of governance. Books Intell. **06**, 69–74 (2021). (In Chinese)
4. Jhan, X., Wong, S., Ebrahimi, E., et al.: Effects of small talk with a crowd of virtual humans on users' emotional and behavioral responses. IEEE Trans. Visual Comput. Graphics **28**(11), 3767–3777 (2022)
5. Schroeder, N.L., Chiou, E.K., Craig, S.D.: Trust influences perceptions of virtual humans, but not necessarily learning. Comput. Educ. **160**, 15 (2021)
6. Gürerk, Ö., Bönsch, A., Kittsteiner, T., et al.: Virtual humans as co-workers: a novel methodology to study peer effects. J. Behav. Exp. Econ. **78**, 17–29 (2019)
7. Qu, C., Brinkman, W., Ling, Y., et al.: Conversations with a virtual human: synthetic emotions and human responses. Comput. Hum. Behav. **34**, 58–68 (2014)
8. Morina, N., Brinkman, W., Hartanto, D., et al.: Sense of presence and anxiety during virtual social interactions between a human and virtual human. PeerJ **2**, 1 (2014)
9. Bouchard, S., Bossé, J., Loranger, C., et al.: Advances in Virtual Reality and Anxiety Disorders. Springer, New York (2014). https://doi.org/10.1007/978-1-4899-8023-6
10. Arghashi, V., Yuksel, C.A.: Interactivity, inspiration, and perceived usefulness! how retailers' AR-apps improve consumer engagement through flow. J. Retail. Consum. Serv. **64**, 102756 (2022)
11. Csikszentmihalhi, M.: Finding Flow: The Psychology of Engagement with Everyday Life. Hachette, London (2020)
12. Forough, S., Zarei, A., Mohsen, S.N.: Consumers' impulse buying behavior on instagram: examining the influence of flow experiences and hedonic browsing on impulse buying. J. Internet Commer. **19**(4), 437–465 (2020)
13. van Noort, G., Voorveld, H.A.M., van Reijmersdal, E.A.: Interactivity in brand web sites: cognitive, affective, and behavioral responses explained by consumers' online flow experience. J. Interact. Mark. **26**(4), 223–234 (2012)
14. Rauschnabel, P.A., Rossmann, A., Tom Dieck, M.C.: An adoption framework for mobile augmented reality games: the case of Pokémon Go. Comput. Hum. Behav. **76**, 276–286 (2017)
15. Mahdi Hosseini, S., Fattahi, R.: Databases' interface interactivity and user self-efficacy: two mediators for flow experience and scientific behavior improvement. Comput. Hum. Behav. **36**, 316–322 (2014)

16. Bressler, D.M., Bodzin, A.M.: A mixed methods assessment of students' flow experiences during a mobile augmented reality science game. J. Comput. Assist. Learn. **29**(6), 505–517 (2013)

17. Hooker, R., Wasko, M., Paradice, D., et al.: Beyond gaming: linking flow, brand attitudes, and purchase intent in realistic and emergent three-dimensional virtual environments. Inf. Technol. People **32**(6), 1397–1422 (2019)

18. Javornik, A.: 'It's an illusion, but it looks real!' consumer affective, cognitive and behavioural responses to augmented reality applications. J. Mark. Manag. **32**(9–10), 987 (2016)

19. Hoffman, D.L., Novak, T.P.: Flow online: lessons learned and future prospects. J. Interact. Mark. **23**(1), 23–34 (2009)

20. Rezaei, S., Valaei, N.: Branding in a multichannel retail environment: online stores vs app stores and the effect of product type. Inf. Technol. People **30**(4), 853–886 (2017)

21. Quinn, R.W.: Flow in knowledge work: high performance experience in the design of national security technology. Adm. Sci. Q. **50**(4), 610–641 (2005)

22. Chen, H.: Flow on the net–detecting Web users' positive affects and their flow states. Comput. Hum. Behav. **22**(2), 221–233 (2006)

23. Guo, Y.M., Poole, M.S.: Antecedents of flow in online shopping: a test of alternative models. Inf. Syst. J. **19**(4), 369–390 (2009)

24. Jackson, S.A., Eklund, R.C.: Assessing flow in physical activity: the flow state scale–2 and dispositional flow scale–2. J. Sport Exerc. Psychol. **24**(2), 133–150 (2002)

25. Barhorst, J.B., McLean, G., Shah, E., et al.: Blending the real world and the virtual world: exploring the role of flow in augmented reality experiences. J. Bus. Res. **122**, 423–436 (2021)

26. Hua, Z.X., Fu, D.M.: Study on the connotation, mechanism, structure and application of learning metaverse – and the learning promotion effect of virtual human. J. Dist. Educ. **40**(01), 26–36 (2022). (In Chinese)

27. Rosengren, S., Eisend, M., Koslow, S., et al.: A meta-analysis of when and how advertising creativity works. J. Mark. **84**(6), 39–56 (2020)

28. Micael, D., Lars, F., Erik, N.: Long live creative media choice: the medium as a persistent brand cue. J. Advert. **38**, 121–129 (2009)

29. Liu, F., Ngai, E., Ju, X.: Understanding mobile health service use: an investigation of routine and emergency willingness to uses. Int. J. Inf. Manag. **45**, 107–117 (2019)

30. Faul, F., Erdfelder, E., Albert-Georg, L., et al.: G*Power 3: a flexible statistical power analysis program for the social, behavioral, and biomedical sciences. Behav. Res. Methods **39**(2), 175–191 (2007)

When Virtual Influencers are Used as Endorsers: Will Match-Up and Attractiveness Affect Consumer Purchase Intention?

Yanling Zhang, Duo Du, and Jiao Ge[✉]

Harbin Institute of Technology Shenzhen, Shenzhen 518055, China
jiaoge@hit.edu.cn

Abstract. In the domain of social media marketing, the influences of virtual influencers on social media have gradually risen and they received widespread attention. The virtual influencers sometimes have similar size of the number of followers with the celebrities. Crises that bring losses to the brand and the company due to the negative news of celebrities often occur. The virtual influencers provide new opportunities for social media marketing, and using virtual influencers as endorsers for advertising to influence consumer behavior has becoming popular in marketing. However, this advertising effect of virtual influencers needs further investigation. The purpose of this study is to investigate whether the product-virtual endorsers match and the attractiveness of virtual influencers have an impact on consumers' purchase intention, and to explore whether consumers' familiarity with virtual influencers has a moderating effect on this impact. The results show that the matching of products and virtual endorsers and the attractiveness of virtual endorsers have positive impacts on consumers' purchase intention, while the familiarity of consumers with virtual people negatively regulates this impact.

Keywords: Virtual influencer · Product-virtual endorser match · Attractiveness · Familiarity · Purchase intention

1 Introduction

Employing celebrity endorsers is one of the commonly used means of marketing, because celebrity endorsers can attract more consumers and media attention with their own influence. The possible influence of endorsers can be summarized as two reasons. First, they have attractive and desirable qualities, and are generally highly active. At the same time, their popularity can attract consumers' attention to the products (Atkin 1983; Kamins 1990). However, there are also great risks in the use of endorsers. Any negative information such as drunk driving or inappropriate speech will have negative impacts on endorsers, thereby reducing consumer awareness (Thwaites et al. 2012). With the application of emerging technologies, virtual influencers have become popular on social platforms. Virtual influencers have similar popularity with celebrities, but their virtuality reduces the risk of negative news and improves the controllability of brands over virtual endorsers. Virtual influencer becomes a new endorser choice favored by many brands.

© Springer Nature Switzerland AG 2023
F. Fui-Hoon Nah and K. Siau (Eds.): HCII 2023, LNCS 14038, pp. 487–496, 2023.
https://doi.org/10.1007/978-3-031-35969-9_33

Influencers are "individuals who have a significant impact on others". This concept has also been introduced into social media (Park et al. 2021). With the rise of social media, "social media influencers", also known as "digital influencers", have emerged. They also have a huge number of fans and influence. Many brands also use them to carry out influencer marketing, and use their popularity and the trust of consumers to promote brand and establish word of mouth. Therefore, influencer marketing is also described as an electronic word of mouth (Breves 2019). Among them, there is a special kind of social media influencers-virtual influencers, virtual influencers are computer-generated influencers whose vivid images have attracted wide attention on social media. For example, the first video of Liu Yexi on Tiktok platform attracts 3.6 million likes and she has 8.4 million fans now (Douyin 2023). Therefore, highly popular virtual influencers have brought new opportunities and forms for social media marketing. Cooperation with brands has also become one of the business models operated by virtual influencers. Virtual influencers have become a brand-new type of celebrity. Many brands have also begun to try to use them as endorsers for brand publicity to attract traffics. Prada and many other brands have cooperated with well-known virtual influencers for drainage, and many brands have launched their own virtual endorsers. For example, Huaxizi launched its own Chinese-style virtual image endorsers "Huaxizi" in June 2021, some virtual characters even appear in the live broadcast with goods on the social media platform to interact directly with consumers. According to the data of iMedia Consulting, China's virtual influencers have driven the overall market size to 107.49 billion yuan, and the core market scale has reached 6.22 billion yuan, and the business opportunities and values of virtual influencers are constantly being explored and utilized.

In the field of marketing, the degree of personalization provided by virtual influencers and humans is similar (Sands et al. 2022).It can also meet many needs of consumers in social media (Arsenyan and Mirowska 2021) and establish a harmonious relationship with mankind (Gratch et al. 2007). But compared with humans, virtual influencers have advantages in terms of consistency and risk (Arsenyan and Mirowska 2021), and more likely to cause word-of-mouth intentions (Sands et al. 2022).Virtual characters will not experience appearance changes and can always maintain their image in the minds of consumers (Hsu et al. 2018), and then improve customers' purchase intention effectively (Holzwarth 2006). However, few studies focus on the impact of the product-virtual endorser match in ads on the consumer's purchase intention, which is very important for the virtual influencers to translate the endorsement into brand value, and it can provide some reference for the brand to choose the virtual influencer to cooperate with. Thus, the objective of this research is to investigate the effect of product-virtual endorser match on consumer purchase intension with the consideration of consumer familiarity towards the virtual influencer.

2 Literature Review and Hypotheses

2.1 Match Effect on Consumer Purchase Intention

Kahle and Homerput (1985) forwarded the celebrity-product matching hypothesis, "the matching hypothesis is that the information conveyed by the celebrity image and the information about the product should be gathered in the effective advertisement". The matching hypothesis has been widely used in the research of the results of traditional celebrity endorsers. The product-endorser match is an important factor for the success of endorsers advertising. However, the results of research on the relationship between product-endorsers match and its effect of advertising are not consistent.

On one hand, some scholars believe that the endorsers-product consistency have positive effect on authenticity and consumer trust, leading to positive results. Previous studies have shown that the personality consistency of celebrities and brands has a positive impact on consumer purchase intention and brand attitude (Pradhan et al. 2014). The effect of advertising will be worse when the product and the spokesperson match poorly (Kamins and Gupta 1994). In the research on the influence of spokesperson-product matching on consumer's purchase intention, the result shows that spokesperson-product matching of athletes will significantly and positively affect consumer's purchase intention (Liu and Brock 2011), and the consistency of virtual animation endorsers and products will also have a significant positive impact on consumers' purchase intention (Liang and Yang 2022). On the other hand, some scholars believe that inconsistent and uncoordinated endorsers will improve the communication effect, which will have a positive impact on consumers' advertising time investment, brand attitude, purchase intention and word of mouth (Törn 2012).

We believe that there is a positive relationship between the product-endorser match and consumers' purchase intention. First of all, research on virtual influencers shows that there is a positive relationship between the highly matching degree of virtual people and products and the positive attitude of consumers towards the recognition of virtual images (Li et al. 2023). Recognition of endorsers is the basis for generating trust, so as to further establish contact with consumers. Secondly, the majority of audiences of social media virtual influencers are Millennials and Generation Z (Moustakas et al. 2020). The existence of endorsers makes millennials more likely to express their views through consumer products (McCormick 2016), therefore, when virtual influencer endorsers and products are more consistent, they are more likely to be convinced. Therefore, we hypothesize that.

H1. Product-virtual-endorser match positively affects consumers' purchase intention.

2.2 Attractiveness Effect on Consumer Purchase Intention

Relevant characteristics of endorsers will affect consumer brand perception, brand loyalty and purchase intention (Osei-Frimpong et al. 2019), and the attraction of the spokesperson is one of the most important characteristics of the spokesperson. First of all, attractive virtual influencers will enhance consumer trust. Research has shown that attractive celebrity endorsers have a positive impact on consumers' purchase intention. The spokesperson is regarded as a credible person by virtue of its high attractiveness to enhance the trust of consumers (Spry et al. 2011), and trust is an important factor that affects the role and value of endorsements (Schouten et al. 2020). Secondly, attractive virtual influencers will have a visual impact on consumers. Virtual influencers used for endorsement and advertising can bring consumers a higher sense of advertising freshness (Liu and Brock 2011), thus arousing consumers' interest in products and Improve consumers' memory of advertisements and products. It can thus make consumers have brand memory when facing product selection, which is conducive to consumers' willingness to buy advertising products. Thus, we propose that.

H2. The attractiveness of virtual endorsers has positive impact on consumers' purchase intention.

2.3 Moderating Effect of Familiarity

In the research of biology, familiarity is considered to be a similar process of fast signal retrieval, which is not conducive to the learning of new association (Yonelinas 2002). Similarly, a study on facial familiarity found that people can extract the same information from different photos of the same person (Kramer and Young et al. 2018). Therefore, when using virtual influencers that consumers are very familiar with for product endorsement and advertising, consumers will more extract the memory of the former virtual influencers than pay attention to the virtual influencers in the new situation. That is, consumers' inherent impression and personal feelings of the virtual influencers will play a more role. Thus, the influence of the attractiveness of virtual influencers on consumer purchase intention will be weakened. At the same time, it is a novel way to use virtual influencers as endorsers, especially when unfamiliar virtual spokesmen appear. It is more likely to bring consumers a fresh sense of advertising, and consumers will pay more attention to virtual influencers, and more likely to be attracted by the charm of virtual influencers to deepen product attention. Therefore, we believe that familiarity negatively moderates the influence of attractiveness of virtual influencers on consumer purchase intention.

When firms use the virtual influencer to endorse and the advertisement reappears in the public's view, it can be considered as a kind of advertisement repetition. When the advertisement reappears, compared with the familiar brand, if the advertisement of the unfamiliar brand is very novel, it will cause consumers to deal with it more widely, and it is easier to stimulate consumers to consider the improper advertisement (Campbell and Keller 2003). Therefore, when unfamiliar virtual influencers are used for endorsement advertising, consumers are likely to have more in-depth thinking about the matching degree of virtual influencers and products, that is, when consumers are less familiar with virtual influencers, they will pay more attention to the matching degree of virtual influencers and products, so that the impact of matching degree on consumers' purchase intention will also be more obvious, When the virtual influencer with high familiarity is applied to marketing, the impact of the match between the product and the virtual influencer on the consumer's purchase intention will decrease. In summary, the familiarity negatively moderates the impact of the match between product and its virtual endorsers on the consumer's purchase intention. Thus, we propose that.

H3a. The familiarity of consumers with virtual endorsers negatively moderates the positive relationship between product-virtual-endorser match and consumer purchase intention.

H3b. The familiarity of consumers with virtual influencers negatively moderates the positive relationship between attractiveness of virtual endorsers and consumer purchase intention.

3 Method and Data

3.1 Measurement of Variables

With regard to the measure of product-endorsers match, the three dimensions of compatible, good-fit and congruent (Lee and Thorson 2008; Liang and Yang 2022) were used. Virtual endorser attractiveness is measured through five dimensions: Attractive, Classy, Beautiful, Elegant and Sexy (Ohanian 1990). With regard to consumer purchase intention, three questions are selected to measure according to the previous research, including "Possibility of purchasing this product" (Liu et al. 2007); "The possibility of actively looking for this product in the store" (McCormick 2016); "To what extent does advertising increase the opportunity to buy this product in the future" (Kamins 1990).

3.2 Data Collection

In this paper, a random questionnaire is selected. Considering the audience group of virtual influencers, this experiment is mainly aimed at generation Z (people born after 1995) to observe and study their attitude to the advertising of virtual endorsers and the impact on their purchase intention for the products in the advertisement.

In the experiments, the seven-point Likert scale was used. Participants will be asked to watch two short advertising videos for different brands with virtual endorsers. Then ask them to rate the familiarity of the virtual person and to fill in questions about product-virtual endorsers matching, virtual endorsers attraction and purchase intention. The last part is participant personal information, mainly including age, gender, frequency of using social media, degree of attention to virtual people. It is also the control variable of this experiment to reduce the impact of personal characteristics and habits on their purchase intention.

We distributed the videos and the associated questionnaire on Questionnaire Star online platform and invited participants to carry out the online experiment with compensation from December of 2022 to January of 2023. Some samples were screened out because of the short filling time, and the final valid sample is 390.

3.3 Model

We summarized the model as shown in the following equation, where i represents consumers and j represents brands.

$$purchase\ intention_{ij} = \beta_0 + \beta_1 match_{ij} + \beta_2 attractiveness_{ij} +$$
$$\beta_3 familiarity_{ij} + \beta_4 match_{ij} * familiarity_{ij} +$$
$$\beta_5 attractiveness_{ij} * familiarity_{ij} + \beta_6 frequency_i +$$
$$\beta_7 intention_i + \beta_8 gender_i + \beta_9 age_i + \beta_{10} brand_j + \varepsilon_{it}$$

where *purchase intension* is consumers' purchase intension after they watch the short video ads with virtual endorsers, *match* represents that product-virtual-endorser matchiness, *attractiveness* is the virtual endorser's attractiveness, *familiarity* indicates how consumers familiar with the advertised virtual endorsers, all other variables are control variables which has been explained in the measurements of variables.

4 Results and Discussion

Among all participants in our sample, women account for 52.8% while men account for 47.2%. Respondents born after 1995 account for 55.3%, and the rest born after 2000. The reliability and validity test results are shown in Table 1, which indicates that the Cronbach Alpha coefficient of product-virtual endorsers match is 0.912, and the Cronbach Alpha coefficients of attractiveness and purchase intention are 0.946 and 0.913 respectively. Therefore, the reliability of the questionnaire is good. According to the validity results, the KMO index is 0.981, so the validity structure of the questionnaire is good.

Table 1. Reliability analysis.

Variables	Cronbach's alpha	Cronbach' Alpha based on standardized items
Product-endorsers Match	0.912	0.912
Attractiveness	0.946	0.946
Purchase intention	0.913	0.913

Regression results are shown in Table 2. Model 1 tested the impact of product-virtual endorsers match on consumers' purchase intention. The results showed that product-virtual endorsers match had a significant positive impact on consumers' purchase intention at the significant of 1% level, and H1 was supported. Model 2 examines the impact of the virtual influencer's attractiveness on consumers' purchase intention. The results show that the higher the virtual influencer's attractiveness, the stronger the consumers' purchase intention. Therefore, there is a significant positive relationship between the virtual influencer's attractiveness and consumers' purchase intention, and H2 is supported. Model 3 includes both product-virtual-endorser match and attractiveness of virtual influencers. We can see that both of them have significant positive impacts on consumers' purchase intention at the significant of 1% level. For each level of virtual influencer's attractiveness increases, consumers' purchase intention increases by 0.442 level, while for each level of matching degree of product-virtual endorsers increases, consumers' purchase intention increases by 0.356 level. Therefore, the product-virtual endorsers match and the attractiveness of virtual influencers both have significant positive impacts on the consumers' purchase intention, and the attractiveness of virtual influencers is more important than the product-virtual-endorser match.

Model 4 and 5 in Table 2 examines the moderating effect of consumers' familiarity with virtual influencers. The results show that familiarity to a certain extent inhibits the impact of the product-virtual endorser match and attractiveness of virtual influencers on consumers' purchase intention, and H3a and H3b are supported. This implies that when consumers' familiarity with products is low, the product-virtual endorser match has more significant positive effect on consumers' purchase intention, However, with the deepening of consumers' familiarity with the virtual influencer, the role of product-virtual endorser match will be weakened. Similarly, when consumers are less familiar with the virtual influencer, the attraction of the virtual influencer has also significant

Table 2. Regression results.

Purchase Intension	Model 1	Model 2	Model 3	Model 4	Model 5	Model 6
Product-virtual endorsers match	0.6068***		0.3570***	0.6697***	0.2467***	0.2833*
	(0.0411)		(0.0479)	(0.0921)	(0.0492)	(0.1590)
Attractiveness		0.6819***	0.4423***	0.3182***	0.8055***	0.7728***
		(0.0440)	(0.0522)	(0.0554)	(0.0902)	(0.1627)
Familiarity				0.4706***	0.5554***	0.5616***
				(0.0908)	(0.0914)	(0.0950)
Match*Familiarity				−0.0756***		−0.0069
				(0.0170)		(0.0286)
Frequency	0.1377***	0.0989***	0.0490	0.0171	0.00313	0.0030
	(0.0311)	(0.0317)	(0.0304)	(0.0302)	(0.0303)	(0.0303)
Intention	0.1864***	0.1564***	0.1012***	0.0625**	0.0581*	0.0576*
	(0.0320)	(0.0322)	(0.0310)	(0.0308)	(0.0305)	(0.0306)
Gender	0.0923	0.0433	0.0693	0.0699	0.0697	0.0696
	(0.0689)	(0.0677)	(0.0634)	(0.0613)	(0.0606)	(0.0607)
Age	−0.0764	0.0156	−0.0179	−0.00523	−0.00203	−0.0019
	(0.0690)	(0.0680)	(0.0638)	(0.0617)	(0.0610)	(0.0611)
Brand	−0.0590	−0.0505	−0.0582	−0.0367	−0.0574	−0.0556
	(0.0686)	(0.0674)	(0.0630)	(0.0610)	(0.0602)	(0.0608)
Attractiveness* Familiarity					−0.0929***	−0.0872***
					(0.0173)	(0.0294)
Constant	0.3263***	0.2897**	0.2372**	−0.609***	−0.7724***	−0.787***
	(0.1174)	(0.1156)	(0.1084)	(0.213)	(0.2104)	(0.219)
Observations	390	390	390	390	390	390
R−squared	0.8641	0.869	0.886	0.894	0.896	0.896
F	405.94	423.36	422.57	355.17	364.35	327.11

Standard errors in parentheses, *** $p < 0.01$, ** $p < 0.05$, * $p < 0.1$

positive impact on the purchase intention, but with the deepening of the familiarity so that impact decreases. It may be that with the deepening of familiarity, consumers will pay more attention to the virtual influencer itself, and reduced attention and concern on matching degree, thus weakening the impact of match-up on purchase intention. Besides, when consumers are not familiar with the virtual influencer, using the virtual influencer for endorsement and advertising will give consumers a sense of fresh stimulation. It is easier to arouse consumers' interest and deepen consumers' product memory. With the deepening of familiarity, attractiveness is no longer the key point of arousing interest for consumers, and the corresponding role will be weakened. Model 6 shows that when all variables are included, the moderating effect of familiarity on product-virtual endorsers match is no longer significant, while the adjustment effect on attractiveness is still significantly negative, that is, the moderating effect is mainly on the attractiveness of the virtual influencer.

5 Conclusion

This paper mainly studies the influence of the product-virtual endorser match and the attractiveness of virtual influencers on consumers' purchase intention when using virtual influencers as advertising endorsers, and tests the moderating effect of consumers' familiarity with virtual endorsers. The results show that both the attractiveness of virtual influencers and the product-virtual endorsers match have significant positive impacts on consumers' purchase intention, while the familiarity of consumers with virtual influencers negatively regulates these impacts. We suggest that the best marketing strategy when selecting virtual influencers for products and endorsements, the company should fully consider the matching degree between virtual influencers and products, and also measure the value that virtual influencers' attractiveness can bring, so as to select more suitable virtual influencers for endorsement and promotion.

Our research contributes the research domain in virtual influencer marketing in three aspects. First, this paper has extended the product-influencer match to the human-like virtual influencers and their match with products and the associated effects on consumers' purchase intention since there are few papers involved how the matching degree of human-like virtual influencers and products affects consumers' purchase intention currently. Second, this paper as the pioneer study of investigating the impact of matching degree and the influencer attractiveness on advertising effect since few literatures focus on the impact of matching degree and attractiveness on advertising effect.

However, the sample is limited to most of respondents are generation Z. People of different ages and experiences may have different views on the attitude of the virtual influencer, so future research can expand the sample range and increase the sample type. For example, will the match between products and virtual influencers and the attractiveness of virtual influencers still have a significant impact on consumers' purchase intention if there is no limitation of age or identities? In addition, due to the limitations of virtual people and brands, the experimental content selected in this paper is limited. The future research could continue to enrich the research of virtual endorsers and products, and can specifically study: Will different types of products affect the research results? What kind of products are more suitable for virtual influencers to advertise? What kind of virtual influencers are more suitable as endorsers? Or we can extend the research on what kind of virtual influencer should be selected for each type of product? In addition, is the experience of virtual influencers a factor to be considered? If he has endorsed other brands of the same type, will it affect the research results. Because of its low-cost, multiple fans and low-risk, virtual influencers will be more and more used in corporate marketing. Therefore, it is very important to continue to study the matching of virtual influencers and products, or other fields, and can also provide better support for the company to create new value and profits.

Funding. The research was financially supported by the National Natural Science Foundation of China under grant [Number 71831005], and Natural Science Foundation of Shenzhen under grant [Number JCYJ20220531095216037].

References

Arsenyan, J., Mirowska, A.: Almost human? a comparative case study on the social media presence of virtual influencers. Int. J. Hum.-Comput. Stud. **155**, 102694 (2021)

Atkin, C., Block, M.: Effectiveness of celebrity endorsers. J. Advert. Res. **23**, 57–61 (1983)

Campbell, M.C., Keller, K.L.: Brand familiarity and advertising repetition effects. J. Consum. Res. **30**(2), 292–304 (2003)

Douyin (2023). https://www.douyin.com/user/MS4wLjABAAAA_hsHsgWtEn6fIAs-1U5uhN NlvjtTDcAQZuApn4s-vCaVQdY6ulIRGeQTyYvJxRSf

Gratch, J., Wang, N., Gerten, J., Fast, E., Duffy, R.: Creating rapport with virtual agents. In: Pelachaud, C., Martin, J.-C., André, E., Chollet, G., Karpouzis, K., Pelé, D. (eds.) IVA 2007. LNCS (LNAI), vol. 4722, pp. 125–138. Springer, Heidelberg (2007). https://doi.org/10.1007/978-3-540-74997-4_12

Hsu, Y., Chen, H., et al.: The effect of virtual spokescharacter type on online advertisements. Int. J. Electron. Commer. Stud. **9**(2), 161–190 (2018)

Kahle, L.R., Homer, P.M.: Physical attractiveness of the celebrity endorser: a social adaptation perspective. J. Consum. Res. **11**(4), 954 (1985)

Kamins, M.A.: An investigation into the "'Match-up'" hypothesis in celebrity advertising: when beauty may be only skin deep. J. Advert. **19**(1), 4–13 (1990)

Kamins, M.A., Gupta, K.: Congruence between spokesperson and product type: a matchup hypothesis perspective. Psychol. Mark. **11**(6), 569–586 (1994)

Kramer, R.S.S., Young, A.W., Mike Burton, A.: Understanding face familiarity. Cognition **172**, 46–58 (2018)

Lee, J., Thorson, E.: The impact of celebrity-product incongruence on the effectiveness of product endorsement. J. Advert. Res. **48**(3), 433–449 (2008)

Li, J., Huang, J., et al.: Examining the effects of authenticity fit and association fit: a digital human avatar endorsement model. J. Retail. Consum. Serv. **71**, 103230 (2023)

Liang, H.L., Yang, F.H.: Are virtual anime endorsers a new type of endorser? examining product involvement as a moderating role. Front. Psychol. **13**, 779519 (2022)

Liu, M.T., Brock, J.L.: Selecting a female athlete endorser in China: the effect of attractiveness, match-up, and consumer gender difference. Eur. J. Mark. **45**(7–8), 1214–1235 (2011)

Holzwarth, M., Janiszewski, C., Neumann, M.M.: The influence of avatars on online consumer shopping behavior. J. Market. **70**(4), 19–36 (2006)

McCormick, K.: Celebrity endorsements: Influence of a product-endorser match on Millennials attitudes and purchase intentions. J. Retail. Consum. Serv. **32**, 39–45 (2016)

Moustakas, E., Lamba, N., Mahmoud, D., Ranganathan, C.: Blurring lines between fiction and reality: perspectives of experts on marketing effectiveness of virtual influencers. IEEE (2020)

Ohanian, R.: Construction and validation of a scale to measure celebrity endorsers' perceived expertise, trustworthiness, and attractiveness. J. Advert. **19**(3), 39–52 (1990)

Osei-Frimpong, K., Donkor, G., Owusu-Frimpong, N.: The impact of celebrity endorsement on consumer purchase intention: an emerging market perspective. J. Market. Theory Pract. **27**(1), 103–121 (2019)

Breves, P.L., Liebers, N., Abt, M., Kunze, A.: The perceived fit between instagram influencers and the endorsed brand: How influencer–brand fit affects source credibility and persuasive effectiveness. J. Advert. Res. **59**(4), 440–454 (2019)

Park, G., Nan, D., Park, E., Kim, K.J., Han, J., del Pobil, A.P.: Computers as social actors? examining how users perceive and interact with virtual influencers on social media. IEEE (2021)

Pradhan, D., Duraipandian, I., et al.: Celebrity endorsement: how celebrity–brand–user personality congruence affects brand attitude and purchase intention. J. Mark. Commun. **22**(5), 456–473 (2014)

Sands, S., Campbell, C.L., Plangger, K., Ferraro, C.: Unreal influence: leveraging AI in influencer marketing. Eur. J. Mark. **56**(6), 1721–1747 (2022)

Schouten, A.P., Janssen, L., et al.: Celebrity vs. Influencer endorsements in advertising: the role of identification, credibility, and Product-Endorser fit. Int. J. Advert. **39**(2), 258–281 (2020)

Spry, A., Pappu, R., Bettina Cornwell, T.: Celebrity endorsement, brand credibility and brand equity. Eur. J. Mark. **45**(6), 882–909 (2011)

Thwaites, D., Lowe, B., et al.: The impact of negative publicity on celebrity ad endorsements. Psychol. Mark. **29**(9), 663–673 (2012)

Liu, M.T., Huang, Y.Y., Jiang, M.: Relations among attractiveness of endorsers, match-up, and purchase intention in sport marketing in China. J. Consum. Mark. **24**(6), 358–365 (2007)

Törn, F.: Revisiting the match-up hypothesis: effects of brand-incongruent celebrity endorsements. J. Curr. Iss. Res. Advert. **33**(1), 20–36 (2012)

Yonelinas, Λ.P.: The nature of recollection and familiarity: a review of 30 years of research. J. Memory Lang. **46**(3), 441–517 (2002)

Author Index

© Springer Nature Switzerland AG 2023
F. Fui-Hoon Nah and K. Siau (Eds.): HCII 2023, LNCS 14038, pp. 497–499, 2023.
https://doi.org/10.1007/978-3-031-35969-9